TEACHER'S EDITION

ACTIVE™ PHYSICAL SCIENCE

An Inquiry Approach to Physical Science

TEACHER'S EDITION

ACTIVE™ PHYSICAL SCIENCE

An Inquiry Approach to Physical Science

Dr. Arthur Eisenkraft

Volume 1 Teacher's Edition

IT's ABOUT TIME®
HERFF JONES EDUCATION DIVISION

84 Business Park Drive, Armonk, NY 10504 Phone (914) 273-2233
Fax (914) 273-2227 Toll Free (888) 698-TIME (8463) www.its-about-time.com

It's About Time, Founder
Laurie Kreindler

Creative/Art Director
John Nordland

Design/Production
Burmar Technical Corporation
Murray Fuchs
Robin Hoffmann
Jennifer Von Holstein

Technical Art
Burmar Technical Corporation
Jennifer Von Holstein
Dennis Falcon

Illustrations
Tomas Bunk

Project Managers
Barbara Zahm
John Nordland

Project Editor
Ruta Demery

Physics Reviewers
John Hubrisz
Dwight Neuenschwander
John Roeder
Patty Rourke

Photo Research
Jennifer Von Holstein

Safety Reviewer
Ed Robeck

Active Physical Science™

Printed and bound in the United States of America

ISBN #1-58591-321-9 3 Volume Set ISBN #1-58591-320-0

1 2 3 4 5 VH 08 07 06 05 04

This project was supported, in part, by the National Science Foundation
Opinions expressed are those of the authors and not necessarily those of the National Science Foundation

Project Director, Active Physics

Arthur Eisenkraft has taught high school physics for over 28 years and is currently the Distinguished Professor of Science Education and a Senior Research Fellow at the University of Massachusetts, Boston. Dr. Eisenkraft is the author of numerous science and educational publications. He holds U.S. Patent #4447141 for a Laser Vision Testing System (which tests visual acuity for spatial frequency).

Dr. Eisenkraft has been recognized with numerous awards including: Presidential Award for Excellence in Science Teaching, 1986 from President Reagan; American Association of Physics Teachers (AAPT) Excellence in Pre-College Teaching Award, 1999; AAPT Distinguished Service Citation for "excellent contributions to the teaching of physics", 1989; Science Teacher of the Year, Disney American Teacher Awards in their American Teacher Awards program, 1991; Honorary Doctor of Science degree from Rensselaer Polytechnic Institute, 1993; Tandy Technology Scholar Award 2000.

In 1999 Dr. Eisenkraft was elected to a 3-year cycle as the President-Elect, President and Retiring President of the National Science Teachers Association (NSTA), the largest science teacher organization in the world. In 2003, he was elected a fellow of the American Association for the Advancement of Science (AAAS).

Dr. Eisenkraft has been involved with a number of projects and chaired many competition programs, including: the Toshiba/NSTA ExploraVisions Awards (1991 to the present); the Toyota TAPESTRY Grants (1990 to the present); the Duracell/NSTA Scholarship Competitions (1984 to 2000). He was a columnist and on the Advisory Board of *Quantum* (a science and math student magazine that was published by NSTA as a joint venture between the United States and Russia; 1989 to 2001). In 1993, he served as Executive Director for the XXIV International Physics Olympiad after being Academic Director for the United States Team for six years. He has served on a number of committees of the National Academy of Sciences including the content committee that helped write the National Science Education Standards.

Dr. Eisenkraft has appeared on *The Today Show*, *National Public Radio*, *Public Television*, *The Disney Channel* and numerous radio shows. He serves as an advisor to the ESPN Sports Figures Video Productions.

He is a frequent presenter and keynote speaker at National Conventions. He has published over 100 articles and presented over 200 papers and workshops. He has been featured in articles in *The New York Times*, *Education Week*, *Physics Today*, *Scientific American*, *The American Journal of Physics* and *The Physics Teacher*.

TABLE OF CONTENTS

Acknowledgements

Active Physical Science

was developed by a team of leading physicists, university educators and classroom teachers with financial support from the National Science Foundation.

NSF Program Officer

Gerhard Salinger
Instructional Materials Development (IMD)

Principal Investigators

Bernard V. Khoury
American Association of Physics Teachers

Dwight Edward Neuenschwander
American Institute of Physics

Project Director

Arthur Eisenkraft
University of Massachusetts

Primary and Contributing Authors

Richard Berg
University of Maryland
College Park, MD

Howard Brody
University of Pennsylvania
Philadelphia, PA

Chris Chiaverina
New Trier Township High School
Crystal Lake, IL

Ron DeFronzo
Eastbay Ed. Collaborative
Attleboro, MA

Ruta Demery
Blue Ink Editing
Stayner, ON

Carl Duzen
Lower Merion High School
Havertown, PA

Jon L. Harkness
Active Physics Regional Coordinator
Wausau, WI

Ruth Howes
Ball State University
Muncie, IN

Douglas A. Johnson
Madison West High School
Madison, WI

Ernest Kuehl
Lawrence High School
Cedarhurst, NY

Robert L. Lehrman
Bayside, NY

Salvatore Levy
Roslyn High School
Roslyn, NY

Tom Liao
SUNY Stony Brook
Stony Brook, NY

Charles Payne
Ball State University
Muncie, IN

Mary Quinlan
Radnor High School
Radnor, PA

Harry Rheam
Eastern Senior High School
Atco, NJ

Bob Ritter
University of Alberta
Edmonton, AB, CA

John Roeder
The Calhoun School
New York, NY

John J. Rusch
University of Wisconsin, Superior
Superior, WI

Patty Rourke
Potomac School
McLean, VA

Ceanne Tzimopoulos
Omega Publishing
Medford, MA

Larry Weathers
The Bromfield School
Harvard, MA

David Wright
Tidewater Comm. College
Virginia Beach, VA

Consultants

Peter Brancazio
Brooklyn College of CUNY
Brooklyn, NY

Robert Capen
Canyon del Oro High School
Tucson, AZ

Carole Escobar

Earl Graf
SUNY Stony Brook
Stony Brook, NY

Jack Hehn
American Association of Physics Teachers
College Park, MD

Donald F. Kirwan
Louisiana State University
Baton Rouge, LA

Gayle Kirwan
Louisiana State University
Baton Rouge, LA

James La Porte
Virginia Tech
Blacksburg, VA

Charles Misner
University of Maryland
College Park, MD

Robert F. Neff
Suffern, NY

Ingrid Novodvorsky
Mountain View High School
Tucson, AZ

John Robson
University of Arizona
Tucson, AZ

Mark Sanders
Virginia Tech
Blacksburg, VA

Brian Schwartz
Brooklyn College of CUNY
New York, NY

Bruce Seiger
Wellesley High School
Newburyport, MA

Clifford Swartz
SUNY Stony Brook
Setauket, NY

Barbara Tinker
The Concord Consortium
Concord, MA

Robert E. Tinker
The Concord Consortium
Concord, MA

Joyce Weiskopf
Herndon, VA

Donna Willis
American Association of
Physics Teachers
College Park, MD

Safety Reviewer

Gregory Puskar
University of West Virginia
Morgantown, WV

Equity Reviewer

Leo Edwards
Fayetteville State University
Fayetteville, NC

Physics at Work

Alex Straus
writer
New York, NY

Mekea Hurwitz
photographer

Physics InfoMall

Brian Adrian
Bethany College
Lindsborg, KS

First Printing Reviewer

John L. Hubisz
North Carolina State University
Raleigh, NC

Unit Reviewers

Robert Adams
Polytech High School
Woodside, DE

George A. Amann
F.D. Roosevelt High School
Rhinebeck, NY

Patrick Callahan
Catasaugua High School
Center Valley, PA

Beverly Cannon
Science and Engineering
Magnet High School
Dallas, TX

Barbara Chauvin

Elizabeth Chesick
The Baldwin School
Haverford, PA

Chris Chiaverina
New Trier Township High School
Crystal Lake, IL

Andria Erzberger
Palo Alto Senior High School
Los Altos Hills, CA

Elizabeth Farrell Ramseyer
Niles West High School
Skokie, IL

Mary Gromko
President of Council of State Science
Supervisors
Denver, CO

Thomas Guetzloff

Jon L. Harkness
Active Physics Regional Coordinator
Wausau, WI

Dawn Harman
Moon Valley High School
Phoenix, AZ

James Hill
Piner High School
Sonoma, CA

Bob Kearney

Claudia Khourey-Bowers
McKinley Senior High School

Steve Kliewer
Bullard High School
Fresno, CA

Ernest Kuehl
Roslyn High School
Cedarhurst, NY

Jane Nelson
University High School
Orlando, FL

Mary Quinlan
Radnor High School
Radnor, PA

John Roeder
The Calhoun School
New York, NY

Patty Rourke
Potomac School
McLean, VA

Gerhard Salinger
Fairfax, VA

Irene Slater
La Pietra School for Girls

Pilot Test Teachers

John Agosta

Donald Campbell
Portage Central High School
Portage, MI

John Carlson
Norwalk Community
Technical College
Norwalk, CT

Veanna Crawford
Alamo Heights High School
New Braunfels, TX

Janie Edmonds
West Milford High School
Randolph, NJ

Eddie Edwards
Amarillo Area Center for Advanced Learning
Amarillo, TX

Arthur Eisenkraft
Fox Lane High School
Bedford, NY

Tom Ford

Bill Franklin

Roger Goerke
St. Paul, MN

Tom Gordon
Greenwich High School
Greenwich, CT

Ariel Hepp

John Herrman
College of Steubenville
Steubenville, OH

Linda Hodges

Ernest Kuehl
Lawrence High School
Cedarhurst, NY

Fran Leary
Troy High School
Schenectady, NY

Harold Lefcourt

Cherie Lehman
West Lafayette High School
West Lafayette, IN

Kathy Malone
Shady Side Academy
Pittsburgh, PA

Bill Metzler
Westlake High School
Thornwood, NY

Elizabeth Farrell Ramseyer
Niles West High School
Skokie, IL

Daniel Repogle
Central Noble High School
Albion, IN

Evelyn Restivo
Maypearl High School
Maypearl, TX

Doug Rich
Fox Lane High School
Bedford, NY

John Roeder
The Calhoun School
New York, NY

Tom Senior
New Trier Township High School
Highland Park, IL

John Thayer
District of Columbia Public Schools
Silver Spring, MD

Carol-Ann Tripp
Providence Country Day
East Providence, RI

Yvette Van Hise
High Tech High School
Freehold, NJ

Jan Waarvick

Sandra Walton
Dubuque Senior High School
Dubuque, IA

Larry Wood
Fox Lane High School
Bedford, NY

Field Test Coordinator

Marilyn Decker
Northeastern University
Acton, MA

Field Test Workshop Staff

John Carlson

Marilyn Decker

Arthur Eisenkraft

Douglas Johnson

John Koser

Ernest Kuehl

Mary Quinlan

Elizabeth Farrell Ramseyer

John Roeder

Field Test Evaluators

Susan Baker-Cohen

Susan Cloutier

George Hein

Judith Kelley

all from Lesley College,
Cambridge, MA

Field Test Teachers and Schools

Rob Adams
Polytech High School
Woodside, DE

Benjamin Allen
Falls Church High School
Falls Church, VA

Robert Applebaum
New Trier High School
Winnetka, IL

Joe Arnett
Plano Sr. High School
Plano, TX

Bix Baker
GFW High School
Winthrop, MN

Debra Beightol
Fremont High School
Fremont, NE

Patrick Callahan
Catasaugua High School
Catasaugua, PA

George Coker
Bowling Green High School
Bowling Green, KY

Janice Costabile
South Brunswick High School
Monmouth Junction, NJ

Stanley Crum
Homestead High School
Fort Wayne, IN

Russel Davison
Brandon High School
Brandon, FL

Christine K. Deyo
Rochester Adams High School
Rochester Hills, MI

Jim Doller
Fox Lane High School
Bedford, NY

Jessica Downing
Esparto High School
Esparto, CA

Douglas Fackelman
Brighton High School
Brighton, CO

Rick Forrest
Rochester High School
Rochester Hills, MI

Mark Freeman
Blacksburg High School
Blacksburg, VA

Jonathan Gillis
Enloe High School
Raleigh, NC

Karen Gruner
Holton Arms School
Bethesda, MD

Larry Harrison
DuPont Manual High School
Louisville, KY

Alan Haught
Weaver High School
Hartford, CT

Steven Iona
Horizon High School
Thornton, CO

Phil Jowell
Oak Ridge High School
Conroe, TX

Deborah Knight
Windsor Forest High School
Savannah, GA

Thomas Kobilarcik
Marist High School
Chicago, IL

Sheila Kolb
Plano Senior High School
Plano, TX

Todd Lindsay
Park Hill High School
Kansas City, MO

Malinda Mann
South Putnam High School
Greencastle, IN

Steve Martin
Maricopa High School
Maricopa, AZ

Nancy McGrory
North Quincy High School
N. Quincy, MA

David Morton
Mountain Valley High School
Rumford, ME

Charles Muller
Highland Park High School
Highland Park, NJ

Fred Muller
Mercy High School
Burlingame, CA

Vivian O'Brien
Plymouth Regional High School
Plymouth, NH

Robin Parkinson
Northridge High School
Layton, UT

Donald Perry
Newport High School
Bellevue, WA

Francis Poodry
Lincoln High School
Philadelphia, PA

John Potts
Custer County District
High School
Miles City, MT

Doug Rich
Fox Lane High School
Bedford, NY

John Roeder
The Calhoun School
New York, NY

Consuelo Rogers
Maryknoll Schools
Honolulu, HI

Lee Rossmaessler
Mott Middle College High School
Flint, MI

John Rowe
Hughes Alternative Center
Cincinnati, OH

Rebecca Bonner Sanders
South Brunswick High School
Monmouth Junction, NJ

David Schilpp
Narbonne High School
Harbor City, CA

Eric Shackelford
Notre Dame High School
Sherman Oaks, CA

Robert Sorensen
Springville-Griffith Institute
and Central School
Springville, NY

Teresa Stalions
Crittenden County High School
Marion, KY

Roberta Tanner
Loveland High School
Loveland, CO

Anthony Umelo
Anacostia Sr. High School
Washington, D.C.

Judy Vondruska
Mitchell High School
Mitchell, SD

Deborah Waldron
Yorktown High School
Arlington, VA

Ken Wester
The Mississippi School for Mathematics
and Science
Columbus, MS

Susan Willis
Conroe High School
Conroe, TX

Meeting *Active Physical Science* for the First Time

Welcome! A Five-Minute Introduction

Active Physical Science is a different species of physical science course. In *Active Physical Science* students are introduced to physics and chemistry concepts on an appropriate basis as they explore issues that deal with sports, transportation, the home, roller coasters, atomic structure, special effects, educational games, and chemistry shows. Each chapter is independent of any other chapter. You can begin the physical science course with any chapter.

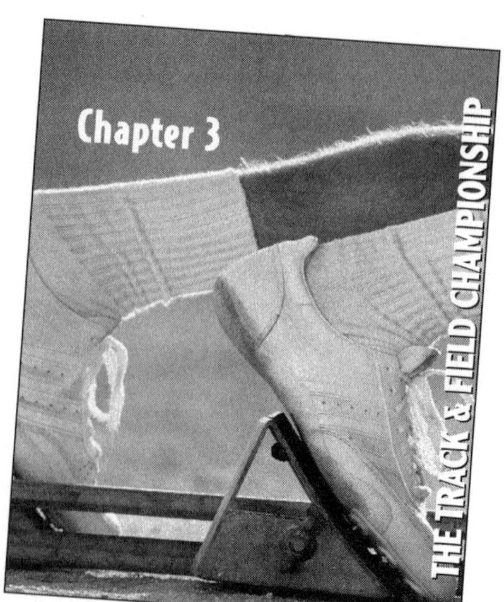

As an example, let's begin with **Chapter 3, The Track and Field Championship**. On Day One, students are introduced to the **Chapter Challenge**. Your school has been invited to send a team to the Penn Relays, the country's largest track and field event. Some schools hire professionals to help improve the performances of their athletes. However, your school cannot afford to hire any professionals. Can the physical science class help prepare the track and field team for the Penn Relays?

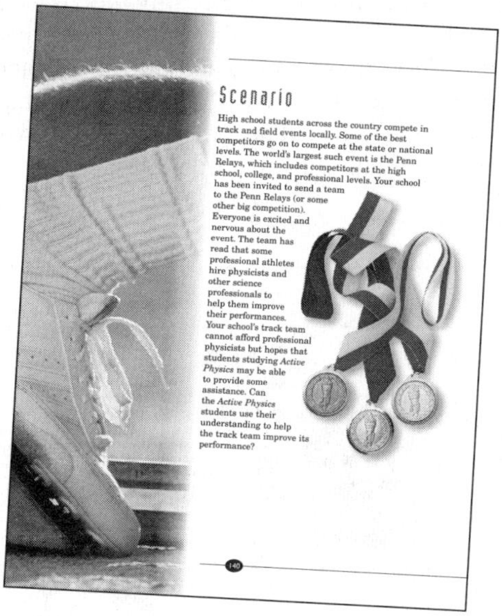

Scenario

High school students across the country compete in track and field events locally. Some of the best competitors go on to compete at the state or national levels. The world's largest such event is the Penn Relays, which includes competitors at the high school, college, and professional levels. Your school has been invited to send a team to the Penn Relays (or some other big competition). Everyone is excited and nervous about the event. The team has read that some professional athletes hire physicists and other science professionals to help them improve their performances. Your school's track team cannot afford professional physicists but hopes that students studying *Active Physics* may be able to provide some assistance. Can the *Active Physics* students use their understanding to help the track team improve its performance?

140

The **Chapter Challenge** has three parts:

1. Help the students on the track and field team compare themselves with their competitors.
2. Provide a description of the physics principles as they relate to track events.
3. Provide specific techniques to improve the performance of the athletes.

How can students get started? How can students complete such a challenge without the requisite physical science knowledge? Before the chapter activities begin, a discussion takes place about the criteria for success. The class discusses what is expected in an excellent response to the challenge. How will they be graded? For instance, should the track and field manual contain physics principles, charts, past records, relevant equations, definitions, and specific techniques? Once the students decide what should be included, they will then have to decide on how each element will be assessed. Should the physics principles be given more weight than past records? Students will have a sense of what will be required from them even before they begin. This grading system will be revisited before work on the **Chapter Challenge** is finalized.

Challenge

Your challenge is to write a physics manual about track and field training for your high-school team to help improve its performance. The manual should:

- **help students compare themselves with their competitors**
- **include a description of physics principles as they relate to track events**
- **provide specific techniques to improve performance**

Criteria

How will the manual be graded? What qualities should a good manual have? Discuss these issues in small groups and with your class. You may decide that some or all of the following qualities should be graded in your track and field manual:

- **physics principles**
- **inclusion of charts**
- **past records**
- **relevant equations**
- **definitions**
- **specific techniques**

Any advice you give should be understandable to athletes who have not studied physics. You can describe any activities you have done to explain how you know that the technique works, but you should not tell so much about each activity that the reader becomes bored.

Once you list the criteria for judging the track and field manuals, you should also decide how many points should be given for each criterion. If the group is going to hand in one manual, you must also agree on the way in which individuals will receive their grade. One method may be that the individual contributions toward the project receive 75% of the individual grade, and the group grade provides the remaining 25% of the individual grade. You should discuss different strategies and choose the one that is best suited to your school.

141 Active Physics CoreSelect

Day Two begins the first of ten activities. Each successive day begins with another activity. *Active Physical Science* is an activity-based curriculum. **Let's look at Activity 5: Acceleration.**

Each activity begins with a **What Do You Think?** feature. This particular activity begins by mentioning, "Accelerating out of the starting blocks is important if a runner is going to win a race." The students are then given a couple of questions to think about. One asks the students what determines who wins a race if all runners have the same top speed?

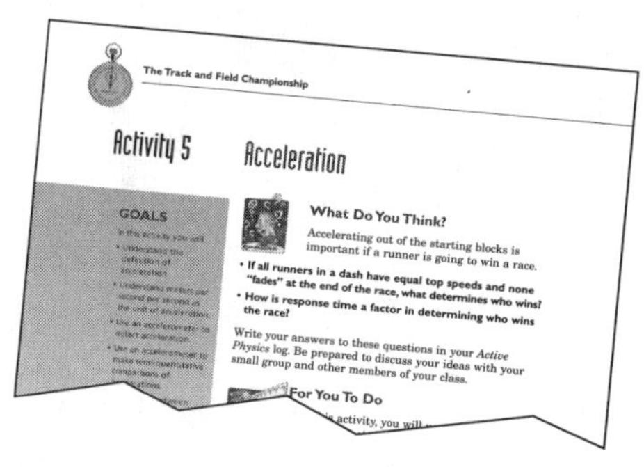

The Track and Field Championship

Activity 5 Acceleration

GOALS

What Do You Think?

Accelerating out of the starting blocks is important if a runner is going to win a race.

- **If all runners in a dash have equal top speeds and none "fades" at the end of the race, what determines who wins?**
- **How is response time a factor in determining who wins the race?**

Write your answers to these questions in your *Active Physics* log. Be prepared to discuss your ideas with your small group and other members of your class.

For You To Do

These **What Do You Think?** questions are intended to find out what students already know about acceleration—to get into the "acceleration" part of their brains. Formally, we say that these questions are to elicit the students' prior understanding and is part of the constructivist approach. Students write a response for one minute and discuss for another two minutes. But we don't reach closure. The questions open the conversation.

Students then begin the **For You To Do** (or **Investigate**) part of the activity.

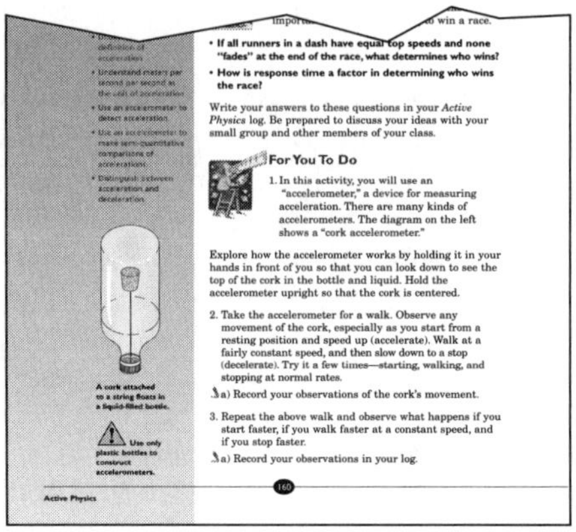

- **If all runners in a dash have equal top speeds and none "fades" at the end of the race, what determines who wins?**
- **How is response time a factor in determining who wins the race?**

Write your answers to these questions in your *Active Physics* log. Be prepared to discuss your ideas with your small group and other members of your class.

For You To Do

1. In this activity, you will use an "accelerometer," a device for measuring acceleration. There are many kinds of accelerometers. The diagram on the left shows a "cork accelerometer."

Explore how the accelerometer works by holding it in your hands in front of you so that you can look down to see the top of the cork in the bottle and liquid. Hold the accelerometer upright so that the cork is centered.

2. Take the accelerometer for a walk. Observe any movement of the cork, especially as you start from a resting position and speed up (accelerate). Walk at a fairly constant speed, and then slow down to a stop (decelerate). Try it a few times—starting, walking, and stopping at normal rates.
   a) Record your observations of the cork's movement.
3. Repeat the above walk and observe what happens if you start faster, if you walk faster at a constant speed, and if you stop faster.
   a) Record your observations in your log.

A cork attached to a string floats in a liquid-filled bottle.

Use only plastic bottles to construct accelerometers.

160

Active Physics

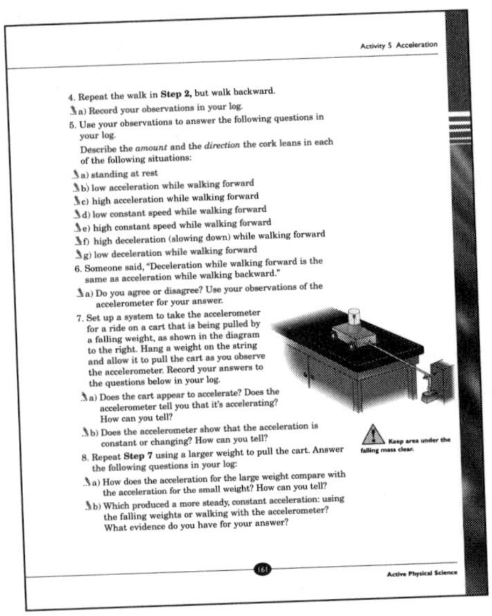

Activity 5 Acceleration

4. Repeat the walk in **Step 2,** but walk backward.
   a) Record your observations in your log.
5. Use your observations to answer the following questions in your log.
   Describe the *amount* and the *direction* the cork leans in each of the following situations:
   a) standing at rest
   b) low acceleration while walking forward
   c) high acceleration while walking forward
   d) low constant speed while walking forward
   e) high constant speed while walking forward
   f) high deceleration (slowing down) while walking forward
   g) low deceleration while walking forward
6. Someone said, "Deceleration while walking forward is the same as acceleration while walking backward."
   a) Do you agree or disagree? Use your observations of the accelerometer for your answer.
7. Set up a system to take the accelerometer for a ride on a cart that is being pulled by a falling weight, as shown in the diagram to the right. Hang a weight on the string and allow it to pull the cart as you observe the accelerometer. Record your answers to the questions below in your log.
   a) Does the cart appear to accelerate? Does the accelerometer tell you that it's accelerating? How can you tell?
   b) Does the accelerometer show that the acceleration is constant or changing? How can you tell?
8. Repeat **Step 7** using a larger weight to pull the cart. Answer the following questions in your log:
   a) How does the acceleration for the large weight compare with the acceleration for the small weight? How can you tell?
   b) Which produced a more steady, constant acceleration: using the falling weights or walking with the accelerometer? What evidence do you have for your answer?

Keep area under the falling mass clear.

161

Active Physical Science

In this activity students investigate acceleration. First they explore how an accelerometer that they have constructed works. They stand still, walk forwards, walk backwards, and change speed to determine how the accelerometer will "read" their movements. Then they use the accelerometer to record the acceleration of a cart attached to a falling weight.

The **Physics Talk** (or **Chem Talk**) focuses the students' attention on the principles of physical science introduced in the activity. In this activity students are presented with an equation that relates acceleration to speed and time. It also provides a sample problem for the students.

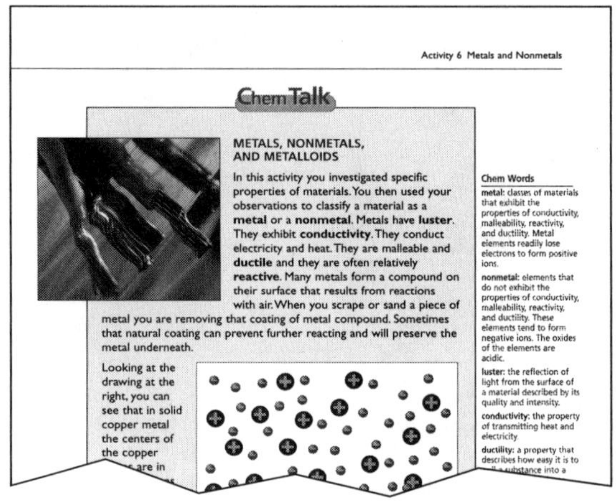

Activity 6 Metals and Nonmetals

Chem Talk

METALS, NONMETALS, AND METALLOIDS

In this activity you investigated specific properties of materials. You then used your observations to classify a material as a **metal** or a **nonmetal**. Metals have **luster**. They exhibit **conductivity**. They conduct electricity and heat. They are malleable and **ductile** and they are often relatively **reactive**. Many metals form a compound on their surface that results from reactions with air. When you scrape or sand a piece of metal you are removing that coating of metal compound. Sometimes that natural coating can prevent further reacting and will preserve the metal underneath.

Looking at the drawing at the right, you can see that in solid copper metal the centers of the copper are in

Chem Words

metal: classes of materials that exhibit the properties of conductivity, malleability, reactivity, and ductility. Metal elements readily lose electrons to form positive ions.

nonmetal: elements that do not exhibit the properties of conductivity, malleability, reactivity, and ductility. These elements tend to form negative ions. The oxides of the elements are acidic.

luster: the reflection of light from the surface of a material described by its quality and intensity.

conductivity: the property of transmitting heat and electricity

ductility: a property that describes how easy it is to … substance into a

A **For You to Read** explains the difference between scalar and vector quantities.

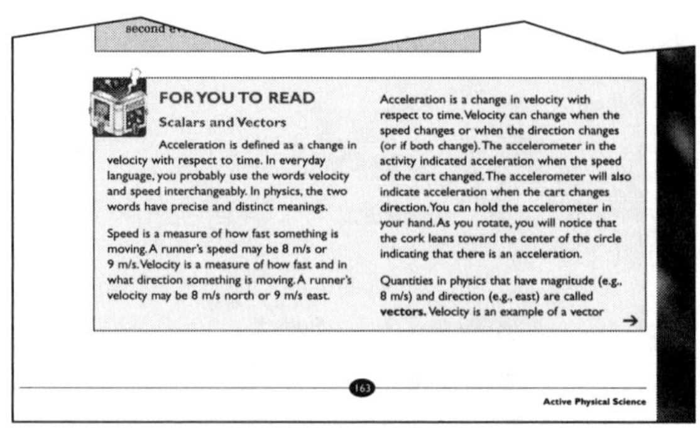

FOR YOU TO READ

Scalars and Vectors

Acceleration is defined as a change in velocity with respect to time. In everyday language, you probably use the words velocity and speed interchangeably. In physics, the two words have precise and distinct meanings.

Speed is a measure of how fast something is moving. A runner's speed may be 8 m/s or 9 m/s. Velocity is a measure of how fast and in what direction something is moving. A runner's velocity may be 8 m/s north or 9 m/s east.

Acceleration is a change in velocity with respect to time. Velocity can change when the speed changes or when the direction changes (or if both change). The accelerometer in the activity indicated acceleration when the speed of the cart changed. The accelerometer will also indicate acceleration when the cart changes direction. You can hold the accelerometer in your hand. As you rotate, you will notice that the cork leans toward the center of the circle indicating that there is an acceleration.

Quantities in physics that have magnitude (e.g., 8 m/s) and direction (e.g., east) are called **vectors.** Velocity is an example of a vector →

163

Active Physical Science

A **Reflecting on the Activity and the Challenge** relates the activity to the larger challenge of creating a training manual for the track and field team. "Since a race always begins with a speed of zero and ends with runners in motion, acceleration is part of every race." Students have been given another piece of the jigsaw puzzle. They can use acceleration at the starting blocks as part of the advice they provide the track and field team.

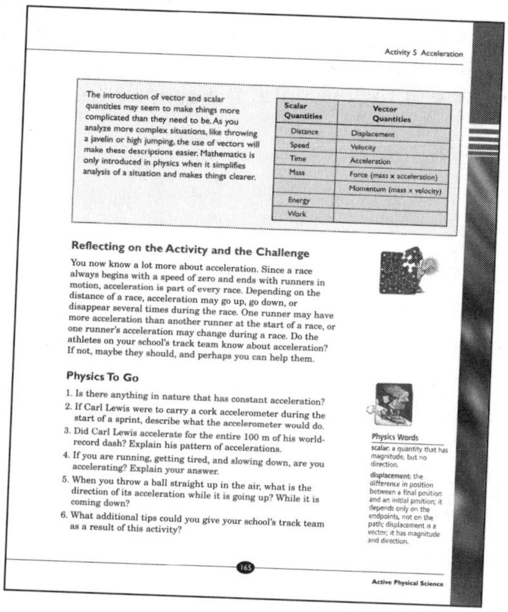

Activity 5 Acceleration

The introduction of vector and scalar quantities may seem to make things more complicated than they need to be. As you analyze more complex situations, like throwing a javelin or high jumping, the use of vectors will make these descriptions easier. Mathematics is only introduced in physics when it simplifies analysis of a situation and makes things clearer.

Scalar Quantities	Vector Quantities
Distance	Displacement
Speed	Velocity
Time	Acceleration
Mass	Force (mass x acceleration)
	Momentum (mass x velocity)
Energy	
Work	

Reflecting on the Activity and the Challenge

You now know a lot more about acceleration. Since a race always begins with a speed of zero and ends with runners in motion, acceleration is part of every race. Depending on the distance of a race, acceleration may go up, go down, or disappear several times during the race. One runner may have more acceleration than another runner at the start of a race, or one runner's acceleration may change during a race. Do the athletes on your school's track team know about acceleration? If not, maybe they should, and perhaps you can help them.

Physics To Go

1. Is there anything in nature that has constant acceleration?
2. If Carl Lewis were to carry a cork accelerometer during the start of a sprint, describe what the accelerometer would do.
3. Did Carl Lewis accelerate for the entire 100 m of his world-record dash? Explain his pattern of accelerations.
4. If you are running, getting tired, and slowing down, are you accelerating? Explain your answer.
5. When you throw a ball straight up in the air, what is the direction of its acceleration while it is going up? While it is coming down?
6. What additional tips could you give your school's track team as a result of this activity?

Physics Words

scalar: a quantity that has magnitude, but no direction.

displacement: the difference in position between a final position and an initial position; it depends only on the endpoints, not on the path; displacement is a vector; it has magnitude and direction.

165

Active Physical Science

The activity concludes with a **Physics to Go** (or **Chemistry to Go**) homework (or in-class) assignment. Here students are asked about the specifics of the activity. In this activity students are required to explain how acceleration may affect an athlete's performance in a race. They are also given problems to solve based on the sample problem provided in the **Physics Talk** section.

This chapter also has activities that help students understand how measurement is important in track and field, how to analyze trends from a graph, ways to increase top speed, and the components of projectile motion.

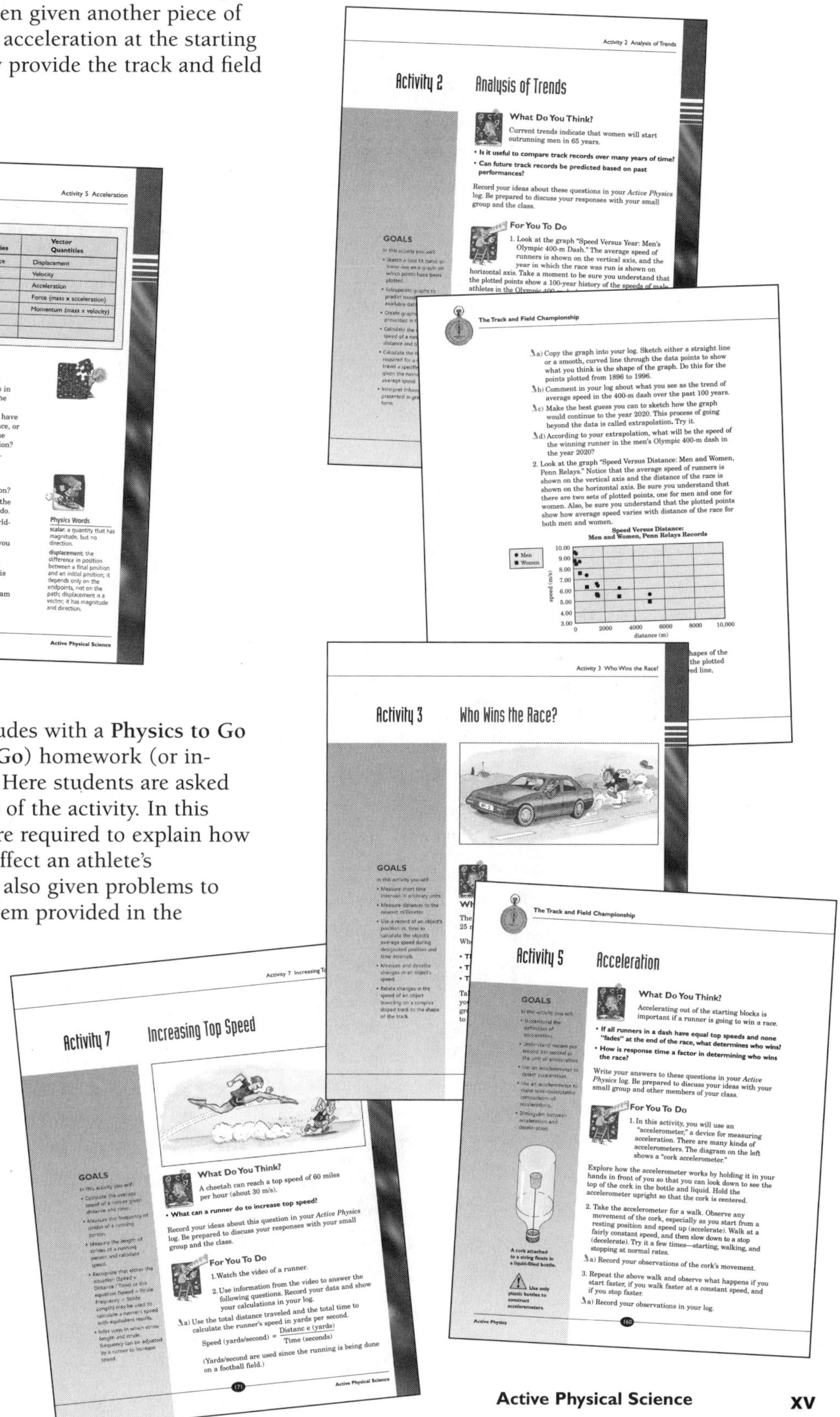

Activity 2 Analysis of Trends

Activity 2 Analysis of Trends

What Do You Think?

Current trends indicate that women will start outrunning men in 65 years.

- Is it useful to compare track records over many years of time?
- Can future track records be predicted based on past performances?

Record your ideas about these questions in your *Active Physics* log. Be prepared to discuss your responses with your small group and the class.

For You To Do

1. Look at the graph "Speed Versus Year: Men's Olympic 400-m Dash." The average speed of runners is shown on the vertical axis, and the year in which the race was run is shown on horizontal axis. Take a moment to be sure you understand that the plotted points show a 100-year history of the speeds of male athletes in the Olympic ...

The Track and Field Championship

a) Copy the graph into your log. Sketch either a straight line or a smooth, curved line through the data points to show what you think is the shape of the graph. Do this for the points plotted from 1896 to 1996.

b) Comment in your log about what you see as the trend of average speed in the 400-m dash over the past 100 years.

c) Make the best guess you can to sketch how the graph would continue to the year 2020. This process of going beyond the data is called extrapolation. Try it.

d) According to your extrapolation, what will be the speed of the winning runner in the men's Olympic 400-m dash in the year 2020?

2. Look at the graph "Speed Versus Distance: Men and Women, Penn Relays." Notice that the average speed of runners is shown on the vertical axis and the distance of the race is shown on the horizontal axis. Be sure you understand that there are two sets of plotted points, one for men and one for women. Also, be sure you understand that the plotted points show how average speed varies with distance of the race for both men and women.

Speed Versus Distance: Men and Women, Penn Relays Records

Activity 3 Who Wins the Race?

Activity 3 Who Wins the Race?

Activity 5 Acceleration

What Do You Think?

Accelerating out of the starting blocks is important if a runner is going to win a race.

- If all runners in a dash have equal top speeds and none "fades" at the end of the race, what determines who wins?
- How is response time a factor in determining who wins the race?

Write your answers to these questions in your *Active Physics* log. Be prepared to discuss your ideas with your small group and other members of your class.

For You To Do

1. In this activity, you will use an "accelerometer," a device for measuring acceleration. There are many kinds of accelerometers. The diagram on the left shows a "cork accelerometer."

Explore how the accelerometer works by holding it in your hands in front of you so that you can look down to see the top of the cork in the bottle and liquid. Hold the accelerometer upright so that the cork is centered.

2. Take the accelerometer for a walk. Observe any movement of the cork, especially as you start from a resting position and speed up (accelerate). Walk at a fairly constant speed, and then slow down to a stop (decelerate). Try it a few times—starting, walking, and stopping at normal rates.

a) Record your observations of the cork's movement.

3. Repeat the above walk and observe what happens if you start faster, if you walk faster at a constant speed, and if you stop faster.

a) Record your observations in your log.

A cork attached to a string floats in a liquid-filled bottle.

Use only plastic bottles to construct accelerometers.

Active Physics 160

Activity 7 Increasing Top Speed

Activity 7 Increasing Top Speed

What Do You Think?

A cheetah can reach a top speed of 60 miles per hour (about 30 m/s).

- What can a runner do to increase top speed?

Record your ideas about this question in your *Active Physics* log. Be prepared to discuss your responses with your small group and the class.

For You To Do

1. Watch the video of a runner.
2. Use information from the video to answer the following questions. Record your data and show your calculations in your log.

a) Use the total distance traveled and the total time to calculate the runner's speed in yards per second.

$$\text{Speed (yards/second)} = \frac{\text{Distance (yards)}}{\text{Time (seconds)}}$$

(Yards/second are used since the running is being done on a football field.)

171

Active Physical Science

The chapter concludes with a **Physics at Work** profile where students are introduced to someone whose job is related to the **Chapter Challenge**. In this chapter Erv Hunt, a former coach of the track and field team at the University of California explains the importance of the physics principles involved in the "pace work and intervals during the different stages of a race."

With the results of all of the activities before them, student teams now complete the **Chapter Challenge**. They put the jigsaw-puzzle pieces of distance, time, speed, acceleration, trajectory motion, and potential and kinetic energy together to create a manual for the track and field team. The teams share their work with the rest of the class and the chapter concludes.

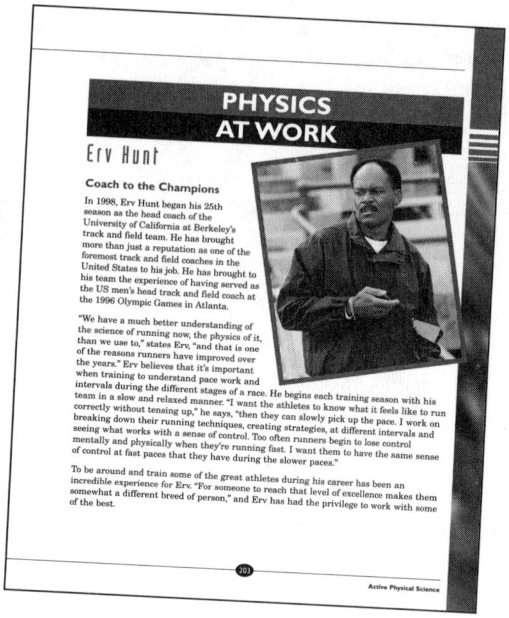

PHYSICS AT WORK

Erv Hunt

Coach to the Champions

In 1998, Erv Hunt began his 25th season as the head coach of the University of California at Berkeley's track and field team. He has brought more than just a reputation as one of the foremost track and field coaches in the United States to his job. He has brought to his team the experience of having served as the US men's head track and field coach at the 1996 Olympic Games in Atlanta.

"We have a much better understanding of the science of running now, the physics of it, than we use to," states Erv, "and that is one of the reasons runners have improved over the years." Erv believes that it's important when training to understand pace work and intervals during the different stages of a race. He begins each training season with his team in a slow and relaxed manner. "I want the athletes to know what it feels like to run correctly without tensing up," he says, "then they can slowly pick up the pace. I work on breaking down their running techniques, creating strategies, at different intervals and seeing what works with a sense of control. Too often runners begin to lose control mentally and physically when they're running fast. I want them to have the same sense of control at fast paces that they have during the slower paces."

To be around and train some of the great athletes during his career has been an incredible experience for Erv. "For someone to reach that level of excellence makes them somewhat a different breed of person," and Erv has had the privilege to work with some of the best.

203

Active Physical Science

4

Thrills and Chills

Scenario

You are excited and scared as you sit back into the seat. You pull the safety restraints into place. The next thing you know, you are beginning a slow but steady ascent into the sky. Then, with a sudden jolt, you reach the top. This is where the thrill or nightmare begins. You hurtle down the track at ever-increasing speeds. You are flung against one side of your seat as you scream around a curve. You shriek as you hang upside down, fortunately, firmly secured to your seat. All the time, your stomach has no idea where it is or you are. Finally, you come to rest where you began. What a ride! Want to go again?

Roller coasters have been enjoyed for many years. However, the roller coaster that may appeal to you, may not appeal to your parents or other friends and relatives.

206

Scenario

Probably the most dangerous thing you will do today is travel to your destination. Transportation is necessary, but the need to get there in a hurry, and the large number of people and vehicles, have made transportation very risky. There is a greater chance of being killed or injured traveling than in any other common activity. Realizing this, people and governments have begun to take action to alter the statistics. New safety systems have been designed and put into use in automobiles and airplanes. New laws and a new awareness are working together with these systems to reduce the danger in traveling.

What are these new safety systems? You are probably familiar with many of them. In this chapter, you will become more familiar with most of these designs. Could you design or even build a better safety device for a car or a plane? Many students around the country have been doing just that, and with great success!

Challenge

Your design team will develop a safety system for protecting automobile, airplane, bicycle, motorcycle, or train passengers. As you study existing safety systems, you and your design team should be listing ideas for improving an existing system or designing a new system for preventing accidents. You may also consider a system that will minimize the harm caused by accidents.

78

Your final product will be a working model or prototype of a safety system. On the day that you bring the final product to class, the teams will display them around the room while class members informally view them and discuss them with members of the design team. During this time, class members will ask questions about each others products. The questions will be placed in envelopes provided to each team by the teacher. The teacher will use some of these questions during the oral presentations on the next day.

The product will be judged according to the following three parts:

1. The quality of your safety feature enhancement and the working model or prototype.
2. The quality of a five-minute oral report that should include:
   - **the need for the system**
   - **the method used to develop the working model**
   - **the demonstration of the working model**
   - **the discussion of the physics concepts involved**
   - **the description of the next-generation version of the system**
   - **the answers to questions posed by the class**
3. The quality of a written and/or multimedia report including:
   - **the information from the oral report**
   - **the documentation of the sources of expert information**
   - **the discussion of consumer acceptance and market potential**
   - **the discussion of the physics concepts applied in the design of the safety system**

Criteria

You and your classmates will work with your teacher to define the criteria for determining grades. You will also be asked to evaluate your own work. Discuss as a class the performance task and the points that should be allocated for each part. A starting point for your discussions may be:

- **Part 1 = 40 points**
- **Part 2 = 30 points**
- **Part 3 = 30 points**

Since group work is made up of individual work, your teacher will assign some points to each individual's contribution to the project. If individual points total 30 points, then parts 1, 2 and 3 must be changed so that the total remains at 100.

Criteria

Work with your classmates to agree on the relative importance of the following assessment criteria. Each item in the list has a point value given after it, but your class must decide what kind of grading system to use.

1. The variety and number of physics concepts used to produce the light and sound effects: 30 points
   - four or more concepts: 30 points
   - three concepts: 25 points
   - two concepts: 20 points
   - one concept: 10 points
2. Your understanding of the physics concepts: 40 points

   Following your production, you will be asked to:
   - a) Name the physics concepts that you used. 10 points
   - b) Explain each concept. 10 points
   - c) Give an example of something that each concept explains or an example of how each concept is used. 10 points
   - d) Explain why each concept is important. 10 points

As a class, you will have to decide if your answers will be in an oral report or a written report.

3. Entertainment value: 30 points

   Your class will need to decide on a way to assign points for creativity. Note that an entertaining and interesting show need not be loud or bright.

You will have a chance later in the chapter to again discuss these criteria. At that time, you may have more information on the concepts and how you might produce your show. You may want to then propose changes in the criteria and the point values.

325

Features of Active Physical Science

1. Scenario

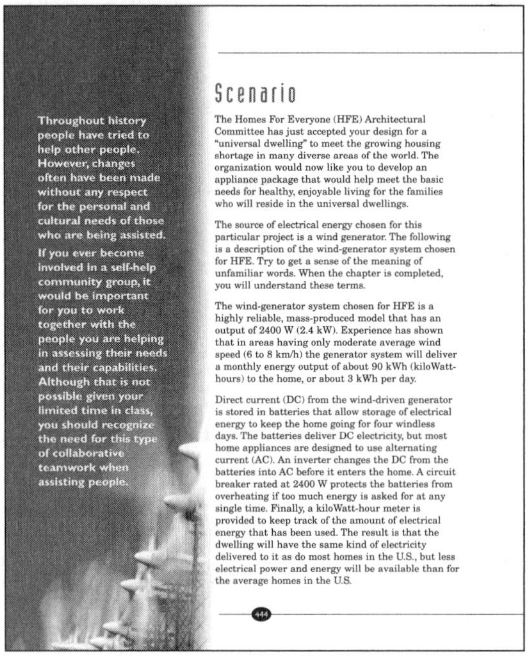
Scenario

Throughout history people have tried to help other people. However, changes often have been made without any respect for the personal and cultural needs of those who are being assisted.

If you ever become involved in a self-help community group, it would be important for you to work together with the people you are helping in assessing their needs and their capabilities. Although that is not possible given your limited time in class, you should recognize the need for this type of collaborative teamwork when assisting people.

The Homes For Everyone (HFE) Architectural Committee has just accepted your design for a "universal dwelling" to meet the growing housing shortage in many diverse areas of the world. The organization would now like you to develop an appliance package that would help meet the basic needs for healthy, enjoyable living for the families who will reside in the universal dwellings.

The source of electrical energy chosen for this particular project is a wind generator. The following is a description of the wind-generator system chosen for HFE. Try to get a sense of the meaning of unfamiliar words. When the chapter is completed, you will understand these terms.

The wind-generator system chosen for HFE is a highly reliable, mass-produced model that has an output of 2400 W (2.4 kW). Experience has shown that in areas having only moderate average wind speed (6 to 8 km/h) the generator system will deliver a monthly energy output of about 90 kWh (kiloWatt-hours) to the home, or about 3 kWh per day.

Direct current (DC) from the wind-driven generator is stored in batteries that allow storage of electrical energy to keep the home going for four windless days. The batteries deliver DC electricity, but most home appliances are designed to use alternating current (AC). An inverter changes the DC from the batteries into AC before it enters the home. A circuit breaker rated at 2400 W protects the batteries from overheating if too much energy is asked for at any single time. Finally, a kiloWatt-hour meter is provided to keep track of the amount of electrical energy that has been used. The result is that the dwelling will have the same kind of electricity delivered to it as do most homes in the U.S., but less electrical power and energy will be available than for the average homes in the U.S.

444

Each *Active Physical Science* chapter opens with an engaging **Scenario**. Students from diverse backgrounds and localities have been interviewed in order to find situations which are not only realistic but meaningful to the high school population. The scenarios (only a paragraph or two in length) set the stage for the **Chapter Challenge** which immediately follows. Many teachers choose to read the scenario aloud to the class as a way of introducing the new chapter.

2. Challenge

The **Chapter Challenge** is the heart and soul of *Active Physical Science*. It provides a purpose for all of the work that will follow. The challenges provide the rationale for learning. One of the common complaints teachers hear from students is, "Why am I learning this?" In *Active Physical Science*, no students raise this criticism. Similarly, no teacher has to answer, "Because one day it will be useful to you." The complaint is avoided because on Day One of the chapter students are presented with a challenge that, in essence, becomes their job for the next few weeks.

In *Safety*, Chapter 2, students are challenged to design and build an improved safety device for an automobile. The study of momentum, forces and Newton's Laws will be integral to their understanding of the required features in a safety device.

In *Electricity for Everyone*, Chapter 7, students must create an appliance package that can be used in developing nations. The appliance package is limited by the wind generator available to the households. Students must also supply a rationale for how each suggested appliance will enhance the well-being of the family using it. This requires students to be able to differentiate between power and energy. It also provides a basis for students to reflect on quality of life issues in parts of the globe that they learn about in their social studies classes.

The beauty of the challenges lies in the variety of tasks and opportunities for students of different talents and skills to excel. Students who express themselves artistically will have an opportunity to shine in some challenges, while the student who can design and build may be the group leader in another challenge. Some challenges have a major component devoted to writing while others require oral or visual presentations. All challenges require the demonstration of solid physics understanding.

The challenges are not contrived situations for high school students. Professional engineers also design and build improved safety devices. Medical writers and illustrators design posters and pamphlets. The challenge in Chapter 4, *Thrills and Chills*, requires students to adapt a roller coaster design for a select group of riders. The group can consist of young children, daredevils, or physically challenged people. This challenge has been successfully completed by 9th grade students, 12th grade physics students, and by mechanical engineers. The expectations may be different for each of these audiences, but the challenge is consistent.

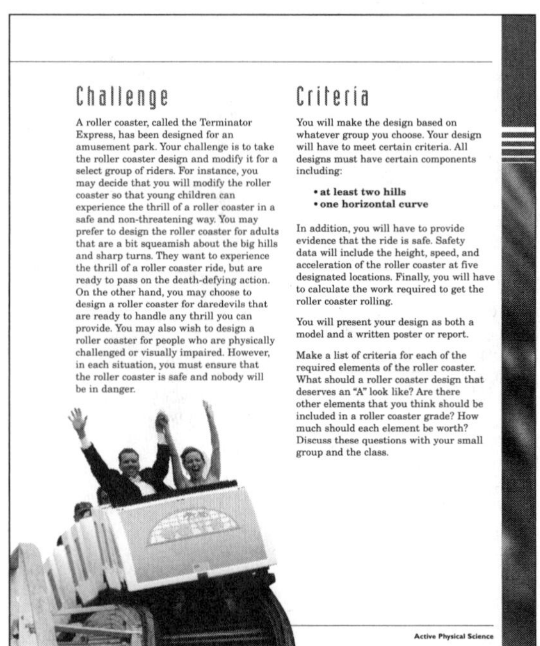
Challenge

A roller coaster, called the Terminator Express, has been designed for an amusement park. Your challenge is to take the roller coaster design and modify it for a select group of riders. For instance, you may decide that you will modify the roller coaster so that young children can experience the thrill of a roller coaster in a safe and non-threatening way. You may prefer to design the roller coaster for adults that are a bit squeamish about the big hills and sharp turns. They want to experience the thrill of a roller coaster ride, but are ready to pass on the death-defying action. On the other hand, you may choose to design a roller coaster for daredevils that are ready to handle any thrill you can provide. You may also wish to design a roller coaster for people who are physically challenged or visually impaired. However, in each situation, you must ensure that the roller coaster is safe and nobody will be in danger.

Criteria

You will make the design based on whatever group you choose. Your design will have to meet certain criteria. All designs must have certain components including:

- **at least two hills**
- **one horizontal curve**

In addition, you will have to provide evidence that the ride is safe. Safety data will include the height, speed, and acceleration of the roller coaster at five designated locations. Finally, you will have to calculate the work required to get the roller coaster rolling.

You will present your design as both a model and a written poster or report.

Make a list of criteria for each of the required elements of the roller coaster. What should a roller coaster design that deserves an "A" look like? Are there other elements that you think should be included in a roller coaster grade? How much should each element be worth? Discuss these questions with your small group and the class.

Active Physical Science

3. Criteria

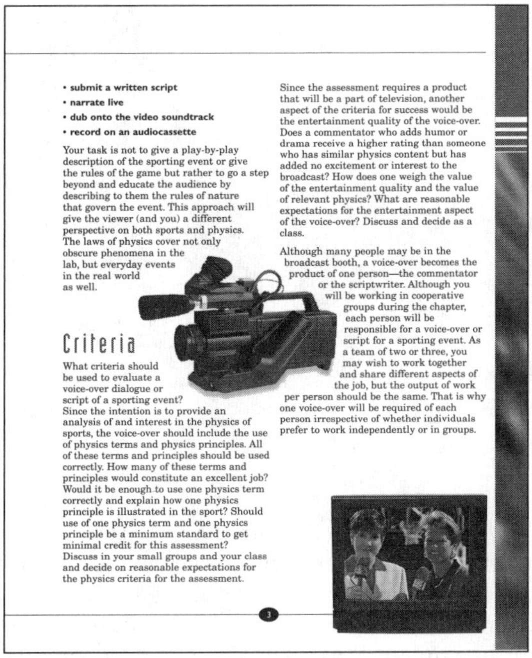
- **submit a written script**
- **narrate live**
- **dub onto the video soundtrack**
- **record on an audiocassette**

Your task is not to give a play-by-play description of the sporting event or give the rules of the game but rather to go a step beyond and educate the audience by describing to them the rules of nature that govern the event. This approach will give the viewer (and you) a different perspective on both sports and physics. The laws of physics cover not only obscure phenomena in the lab, but everyday events in the real world as well.

Criteria

What criteria should be used to evaluate a voice-over dialogue or script of a sporting event? Since the intention is to provide an analysis of and interest in the physics of sports, the voice-over should include the use of physics terms and physics principles. All of these terms and principles should be used correctly. How many of these terms and principles would constitute an excellent job? Would it be enough to use one physics term correctly and explain how one physics principle is illustrated in the sport? Should use of one physics term and one physics principle be a minimum standard to get minimal credit for this assessment? Discuss in your small groups and your class and decide on reasonable expectations for the physics criteria for the assessment.

Since the assessment requires a product that will be a part of television, another aspect of the criteria for success would be the entertainment quality of the voice-over. Does a commentator who adds humor or drama receive a higher rating than someone who has similar physics content but has added no excitement or interest to the broadcast? How does one weigh the value of the entertainment quality and the value of relevant physics? What are reasonable expectations for the entertainment aspect of the voice-over? Discuss and decide as a class.

Although many people may be in the broadcast booth, a voice-over becomes the product of one person—the commentator or the scriptwriter. Although you will be working in cooperative groups during the chapter, each person will be responsible for a voice-over or script for a sporting event. As a team of two or three, you may wish to work together and share different aspects of the job, but the output of work per person should be the same. That is why one voice-over will be required of each person irrespective of whether individuals prefer to work independently or in groups.

3

In creating *Active Physical Science*, we had thought that the generation of the challenge was good enough. Upon reflection, we soon realized that criteria for success must also be included. When students agree to the matrix by which they will be measured, the research has shown that the students will perform better and achieve more. It makes sense. In the simplest situation of cleaning a lab room, the teacher may simply state, "Please clean up the lab." The results are often a minimal cleanup. If the teacher begins by asking, "What does a clean lab room look like?" and students and teacher jointly list the attributes of a clean lab room (i.e., no paper on the floor, all beakers put away, all materials on the back of the lab tables, all power supplies unplugged and all water removed), the students respond differently and the cleanup is better. When students are asked to include physical science principles in an explanation, the students should know whether the expectation is for three physical science principles or five.

The discussion of grading criteria and the creation of a grading rubric is a crucial ingredient for student success. *Active Physical Science* requires a class discussion, after the introduction of the challenge, about the grading criteria. How much is required? What does an "A" presentation look like? Should creativity be weighed more than delivery? The criteria can be visited again at the end of the chapter, but at this point it provides a clarity to the challenge and the expectation level that the students should set for themselves.

4. What Do You Think?

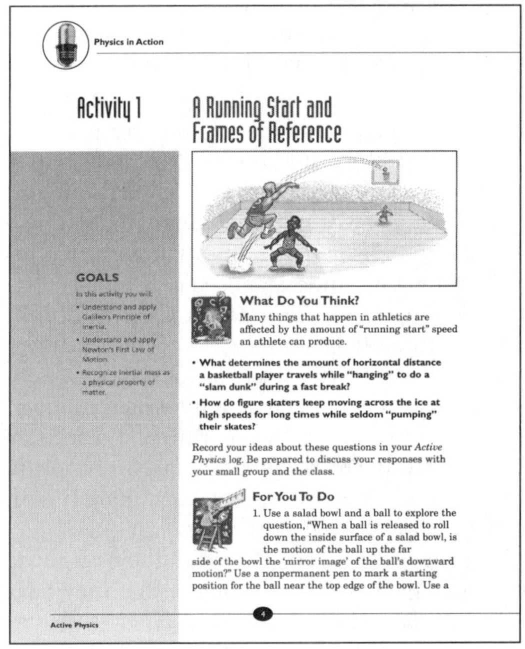
Physics in Action

Activity 1

A Running Start and Frames of Reference

GOALS

In this activity you will:

- Understand and apply Galileo's Principle of Inertia.
- Understand and apply Newton's First Law of Motion.
- Recognize inertial mass as a physical property of matter.

What Do You Think?

Many things that happen in athletics are affected by the amount of "running start" speed an athlete can produce.

- **What determines the amount of horizontal distance a basketball player travels while "hanging" to do a "slam dunk" during a fast break?**
- **How do figure skaters keep moving across the ice at high speeds for long times while seldom "pumping" their skates?**

Record your ideas about these questions in your *Active Physics* log. Be prepared to discuss your responses with your small group and the class.

For You To Do

1. Use a salad bowl and a ball to explore the question, "When a ball is released to roll down the inside surface of a salad bowl, is the motion of the ball up the far side of the bowl the 'mirror image' of the ball's downward motion?" Use a nonpermanent pen to mark a starting position for the ball near the top edge of the bowl. Use a

Active Physics 4

During the past few years much has been written about a constructivist approach to learning. Videos of Harvard graduates, in caps and gowns, show that the students are not able to explain correctly why it is colder in the winter than it is in the summer. These students have previously answered these questions correctly in 4th grade, in middle school, and then again in high school. How else would they have gotten into Harvard? We believe that they never internalized the logic and understanding of the seasons. One reason for this problem is that they were never confronted by what they did believe, and were never adequately shown why they should give up that belief system. Certainly, it is worth writing down a "book's perfect answer" on a test to secure a good grade, but to actually believe requires a more thorough examination of competing explanations.

The best way to ascertain a student's prior understanding is through extensive interviewing. Much of the research literature in this area includes the results of these interviews. In a classroom, this one-on-one dialogue is rarely possible. The **What Do You Think?** questions introduce each activity in a way in which to elicit prior understandings. It gives students an opportunity to verbalize what they think about speed, or energy, or elements and compounds, before they embark on an activity. The brief discussion of the range of answers brings the student a little closer in touch with that part of his/her brain which understands speed, energy, or elements and compounds. The **What Do You Think?** question is not intended to produce a correct answer or a discussion of the features of the questions. It is not intended to bring closure. The activity which follows will provide that discussion as experimental results are analyzed. The **What Do You Think?** question should take no more than a few minutes of class time. It is the lead into the physical science investigation. Students should be strongly encouraged to write their responses to the questions in their logs, to ensure that they have in fact addressed their prior conceptions. After students have discussed their responses in their small groups, activate a class discussion. Ask students to volunteer other students' answers which they found interesting. This may encourage students to exchange ideas without the fear of personally giving a "wrong" answer.

5. For You To Do (or Investigate)

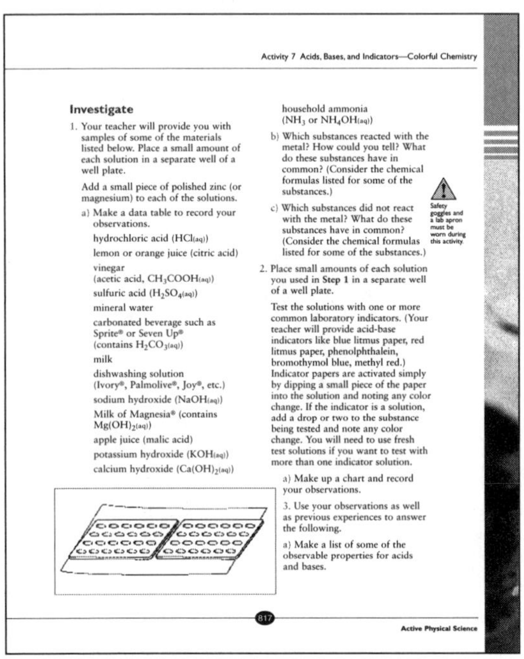

Activity 7 Acids, Bases, and Indicators—Colorful Chemistry

Investigate

1. Your teacher will provide you with samples of some of the materials listed below. Place a small amount of each solution in a separate well of a well plate.

   Add a small piece of polished zinc (or magnesium) to each of the solutions.

   a) Make a data table to record your observations.

   hydrochloric acid ($HCl_{(aq)}$)
   lemon or orange juice (citric acid)
   vinegar (acetic acid, $CH_3COOH_{(aq)}$)
   sulfuric acid ($H_2SO_{4(aq)}$)
   mineral water
   carbonated beverage such as Sprite® or Seven Up® (contains $H_2CO_{3(aq)}$)
   milk
   dishwashing solution (Ivory®, Palmolive®, Joy®, etc.)
   sodium hydroxide ($NaOH_{(aq)}$)
   Milk of Magnesia® (contains $Mg(OH)_{2(aq)}$)
   apple juice (malic acid)
   potassium hydroxide ($KOH_{(aq)}$)
   calcium hydroxide ($Ca(OH)_{2(aq)}$)
   household ammonia (NH_3 or $NH_4OH_{(aq)}$)

   b) Which substances reacted with the metal? How could you tell? What do these substances have in common? (Consider the chemical formulas listed for some of the substances.)

   c) Which substances did not react with the metal? What do these substances have in common? (Consider the chemical formulas listed for some of the substances.)

Safety goggles and a lab apron must be worn during this activity.

2. Place small amounts of each solution you used in **Step 1** in a separate well of a well plate.

   Test the solutions with one or more common laboratory indicators. (Your teacher will provide acid-base indicators like blue litmus paper, red litmus paper, phenolphthalein, bromothymol blue, methyl red.) Indicator papers are activated simply by dipping a small piece of the paper into the solution and noting any color change. If the indicator is a solution, add a drop or two to the substance being tested and note any color change. You will need to use fresh test solutions if you want to test with more than one indicator solution.

   a) Make up a chart and record your observations.

3. Use your observations as well as previous experiences to answer the following.

   a) Make a list of some of the observable properties for acids and bases.

817 Active Physical Science

Active Physical Science is a hands-on, minds-on curriculum. Students *do* physical science; they do not *read* about doing physical science. Each activity has instructions for each part of the investigation. The pencil icons in the Student Edition are provided to remind students that data, hypotheses, or conclusions should be recorded in their log or laboratory manual.

Activities are the opportunity for students to garner the knowledge that they will need to complete the **Chapter Challenge**. Students will understand the physical science principle involved because they have investigated it. In *Active Physical Science*, if a student is asked, "How do you know?" the response is, "Because I did an experiment!"

Recognizing that many students know how to read, but do not like reading, Background information is provided within the context of the activity. Students have demonstrated that they will read when the information is required for them to continue withtheir exploration.

Occasionally, the activity will require the entire class to participate in a large, single demonstration simultaneously. The teacher, on other occasions, may decide that a specific activity is best done as a demonstration. This would be appropriate if there is limited equipment for that one activity, or the facilities are not available. Viewing demonstrations on an ongoing basis, however, is not what these 12 chapters are about.

There are specific **For You To Do** (or **Investigate**) activities where computer spreadsheets, force transducers, or specific electronic equipment is required. Most of these activities have 'low-tech' alternatives provided in the Teacher's Edition. In the initial teaching of *Active Physical Science*, the low-tech alternative may be the only reasonable approach. As the course becomes a staple of the school offerings, it is hoped that funds can be set aside to improve the students' access to equipment.

Most of the **For You To Do** (or **Investigate**) activities require between one and two class periods. With the present trend toward block scheduling, there are so many time structures that it is difficult to predict how *Active Physical Science* will best fit with your schedule. The other impact on time is the achievement and preparation level of the students. In a given activity, students may be required to complete a graph of their data. This is considered one small part of the activity. If the students have never been exposed to graphing, this could require a two-period lesson to teach the rudiments of graphing with suitable practice in interpretation. *Active Physical Science* is accessible to all students. The teacher is in the best position to make accommodations in time reflecting the needs of the students.

6. Physics Talk and Chem Talk

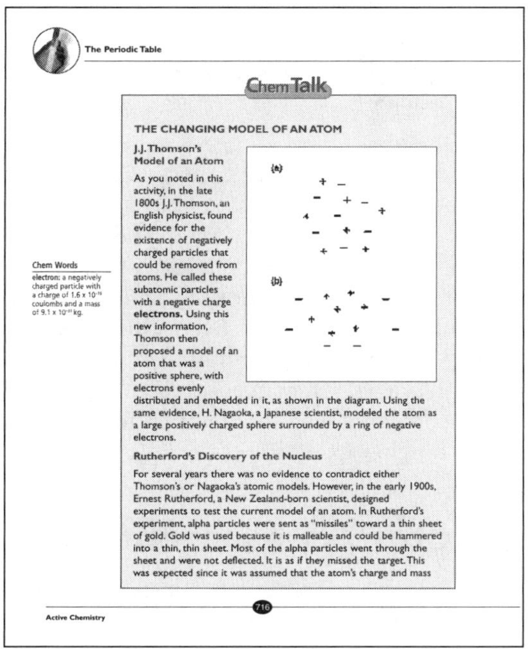
The Periodic Table

Chem Talk

THE CHANGING MODEL OF AN ATOM

J.J. Thomson's Model of an Atom

As you noted in this activity, in the late 1800s J.J. Thomson, an English physicist, found evidence for the existence of negatively charged particles that could be removed from atoms. He called these subatomic particles with a negative charge **electrons.** Using this new information, Thomson then proposed a model of an atom that was a positive sphere, with electrons evenly distributed and embedded in it, as shown in the diagram. Using the same evidence, H. Nagaoka, a Japanese scientist, modeled the atom as a large positively charged sphere surrounded by a ring of negative electrons.

Chem Words

electron: a negatively charged particle with a charge of 1.6×10^{-19} coulombs and a mass of 9.1×10^{-31} kg.

Rutherford's Discovery of the Nucleus

For several years there was no evidence to contradict either Thomson's or Nagaoka's atomic models. However, in the early 1900s, Ernest Rutherford, a New Zealand-born scientist, designed experiments to test the current model of an atom. In Rutherford's experiment, alpha particles were sent as "missiles" toward a thin sheet of gold. Gold was used because it is malleable and could be hammered into a thin, thin sheet. Most of the alpha particles went through the sheet and were not deflected. It is as if they missed the target. This was expected since it was assumed that the atom's charge and mass

716

Active Chemistry

Equations and formulas are often the simplest, most straightforward, most concise, and clearest way of expressing physical science principles. *Active Physical Science* limits the mathematics to the 9th grade curriculum. Students who have shied away from studying physics because of the mathematics prerequisites find that they are welcomed into *Active Physical Science*. **Physics** and **Chem Talk** is a means by which specific attention can be given to the mathematical equations. It also provides an opportunity to illustrate a problem solution or to derive a complex equation. For some students, there is a need to guide them through the algebraic manipulation which shows the equivalence of $F = ma$ and $a = F/m$. Where appropriate, this manipulation is explicitly shown. Finally, sample problems required for the **Chapter Challenge** will also be in **Physics Talk**.

7. For You To Read

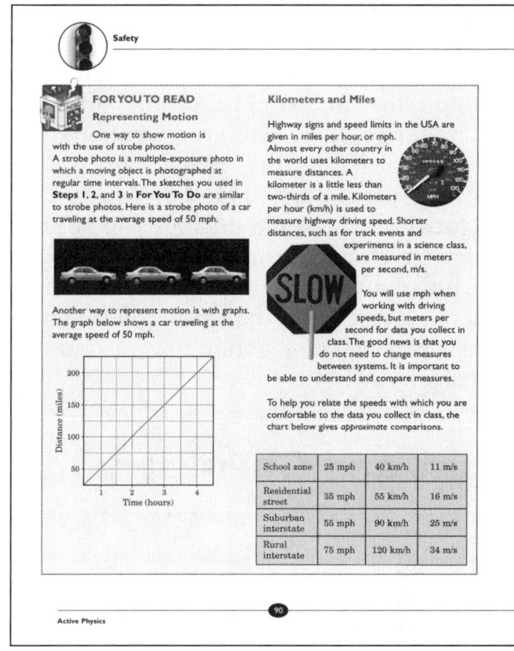
Safety

FOR YOU TO READ

Representing Motion

One way to show motion is with the use of strobe photos. A strobe photo is a multiple-exposure photo in which a moving object is photographed at regular time intervals. The sketches you used in **Steps 1, 2,** and **3** in **For You To Do** are similar to strobe photos. Here is a strobe photo of a car traveling at the average speed of 50 mph.

Another way to represent motion is with graphs. The graph below shows a car traveling at the average speed of 50 mph.

Kilometers and Miles

Highway signs and speed limits in the USA are given in miles per hour, or mph. Almost every other country in the world uses kilometers to measure distances. A kilometer is a little less than two-thirds of a mile. Kilometers per hour (km/h) is used to measure highway driving speed. Shorter distances, such as for track events and experiments in a science class, are measured in meters per second, m/s.

You will use mph when working with driving speeds, but meters per second for data you collect in class. The good news is that you do not need to change measures between systems. It is important to be able to understand and compare measures.

To help you relate the speeds with which you are comfortable to the data you collect in class, the chart below gives *approximate* comparisons.

School zone	25 mph	40 km/h	11 m/s
Residential street	35 mph	55 km/h	16 m/s
Suburban interstate	55 mph	90 km/h	25 m/s
Rural interstate	75 mph	120 km/h	34 m/s

90

Active Physics

The **For You To Read** inserts provide students with some reading at the 9th grade level. This section may be used to tie together concepts from the present activity or a set of activities. It may also be used to provide a glimpse into the history of the physics principle being investigated. Finally, **For You To Read** may provide background information which will help clarify the meaning of the physics principle investigated in **For You To Do** (or **Investigate**).

8. Reflecting on the Activity and the Challenge

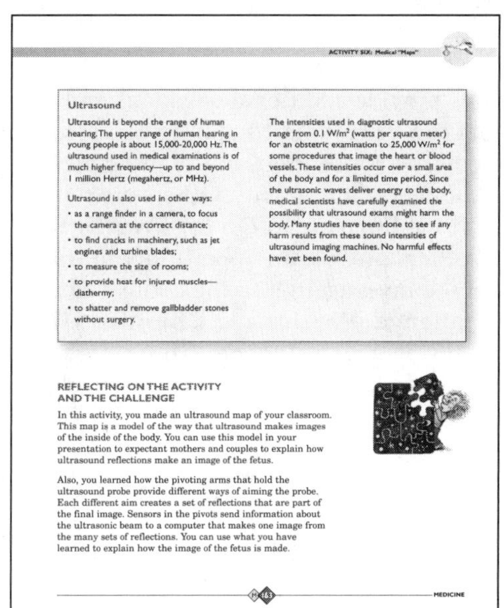
ACTIVITY SIX: Medical "Maps"

Ultrasound

Ultrasound is beyond the range of human hearing. The upper range of human hearing in young people is about 15,000-20,000 Hz. The ultrasound used in medical examinations is of much higher frequency—up to and beyond 1 million Hertz (megahertz, or MHz).

Ultrasound is also used in other ways:

- as a range finder in a camera, to focus the camera at the correct distance;
- to find cracks in machinery, such as jet engines and turbine blades;
- to measure the size of rooms;
- to provide heat for injured muscles—diathermy;
- to shatter and remove gallbladder stones without surgery.

The intensities used in diagnostic ultrasound range from 0.1 W/m^2 (watts per square meter) for an obstetric examination to 25,000 W/m^2 for some procedures that image the heart or blood vessels. These intensities occur over a small area of the body and for a limited time period. Since the ultrasonic waves deliver energy to the body, medical scientists have carefully examined the possibility that ultrasound exams might harm the body. Many studies have been done to see if any harm results from these sound intensities of ultrasound imaging machines. No harmful effects have yet been found.

REFLECTING ON THE ACTIVITY AND THE CHALLENGE

In this activity, you made an ultrasound map of your classroom. This map is a model of the way that ultrasound makes images of the inside of the body. You can use this model in your presentation to expectant mothers and couples to explain how ultrasound reflections make an image of the fetus.

Also, you learned how the pivoting arms that hold the ultrasound probe provide different ways of aiming the probe. Each different aim creates a set of reflections that are part of the final image. Sensors in the pivots send information about the ultrasonic beam to a computer that makes one image from the many sets of reflections. You can use what you have learned to explain how the image of the fetus is made.

M 183

MEDICINE

At the close of each activity, the student is often so involved with the completion of the single experiment that the larger context of the investigation is lost. **Reflecting on the Activity and the Challenge** is the opportunity for students to place the new insights and information into the context of the chapter and the **Chapter Challenge**. If the **Chapter Challenge** is considered a completed picture, each activity is a jigsaw piece. By completing enough of the **For You To Do** (or **Investigate**) activities, the students will be able to fit the jigsaw pieces together and complete the challenge. This summary section ensures that the students do not forget about the larger context and continue their personal momentum toward completion of the challenge.

9. Physics and Chemistry To Go

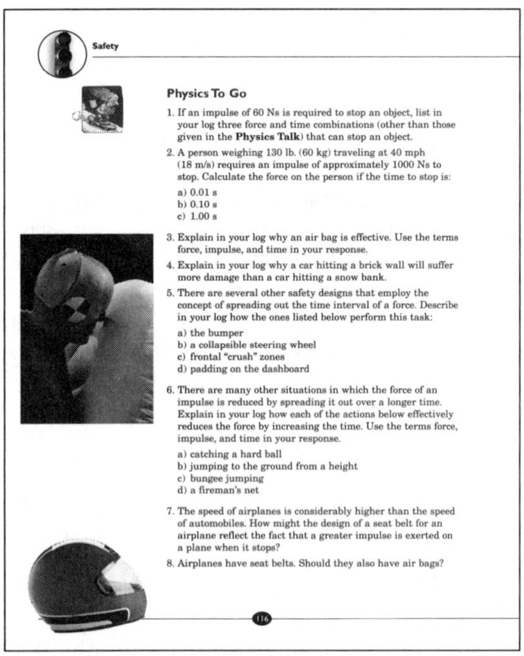

Safety

Physics To Go

1. If an impulse of 60 Ns is required to stop an object, list in your log three force and time combinations (other than those given in the **Physics Talk**) that can stop an object.
2. A person weighing 130 lb. (60 kg) traveling at 40 mph (18 m/s) requires an impulse of approximately 1000 Ns to stop. Calculate the force on the person if the time to stop is:
   a) 0.01 s
   b) 0.10 s
   c) 1.00 s
3. Explain in your log why an air bag is effective. Use the terms force, impulse, and time in your response.
4. Explain in your log why a car hitting a brick wall will suffer more damage than a car hitting a snow bank.
5. There are several other safety designs that employ the concept of spreading out the time interval of a force. Describe in your log how the ones listed below perform this task:
   a) the bumper
   b) a collapsible steering wheel
   c) frontal "crush" zones
   d) padding on the dashboard
6. There are many other situations in which the force of an impulse is reduced by spreading it out over a longer time. Explain in your log how each of the actions below effectively reduces the force by increasing the time. Use the terms force, impulse, and time in your response.
   a) catching a hard ball
   b) jumping to the ground from a height
   c) bungee jumping
   d) a fireman's net
7. The speed of airplanes is considerably higher than the speed of automobiles. How might the design of a seat belt for an airplane reflect the fact that a greater impulse is exerted on a plane when it stops?
8. Airplanes have seat belts. Should they also have air bags?

116

These sections provide additional questions and problems that can be completed outside of class. Some of the problems are applications of the principles involved in the preceding activity. Others are replication of the work in the **For You To Do** (or **Investigate**) activity. Still others provide an opportunity to transfer the results of the investigation to the context of the **Chapter Challenge**. **Physics** and **Chemistry To Go** provide a means by which students can be working on the larger **Chapter Challenge** in smaller chunks during the chapter.

10. Stretching Exercises (or Inquiring Further)

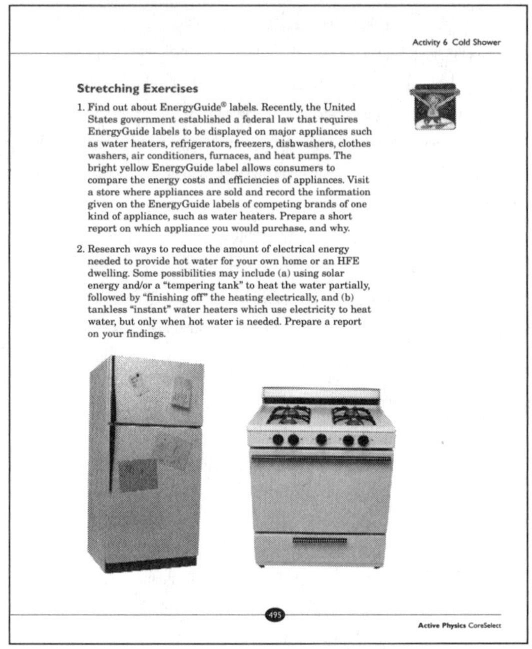

Activity 6 Cold Shower

Stretching Exercises

1. Find out about EnergyGuide® labels. Recently, the United States government established a federal law that requires EnergyGuide labels to be displayed on major appliances such as water heaters, refrigerators, freezers, dishwashers, clothes washers, air conditioners, furnaces, and heat pumps. The bright yellow EnergyGuide label allows consumers to compare the energy costs and efficiencies of appliances. Visit a store where appliances are sold and record the information given on the EnergyGuide labels of competing brands of one kind of appliance, such as water heaters. Prepare a short report on which appliance you would purchase, and why.
2. Research ways to reduce the amount of electrical energy needed to provide hot water for your own home or an HFE dwelling. Some possibilities may include (a) using solar energy and/or a "tempering tank" to heat the water partially, followed by "finishing off" the heating electrically, and (b) tankless "instant" water heaters which use electricity to heat water, but only when hot water is needed. Prepare a report on your findings.

495 Active Physics CoreSelect

Some students express additional interest in a specific topic or an extension to a topic. The **Stretching Exercises** (or **Inquiring Further**) provide an avenue in which to pursue that interest. **Stretching Exercises** (or **Inquiring Further**) often require additional readings or interviews. They may be given for extra credit to students who wish to attempt a more in-depth problem or a tougher exercise.

11. Chapter Assessment

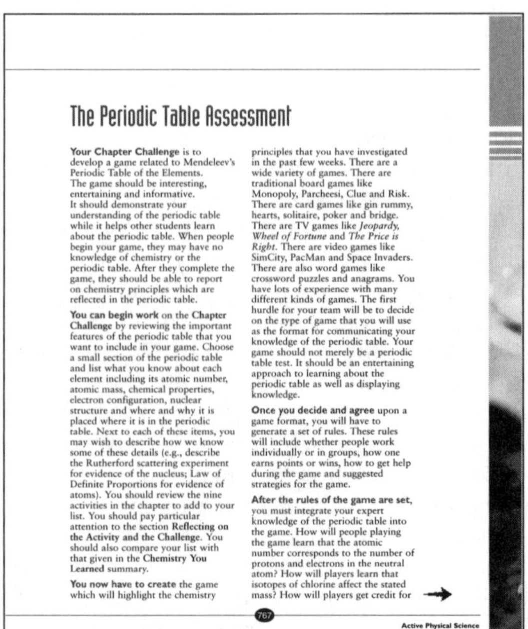

The Periodic Table Assessment

Your Chapter Challenge is to develop a game related to Mendeleev's Periodic Table of the Elements. The game should be interesting, entertaining and informative. It should demonstrate your understanding of the periodic table while it helps other students learn about the periodic table. When people begin your game, they may have no knowledge of chemistry or the periodic table. After they complete the game, they should be able to report on chemistry principles which are reflected in the periodic table.

You can begin work on the **Chapter Challenge** by reviewing the important features of the periodic table that you want to include in your game. Choose a small section of the periodic table and list what you know about each element including its atomic number, atomic mass, chemical properties, electron configuration, nuclear structure and where and why it is placed where it is in the periodic table. Next to each of these items, you may wish to describe how we know some of these details (e.g., describe the Rutherford scattering experiment for evidence of the nucleus; Law of Definite Proportions for evidence of atoms). You should review the nine activities in the chapter to add to your list. You should pay particular attention to the section **Reflecting on the Activity and the Challenge**. You should also compare your list with that given in the **Chemistry You Learned** summary.

You now have to create the game which will highlight the chemistry principles that you have investigated in the past few weeks. There are a wide variety of games. There are traditional board games like Monopoly, Parcheesi, Clue and Risk. There are card games like gin rummy, hearts, solitaire, poker and bridge. There are TV games like *Jeopardy, Wheel of Fortune* and *The Price is Right*. There are video games like SimCity, PacMan and Space Invaders. There are also word games like crossword puzzles and anagrams. You have lots of experience with many different kinds of games. The first hurdle for your team will be to decide on the type of game that you will use as the format for communicating your knowledge of the periodic table. Your game should not merely be a periodic table test. It should be an entertaining approach to learning about the periodic table as well as displaying knowledge.

Once you decide and agree upon a game format, you will have to generate a set of rules. These rules will include whether people work individually or in groups, how one earns points or wins, how to get help during the game and suggested strategies for the game.

After the rules of the game are set, you must integrate your expert knowledge of the periodic table into the game. How will people playing the game learn that the atomic number corresponds to the number of protons and electrons in the neutral atom? How will players learn that isotopes of chlorine affect the stated mass? How will players get credit for →

767 Active Physical Science

The **Chapter Assessment** is the return to the **Chapter Challenge and Criteria**. The students are ready to complete the challenge. They are able to view the challenge with a clarity that has emerged from the completion of the **For You To Do** (or **Investigate**) activities of the chapter. Students are able to review the chapter as they discuss the synthesis of the information into the required context of the challenge. The students should have some class time to work together to complete the challenge and to present their project. In many physical science courses, all students are expected to converge on the same solution. In *Active Physical Science*, each group is expected to have a unique solution. All solutions must have correct physical science, but there is ample room for creativity on the students' part. This is one of the features that captures the imagination of students who have often previously chosen not to enroll in physical science classes.

12. Physics and Chemistry You Learned

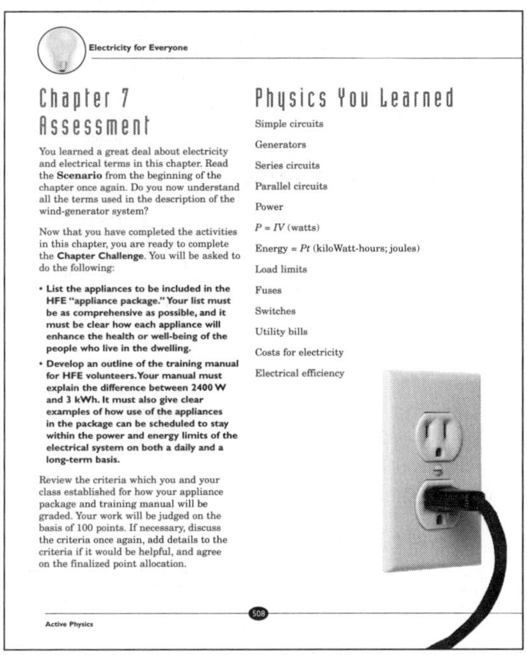
Electricity for Everyone

Chapter 7 Assessment

You learned a great deal about electricity and electrical terms in this chapter. Read the **Scenario** from the beginning of the chapter once again. Do you now understand all the terms used in the description of the wind-generator system?

Now that you have completed the activities in this chapter, you are ready to complete the **Chapter Challenge**. You will be asked to do the following:

- **List the appliances to be included in the HFE "appliance package." Your list must be as comprehensive as possible, and it must be clear how each appliance will enhance the health or well-being of the people who live in the dwelling.**
- **Develop an outline of the training manual for HFE volunteers. Your manual must explain the difference between 2400 W and 3 kWh. It must also give clear examples of how use of the appliances in the package can be scheduled to stay within the power and energy limits of the electrical system on both a daily and a long-term basis.**

Review the criteria which you and your class established for how your appliance package and training manual will be graded. Your work will be judged on the basis of 100 points. If necessary, discuss the criteria once again, add details to the criteria if it would be helpful, and agree on the finalized point allocation.

Physics You Learned

Simple circuits

Generators

Series circuits

Parallel circuits

Power

$P = IV$ (watts)

Energy $= Pt$ (kiloWatt-hours; joules)

Load limits

Fuses

Switches

Utility bills

Costs for electricity

Electrical efficiency

508

Active Physics

This small section at the end of the chapter provides a list of physical science concepts and equations which were studied in the context of the **For You To Do** (or **Investigate**) activities. It provides students with a sense of accomplishment and serves as a quick review of all that was learned during the preceding weeks.

13. Physics and Chemistry At Work

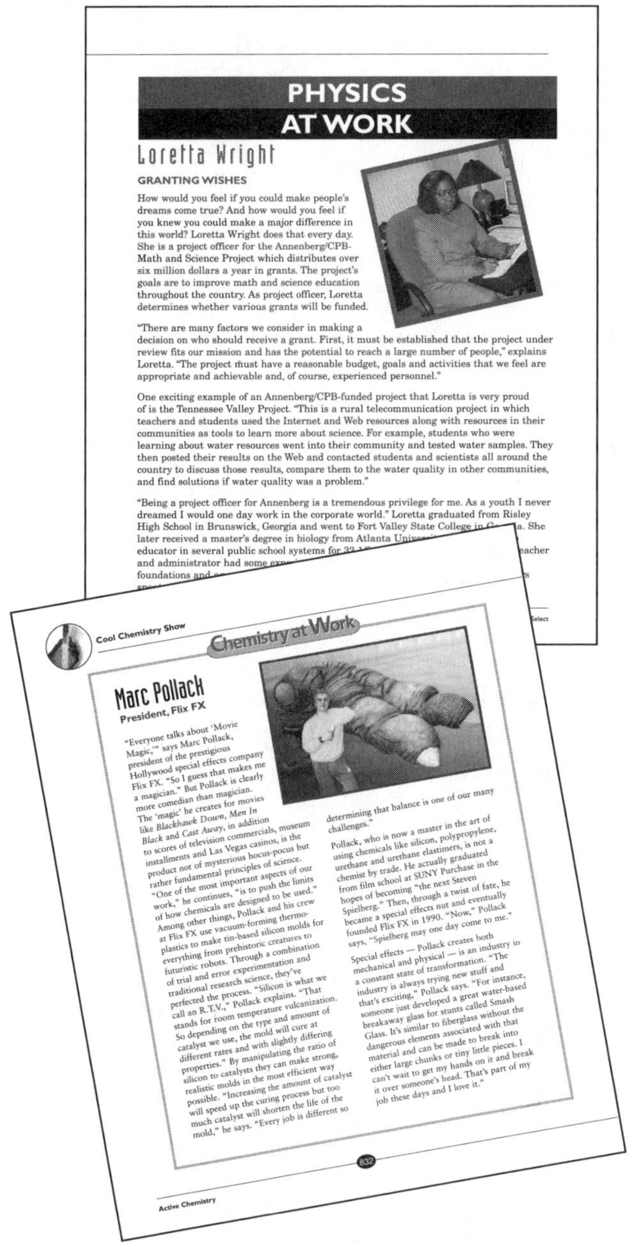
PHYSICS AT WORK

Loretta Wright

GRANTING WISHES

How would you feel if you could make people's dreams come true? And how would you feel if you knew you could make a major difference in this world? Loretta Wright does that every day. She is a project officer for the Annenberg/CPB-Math and Science Project which distributes over six million dollars a year in grants. The project's goals are to improve math and science education throughout the country. As project officer, Loretta determines whether various grants will be funded.

"There are many factors we consider in making a decision on who should receive a grant. First, it must be established that the project under review fits our mission and has the potential to reach a large number of people," explains Loretta. "The project must have a reasonable budget, goals and activities that we feel are appropriate and achievable and, of course, experienced personnel."

One exciting example of an Annenberg/CPB-funded project that Loretta is very proud of is the Tennessee Valley Project. "This is a rural telecommunication project in which teachers and students used the Internet and Web resources along with resources in their communities as tools to learn more about science. For example, students who were learning about water resources went into their community and tested water samples. They then posted their results on the Web and contacted students and scientists all around the country to discuss those results, compare them to the water quality in other communities, and find solutions if water quality was a problem."

"Being a project officer for Annenberg is a tremendous privilege for me. As a youth I never dreamed I would one day work in the corporate world." Loretta graduated from Risley High School in Brunswick, Georgia and went to Fort Valley State College in G... She later received a master's degree in biology from Atlanta Univ... educator in several public school systems for 32... and administrator had some ex... foundations and ...

Cool Chemistry Show

Chemistry at Work

Marc Pollack

President, Flix FX

"Everyone talks about 'Movie Magic,'" says Marc Pollack, president of the prestigious Hollywood special effects company Flix FX. "So I guess that makes me a magician." But Pollack is clearly more comedian than magician. The 'magic' he creates for movies like *Blackhawk Down*, *Men In Black* and *Cast Away*, in addition to scores of television commercials, museum installments and Las Vegas casinos, is the product not of mysterious hocus-pocus but rather fundamental principles of science. "One of the most important aspects of our work," he continues, "is to push the limits of how chemicals are designed to be used." Among other things, Pollack and his crew at Flix FX use vacuum-forming thermo-plastics to make tin-based silicon molds for everything from prehistoric creatures to futuristic robots. Through a combination of trial and error experimentation and traditional research science, they've perfected the process. "Silicon is what we call an R.T.V.," Pollack explains. "That stands for room temperature vulcanization. So depending on the type and amount of catalyst we use, the mold will cure at different rates and with slightly differing properties." By manipulating the ratio of silicon to catalysts they can make strong, realistic molds in the most efficient way possible. "Increasing the amount of catalyst will speed up the curing process but too much catalyst will shorten the life of the mold," he says. "Every job is different so determining that balance is one of our many challenges."

Pollack, who is now a master in the art of using chemicals like silicon, polypropylene, urethane and urethane elastimers, is not a chemist by trade. He actually graduated from film school at SUNY Purchase in the hopes of becoming "the next Steven Spielberg." Then, through a twist of fate, he became a special effects nut and eventually founded Flix FX in 1990. "Now," Pollack says, "Spielberg may one day come to me."

Special effects — Pollack creates both mechanical and physical — is an industry in a constant state of transformation. "The industry is always trying new stuff and that's exciting," Pollack says. "For instance, someone just developed a great water-based breakaway glass for stunts called Smash Glass. It's similar to fiberglass without the dangerous elements associated with that material and can be made to break into either large chunks or tiny little pieces. I can't wait to get my hands on it and break it over someone's head. That's part of my job these days and I love it."

832

Active Chemistry

These sections highlight an individual whose work or hobby is illustrative of the **Chapter Challenge. Physics** and **Chemistry At Work** speaks to the authenticity of the **Chapter Challenges**. The profiles illustrate how knowledge of physical science is important and valuable in different walks of life. The choice of profiles span the ethnic, racial, and gender diversity that we find in our nation.

Active Physical Science and the National Science Education Standards

Active Physical Science was designed and developed to provide teachers with instructional strategies that model the following from the Standards:

Guide and Facilitate Learning

- Focus and support inquiries while interacting with students.
- Orchestrate discourse among students about scientific ideas.
- Challenge students to accept and share responsibility for their own learning.
- Recognize and respond to student diversity; encourage all to participate fully in science learning.
- Encourage and model the skills of scientific inquiry as well as the curiosity openness to new ideas and data and skepticism that characterize science.

Engage in ongoing assessment of their teaching and student learning

- Use multiple methods and systematically gather data about student understanding and ability.
- Analyze assessment data to guide teaching.
- Guide students in self-assessment.

Design and manage learning environments that provide students with time, space and resources needed for learning science

- Structure the time available so students are able to engage in extended investigations.
- Create a setting for student work that is flexible and supportive of science inquiry.
- Make available tools, materials, media, and technological resources accessible to students.
- Identify and use resources outside of school.

Develop communities of science learners that reflect the intellectual rigor of scientific attitudes and social values conducive to science learning

- Display and demand respect for diverse ideas, skills, and experiences of students.
- Enable students to have significant voice in decisions about content and context of work and require students to take responsibility for the learning of all members of the community.
- Nurture collaboration among students.
- Structure and facilitate ongoing formal and informal discussion based on shared understanding of rules.
- Model and emphasize the skills, attitudes and values of scientific inquiry.

Assessment Standards

- Features claimed to be measured are actually measured.
- Students have adequate opportunity to demonstrate their achievement and understanding.
- Assessment tasks are authentic and developmentally appropriate, set in familiar context, and engaging to students with different interests and experiences.
- Assesses student understanding as well as knowledge.
- Improve classroom practice and plan curricula.
- Develop self-directed learners.

Active Physical Science Addresses Key NSES Recommendations

Scenario-Driven

There are 12 chapters in *Active Physical Science*. Each chapter begins with an engaging **Scenario** or project assignment that challenges the students and sets the stage for the learning activities and chapter assessments to follow. Chapter contents and activities are selectively aimed at providing the students with the knowledge and skills needed to address the introductory challenge, thus providing a natural content filter in the "less is more" curriculum.

Flexibly Formatted

Units are designed to stand alone, so teachers have the flexibility of changing the sequence of presentation of the units, omitting the entire unit, or not finishing all of the chapters within a unit. Although intended to serve as a full-year physics course, the units of *Active Physical Science* could be adapted to spread across a four-year period in an integrated high school curriculum.

Multiple Exposure Curriculum

The thematic nature of the course requires students to continually revisit fundamental physics principles throughout the year, extending and deepening their understanding of these principles as they apply them in new contexts. This repeated exposure fosters the retention and transferability of learning, and promotes the development of critical thinking skills.

Constructivist Approach

Students are continually asked to explore how they think about certain situations. As they investigate new situations, they are challenged to either explain observed phenomena using an existing paradigm or to develop a more consistent one. This approach can be helpful in including situations to abandon previously held notions in favor of the more powerful ideas and explanations offered by scientists.

Authentic Assessment

For the culmination of each chapter, students are required to demonstrate the usefulness of their newly acquired knowledge by adequately meeting the challenge posed in the chapter introduction. Students are then evaluated on the degree to which they accomplish this performance task. The curriculum also includes other methods and instruments for authentic assessments as well as non-traditional procedures for evaluating and rewarding desirable behaviors and skills.

Cooperative Grouping Strategies

Use of cooperative groups is integral to the course as students work together in small groups to acquire the knowledge and information needed to address the series of challenges presented through the chapter scenarios. Ample teacher guidance is provided to assure that effective strategies are used in group formation, function, and evaluation.

Math Skills Development/Graphing Calculators and Computer Spreadsheets

The presentation and use of math in *Active Physical Science* varies substantially from traditional high school physics courses. Math, primarily algebraic expressions, equations, and graphs is approached as a way of representing ideas symbolically. Students begin to recognize the usefulness of math as an aid in exploring and understanding the world around them. Finally, since many of the students in the target audience are insecure about their math backgrounds, the course engages and provides instruction for the use of graphing calculators and computer spreadsheets to provide math assistance.

Minimal Reading Required

Because it is assumed that the target audience reads only what is absolutely necessary, the entire course is activity-driven. Reading passages are presented mainly within the context of the activities, and are written at the ninth grade level.

Use of Educational Technologies

Videos which capture students' attention explore a variety of the *Active Physical Science* topics. Opportunities are also provided for students to produce their own videos in order to record and analyze events. Computer software programs make use of various interfacing devices.

Problem Solving

For the curriculum to be both meaningful and relevant to the target population, problem-solving related to technological applications and related issues is an essential component of the course. Problem solving ranges from simple numerical solutions where one result is expected, to more involved decision-making situations where multiple alternatives must be compared.

Challenging Learning Extensions

Throughout the text, a variety of **Stretching Exercises** are provided for more motivated students. These extensions range from more challenging design tasks, to enrichment readings, to intriguing and unusual problems. Many of the extensions take advantage of the frequent opportunities the curriculum provides for oral and written expression of student ideas.

Cooperative Learning

Benefits of Cooperative Learning

Cooperative learning requires you to organize and structure a lesson so that students work with other students to jointly accomplish a task. Group learning is an essential part of balanced methodology. It should be blended with whole-class instruction and individual study to meet a variety of learning styles and expectations as well as maintain a high level of student involvement.

Cooperative learning has been thoroughly researched and agreement has been reached on a number of results. Cooperative learning:

- promotes trust and risk-taking
- elevates self-esteem
- encourages acceptance of individual differences
- develops social skills
- permits a combination of a wide range of backgrounds and abilities
- provides an inviting atmosphere
- promotes a sense of community
- develops group and individual responsibility
- reduces the time on a task
- results in better attendance
- produces a positive effect on student achievement
- develops key employability skills

As with any learning approach, some students will benefit more than others from cooperative learning. Therefore, you may question as to what extent you should use cooperative learning strategies. It is important to involve the student in helping decide which type of learning approaches they prefer, and to what extent each is used in the classroom. When students have a say in their learning, they will accept to a greater extent any method which you choose to use.

Phases of Cooperative Learning Lessons

Organizational Pre-lesson Decisions

What academic and social objectives will be emphasized? In other words, what content and skills are to be learned and what interaction skills are to be emphasized or practiced?

What will be the group size? Or, what is the most appropriate group size to facilitate the achievement of the academic and social objectives? This will depend on the amount of individual involvement expected (small groups promote more individual involvement), the task (diverse thinking is promoted by larger groups), nature of the task or materials available and the time available (shorter time demands smaller groupings to promote involvement).

Who will make up the different groups? Teacher-selected groups usually have the best mix, but this can only happen after the teacher gets to know his/her students well enough to know who works well together. Heterogeneous groupings are most successful in that all can learn through active participation. The duration of the groups' existence may have some bearing on deciding the membership of groups.

How should the room be arranged? Practicing routines where students move into their groups quickly and quietly is an important aspect. Having students face-to-face is important. The teacher should still be able to move freely among the groups.

What Materials and/or Rewards Might be Prepared in Advance?

Setting the Lesson

Structure for Positive Interdependence: When students feel they need one another, they are more likely to work together—goal interdependence becomes important. Class interdependence can be promoted by setting class goals which all teams must achieve in order for class success.

Explanation of the Academic Task: Clear explanations and sometimes the use of models can help the students. An explanation of the relevance of the activity is importance. Checks for clear understanding can be done either before the groups form or after, but they are necessary for delimiting frustrations.

Explanation of Criteria for Success: Groups should know how their level of success will be determined.

Structure for Individual Accountability: The use of individual follow-up activities for tasks or social skills will provide for individual accountability.

Specification of Desired Social Behaviors: Definition and explanations of the importance of values of social skills will promote student practice and achievement of the different skills.

Monitoring/Intervening During Group Work

Through monitoring students' behaviors, intervention can be used more appropriately. Students can be involved in the monitoring by being "a team observer," but only when the students have a very clear understanding of the behavior being monitored.

Interventions to increase chances for success in completing the task or activity and for the teaching of collaborative skills should be used as necessary—they should not be interruptions. This means that the facilitating teacher should be moving among the groups as much as possible. During interventions, the problem should be turned back to the students as often as possible, taking care not to frustrate them.

Evaluating the Content and Process of Cooperative Group Work

Assessment of the achievement of content objectives should be completed by both the teacher and the students. Students can go back to their groups after an assignment to review the aspects in which they experienced difficulties.

When assessing the accomplishment of social objectives, two aspects are important: how well things proceeded and where/how improvements might be attempted. Student involvement in this evaluation is a very basic aspect of successful cooperative learning programs.

Organizing and Monitoring Groups

An optimum size of group for most activities appears to be four; however, for some tasks, two may be more efficient. Heterogeneous groups organized by the teacher are usually the most successful. The teacher will need to decide what factors should be considered in forming the heterogeneous groups. Factors which can be considered are: academic achievement, cultural background, language proficiency, sex, age, learning style, and even personality type.

Level of academic achievement is probably the simplest and initially the best way to form groups. Sort the students on the basis of marks on a particular task or on previous year's achievement. Then choose a student from each quartile to form a group. Once formed, groups should be flexible. Continually monitor groups for compatibility and make adjustments as required.

Students should develop an appreciation that it is a privilege to belong to a group. Remove from group work any student who is a poor participant or one who is repeatedly absent. These individuals can then be assigned the same tasks to be completed in the same time line as a group. You may also wish to place a ten percent reduction on all group work that is completed individually.

The chart on the next page presents some possible group structures and their functions.

What Does Cooperative Learning Look Like?

During a cooperative learning situation, students should be assigned a variety of roles related to the particular task at hand. Following is a list of possible roles that students may be given. It is important that students are given the opportunity of assuming a number of different roles over the course of a semester.

Leader:

Assigns roles for the group. Gets the group started and keeps the group on task.

Organizer:

Helps focus discussion and ensures that all members of the group contribute to the discussion. The organizer ensures that all of the equipment has been gathered and that the group completes all parts of the activity.

Recorder:

Provides written procedures when required, diagrams where appropriate and records data. The recorder must work closely with the organizer to ensure that all group members contribute.

Researcher:

Seeks written and electronic information to support the findings of the group. In addition, where appropriate, the researcher will develop and test prototypes. The researcher will also exchange information gathered among different groups.

Encourager:

Encourages all group members to participate. Values contributions and supports involvement.

Checker:

Checks that the group has answered all the questions and the group members agree upon and understand the answers.

Diverger:

Seeks alternative explanations and approaches. The task of the diverger is to keep the discussion open. "Are other explanations possible?"

Some Possible Group Structures and Their Functions*

	Structure	Brief Description	Academic and Social Functions
Team Building	Round-robin	Each student in turn shares something with his/her teammates.	Expressing ideas and opinions, creating stories. Equal participation, getting acquainted with each other.
Class Building	Corners	Each student moves to a group in a corner or location as determined by the teacher through specified alternatives. Students discuss within groups, then listen to and paraphrase ideas from other groups.	Seeing alternative hypotheses, values, and problem solving approaches. Knowing and respecting differing points of view.
Mastery	Numbered heads together	The teacher asks a question, students consult within their groups to make sure that each member knows the answer. Then one student answers for the group in response to the number called out by the teacher.	Review, checking for knowledge comprehension, analysis, and divergent thinking. Tutoring.
	Color coded co-op cards	Students memorize facts using a flash card game or an adaption. The game is structured so that there is a maximum probability for success at each step, moving from short to long-term memory. Scoring is based on improvement.	Memorizing facts. Helping, praising.
	Pairs check	Students work in pairs within groups of four. Within pairs students alternate — one solves a problem while the other coaches. After every problem or so, the pair checks to see if they have the same answer as the other pair.	Practicing skills. Helping, praising.
Concept Development	Three-step interview	Students interview each other in pairs, first one way, then the other. Each student shares information learned during interviews with the group.	Sharing personal information such as hypotheses, views on an issue, or conclusions from a unit. Participation, involvement.
	Think-pair-share	Students think to themselves on a topic provided by the teacher; they pair up with another student to discuss it; and then share their thoughts with the class.	Generating and revising hypotheses, inductive and deductive reasoning, and application. Participation and involvement.
	Team word-webbing	Students write simultaneously on a piece of paper, drawing main concepts, supporting elements, and bridges representing the relation of concepts/ideas.	Analysis of concept into components, understanding multiple relations among ideas, and differentiating concepts. Role-taking.
Multifunctional	Roundtable	Each student in turn writes one answer as a paper and a pencil are passed around the group. With simultaneous roundtables, more than one pencil and paper are used.	Assessing print knowledge, practicing skills, recalling information, and creating designs. Team building, participation of all.
	Partners	Students work in pairs to create or master content. They consult with partners from other teams. Then they share their products or understandings with the other partner pair in their team.	Mastery and presentation of new material, concept development. Presentation and communication skills.
	Jigsaw	Each student from each team becomes an "expert" on one topic area by working with members from other teams assigned to the same topic area. On returning to their own teams, each one teaches the other members of the group and students are assessed on all aspects of the topic.	Acquisition and presentation of new material review and informed debate. Independence, status equalization.

* Adapted from Spencer Kagan (1990), "*The Structural Approach to Cooperative Learning,*" Educational Leadership, December 1989/January 1990.

Active Listener:

Repeats or paraphrases what has been said by the different members of the group.

Idea Giver:

Contributes ideas, information, and opinions.

Materials Manager:

Collects and distributes all necessary material for the group.

Observer:

Completes checklists for the group.

Questioner:

Seeks information, opinions, explanations, and justifications from other members of the group.

Reader:

Reads any textual material to the group.

Reporter:

Prepares and/or makes a report on behalf of the group.

Summarizer:

Summarizes the work, conclusions, or results of the group so that they can be presented coherently.

Timekeeper:

Keeps the group members focused on the task and keeps time.

Safety Manager:

Responsible for ensuring that safety measures are being followed, and the equipment is clean prior to and at the end of the activity.

Group Assessment

Assessment should not end with a group mark. Students and their parents have a right to expect marks to reflect the students' individual contributions to the task. It is impossible for you as the instructor to continuously monitor and record the contribution of each individual student. Therefore, you will need to rely on the students in the group to assign individual marks as merited.

There are a number of ways that this can be accomplished. The group mark can be multiplied by the number of students in the group, and then the total mark can be divided among the students, as shown in the graphics that follow.

Activity:______________________

Group Mark: 8/10

Number in Group: 4

Total Marks: 32/40

Distribution of Marks

Student's Name	Mark	Signature
Ahmed	8/10	______________________
Jasmin	8/10	______________________
Mike	7/10	______________________
Tabitha	9/10	______________________

Another way to share group marks is to assign a factor to each student. The mark factors must total the number of students in the group. The group mark is then multiplied by this factor to arrive at each student's individual mark which best represents their contribution to the task, as shown below.

Activity:______________________

Group Mark: 8/10

Number in Group: 4

Mark Factors and Individual Marks

Student's Name	Mark Factor	Individual Mark	Signature
Ahmed	1.0	8/10	______________________
Jasmin	1.0	8/10	______________________
Mike	0.9	7.2/10	______________________
Tabitha	1.1	8.8/10	______________________
Total Mark Factor	4		

In any case, students must sign to show that they are in agreement with the way the individual marks were assigned.

You may also wish to provide students with an **Assessment Rubric** similar to the one shown which they can use to assess the manner in which the group worked together.

Assessment Rubric for Group Work: Individual Assessment of the Group

Individual's name: ____________________

Names of group members: ____________________

Name of activity: ____________________

Circle the appropriate number: #1 is excellent, #2 is good, #3 is average, and #4 is poor.

1. The group worked cooperatively. Everyone assumed a role and carried it out.	1	2	3	4
2. Everyone contributed to the discussion. Everyone's opinion was valued.	1	2	3	4
3. Everyone assumed the roles assigned to them.	1	2	3	4
4. The group was organized. Materials were gathered, distributed, and collected.	1	2	3	4
5. Problems were addressed as a group.	1	2	3	4
6. All parts of the task were completed within the time assigned.	1	2	3	4

Comments:

If you were to repeat the activity, what things would you change?

Chapter 1

PHYSICS IN ACTION

Chapter 1-Physics in Action

National Science Education Standards

Chapter Challenge

PBS has decided to televise sporting events in a program that has educational value. Students are challenged to provide the voice-over on a sports video that explains the physics of the action as an audition for the job of sports broadcaster.

Chapter Summary

To gain knowledge and understanding of physics principles necessary to meet this challenge, students work collaboratively on activities that investigate a variety of the forces that affect the changes in motion that are commonly observed in sports. These experiences engage students in the following content identified in the National Science Education Standards.

Content Standards

Unifying Concepts

- Systems, order, and organization
- Evidence, models and explanations
- Constancy, change and measurement

Science as Inquiry

- Identify questions and concepts that guide scientific investigations
- Use technology and mathematics to improve investigations
- Communicate and defend a scientific argument
- Formulate and revise scientific explanations and models using logic and evidence

Physical Science

- Motions and Forces
- Conservation of energy and increase in disorder

History and Nature of Science

- Science as a human endeavor
- Nature of scientific knowledge
- Historical perspectives

Key Physics Concepts and Skills

Activity Summaries	Physics Principles
Activity 1: A Running Start and Frames of Reference Students measure the motion of a ball rolling down then up the sides of a bowl and find the ratio of the "running start" to the vertical distance. From this, they are introduced to the concept of inertia.	• Acceleration • Gravity • Galileo's Principle of Inertia • Newton's First Law of Motion
Activity 2: Push or Pull–Adding Vectors Students construct, calibrate, and use a simple force meter to explore the variables involved in throwing a shot put. They then connect their observations and data to a study of the laws of motion.	• Newton's Second Law of Motion • Relationship of mass and force to acceleration • Gravity
Activity 3: Center of Mass By finding the balance points on objects with a variety of shapes, students are introduced to the effect motion of the athlete's center of mass has on balance and performance.	• Center of Mass • Gravity
Activity 4: Defy Gravity Students learn to measure hang time and analyze vertical jumps of athletes using slow-motion videos. This introduces the concept that work when jumping is force applied against gravity.	• Gravity • Potential and kinetic energy • Work • Vertical accelerated motion
Activity 5: Run and Jump Thinking about the direction in which they apply force to move in a desired way introduces students to the concept that a force has an equal and opposite force. They test this concept, then apply it to a variety of motions observed in sports.	• Force vectors • Weight and gravity as forces • Newton's Third Law of Motion
Activity 6: The Mu of the Shoe Students measure the amount of force necessary to slide athletic shoes on a variety of surfaces. From this and the weight of the shoe, they learn to calculate friction coefficients. They then consider the effect of friction on an athlete's performance.	• Gravity • Frictional force • Normal force • Coefficient of Sliding friction
Activity 7: Concentrating on Collisions Students investigate the affect of a ball's velocity on its motion after a collision. They then apply these observations and what they now know about opposing forces in motion to describe collisions of balls and athletes in sporting events.	• Newton's Third Law of Motion • Mass • Velocity • Momentum
Activity 8: Conservation of Momentum Additional collisions between objects allow students to investigate what happens when the objects stay together or "stick" after the collision.	• Newton's Third Law of Motion • Momentum = Mass × Velocity • Velocity • Law of Conservation of Momentum
Activity 9: Circular Motion Students use an accelerometer to test the direction of acceleration when spinning in a chair. From this, they investigate the forces involved in the movement of turning objects and athletes.	• Inertia • Centripetal acceleration • Centripetal force

GETTING STARTED WITH EQUIPMENT NEEDED TO CONDUCT THE ACTIVITIES.

Items needed—not supplied in Material Kits

Preparing the equipment needed for each activity in this chapter is an important procedure. There are some items, however, needed for the chapter that are not supplied in the It's About Time material kitpackage. Many of these items may already be in your school and would be an unnecessary expense to duplicate. Please read carefully the list of items to the right which are not found in the supplied kits and locate them before beginning activities.

Items needed—not supplied by It's About Time:

- **Dried Peas**
- **Plastic beads**
- **Raw rice**
- **Eraser**
- **Teeter-totter**
- **VCR & Monitor**
- **Sports Content Video**
- **Computer**
- **MBL or CBL with Motion sensor**
- **Safety Helmet**
- **Knee Pads**
- **Elbow pads**
- **Chair with wheels**
- **Shoe**
- **Smooth surface (desk top, tile)**
- **Stack of books**
- **Wood board, 4 ft**
- **Rotating stool or chair**

Equipment List For Chapter 1

PART	ITEM	QTY	ACTIVITY	TO SERVE
AH-9251	Accelerometer	1	9	Group
AH-9252-S2	Accelerometer, Cork	1	9	Group
AH-9250-S1	Accelerometer-To-Cart	1	9	Group
PH-1133	Air Puck	1	1A	Class
BS-7203-S2	Balls, Bocci		7,9	Group
BS-1400-S2	Ball, Golf	1	7	Group
NB-0004	Ball, Nerf	1	7, 9	Group
BS-6423	Ball, Practice Golf	1	7	Group
BS-7207-S2	Balls, Set of 2	1	1, 2	Group
BS-0787-S2	Ball, Super, 1" Diameter	1	1	Group
BS-0790-S2	Ball, Tennis	1	2,7	Group
BS-5856-S2	Bowl, Salad,	1	1	Group
CM-1108-S2	Calculator, Basic	1	6A	Group
CS-3246-S2	C-Clamp	1	2	Group
DS-7201-S2	Dots, Adhesive, 3/4"	28	3	Group
GC-0001	Dynamics Cart	2	8	Group
MS-1425-S2	Marking Pen, Felt Tip	1	1	Group
MC-5418-S2	Modeling Clay	1	8	Group
PM-0220-S2	Plumb Bob W/String	1	3	Group
PP-6128-S2	Push Pins	100	3	Group
SS-7209-S2	Rough Horizontal Surface, 6" x 36"	1	6	Group
RS-2723-S2	Ruler, Flexible Plastic, 30 cm	1	1,2,9	Group
RS-2826-S2	Ruler, Metric, mm Marking	1	1, 2, 6A, 9	Group
SS-0491	Scale, Bathroom	1	5A	Class
SS-7210-S2	Set of Shapes, A,B,C,D	1	3	Group
SH-7708-S2	Skateboard	1	5	Group
SM-1676-S2	Stick, Meter, 100 cm, Hardwood	1	3, 4, 5	Group
SS-2304-S2	Spring Scale, 0-20 Newton Range	1	6,6A	Group
SS-2303-S2	Spring Scale, 0-10 Newton Range	1	8	Group
RS-7211-S2	Starting Ramp For Bocci Ball	2	7	Group
TS-2662-S2	Tape, Masking, 3/4" x 60 yds.	2	1,4,7,8	Group
TT-6100-S2	Ticker Tape Timer	1	8	Group
RS-4462-S2	Track W/Adjustable Outrun Slope	1	1	Group
WS-0376-S2	Weight, 100 g Slotted Mass	10	5,6,8	Group
WS-6910-S2	Washers, Metal, 3/4"	4	2,5	Group
ITEMS NEEDED – NOT SUPPLIED BY IT'S ABOUT TIME				
	Dried Peas	1	1A	
	Plastic beads	1	1A	
	Raw rice	1	1A	
	Eraser	1	1A	
	Teeter-totter	1	3A	
	VCR & Monitor	1	4	
	Sports Content Video	1	4,7	
	Computer	1	4A	
	MBL or CBL with Motion sensor	1	4A	
	Safety Helmet	1	5	
	Knee Pads	1	5	
	Elbow pads	1	5	
	Chair with wheels	1	5	
	Shoe	1	6,6A	
	Smooth surface (desk top, tile)	1	6	
	Stack of books	1	6,6A	
	Wood board, 4 ft	1	6A	
	Rotating stool or chair	1	9	

Organizer for Materials Available in Teacher's Edition

Activity in Student Text	Additional Material	Alternative / Optional Activities
ACTIVITY 1: A Running Start and Frames of Reference, p. 4	Performance Assessment Rubrics, pgs. 28-29	Activity 1 A: Can Objects Move Forever?, p. 27
ACTIVITY 2: Push or Pull–Adding Vectors, p. 15		
ACTIVITY 3: Center of Mass, p. 26	Templates for Shapes A, B, C, and D, pgs. 52-53	Activity 3 A: Alternative Method for Determining Center of Gravity, p. 51
ACTIVITY 4: Defy Gravity p. 31	Calculating Hang Time and Force During a Vertical Jump (Worksheet) pgs. 70-71	Activity 4 A: High-Tech Alternative for Monitoring Vertical Jump Height, p. 69
ACTIVITY 5: Run and Jump, p. 45		Activity 5 A: Using a Bathroom Scale to Measure Forces, pgs. 80-81
ACTIVITY 6: The Mu of the Shoe, p. 50	Assessment Rubric, Physics to Go Question 8, p. 90 Background Information for Activity 6 A: Alternative Activity for Measuring the Mu of the Shoe, pgs. 93-94	Activity 6 A: Alternative Activity for Measuring the Mu of the Shoe, p. 92
ACTIVITY 7: Concentrating on Collisions, p. 56		
ACTIVITY 8: Conservation of Momentum, p. 61		
ACTIVITY 9: Circular Motion, p. 69		

Scenario

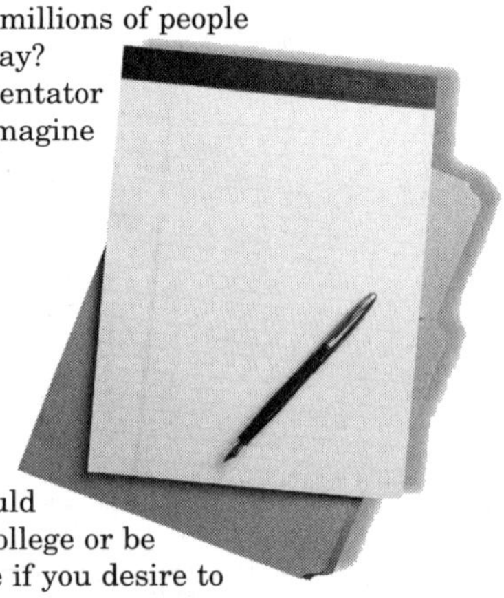

Have you ever dreamed of being a sports analyst for Monday Night Football with millions of people listening to every word you say? What about the sports commentator for the Summer Olympics? Imagine interviewing the Most Valuable Player (MVP) after an NBA championship game or interviewing the Olympic Gold Medalist in women's figure skating. What type of credentials are needed to have such a glamorous career in sportscasting? Should you major in journalism in college or be a retired professional athlete if you desire to land such a lucrative and exciting job? Could the study of physics be a key to becoming a sports analyst? Could a student with a knowledge of physics bring to the TV viewer a different perspective that might provide a new outlook on sporting events?

Challenge

PBS has decided that it wants to televise certain sporting events and that they would like these programs to have some educational as well as entertainment value. As a test of this idea, you are to provide the voice-over on a sports video and to explain the physics of the action appearing on the screen. Here is your chance to audition for a job in sportscasting. Each student (or group of students) will do a "science commentary" on a short (2–3 minute) sports video.

To assess how well you understand this material, you (or your group) are to do one of these:

Chapter and Challenge Overview

In this chapter Newton's Laws of Motion and also the concepts of force, inertia (mass), momentum, and the physics of rotation are introduced.

Students will be asked to produce a voice-over (or a script for a voice-over) to explain the physics behind a short sports video. You will select the video, but if the students are very ambitious, they can either find some footage themselves, shoot some scenes with a camcorder, or tape some sporting events from TV. The entire chapter will build toward this end, and the final evaluation of the student's progress will be based on the video voice-over.

There are a number of objectives in this chapter, one of which is to show the students that the laws of physics hold true not only in their science class and lab, but out in the world as well. The students should be able to look at a sporting event and realize what physical principle is involved. Hopefully this will carry over to everyday life, and the student will then be able to see the physics in the world around them.

Each class might start with a short video segment showing sports bloopers. They are commercially available and many of the students may have their own. After the class has covered some of the material, it is increasingly appropriate to discuss the physics that is being displayed in the blooper. Many of these bloopers are very humorous and the students look forward to the beginning of the class.

As you review the **Chapter Challenge** assignments, reassure the students that while they may feel incompetent now, by the end of the chapter they will have the necessary skills and vocabulary to respond adequately.

On the following pages of the Teacher's Edition there are suggestions on how to evaluate students on this material. It is very important at this time that the students be made aware of the method you are going to use and how you will evaluate their work. Have the students actively participate in deciding the criteria for evaluation.

The **Physics To Go** at the end of each section often contains more questions than should ever be assigned for homework. This section has been written in such a way as to give you a choice as to how much work, and the nature of the work the students will be expected to do each day out of class.

As you work with *Active Physical Science*, be aware that the same physics concepts appear repeatedly in different contexts. It is not necessary for the students to achieve total understanding the first time that they encounter Newton's Laws of Motion, and the physics of rotation.

- **submit a written script**
- **narrate live**
- **dub onto the video soundtrack**
- **record on an audiocassette**

Your task is not to give a play-by-play description of the sporting event or give the rules of the game but rather to go a step beyond and educate the audience by describing to them the rules of nature that govern the event. This approach will give the viewer (and you) a different perspective on both sports and physics. The laws of physics cover not only obscure phenomena in the lab, but everyday events in the real world as well.

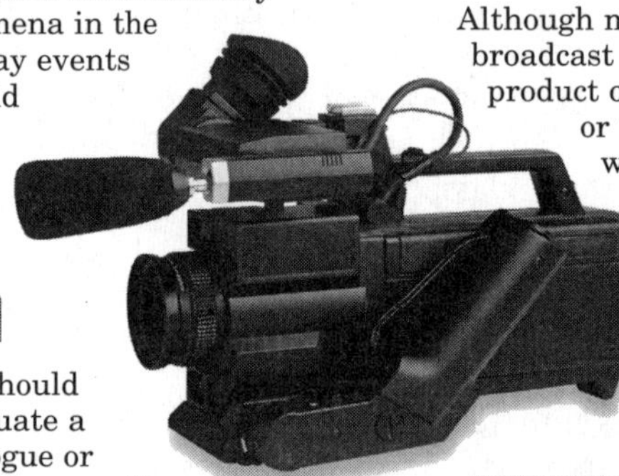

Criteria

What criteria should be used to evaluate a voice-over dialogue or script of a sporting event? Since the intention is to provide an analysis of and interest in the physics of sports, the voice-over should include the use of physics terms and physics principles. All of these terms and principles should be used correctly. How many of these terms and principles would constitute an excellent job? Would it be enough to use one physics term correctly and explain how one physics principle is illustrated in the sport? Should use of one physics term and one physics principle be a minimum standard to get minimal credit for this assessment? Discuss in your small groups and your class and decide on reasonable expectations for the physics criteria for the assessment.

Since the assessment requires a product that will be a part of television, another aspect of the criteria for success would be the entertainment quality of the voice-over. Does a commentator who adds humor or drama receive a higher rating than someone who has similar physics content but has added no excitement or interest to the broadcast? How does one weigh the value of the entertainment quality and the value of relevant physics? What are reasonable expectations for the entertainment aspect of the voice-over? Discuss and decide as a class.

Although many people may be in the broadcast booth, a voice-over becomes the product of one person—the commentator or the scriptwriter. Although you will be working in cooperative groups during the chapter, each person will be responsible for a voice-over or script for a sporting event. As a team of two or three, you may wish to work together and share different aspects of the job, but the output of work per person should be the same. That is why one voice-over will be required of each person irrespective of whether individuals prefer to work independently or in groups.

Assessment Rubric for Voice-Over Dialogue or Script

Meets the standard of excellence. **5**	• A significant number of physics principles are consistently and correctly addressed. • Physics concepts from the chapter are repeatedly integrated in the appropriate places. • Physics terminology and equations are consistently incorporated as applicable. • Correct estimates of the magnitude of physical quantities are frequently used. • Additional research, beyond basic concepts presented in the chapter, is evident. Knowledge of the rules of the game are evident. • The voice-over has great entertainment value. It contains humor and excitement.
Approaches the standard of excellence. **4**	• A significant number of physics principles are often correctly addressed. • Physics concepts from the chapter are integrated in the appropriate places. • Physics terminology and equations are incorporated as applicable. • Correct estimates of the magnitude of physical quantities are frequently used. • Knowledge of the rules of the game are evident. • The voice-over has entertainment value. It contains some humor and excitement.
Meets an acceptable standard. **3**	• A sufficient number of physics principles are correctly addressed. • Physics concepts from the chapter are integrated in the appropriate places. • A limited amount of physics terminology and equations are incorporated as applicable. • Correct estimates of the magnitude of physical quantities are occasionally used. • Knowledge of the rules of the game are general. • The voice-over has some entertainment value.
Below acceptable standard and requires remedial help. **2**	• Very few physics principles are addressed. • Physics concepts from the chapter are not always integrated in the appropriate places. • A limited amount of physics terminology is incorporated as applicable. • Estimates of the magnitude of physical quantities are seldom used. • Knowledge of the rules of the game is weak. • The voice-over has a limited entertainment value.
Basic level that requires remedial help or demonstrates a lack of effort. **1**	• Physics principles are not addressed correctly. • Physics concepts from the chapter are not integrated in the appropriate places. • Physics terminology is not used. • No attempt is made to include the magnitude of physical quantities. • Knowledge of the rules of the game is lacking. • The voice-over is difficult to follow and portions are missing.

For use with *Physics In Action*, Chapter 1

Assessment Rubric for Voice-Over Dialogue or Script

Meets the standard of excellence. **5**	• Scientific vocabulary is used consistently and precisely. • Sentence structure is consistently controlled. • Spelling, punctuation, and grammar are consistently used in an effective manner. • Scientific symbols for units of measurement are used appropriately in all cases.
Approaches the standard of excellence. **4**	• Scientific vocabulary is used appropriately in most situations. • Sentence structure is usually consistently controlled. • Spelling, punctuation, and grammar are generally used in an effective manner. • Scientific symbols for units of measurement are used appropriately in most cases.
Meets an acceptable standard. **3**	• Some evidence that the student has used scientific vocabulary, although usage is not consistent or precise. • Sentence structure is generally controlled. • Spelling, punctuation, and grammar do not impede the meaning. • Some scientific symbols for units of measurement are used. Generally, the usage is appropriate.
Below acceptable standard and requires remedial help. **2**	• Limited evidence that the student has used scientific vocabulary. Generally, the usage is not consistent or precise. • Sentence structure is poorly controlled. • Spelling, punctuation, and grammar impedes the meaning. • Some scientific symbols for units of measurement are used, but most often, the usage is inappropriate.
Basic level that requires remedial help or demonstrates a lack of effort. **1**	• Limited evidence that the student has used scientific vocabulary and usage is not consistent or precise. • Sentence structure is poorly controlled. • Spelling, punctuation, and grammar impedes the meaning. • No attention to using scientific symbols for units of measurement.

Maximum = 10 points

For use with *Physics In Action*, Chapter 1

What is in the Physics InfoMall for Chapter 1?

Chapter 1 deals with the physics of sports

If you have had much experience with the *Physics InfoMall CD-ROM*, you have probably done a few searches, and no doubt some of the searches have resulted in "Too many hits." Surprisingly, searching the entire CD-ROM with the keyword "sport*" does not give "too many" hits, but provides some interesting hits. Note that the asterisk is a wild character; this searches for any word beginning with "sport."

If you do the search just mentioned, the first hit is a resource letter

("Resource letter PS-1: Physics of sports," *American Journal of Physics, vol. 54, issue 7*) that discusses the published discussions on the physics of sports. According to this letter, "there is surprisingly little published information about the basic physics underlying most sports, even though the relevant physics is all classical." Included is a list of places you might find such information, including journals and books. The letter contains a list of specific references grouped by sport, such as Physics of basketball, *American Journal of Physics, vol. 49, issue 4*. Another interesting article is "Students do not think physics is 'relevant.' What can we do about it?," in the *American Journal of Physics, vol. 36, issue 12*.

Given that the physics in sports is classical, you might search for student difficulties learning classical physics in general. One article you might find is "Factors influencing the learning of classical mechanics," *American Journal of Physics 48, issue 12*. Knowledge of such factors affecting learning can be a valuable tool. Perform other searches that meet your needs, and the InfoMall is very likely to provide good information. And we have not even opened the Textbook Trove yet!

ACTIVITY 1
A Running Start and Frames of Reference

Background Information

Two major ideas are introduced in this activity:

- Galileo's Principle of Inertia
- Newton's Second Law of Motion

Before attempting to identify causes and effects for generating, sustaining, and arresting motion, a pivotal question first must be answered: What kinds of motion require explanation?

Two distinct kinds of motion along a straight line often are encountered in nature: (1) motion with constant speed and (2) motion with uniform, or constant, acceleration.

Since the contributions of Galileo, physics has operated from the perspective that the first of these kinds of motion, constant speed, has no cause. Galileo devised a number of arguments and demonstrations, some of which are replicated in this activity, to support this notion.

The cause of all accelerated motion is force; some agent(s) must be pushing or pulling—exerting a force—on any object observed to be accelerating. Sources or kinds of forces abound. Every situation that involves acceleration has an associated net force. Observation: If an orange is dropped, it accelerates; assigned cause: the downward force due to gravity. When the orange hits the floor it stops; another acceleration, another force. The force which stops the orange is provided by the floor, upward. A magnet brought near another magnet will cause an acceleration; therefore, there must be a magnetic force.

Sometimes, we can also discover forces hiding in constant-speed linear motion. Drop a coffee filter: it accelerates downward for a bit, but the amount of acceleration falls off to zero, so that the coffee filter falls most of the way at constant speed. Did the force of gravity decrease or disappear? No, a coffee filter seems to weigh (a measure of the force of gravity) the same at every point in the descent path. Conclusion: there must be another force, the force of air resistance, acting in the opposite direction to gravity. The force of air resistance eventually balances out the gravitational force. It is possible for a combination of forces to have a net effect of zero.

So, it is the net force on an object that imparts the acceleration. Newton's First Law of Motion states the case: An object at rest tends to remain at rest, and an object in motion (in a straight line) tends to remain in motion unless acted upon by an outside (net, non-zero) force. This statement is more complete than the one provided to the students in **Activity 1**. Whenever speed, direction, or both speed and direction, are observed to change, a net force is the cause.

The First Law does not attempt to quantify the relationship between accelerations and the forces that cause them. Establishing the quantitative relationship requires experimental evidence which is the purpose of the next activity.

Active-ating the Physics InfoMall

Note that this activity has students perform a simple experiment, and gradually leads them to concentrate on one aspect of the motion, then leads to predictions and generalizations. The importance of the prediction should not be overlooked; indeed, predictions force students to examine their understanding of a phenomena and actively engage thought. If you were to search the InfoMall to find more about the importance of predictions in learning, you would find that you need to limit your search. For example, a search for "prediction*" AND "inertia" resulted in several hits; the first hit is from *A Guide to Introductory Physics Teaching: Elementary Dynamics,* Arnold B. Arons' Book Basement entry. Here is a quote from that book: "Because of the obvious conceptual importance of the subject matter, the preconceptions students bring with them when starting the study of dynamics, and the difficulties they encounter with the Law of Inertia and the concept of force, have attracted extensive investigation and generated a substantial literature. A sampling of useful papers, giving far more extensive detail than can be incorporated here, is cited in the bibliography [Champagne, Klopfer, and Anderson (1980); Clement (1982); di Sessa (1982); Gunstone, Champagne, and Klopfer (1981); Halloun and Hestenes (1985); McCloskey, Camarazza, and Green (1980); McCloskey (1983); McDermott (1984); Minstrell (1982); Viennot (1979); White (1983), (1984)]." Note that students' preconceptions can have a large effect on how they learn something. It is important that they are forced to consciously acknowledge their preconceptions by making predictions.

Not surprisingly, among the list of hits from the search just mentioned is an article on Galileo,

"Galileo, yesterday and today," *American Journal of Physics* vol. 33, issue 9, 1965. This article provides an interesting insight into Galileo and his work, as well as several historical accounts of his work. Check it out; it might provide interesting additional reading for your students.

Of course, Newton had something to say about inertia, and another hit from the same search provides *Physics for Science and Engineering* in the Textbook Trove. See Chapter 4, Newton's Principles of Motion.

The search above was conducted initially to explore the importance of predictions, especially as related to the concept of inertia. As we can see, additional information was provided that was easily relevant to this topic. This is not unusual when searching the InfoMall — you will often find many interesting bits of information that may take you on unexpected, but enlightening, tangents.

In **Physics To Go, Question 2**, you are encouraged to find something about "curling." Sadly, the InfoMall has only one reference to this sport, and it is a short passage indicating that Lord Kelvin broke his leg while curling, and limped badly thereafter. While not directly related to anything in this section, it is another of those interesting articles one can find on the InfoMall.

Planning for the Activity

Time Requirements

- One class period.

Materials Needed

For each group:

- ball, set of 2
- ball, super, 1" diameter
- bowl, salad
- marking pen, felt tip
- ruler, flexible plastic, 30 cm
- ruler, metric, mm marking
- tape, masking, 3/4" x 60 yds.
- track w/adjustable outrun slope

Advance Preparation and Setup

You may wish to consider using ball bearings or glass marbles rolling within flexible, transparent plastic tubing for the second part of **For You To Do** instead of a ball rolling on an adjustable track. If so, you may wish to procure the tubing and bearings in advance.

Teaching Notes

Active Physical Science uses a modified constructivist model. By confronting students' misconceptions and by having them do hands-on exploration of ideas, we seek to replace their misconceptions with correct perceptions of reality. In order to do this, a consistent scheme is integrated into the course activities to elicit the students' misconceptions early in any activity.

Students' current mental models are sampled by one or more **What Do You Think?** questions. Students are not expected to know a "right" answer. These questions are supposed to elicit from students their beliefs regarding a very specific prediction or outcome, and students should commit to a written specific answer in their logs.

When students have completed **For You To Do,** convene the entire class for a demonstration of objects moving at constant speed on low-friction surfaces. A demonstration is included at the end of this activity. Possible materials for demonstrating motion at constant speed for an object given a push start on a low-friction, flat surface such as a smooth counter top or the glass surface include:

- balloon air puck on a smooth, hard surface
- a piece of dry ice on a smooth, hard surface
- glider on air track
- puck on air table
- puck on raw rice in a ripple tank
- puck on plastic bead bearings in a ripple tank.

Then direct students to read the **For You To Read** and **Physics Talk** sections. Reserve some time for closure after students have completed the reading.

You may wish to direct the students' attention to the fact that several sports involve motion for which an initial speed does not involve running in the literal sense, but may involve an object, such as a shot put ball, being given an initial speed by an athlete.

Activity Overview

Student Objectives

Students will:

- Understand and apply Galileo's Principle of Inertia.
- Understand and apply Newton's First Law of Motion.
- Recognize inertial mass as a physical property of matter.

ANSWERS FOR THE TEACHER ONLY

What Do You Think?

The horizontal distance a basketball player travels while "hanging" is determined by the speed upon jumping; since the speed often is high, the trajectory is quite flat near the peak of flight, giving the illusion that the player "hangs" in the air.

Skaters maintain speed on ice due to very low friction between the blades and the ice.

Physics in Action

Activity 1 A Running Start and Frames of Reference

GOALS

In this activity you will:

- Understand and apply Galileo's Principle of Inertia.
- Understand and apply Newton's First Law of Motion.
- Recognize inertial mass as a physical property of matter.

What Do You Think?

Many things that happen in athletics are affected by the amount of "running start" speed an athlete can produce.

- **What determines the amount of horizontal distance a basketball player travels while "hanging" to do a "slam dunk" during a fast break?**
- **How do figure skaters keep moving across the ice at high speeds for long times while seldom "pumping" their skates?**

Record your ideas about these questions in your *Active Physics* log. Be prepared to discuss your responses with your small group and the class.

For You To Do

1. Use a salad bowl and a ball to explore the question, "When a ball is released to roll down the inside surface of a salad bowl, is the motion of the ball up the far side of the bowl the 'mirror image' of the ball's downward motion?" Use a nonpermanent pen to mark a starting position for the ball near the top edge of the bowl. Use a

Active Physics 4

Activity 1 A Running Start and Frames of Reference

flexible ruler to measure, in centimeters, the distance along the bowl's curved surface from the bottom-center of the bowl to the mark.

a) Make a table similar to the one below in your log.

Start	Trial	Starting Distance (cm)	Recovered Distance (cm)	$\frac{\text{Recovered Distance}}{\text{Starting Distance}}$
High	1			
	2			
	3			
Medium	1			
	2			
	3			
Low	1			
	2			
	3			

b) Record the measured distance in your table as the High Starting Distance.

Do not use a glass bowl, if possible.

2. Prepare to observe and mark the position on the far side of the bowl where the ball stops when it is released from the starting position. Release the ball from the starting position and mark the position where it stops. Measure the distance from the bottom-center to the stop mark.

a) Record the distance in your table as the High Recovered Distance.

3. Repeat **Step 2** above two more times to see if the results are consistent.

a) Record the data for all three trials.

4. Mark two more starting positions on the surface of the bowl, one for Medium Starting Distance and another for Low Starting Distance.

a) Measure and record each of the new distances.

ANSWERS

For You To Do

1. a) Students copy tables into their logs.
   b) Students record data.

2.–4. Students record data. When rolling the ball within the salad bowl, students should find the Recovered Distance to be very nearly equal to the Starting Distance.

b) Observe, mark, measure and record the recovered distances for three trials at each of the medium and low starting positions.

c) Complete the table by calculating and recording the value of the ratio of the Recovered Distance to the Starting Distance for each trial. (The ratio is the Recovered Distance divided by the Starting Distance.)

Example:
If the Recovered Distance is 6.0 cm for a Starting Distance of 10.0 cm, the value of the ratio is $\frac{6.0 \text{ cm}}{10.0 \text{ cm}} = 0.6$.

d) For each of the three starting distances, to what extent is the motion of the ball up the far side of the bowl the "mirror image" of the downward motion? Use data as evidence for your answer.

e) Does the fraction of the starting distance "recovered" when going up the far side of the bowl depend on the amount of starting distance? Describe any pattern of data that supports your answer.

5. Repeat the activity but roll the ball along varying slopes during its upward motion. Make a track that has the same slope on both sides, as shown below. Your teacher will suggest how high the ends of the track sections should be elevated. This time, concentrate on comparing the vertical height of the ball's release position to the vertical height of the position where the ball stops.

a) Measure and record the vertical height (not the distance along the track) from which the ball will be released at the top end of the left-hand section of track.

b) Prepare to observe and mark the position on the right-hand section of track where the ball stops when it is released from the starting position. Release the ball from the top end of the left-hand section of track and mark the position where it stops. Measure and record the vertical height of the position where the ball stops.

ANSWERS

For You To Do *(continued)*

4. c)-e) The ratio of Recovered Distance to Starting Distance should be only slightly less than 1.00 and typically 0.90 or more. The actual value will, of course, depend on the coefficient of friction for the particular kind of ball and bowl used. You may expect that the error of measurement will be nearly as much as the observable difference in distances, indicating nearly complete "conservation" of distance. The ratio should remain essentially constant regardless of the starting height.

5. a)-c) Students record data and calculate the ratios of Recovered to Starting Distance.

Activity 1 A Running Start and Frames of Reference

c) Calculate the ratio of the recovered height to the starting height. How is this case, and the result, similar to what you did when using the salad bowl? How is it different?

6. Leave the left-hand starting section of track unchanged, but change the right-hand section of track so that it has less slope and is at least long enough to allow the ball to recover the starting height. The track should be arranged approximately as shown below.

a) Predict the position where the ball will stop on the right-hand track if it is released from the same height as before on the left-hand track. Mark the position of your guess on the right-hand track and explain the basis for your prediction in your log.

7. Release the ball from the same height on the left-hand section of track as before and mark the position where the ball stops on the right-hand section of track.

a) How well did you guess the position? Why do you think your guess was "on" or "off"?

b) Measure the vertical height of the position where the ball stopped and again calculate the ratio of the recovered height to the starting height. Did the ratio change? Why, do you think, did the ratio change or not change?

8. Imagine what would happen if you again did not change the left-hand starting section of track, but changed the right-hand section of track so that it would be horizontal, as shown below.

a) How far along the horizontal track would the ball need to roll to recover its starting height (or most of it)? How far do you think the ball would roll?

b) When rolling on the horizontal track, what would "keep the ball going"?

Answers

For You To Do *(continued)*

6. a) Make certain that students record their predictions.

7. a)-b) When rolling the ball down the adjustable track, students must shift attention from comparing distances traveled along the "down" and "up" paths to vertical distances "down" and "up." For symmetrical slopes, the former measurement—distance along either track—would serve, but, as the "up" slope is made less, the ball will roll farther on the "up" slope to gain nearly the same vertical height as the height from which it was released on the "down" slope.

8. a)-b) The intent is for students to realize, in accord with Galileo's ideal, that when the "up" track has no slope, the ball will roll "forever" in its attempt to gain the height from which it was released.

Physics in Action

FOR YOU TO READ

Inertia

Italian philosopher Galileo Galilei (1564–1642), who can be said to have introduced science to the world, noticed that a ball rolled down one ramp seems to seek the same height when it rolls up another ramp. He also did a "thought experiment" in which he imagined a ball made of extremely hard material set into motion on a horizontal, smooth surface, similar to the final track in **For You To Do**. He concluded that the ball would continue its motion on the horizontal surface with constant speed along a straight line "to the horizon" (forever). From this, and from his observation that an object at rest remains at rest unless something causes it to move, Galileo formed the Principle of **Inertia**:

Inertia is the natural tendency of an object to remain at rest or to remain moving with constant speed in a straight line.

Isaac Newton, born in England on Christmas day in 1642 (within a year of Galileo's death), used Galileo's Principle of Inertia as the basis for developing his First Law of Motion, presented in **Physics Talk**. Crediting Galileo and others for their contributions to his thinking, Newton said, "If I have seen farther than others, it is because I have stood on the shoulders of giants."

Running Starts

Running starts take place in many sporting activities. Since there seems to be this prior motion in many sports, there must be some advantage to it.

In sports where the objective is to maximize the speed of an object or the distance traveled in air, the prior motion may be essential. When a javelin is thrown, at the instant of release it has the same speed as the hand that is propelling it.

- The hand has a forward speed relative to the elbow, the elbow has a forward speed relative to the shoulder (because the arm is rotating around the elbow and shoulder joints), and the shoulder has a forward speed relative to the ground because the body is rotating and the body is also moving forward.
- The javelin speed then is the sum of each of the above speeds. If the thrower is not running forward, that speed does not add into the equation.

You can write a **velocity** equation to show the speeds involved.

$$v_{javelin} = v_{hand} + v_{elbow} + v_{shoulder} + v_{ground}$$

Motion captures everyone's attention in sports. Starting, stopping, and changing direction (**accelerations**) are part of the motion story, and they are exciting components of many sports. Ordinary, straight-line motion is just as important but is easily overlooked.

8

Active Physics

Activity 1 A Running Start and Frames of Reference

PHYSICS TALK

Newton's First Law of Motion

Isaac Newton included Galileo's Principle of Inertia as part of his **First Law of Motion**:

In the absence of an unbalanced force, an object at rest remains at rest, and an object already in motion remains in motion with constant speed in a straight-line path.

Newton also explained that an object's mass is a measure of its inertia, or tendency to resist a change in motion.

Here is an example of how Newton's First Law of Motion works:

Inertia is expressed in kilograms of mass. If an empty grocery cart has a mass of 10 kg and a cart full of groceries has a mass of 100 kg, which cart would be more difficult to move (have a greater tendency to remain at rest)? If both carts already were moving at equal speeds, which cart would be more difficult to stop (would have a greater tendency to keep moving)? Obviously in both cases, the answer is the cart with more mass.

Physics Words

inertia: the natural tendency of an object to remain at rest or to remain moving with constant speed in a straight line.

acceleration: the change in velocity per unit time.

frame of reference: a vantage point with respect to which position and motion may be described.

FOR YOU TO READ

Frames of Reference

In this activity, you investigated Newton's First Law. In the absence of external forces, an object at rest remains at rest and an object in motion remains in motion. If you were challenged to throw a ball as far as possible, you would probably now be sure to ask if you could have a running start. If you run with the ball prior to throwing it, the ball gets your speed before you even try to release it. If you can run at 5 m/s, then the ball will get the additional speed of 5 m/s when you throw it. When you do throw the ball, the ball's speed is the sum of your speed before releasing the ball, 5 m/s, and the speed of the release.

→

Physics in Action

It may be easier to understand this if you think of a toy cannon that could be placed on a skateboard. The toy cannon always shoots a small ball forward at 7 m/s. This can be checked with multiple trials. The toy cannon is then attached to the skateboard. A release mechanism is set up so that the cannon continues to shoot the ball forward at 7 m/s when the skateboard is at rest. When the skateboard is given an initial push, the skateboard is able to travel at 3 m/s. If the cannon releases the ball while the skateboard is moving, the ball's speed is now measured to be 10 m/s. From where did the additional speed come? The ball's speed is the sum of the ball's speed from the cannon plus the speed of the skateboard. 7 m/s + 3 m/s = 10 m/s.

You may be wondering if the ball is moving at 7 m/s or 10 m/s. Both values are correct — it depends on your **frame of reference**. The ball is moving at 7 m/s relative *to* the skateboard. The ball is moving at 10 m/s relative *to* the Earth.

Imagine that you are on a train that is stopped at the platform. You begin to walk toward the front of the train at 3 m/s. Everybody in the train will agree that you are moving at 3 m/s toward the front of the train. This is your speed *relative to* the train. Everybody looking into the train from the platform will also agree that you are moving at 3 m/s toward the front of the train. This is your speed *relative to* the platform.

Imagine that you are on the same train, but now the train is moving past the platform at 9 m/s. You begin to walk toward the front of the train at 3 m/s. Everybody in the train will agree that you are moving at 3 m/s toward the

front of the train. This is your speed *relative to* the train. Everybody looking into the train from the platform will say that you are moving at 12 m/s (3 m/s + 9 m/s) toward the front of the train. This is your speed relative *to* the platform.

Whenever you describe speed, you must always ask, "*Relative to what?*" Often, when the speed is relative to the Earth, this is assumed in the problem. If your frame of reference is the Earth, then it all seems quite obvious. If your frame of reference is the moving train, then different speeds are observed.

In sports where you want to provide the greatest speed to a baseball, a javelin, a football, or a tennis ball, that speed could be increased if you were able to get on a moving platform. That being against the rules and inappropriate for many reasons, an athlete will try to get the body moving with a running start, if allowed. If the running start is not permitted, the athlete tries to move every part of his or her body to get the greatest speed.

Sample Problem 1

A sailboat has a constant velocity of 22 m/s east. Someone on the boat prepares to toss a rock into the water.

a) Before being tossed, what is the speed of the rock with respect to the boat?

b) Before being tossed, what is the speed of the rock with respect to the shore?

c) If the rock is tossed with a velocity of 16 m/s east, what is the rock's velocity with respect to shore?

d) If the rock is tossed with a velocity of 16 m/s west, what is the rock's velocity with respect to shore?

Strategy: Before determining a velocity, it is important to check the frame of reference. The rock's velocity with respect to the boat is different from the velocity with respect to the shore. The direction of the rock also impacts the final answer.

Givens:

v_b = 22 m/s east

v_r = 16 m/s (direction varies)

Solution:

a) With respect to the boat, the rock's velocity is 0 m/s.
The rock is moving at the same speed as the boat, but you wouldn't notice this velocity if you were in the boat's frame of reference.

b) With respect to shore, the rock's velocity is 22 m/s east.
The rock is on the boat, which is traveling at 22 m/s east. Relative to the shore, the boat and everything on it act as a system traveling at the same velocity.

c) With respect to the shore, the rock's velocity is now 38 m/s east.
It is the sum of the velocity values. Since each is directed east, the relative velocity is the sum of the two.

$$v = v_b + v_r$$
$$= 22 \text{ m/s east} + 16 \text{ m/s east}$$
$$= 38 \text{ m/s east}$$

→

Physics in Action

d) With respect to shore, the rock's velocity is now 6 m/s east.
Since the directions are opposite, the relative velocity is the difference between the two.

$$v = v_b - v_r$$
$$= 22 \text{ m/s east} - 16 \text{ m/s west}$$
$$= 6 \text{ m/s east}$$

Sample Problem 2

A quarterback on a football team is getting ready to throw a pass. If he is moving backward at 1.5 m/s and he throws the ball forward at 10.0 m/s, what is the velocity of the ball relative to the ground?

Strategy: Use a negative sign to indicate the backward direction. Add the two velocities to find the velocity relative to the ground.

Givens:

Solution:

Add the velocities.

$10.0 \text{ m/s} + (-1.5 \text{ m/s}) = 8.5 \text{ m/s}$

The ball is moving forward at 8.5 m/s relative to the ground.

Reflecting on the Activity and the Challenge

Running starts can be observed in many sports. Many observers may not realize the important role that inertia plays in preserving the speed already established when an athlete engages in activities such as jumping, throwing, or skating from a running start. "Immovable objects," such as football linemen, illustrate the tendency of highly massive objects to remain at rest and can be observed in many sports. You should have no problem finding a great variety of video segments that illustrate Newton's First Law.

Physics To Go

1. Provide three illustrations of Newton's First Law in sporting events. Describe the sporting event and which object when at rest stays at rest, or when in motion stays in motion. Describe these same three illustrations in the manner of an entertaining sportscaster.
2. Find out about a sport called curling (it is an Olympic competition that involves some of the oldest Olympians) and how this sport could be used to illustrate Newton's First Law of Motion.
3. When a skater glides across the ice on only one skate, what kind of motion does the skater have? Use principles of physics as evidence for your answer.
4. Use what you have learned in **Activity 1** to describe the motion of a hockey puck between the instant the puck leaves a player's stick and the instant it hits something. (No "slap shot" allowed; the puck must remain in contact with the ice.)
5. Why do baseball players often slide into second base and third base, but almost never slide into first base after hitting the ball? (The answer depends on both the rules of baseball and the laws of physics.)
6. Do you think it is possible to arrange conditions in the "real world" to have an object move, unassisted, in a straight line at constant speed forever? Explain why or why not.

ANSWERS

Physics To Go

1. Answers will vary.
   Possible examples:
   An outfielder diving for a line drive. The outfielder continues in motion, sliding along the ground, his hat also continues in motion.

   A slap shot in hockey. The puck continues to move in a constant horizontal motion, once it has been set in motion by the player.
2. Curling, an Olympic competition, is similar to shuffleboard and involves sliding "stones" on ice.
3. A skater has either nearly constant speed, or very small uniform deceleration.
4. A hockey puck on ice has either nearly constant speed, or very small uniform deceleration.
5. A baseball player slides into second or third to decelerate to a stop at the base because if the base is overrun, the player would be "out" if tagged; at first base, a player can overrun the base without danger of being tagged out and the fastest way of beating a throw to first base is to run without sliding into the base.
6. It does not seem possible to eliminate friction to arrive at perpetual motion in the real world.

ANSWERS

Physics To Go
(continued)

7. a) The ball will appear to go straight up and down.

   b) The little girl will see the ball travel in a parabola.

   c) The speed relative to the girl will be 2.5 m/s + 4.5 m/s = 7.0 m/s.

8. The relative velocity will be 4.2 m/s + 10.3 m/s = 14.5 m/s.

9. a) The velocity relative to the tracks is 5.6 m/s + 2.4 m/s = 8.0 m/s.

   b) The velocity relative to the tracks is 5.6 m/s − 2.4 m/s = 3.2 m/s.

   c) Since the two velocities are perpendicular, we must use the Pythagorean Theorem

$$(5.6 \text{ m/s})^2 + (2.4 \text{ m/s})^2 = v^2$$

$$v = 6.1 \text{ m/s}$$

   Using the tangent button on the calculator or a vector diagram, the angle is 67°

   (Students at this point should be able to make the diagram. Some will be able to use the Pythagorean Theorem. More emphasis on this will come in **Activity 2.**)

10. a) One vaulter is moving 4.3 m/s – 3.8 m/s = 0.5 m/s faster than the other. This is their relative speed.

    b) The one going faster has more kinetic energy. With all else being equal (skill, strength, etc.) the one going faster will also go higher.

11. The speed is 85 m/s – 18 m/s = 67 m/s.

Physics in Action

7. You are pulling your little brother in his red wagon. He has a ball, and he throws it straight up into the air while you are pulling him forward at a constant speed.
   a) What will the path of the ball look like to your little brother in the wagon?
   b) What will the path of the ball look like to a little girl who is standing on the sidewalk watching you?
   c) If your brother throws the ball forward at a velocity of 2.5 m/s while you are pulling the wagon at a velocity of 4.5 m/s, at what speed does the girl see the ball go by?
8. A track and field athlete is running forward with a javelin at a velocity of 4.2 m/s. If he throws the javelin at a velocity relative to him of 10.3 m/s, what is the velocity of the javelin relative to the ground?
9. You are riding the train to school. Since the train car is almost empty, you and your friend are throwing a ball back and forth. The train is moving at a velocity of 5.6 m/s. Suppose you throw the ball to each other at the same speed, 2.4 m/s.
   a) What is the velocity of the ball relative to the tracks when the ball is moving toward the front of the car?
   b) What is the velocity of the ball relative to the tracks when it is moving toward the back of the car?
   c) What if you and your friend throw the ball perpendicular to the aisle of the train? What is the ball's velocity then?
10. Two athletes are running toward the pole vault. One is running at 3.8 m/s and the other is running at 4.3 m/s.
    a) What is their velocity relative to each other?
    b) If they leave the ground at their respective velocities, which one has the energy to go higher in the vault? Explain.
11. While riding a horse, a competitor shoots an arrow toward a target. The speed of the arrow as it reaches the target is 85 m/s. If the horse was traveling at 18 m/s, at what speed did the arrow leave the bow? (Assume the horse and arrow are traveling in the same direction.)

14

Chapter 1

Activity 1 A

Can Objects Move Forever?

FOR YOU TO DO

In this exercise, you will observe the motion of various objects on a variety of surfaces to see whether an object might be able to move forever as Galileo concluded. Since you do not have an infinite time to work, nor an infinitely large room to work in, this question can only be approached by looking at limited examples and by trying to imagine the limitless consequences of what you see. Also, since space is limited, it is important to use as little space as possible for the process of getting the object started on the motion that is to be investigated. That leaves more room to see what happens when the object is "on its own."

1. The first object to start in motion is an eraser on a table top or on the floor. The spine of the eraser is to be in contact with the floor.

a) Does the eraser sustain its motion on its own? Describe what happens.

2. Replace the table top with smooth glass or plastic.

a) Describe the eraser's motion.

3. Put about 1/8 cup of raw rice on the same surface and try the eraser again.

a) Describe the eraser's motion.

4. Remove the rice, and replace it with dried peas. Try the eraser again.

a) Describe the eraser's motion.

5. Remove the peas, and pour on a thin, sparse layer of minute plastic beads. Be careful not to spill the beads on the floor, as they are very difficult to pick up and clean up. Try the eraser again.

a) Describe the eraser's motion.

b) Try other objects on the beaded surface (coins, small blocks of wood, objects with weights on them, etc.). Report the results.

6. Set a non-inflated air puck in motion along a table top.

a) Describe what happens.

7. Fill the air reservoir and set the puck in motion again.

a) Describe and explain what happens in each case.

For use with *Physics In Action*, Chapter 1, Activity 1: A Running Start and Frames of Reference

Performance Assessment Rubrics

Part 1 = maximum 4
Part 2 = maximum 2
Part 3 = maximum 5
Part 4 = maximum 6

1. Student records experimental data demonstrating that the distance a ball travels down a salad bowl is nearly the same distance that the ball travels up the same salad bowl.

Descriptor	Task accomplished	Task not accomplished
a) Measurement is taken with ruler.		
b) Units of measurement are recorded in centimeters.		
c) Three trials are used for high, medium, and low distances.		
d) Release height is compared with recovery height.		

Maximum 4 marks if each of the sub tasks is accomplished.

Total marks: _________

2. Student uses deductive reasoning to make a generalization about how the downward motion of the ball mirrors the upward motion of the ball. The greater the release height, the greater the recovery distance.

Descriptor	Task accomplished	Task not accomplished
a) Student explains that the recovery distance is dependent upon the start distance.		
b) Student notes the constant ratio by comparing start distance of three places (high, medium, and low) to recovery distance.		

Maximum 2 marks if each of the sub tasks is accomplished.

Total marks: _________

For use with *Physics In Action*, Chapter 1, Activity 1: A Running Start and Frames of Reference

3. Student uses experimental data to find a constant ratio of release distance to recovery distance.

Descriptor	Task accomplished	Task not accomplished
a) Vertical measurement is taken with a ruler.		
b) Units of measurement are recorded in centimeters.		
c) Release height is compared with recovery height.		
d) Ratio of the recovery distance to start distance is calculated correctly.		
e) Comparison is made between the ramp and salad bowl. Similarities and differences are identified.		

Maximum 5 marks if each of the sub tasks is accomplished.

Total marks: _________

4. Student correctly decreases the slope on the right of the incline and notes that the release height affects the distance that the ball rolls along a vertical plane.

Descriptor	Task accomplished	Task not accomplished
a) Prediction of recovery distance.		
b) Student notes that the ball travels a greater distance along a horizontal plane.		
c) Student measures the vertical distance that the ball traveled up the ramp.		
d) The ratio of the vertical start height to horizontal recovery height is calculated correctly. The ratio remains constant.		
e) Gravity is identified as the force that caused the ball to move downward and slowed the ball as it moved upward.		
f) Student concludes that the force of gravity remains constant for both downward and upward movement of the ball.		

Maximum 6 marks if each of the sub tasks is accomplished.

Total marks: _________

ACTIVITY 2
Push or Pull – Adding Vectors

Background Information

It is suggested that the "**Physics Talk:** Newton's Second Law of Motion" and "**For You To Read:** Weight and Newton's Second Law" sections of the student text for **Activity 1** be read again before proceeding in this section.

The unit of mass, or quantity of matter, in the International System of Units is the kilogram. One of seven base units from which all other units are derived, the kilogram originally was conceived as the quantity of matter represented by 1 liter of water at the temperature of maximum density, 4°C; today, the kilogram is defined by a carefully protected metal standard called the International Prototype Kilogram. When a balance which employs the force of gravity is used to measure the mass of an object by comparison to prototype masses, the resulting measurement is known as the "gravitational mass" of the object. Mass is also internationally recognized as a measure of the inertial resistance of an object to acceleration. When a standard force is used to compare an object's acceleration to the acceleration of a prototype mass as a means of measuring the mass of the object, the resulting measurement is known as the "inertial mass" of the object. It can be shown that 1 kilogram of gravitationally determined mass is equivalent to 1 kilogram of inertial mass.

A derived unit of force, the newton, is defined in terms of base units of mass, length and time using Newton's Second Law of Motion, $F = ma$. 1 newton (N) is the force which will cause 1 kilogram to accelerate at $1 m/s^2$, or $1 N = 1 (kg)m/s^2$.

The word "weight" denotes a force; the weight of an object is the product of its mass and the acceleration due to gravity, $9.81 m/s^2$. Since weight is the force due to gravity, weight is measured in newtons. One newton is roughly 1/4 lb., prompting the identification of the familiar 1/4 lb-burger as a "newton burger."

In summary, matter seems to have two distinct properties:
1. It exhibits a resistance to acceleration, property called "inertia."
2. It has the property of gravitation; matter is attracted to other matter.

It is clear why it is that all objects, irrespective of mass, have the same free fall acceleration at a given location. The more mass, the more gravitational force; but the more mass, the more difficult it is to accelerate the object. These two factors exactly compensate to produce the same acceleration for every freely falling object at a given location.

Active-ating the Physics InfoMall

A big concept in this activity is the concept of force. Students' understanding of this concept has been studied extensively. An InfoMall search using "force" AND "misconception*" in only the Articles and Abstracts Attic produced many great references. The first such hit is the article containing the Force Concept Inventory. The second is "Common sense concepts about motion," *American Journal of Physics,* vol. 53, issue 11, 1985. in which it is mentioned that "(a) On the pretest (post-test), 47% (20%) of the students showed, at least once, a belief that under no net force, an object slows down. However, only 1% (0%) maintained that belief across similar tasks. (b) About 66% (54%) of the students held, at least once, the belief that under a constant force an object moves at constant speed. However, only 2% (1%) held that belief consistently." More results are reported in this article.

The third hit in this search is "Physics that textbook writers usually get wrong," in *The Physics Teacher*, vol. 30, issue 7, 1992. This article is good reading for any introductory physics teacher. The list of hits from this search is long. In fact, it had to be limited to just the Articles and Abstracts Attic to prevent the "Too many hits" warning. If you search the rest of the CD-ROM, you will find many other great hits, such as this quote from Chapter 3 of Arons' *A Guide to Introductory Physics Teaching: Elementary Dynamics*: "In the study of physics, the Law of Inertia and the concept of force have, historically, been two of the most formidable stumbling blocks for students, and, as of the present time, more cognitive research has been done in this area than in any other."

Newton's Second Law is discussed in virtually every physics textbook in existence, not to mention the InfoMall. Depending on the level at which you wish to present this Law, you may wish to examine the conceptual-level texts, the algebra-based texts, or even the calculus-bases textbooks on the InfoMall.

If you want more exercises to give to your students, searching the InfoMall is a bad idea — there are too many problems on the CD-ROM. Searching with keywords "force" AND "acceleration" AND "mass" in the Problems Place alone produces "Too many hits." However, you will find more than enough by simply going to the Problems Place and browsing a few of the resources you will find there. For example, *Schaum's 3000 Solved Problems in Physics* has a section on Newton's Laws of Motion. You will surely find enough problems there to keep any student busy for some time!

Planning for the Activity

Time Requirements

- One class period.

Materials Needed

For each group:

- balls, set of 2
- ball, tennis
- c-clamp
- ruler, flexible plastic, 30 cm
- ruler, metric, mm marking
- washers, metal, 3/4"

Advance Preparation and Setup

Identify the particular combination of flexible rulers (or plastic strips) and weights (coins or metal washers) which will serve as force meters. Identify a means of preventing the weights from slipping off the bent plastic strip, such as a lightweight cardboard "lip" taped to the plastic strip.

Also identify the set of objects to be accelerated; either balls of about the same diameter but having different masses or laboratory carts which can be loaded to vary the mass would work. If possible, have at least three objects of different masses available for each group.

Try for yourself the calibration procedure and the use of the force meter to accelerate objects in advance of class. You may need to try different sizes of coins or metal washers to find a kind which will produce a reasonable amount of bend in the ruler for a 4-washer load while at the same time providing a reasonable amount of acceleration when the smallest and largest objects are pushed using the smallest and largest forces.

Teaching Notes

Students can be expected to need practice to exert constant amounts of force on moving objects. Only semiquantitative comparisons of the amounts of acceleration (e.g., low, higher, even higher) which result from varying the amount of force (while mass is held constant) and from varying the amount of mass (when force is held constant) are intended.

Direct all students to silently read the **For You To Read** section. Then conduct a brief discussion of the assumption presented in the section. You may wish to point out that assumptions represent beliefs which may be argued, but not proven as "right" or "wrong." Another example of an assumption which could be used for the discussion is "There is a tooth fairy."

You may wish to see if students really believe that gravity treats all athletes equally by probing students about the "hang time" of basketball stars.

NOTES

Activity 2 Push or Pull—Adding Vectors

GOALS

In this activity you will:

- Recognize that a force is a push or a pull.
- Identify the forces acting on an object.
- Determine when the forces on an object are either balanced or unbalanced.
- Calibrate a force meter in arbitrary units.
- Use a force meter to apply measured amounts of force to objects.
- Compare amounts of acceleration semiquantitatively.
- Understand and apply Newton's Second Law of Motion.
- Understand and apply the definition of the newton as a unit of force.
- Understand weight as a special application of Newton's Second Law.

What Do You Think?

Moving a football one yard to score a touchdown requires strategy, timing, and many forces.

- **What is a force?**
- **Can the same force move a bowling ball and a ping-pong ball?**

Record your ideas about these questions in your *Active Physics* log. Be prepared to discuss your responses with your small group and the class.

For You To Do

1. Use a flexible ruler as a "force meter". Use washers to make a scale of measurement for (that is, to calibrate) the meter in pennyweights. The force you are using to calibrate the meter is gravity, the force with which Earth pulls downward on every

Activity Overview

In this activity students calibrate a crude "force meter" by deforming (bending) a plastic strip using washers. Students then use the force meter to accelerate the same object using different forces, and different objects using the same force.

Student Objectives

Students will:

- Recognize that a force is a push or a pull.
- Identify the forces acting on an object.
- Determine when the forces on an object are either balanced or unbalanced.
- Calibrate a force meter in arbitrary units.
- Use a force meter to apply measured amounts of force to objects.
- Compare amounts of acceleration semiquantitatively.
- Understand and apply Newton's Second Law of Motion, $F=ma$.
- Understand and apply the definition of the newton as a unit of force,

  $1\ N = 1\ (kg)m/s^2$.
- Understand weight as a special application of Newton's Second Law,

  Weight = mg.

ANSWERS

For You To Do

1. Student activity.

ANSWERS FOR THE TEACHER ONLY

What Do You Think?

In simple terms, a force is a push or a pull. Some forces, such as gravitational and magnetic forces can act on objects without having to be in contact with them. Many other forces, called mechanical forces, act when particles or objects contact each other Forces are very important in physics because they determine how matter interacts with other matter.

The same force could be used to move both a bowling ball and a ping-pong ball. The difference would be the amount by which each is accelerated by the force. The greater the mass, the less the acceleration experienced by an object when the same force is applied to it. Mass affects acceleration.

object near its surface. *Carefully* clamp the plastic strip into position as shown in the diagram on page 15.

2. Draw a line on a piece of paper. Hold the paper next to the plastic strip so that the line is even with the edge of the strip. Mark the position of the end of the strip on the reference line and label the position as the "zero" mark.

3. Place one coin on the top surface of the strip near the strip's outside end. Notice that the strip bends downward and then stops. Hold the paper in the original position and mark the new position of the end of the strip. Label the mark as "1 pennyweight."

4. Repeat **Step 3** for two, three, and four coins placed on the strip. In each case mark and label the new position of the end of the strip.
   a) Copy the reference line and the calibration marks from the piece of paper into your log.

5. Practice holding one end of the "force meter" (plastic strip) in your hand and pushing the free end against an object until you can bend the strip by forces of 1, 2, 3, and 4-pennyweight amounts. To become good at this, you will need to check the amount of bend in the strip against your calibration marks as you practice.

6. Use the force meter to push an object such as a tennis ball with a continuous 1-pennyweight force. You will need to keep up with the object as it moves and to keep the proper bend in the force meter. You may need to practice a few times to be able to do this.
   a) In your log, record the amount of force used, a description of the object, and the kind of motion the object seemed to have.

7. Repeat **Step 6** three more times, pushing on the same object with steady (constant) 2, 3, and 4-pennyweight amounts of force.
   a) Record the results in your log for each amount of force.

ANSWERS

For You To Do *(continued)*

2. – 3. Student activity.

4. a) Students record calibration in their logs. As each washer is added, the ruler deflects more. The force due to gravity of each washer is responsible for bending the ruler.

5. Student activity.

6. a) Students record observations in their logs.

7. a) The greater the force applied to the tennis ball, the greater its acceleration, as demonstrated by the increasing difficulty in keeping up with the ball to maintain the force on it.

8. Based on your observations, complete the statement: "The greater the constant, unbalanced force pushing on an object,..."
   a) Write the completed statement in your log.
9. Select an object that has a small mass. Use the force meter to push on the object with a rather large, steady force such as 3- or 4-pennyweight amounts.
   a) Record the amount of force used, a description of the object pushed (especially including its mass, compared to the other objects to be pushed) and the kind of motion the object seemed to have.
10. Repeat **Step 9** using the same amount of force to push objects of greater and greater mass.
    a) Record the results in your log for each object.
11. Based on your observations, complete the statement: "When equal amounts of constant, unbalanced force are used to push objects having different masses, the more massive object..."
    a) Write the completed statement in your log.

ANSWERS

For You To Do *(continued)*

8\. a) The greater the constant, unbalanced force pushing on an object, the greater the acceleration of that object.

9\. a) Students record results in their logs.

10\. a) As the mass of the objects increases, the acceleration decreases.

11\. a) When equal amounts of constant, unbalanced forces are used to push objects having different masses, the more massive objects are accelerated less.

Physics Words

Newton's Second Law of Motion: if a body is acted upon by an external force, it will accelerate in the direction of the unbalanced force with an acceleration proportional to the force and inversely proportional to the mass.

weight: the vertical, downward force exerted on a mass as a result of gravity.

PHYSICS TALK

Newton's Second Law of Motion

Based on observations from experiments similar to yours, Isaac Newton wrote his **Second Law of Motion:**

The acceleration of an object is directly proportional to the unbalanced force acting on it and is inversely proportional to the object's mass. The direction of the acceleration is the same as the direction of the unbalanced force.

If 1 N (newton) is defined as the amount of unbalanced force that will cause a 1-kg mass to accelerate at 1 m/s^2 (meter per second every second), the law can be written as an equation:

$$F = ma$$

where F is expressed in newtons (symbol N), mass is expressed in kilograms (kg), and acceleration is expressed in meters per second every second (m/s^2).

By definition, the unit "newton" can be written in its equivalent form: $(kg)m/s^2$.

Newton's Second Law can be arranged in three possible forms:

$$F = ma \qquad a = \frac{F}{m} \qquad m = \frac{F}{a}$$

FOR YOU TO READ

Weight and Newton's Second Law

Newton's Second Law explains what "weight" means, and how to measure it. If an object having a mass of 1 kg is dropped, its free fall acceleration is roughly 10 m/s^2.

Using Newton's Second Law,

$$F = ma$$

the force acting on the falling mass can be calculated as

$$F = ma$$
$$= 1 \text{ kg} \times 10 \text{ m/s}^2 \text{ or } 10 \text{ N}$$

The 10-N force causing the acceleration is known to be the gravitational pull of Earth on the 1-kg object. This gravitational force is given the special name **weight**. Therefore, it is correct to say, "The weight of a 1-kg mass is ten newtons."

What is the weight of a 2-kg mass? If dropped, a 2-kg mass also would accelerate due to gravity (as do all objects in free fall) at about 10 m/s^2. Therefore, according to Newton's Second Law, the weight of a 2-kg mass is equal to

$$2 \text{ kg} \times 10 \text{ m/s}^2 \text{ or } 20 \text{ N}$$

In general, to calculate the numerical value of an object's weight in newtons, it is necessary only to multiply the numerical value of its mass by the numerical value of the g (acceleration due to gravity), which is about 10m/s^2.

$$\text{Weight} = mg$$

The preceding equation is the "special case" of Newton's Second Law that must be applied to any situation in which the force causing an object to accelerate is Earth's gravitational pull.

Where There's Acceleration, There Must Be an Unbalanced Force

There are lots of different everyday forces. You just read about the force of gravity. There is also the force of a spring, the force of a rubber band, the force of a magnet, the force of your hand, the force of a bat hitting a ball, the force of friction, the buoyant force of water, and many more. Newton's Second Law tells you that accelerations are caused by unbalanced external forces. It doesn't matter what kind of force it is or how it originated. If you observe an acceleration (a change in velocity), then there must be an unbalanced force causing it.

When you apply a force, if the object has a small mass, the acceleration may be quite large for a given force. If the object has a large mass, the acceleration will be smaller for the same applied force. Occasionally, the mass is so large that you are not even able to measure the acceleration because it is so small.

If you push on a go-cart with the largest force possible, the cart will accelerate a great deal. If you push on a car with that same force, you

→

Physics in Action

will measure a much smaller acceleration. If you were to push on the Earth, the acceleration would be too small to measure. Can you convince someone that a push on the Earth moves the Earth? Why should you believe something that you can't measure? If you were to assume that the Earth does not accelerate when you push on it, then you would have to believe that Newton's Second Law stops working when the mass gets too big. If that were so, you would want to determine how big is "too big." When you conduct such experiments, you find that the acceleration gets less and less as the mass gets larger and larger. Eventually, the acceleration gets so small that it is difficult to measure. Your inability to measure it doesn't mean that it is zero. It just means that it is smaller than your best measurement. In this way, you can assume that Newton's Second Law is always valid.

All of these statements are summarized in Newton's Second Law as you read in **Physics Talk**:

$$F = ma$$

or in forms that emphasize the acceleration and the mass

$$a = \frac{F}{m} \text{ and } m = \frac{F}{a}$$

Sample Problem 1

A tennis racket hits a ball with a force of 150 N. While the 275-g ball is in contact with the racket, what is its acceleration?

Strategy: Newton's Second Law relates the force acting on an object, the mass of the object, and the acceleration given to it by the force. Use the form of the equation that emphasizes acceleration to find the acceleration. The force unit, the newton, is defined as the amount of force needed to give a mass of 1.0 kg an acceleration of 1.0 m/s^2. Therefore, you will need to change the grams to kilograms.

Givens:

$F = 150.0$ N
$m = 275$ g

Solution:

$$275 \text{ g} = 0.275 \text{ kg}$$

$$a = \frac{F}{m}$$

$$= \frac{150 \text{ N}}{0.275 \text{ kg}}$$

$$= 545 \text{ m/s}^2$$

Sample Problem 2

As the result of a serve, a tennis ball (m_t = 58 g) accelerates at 43 m/s^2.

a) What force is responsible for this acceleration?

b) Could an identical force accelerate a 5.0-kg bowling ball at the same rate?

Strategy: Newton's Second Law states that the acceleration of an object is directly proportional to the applied force and indirectly proportional to the mass ($F = ma$).

Givens:

$a = 43$ m/s^2
$m_t = 58$ g $= 0.058$ kg
$m_b = 5.0$ kg

Solution:

a)

$$F = m_t a$$

$$= 0.058 \text{ kg} \times 43 \text{ m/s}^2$$

$$= 2.494 \text{ N or } 2.5 \text{ N}$$

b) Since the mass of the bowling ball is much greater than that of the tennis ball, an identical force will result in a smaller acceleration.
(You can calculate the acceleration.)

$$a = \frac{F}{m_b}$$

$$= \frac{2.5\ \text{N}}{5.0\ \text{kg}}$$

$$= 0.50\ \text{m/s}^2$$

Adding Vectors

A vector is a quantity that has both magnitude and direction. Velocity is a vector. In the previous activity you found that the direction in which an object was traveling and the speed at which it was moving are equally important.

Force is also a vector because you can measure how big it is (its magnitude) and its direction. Acceleration is also a vector. The equation for acceleration reminds you that the force and the acceleration must be in the same direction.

Often, more than one force acts on an object. If the two forces are in the same direction, the sum of the forces is simply the addition of the two forces. A 30-N force by one person and a force of 40 N by a second person (pushing in the same direction) on the same desk provides a 70-N force on the desk. If the two forces are in opposite directions, then you give one of the forces a negative value and add them again. If one student pushes on a desk to the right with a force of 30 N and a second student pushes on the same desk to the left with a force of 40 N, the net force on the desk will be 10 N to the left. Mathematically, you would state that 30 N + (–40 N) = – 10 N where the negative sign denotes "to the left."

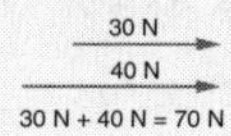

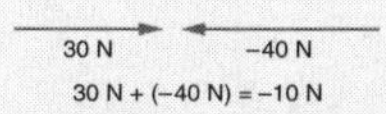

Occasionally, the two forces acting on an object are at right angles. For instance, one student may be kicking a soccer ball with a force of 30 N ahead toward the goal, while the second student kicks the same soccer ball with a force of 40 N toward the sideline. To find the net force on the ball and the direction the ball will travel, you must use vector addition. You can do this by using a vector diagram or the Pythagorean Theorem.

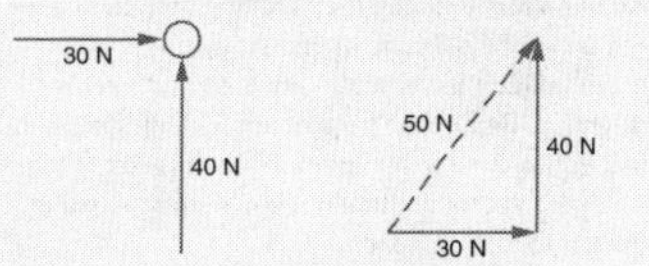

In the vector diagram shown above, the two force vectors are shown as arrows acting on the soccer ball. The magnitudes of the vectors are drawn to scale. The 30-N force may be drawn as 3.0 cm and the 40-N force may be drawn as 4.0 cm, if the scale is 10 N = 1 cm. To add the vectors, slide them so that the tip of the 30-N vector can be placed next to the tail of the 40-N vector (tip to tail method). The sum of the two vectors is then drawn from the tail of the 30-N vector to the tip of the 40-N vector. This *resultant* vector is measured and is found to be 5.0 cm, which is equivalent to 50 N. The angle is measured with a protractor and is found to be 53°.

→

Physics in Action

A second method of finding the *resultant* vector is to recognize that the 30-N and 40-N force vectors form a right triangle. The resultant is the hypotenuse of this triangle. Its length can be found by the Pythagorean Theorem.

$$a^2 + b^2 = c^2$$
$$30N^2 + 40N^2 = c^2$$
$$900N^2 + 1600N^2 = c^2$$
$$c = \sqrt{2500}\ N^2$$
$$c = 50\ N$$

The angle can be found by using the tangent function.

$$\tan\theta = \frac{\text{opposite}}{\text{adjacent}} = \frac{40N}{30N} = 1.33$$
$$\theta = 53°$$

Adding vector forces that are not perpendicular is a bit more difficult mathematically, but no more difficult using scale drawings and vector diagrams. Two other players are kicking a soccer ball in the direction shown in the diagram. The resultant vector force can be determined using the tip to tail approach.

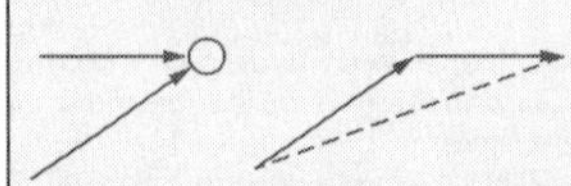

The two arrows in the left diagram correspond to the two players kicking the ball at different angles. The diagram at the right shows the two vectors being added "tip to tail." The resultant vector (shown as a dotted line) represents the net force and is the direction of the acceleration of the soccer ball.

Sample Problem 3

One player applies a force of 125 N in a north direction. Another player pushes with a force of 125 N west. What is the magnitude and direction of the resultant force?

Strategy: Since the forces are acting at right angles, you can use the Pythagorean theorem to find the resultant force. The direction of the force can be found using the tangent function.

Givens:

$F_1 = 125\ N$

$F_2 = 125\ N$

Solution:

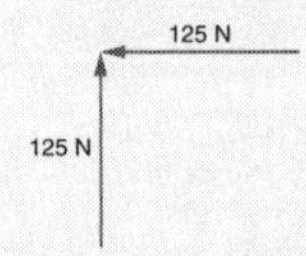

$$F_R^2 = F_1^2 + F_2^2$$
$$F_R = \sqrt{125\ N^2 + 125\ N^2}$$
$$= \sqrt{31250\ N^2}$$
$$= 177\ N$$
$$\tan\theta = \frac{\text{opposite}}{\text{adjacent}} = \frac{125N}{125N} = 1$$
$$\theta = 45°$$

The resultant force is 177 N, 45° west of north.

Active Physics 22

Reflecting on the Activity and the Challenge

What you learned in this activity really increases the possibilities for interpreting sports events in terms of physics. Now you can explain why accelerations occur in terms of the masses and forces involved. You know that forces produce accelerations. Therefore, if you see an acceleration occur, you know to look for the forces involved. You can apply this to the sport that you will describe.

Also, you can explain, in terms of mass and weight, why gravity has no "favorite" athletes; in every case of free fall in sports, g has the same value, about 10 m/s^2.

Physics To Go

1. Copy and complete the following table using Newton's Second Law of Motion. Be sure to include the unit of measurement for each missing item.

Newton's Second Law:	F =	m ×	a
Sprinter beginning 100-meter dash	?	70 kg	5 m/s^2
Long jumper in flight	800 N	?	10 m/s^2
Shot put ball in flight	70 N	7 kg	?
Ski jumper going down hill before jumping	400 N	?	5 m/s^2
Hockey player "shaving ice" while stopping	−1500 N	100 kg	?
Running back being tackled	?	100 kg	−30 m/s^2

2. The following items refer to the table in **Question 1**:

   a) In which cases in the table does the acceleration match "g," the acceleration due to gravity 10 m/s^2?
   Are the matches to g coincidences or not? Explain.

   b) The force on the hockey player stopping is given in the table as a negative value. Should the player's acceleration also be negative? What do you think it means for a force or an acceleration to be negative?

   c) The acceleration of the running back being tackled also is given as negative. Should the unbalanced force acting on him also be negative? Explain.

ANSWERS

Physics To Go

1. See chart below.

2. a) The long jumper and the shot put ball both are cases of free fall; therefore the acceleration is g, the acceleration due to gravity.

   b) The negative sign is used to denote that the force and acceleration are in a direction opposite the motion.

   c) Since acceleration occurs in the direction of the causal force, yes the force should be shown as negative.

Newton's Second Law:	f =	m ×	a
Sprinter beginning 100-meter dash	350 N	70 kg	5 m/s^2
Long jumper in flight	800 N	80 kg	10 m/s^2
Shot put ball in flight	70 N	7 kg	10 m/s^2
Ski jumper going down hill before jumping	400 N	80 kg	5 m/s^2
Hockey player "shaving ice" while stopping	−1500 N	100 kg	-15 m/s^2
Running back being tackled	−3000 N	100 kg	−30 m/s^2

ANSWERS

Physics To Go
(continued)

2. d) Students should be able to provide a plausible "voice-over" narration for an imagined video clip showing each event in the table.

3. 4.2 N / 0.30 kg = 14 m/s^2

4. 0.040 kg × 20 m/s^2 = 0.8 N

5. a) A bowling ball has greater inertia (mass) than a baseball; therefore, a bowling ball has a greater tendency to either remain at rest or remain in motion than does a baseball.

   b) More force is required to cause a bowling ball to accelerate than a baseball; therefore, throwing (accelerating) or catching (decelerating) a bowling ball involves much greater forces than throwing or catching a baseball when equal speeds are involved.

6. The sandwich would weigh 0.1 kg × 10 m/s^2 = 1 N.

7. Example: Weight
= 150 lb. × 4.38 N/lb.
= 657 N

   Mass
= 657 N ÷ 10 m/s^2 = 65.7 kg

8. The component, or effectiveness, of the weight in the downhill direction, parallel to the slope of the hill, is 0.71 times the downward force of gravity (weight), or 7.1 N/kg; therefore, the acceleration is 7.1 N/kg = 7.1 m/s^2. This can be analyzed using either a scale drawing or trigonometry and should not be expected of all students at this level.

d) In your mind, "play" an imagined video clip that illustrates the event represented by each horizontal row of the preceding table. Write a brief voice-over script for each video clip that explains how Newton's Second Law of Motion is operating in the event. Use appropriate physics terms, equations, numbers, and units of measurement in the scripts.

3. What is the acceleration of a 0.30-kg volleyball when a player uses a force of 42 N to spike the ball?

4. What force would be needed to accelerate a 0.040-kg golf ball at 20.0 m/s^2?

5. Most people can throw a baseball farther than a bowling ball, and most people would find it less painful to catch a flying baseball than a bowling ball flying at the same speed as the baseball. Explain these two apparent facts in terms of:
   a) Newton's First Law of Motion.
   b) Newton's Second Law of Motion.

6. Calculate the weight of a new fast food sandwich that has a mass of 0.1 kg. Think of a clever name for the sandwich that would incorporate its weight.

7. In the United States, people measure body weight in pounds. Write down the weight, in pounds, of a person who is known to you. (This could be your weight or someone else's.)
   a) Convert the person's weight in pounds to the international unit of force, newtons. To do so, use the following conversion equation:
   Weight in newtons = Weight in pounds × 4.38 newtons/pound
   b) Use the person's body weight, in newtons, and the equation
   Weight = mg
   to calculate the person's body mass, in kilograms.

8. Imagine a sled (such as a bobsled or luge used in Olympic competitions) sliding down a 45° slope of extremely slippery ice. Assume there is no friction or air resistance (not really possible). Even under such ideal conditions, it is a fact that gravity could cause the sled to accelerate at a maximum of only 7.1 m/s^2. Why would the "ideal" acceleration of the sled not be g, 10 m/s^2? Your answer is expected only to suggest reasons why, on a 45° hill, the ideal free fall acceleration is "diluted" from 10 m/s^2 to about 7 m/s^2; you are not expected to give a complete explanation of why the "dilution" occurs.

9. If you were doing the voice-over for a tug-of-war, how would you explain what was happening? Write a few sentences as if you were the science narrator of that athletic event.
10. You throw a ball. When the ball is many meters away from you, is the force of your hand still acting on the ball?
11. Carlo and Sara push on a desk in the same direction. Carlo pushes with a force of 50 N, and Sara pushes with a force of 40 N. What is the total resultant force acting on the desk?
12. A car is stuck in the mud. Four adults each push on the back of the car with a force of 200 N. What is the total force on the car?
13. During a football game, two players try to tackle another player. One player applies a force of 50.0 N to the east. A second player applies a force of 120.0 N to the north. What is the total applied force? (Since force is a vector, you must give both the magnitude and direction of the force.)
14. In auto racing, a crash occurs. A red car hits a blue car from the front with a force of 4000 N. A yellow car also hits the blue car from the side with a force of 5000 N. What is the total force on the blue car? (Since force is a vector, you must give both the magnitude and direction of the force.)
15. A baseball player throws a ball. While the 700.0-g ball is in the pitcher's hand, there is a force of 125 N on it. What is the acceleration of the ball?
16. If the acceleration due to gravity at the surface of the Earth is approximately 9.8 m/s^2, what force does the gravitational attraction of the Earth exert on a 12.8-kg object?
17. A force of 30.0 N acts on an object. At right angles to this force, another force of 40.0 N acts on the same object.
    a) What is the net force on the object?
    b) What acceleration would this give a 5.6-kg wagon?
18. Bob exerts a 30.0 N force to the left on a box (m = 100.0 kg). Carol exerts a 20.0 N force on the same box, perpendicular to Bob's.
    a) What is the net force on the box?
    b) Determine the acceleration of the box.
    c) At what rate would the box accelerate if both forces were to the left?

Answers

Physics To Go

(continued)

9. Students provide voice-over for tug-of-war.
10. No.
11. The resultant is 40 N + 50 N = 90 N.
12. The total force is 4 x 200 N = 800 N.
13. Application of the Pythagorean Theorem yields:

    $(50\ N)^2 + (120\ N)^2 = F^2$

    $F = 130\ N$

    Using the tangent button on the calculator or a vector diagram, the angle is 23° east of north.
14. Application of the Pythagorean Theorem yields:

    $(4000\ N)^2 + (5000\ N)^2 = F^2$

    $F = 6403\ N$

    Using the tangent button on the calculator or a vector diagram, the angle is 39°.
15. $F = ma$

    $a = F/m = (125\ N)/ 0.7\ kg) = 179\ m/s^2$
16. $F = ma = (12.8\ kg)\ (9.8\ m/s^2) = 125\ N$
17. a) Application of the Pythagorean Theorem yields:

    $(40\ N)^2 + (30\ N)^2 = F^2$

    $F = 50\ N$

    Using the tangent button on the calculator or a vector diagram, the angle is 53°.

    b) $a = F/m = 50\ N/5.6\ kg = 8.9\ m/s^2$
18. a) Application of the Pythagorean Theorem yields:

    $(30\ N)^2 + (20\ N)^2 = F^2$

    $F = 36\ N$

    Using the tangent button on the calculator or a vector diagram, the angle is 34°.

    b) $a = F/m = 36\ N/100\ kg = 0.36\ N/kg = 0.36\ m/s^2$

    c) If both boxes were pushed toward the right, the new force would be 50 N. (30 N + 20 N = 50 N).

    The acceleration can then be calculated:

    $a = F/m = 50\ N/100\ kg = 0.5\ N/kg = 0.5\ m/s^2$

ACTIVITY 3
Center of Mass

Background Information

The center of mass of an object is the only idea introduced in this activity.

Definition: The center of mass is the point at which the entire mass of an object may be thought of as being concentrated for purposes of analyzing the translational motion (motion along a path) or rotational motion (spinning motion) of the object.

For practical purposes, the location of the center of mass of an object having only one significant dimension—such as a straight stick, loaded teeter-totter, twirler's baton, screwdriver or wrench—corresponds to the object's balance point. For a two-dimensional object—such as a sheet of plywood cut into any shape—the location of the center of mass corresponds to the balance point located on either of the two large, flat surfaces of the object; to the extent that a two-dimensional object—such as a triangle cut from a sheet of plywood—may have significant thickness and, therefore, actually be three-dimensional, the center of mass would be located within the object, "in line" with the balance point, at the center of the thickness dimension.

For objects having simple three-dimensional shapes—such as homogeneous or symmetrically layered spheres (examples, in respective order: bowling ball, basketball), cubes, rectangular solids and cylinders—the center of mass is located within the object, at its center.

An alternative to balancing an object to locate the center of mass is to suspend the object from any point which is not the center of mass. When suspended, gravity serves to orient the object so that its center of mass is located directly below the point of suspension (this is an example that the Earth "views" an object near it as a "point mass" (located at the object's center of mass) and pulls the point mass as close to Earth as possible). A line extended straight downward from the point of suspension passes through the object's center of mass. The intersection of two such lines, corresponding to two points of suspension, locates the object's center of mass.

It is possible that the center of mass may not be located within the material of the object for some shapes. The "boomerang" shape is an example of such an object.

For purposes of applying Newton's Laws of Motion, an object is treated as if all of its mass is concentrated at the center of mass. The fact that objects behave this way in nature is verified by the observation that when a baton is thrown through the air as a twirling projectile, the baton's center of mass, if marked for high visibility, is seen to trace the familiar parabolic trajectory of a projectile. A twirling baton brings up another aspect of center of mass: when a force acting on an object is aligned with the object's center of mass, the object accelerates in accordance with Newton's second law; however, if the applied force is not aligned with the center of mass, the object also will rotate, or spin. The latter kind of case is not treated in *Active Physical Science*.

Considerable emphasis in future activities will be placed on the center of mass of the human body. Except for contorted positions of body parts (e.g., the arched "Fosbury Flop" position in the high jump), the normal location of the body's center of mass is within the body at about the level of the navel.

Active-ating the Physics InfoMall

The methods outlined in the *Active Physical Science* text are standard for finding the center of mass for objects. However, you may want demonstrations. A search of the Demo & Lab Shop produces many great, and tested, demonstrations. Just use keywords "Center of mass" and search only the Demo & Lab Shop. Of course, you can also find many problems in the Problems Place, if you wish, using the same keywords.

Planning for the Activity

Time Requirements

- One class period.

Materials Needed

For the class:

- hammer and catch box (demonstration, **Step 8**)

For each group:

- dots, adhesive, 3/4"
- plumb bob w/string
- push pins
- set of shapes, A,B,C,D
- stick, meter, 100 cm, hardwood

Advance Preparation and Setup

Cutouts of shapes A, B, C, and D need to be made for each group in the shapes of templates provided in the Additional Materials for this activity. (The templates are provided only for your convenience. Other shapes, in greater variety, may be used and the size may be scaled differently, if desired. If you depart from the shapes provided, be sure to include a "boomerang" shape for which the C of M will be outside the object.) The shapes may be cut from any thin, flat material such as corrugated cardboard or (more durable) plywood, plastic or metal; it would also be convenient to cut the shapes from a sheet of pegboard material to avoid need to drill holes for suspension.

Drill holes to serve to suspend the shapes, and plan how the shapes will be suspended from pins or nails from areas such as a bulletin board or pieces of wood mounted on laboratory table rods.

Teaching Notes

Prepare a demonstration of one or more objects having complex shapes moving as spinning projectiles. For example, the centers of mass of shapes A, B, C, and D could be brightly marked and observed from a distance as two persons play catch with each of the objects. Even if an object spins, the center of mass will trace a parabolic trajectory. A baton having the center of mass marked with bright tape also could be used. When all students have completed **Steps 1** to **5** of **For You To Do**, convene the entire class to observe the motion of the center of mass as two persons play catch with the objects planned for the demonstration.

You may wish to recommend the opening montage on the *Active Physical Science* video as a possibility for tracing the motion of the center of mass moving as a projectile.

Activity Overview

In this activity students locate the center of mass of various shaped objects by locating the center of gravity. This is done by balancing the object on a finger as well as by suspending the object and using a plumb bob. Students also estimate the location of their own center of mass.

Student Objectives

- Locate the center of mass of oddly shaped two-dimensional objects.
- Infer the location of the center of mass of symmetrical three-dimensional objects.
- Measure the approximate location of the center of mass of the student's body.
- Understand that the entire mass of an object may be thought of as being located at the object's center of mass.

ANSWERS FOR THE TEACHER ONLY

What Do You Think?

The center of mass is the point at which all of the mass of an object may be thought of as being concentrated. (As defined above: The center of mass is the point at which the entire mass of an object may be thought of as being concentrated for purposes of analyzing the translational motion (motion along a path) or rotational motion (spinning motion) of the object.)

The normal location of the body's center of mass is within the body at about the level of the navel.

Physics in Action

Activity 3 Center of Mass

GOALS

In this activity you will:

- **Locate the center of mass of oddly shaped two-dimensional objects.**
- **Infer the location of the center of mass of symmetrical three-dimensional objects.**
- Measure **the approximate location of the center of mass of the student's body.**
- **Understand that the entire mass of an object may be thought of as being located at the object's center of mass.**

What Do You Think?

The center of mass of a high jumper using the "Fosbury Flop" (arched back) technique passes below the bar as the jumper's body successfully passes over the bar.

- **What is "center of mass"? What does it mean?**
- **Where is your body's center of mass?**

Record your ideas about these questions in your *Active Physics* log. Be prepared to discuss your responses with your small group and the class.

Active Physics 26

Activity 3 Center of Mass

For You To Do

1. You will be provided four objects made from thin sheets of material in the shapes shown below.

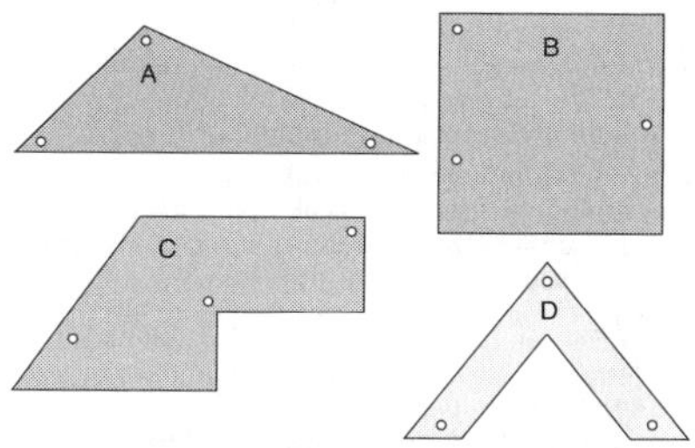

a) Make a sketch in your log to show each shape in a reduced scale of size.

2. Use your intuition and trial and error to locate a point on the flat surfaces of objects *A*, *B*, and *C* where the object will balance on your fingertip. Mark the "balance point" of each object using a nonpermanent method.

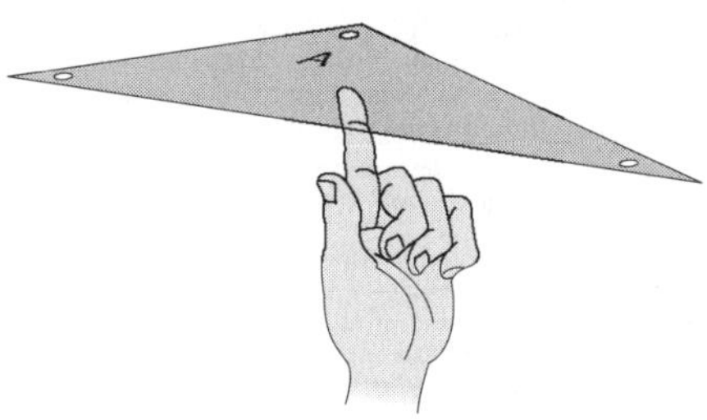

a) Mark the balance points on the sketches in your log.

ANSWERS

For You To Do

1. a) You may wish to provide students with a copy of the templates at the end of this activity, rather than have them redraw each in their log.

2. a) Students should be able to locate the balance points for each shape.

Physics in Action

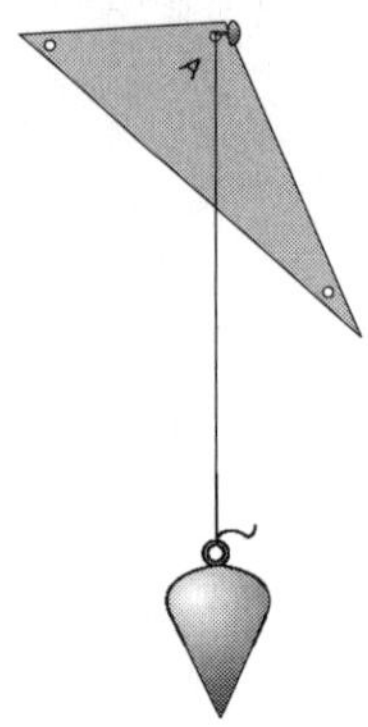

3. To check on the balance points found by the above method for objects *A*, *B* and *C*, use one of the small holes in object *A* to hang it from a pin as shown, and, also as shown, hang a "plumb bob" (a weight on a string) from the same pin.

   a) Does the string pass over the balance point you marked for object *A* when you used your finger to balance the object? Should this happen? Write why you think it should or should not happen in your log.

   b) Use a different hole in object *A* to suspend it from the pin and again hang the plumb bob from the pin. Does the string pass over the balance point marked before? Should it? Write your responses in your log.

4. The intersection of the two lines made where the string passed over the surface of object *A* could have been used to predict the balance point without first trying to balance the object on your finger. Use the suspension and plumb bob method to check the correspondence of the two methods of finding the balance point for objects *B* and *C*.

   a) Record your findings about how well the two methods agree in your log.

5. Locate an "imaginary" balance point for object *D*. Tape a lightweight piece of paper between the "open arms" of object *D* and suspend the object and the plumb bob from the pin. Trace the path of the string across the piece of paper. Suspend the object from a different hole and trace the path of the string across the paper again.

   a) Do you agree that the intersection of the two lines on the paper mark the balance point of object *D*? What is special about this balance point? Write your answers in your log.

6. The above "balance points" that you found for two-dimensional, or "flat," objects *A*, *B*, *C*, and *D* were, in each case, the location of the object's "center of mass." Do a "thought experiment" (an experiment in your mind) to determine the location of the center of mass of each of the following objects:

Shot put ball (solid steel)	Basketball
Banana	Planet Earth
Baseball bat	Hockey stick

ANSWERS

For You To Do *(continued)*

3. a) Yes, the plumb bob should pass over the balance point. When suspended, gravity serves to orient the object so that its center of mass is located directly below the point of suspension. A line extended straight downward from the point of suspension passes though the object's center of mass.

   b) The string will pass over the balance point. The spot where the lines from **a)** and **b)** intersect represents the C of M.

4. a) Students will find that the two methods will produce similar results.

5. a) The C of M is not located on shape D. Students will find that the two lines cross as a point outside the shape.

6. a) Students answer will vary.

Activity 3 Center of Mass

a) For each object, describe in your log how you decided upon the location of the center of mass.

7. The technique that was used to find the center of mass (C of M) relied on the fact that the C of M always lies beneath the point of support when an object is hanging. Similarly, when an object is balanced, the C of M is always above the point of support. To find your C of M, carefully balance on one foot and then the other. Try to keep your arms and legs in roughly identical positions as you shift your weight. Your C of M is located where a vertical meter stick from one foot and the other intersect. Locate this point. The actual C of M is inside your body, since nobody has zero thickness.

a) Record the location of your C of M.

8. Your teacher will balance a hammer on a finger to locate the hammer's C of M and make an obvious mark on the hammer at the C of M. As your teacher drops the hammer into a catch box on the floor, and it twists and turns, notice the movement of the C of M.

a) How does the movement of the C of M compare to the motion of the entire hammer?

Physics Words

center of mass: the point at which all the mass of an object is considered to be concentrated for calculations concerning motion of the object.

Reflecting on the Activity and the Challenge

The **center of mass** is an important concept in any sports activity. The motion of the center of mass of a diver or gymnast is much easier to observe than the movements of the entire body. The sure-fire way of having a football player fall is to move his center of mass away from his support.

Think about the possibilities for using a transparent plastic cover on a TV monitor and using a pen to trace the motion of the center of mass of an athlete executing a free fall jump or dive. This could be used to simulate the light-pen technique used by TV commentators when they comment on football replays. This would seem a good way to add an interesting feature to your TV sports commentary.

ANSWERS

For You To Do *(continued)*

7. a) The center of mass of a body is located inside the body at about the level of the navel.

8. a) The C of M moves directly down in a straight line, whereas the hammer twists and turns as it falls.

Answers

Physics To Go

1. If not directed toward the center of mass, part of the force will be used to make the object rotate, not accelerate along a line.
2. Referring to the above answer to **Question 1**, a player having a low center of gravity must be "hit" low, at the level of the center of mass, to have his state of rest or motion changed.
3. The body's center of mass has no support directly beneath it, so it falls.
4. Fosbury Flop: the center of mass is located behind the back, in the air outside the body.
5. The pushoff force is directed at an angle to the intended path of travel.
6. If the car were suspended from a crane twice, each time from a different point of attachment of the cable to the car, the intersection of lines representing, in each case of suspension, an extension of the cable through the car would locate the center of mass.
7. Students will probably find the center of mass by balancing the bat on their finger. Ask students to record what they did, and any problems they may have encountered.
8. When the support is moved away from the center of mass, the book will fall. By tackling below the center of mass, the support is moved away from under the center of mass, and the player will fall.

Physics To Go

1. When applying a force to make an object move, why is it most effective to have the applied force "aimed" directly at the object's center of mass?
2. "Center of gravity" means essentially the same thing as "center of mass." Why is it often said to be desirable for football players to have a low center of gravity?
3. Stand next to a wall facing parallel to the wall. With your right arm at your side pushing against the wall and with the right edge of your right foot against the wall at floor level, try to remain standing as you lift your left foot. Why is this impossible to do?
4. Think of positions for the human body for which the center of mass might be located outside the body. Describe each position and where you think the center of mass would be located relative to the body for each position.
5. An object tends to rotate (spin) if it is pushed on by a force that is not aimed at the center of mass. How do athletes use this fact to initiate spins before they fly through the air, as in gymnastics, skating, and diving events?
6. Could the suspension technique for finding the center of mass used in **For You To Do** be adapted to locate the center of mass of a three-dimensional object? If you had a crane that you could use to suspend an automobile from various points of attachment, how could you locate the auto's center of mass?
7. Find the center of mass of a baseball bat using the technique that you learned in class.
8. Carefully balance a light object (not too massive) over a table or catch box. Notice that the C of M is directly over the point of support. Move the support a little bit. Explain how this technique can be adapted to tackling in football.
9. Cut out a piece of cardboard in the shape of your state or a country. Find the geographic center of mass of your shape.

Activity 3 A

Alternative Method for Determining Center of Gravity

FOR YOU TO DO

1. Locate the center of mass (often abbreviated C of M) of your body. For this you will need an equal arm "teeter-toter," meter stick, and two assistants. Your teacher will give you safety precautions.

Lay on your back on the teeter-toter as an assistant stabilizes each end to prevent extreme tipping. Adjust your position until balance is achieved without the assistants touching the system.

2. When at balance, have an assistant measure the distance, in meters, from the bottom of the heel of your shoe to the fulcrum (middle support) of the teeter-toter.

3. As the assistants again stabilize the teeter-toter, get off the teeter-toter.

a) Record the distance measured by the assistant.

Distance from heel to C of M = _____ m

4. Standing erect, measure the above distance from the floor upward to locate the height of your body's center of mass relative to a constant reference point such as your navel.

a) Record the location in your log so that you will be able to recover it easily (Example: three fingerwidths below navel). Actually, your center of mass is located inside your body (within your belly) when your body is in most positions. Usually you will need to know only how high above floor level it is located.

Template for Shapes A and B

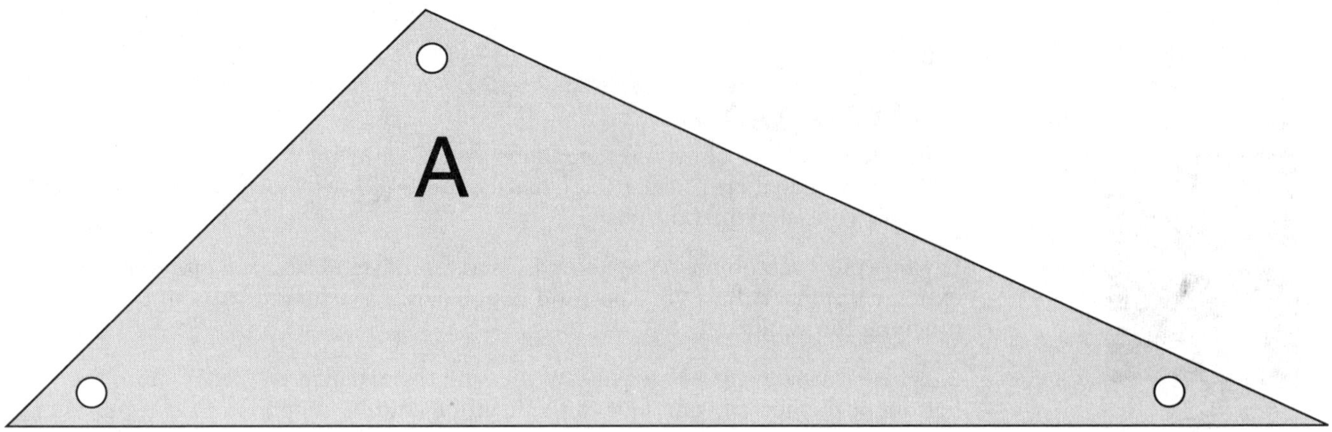

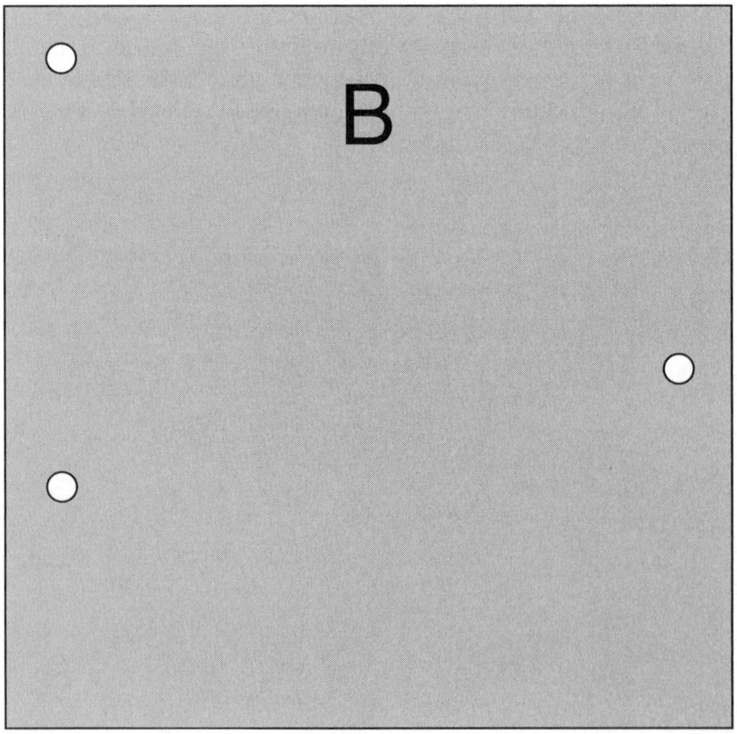

For use with *Physics In Action*, Chapter 1, Activity 3: Center of Mass

Template for Shapes C and D

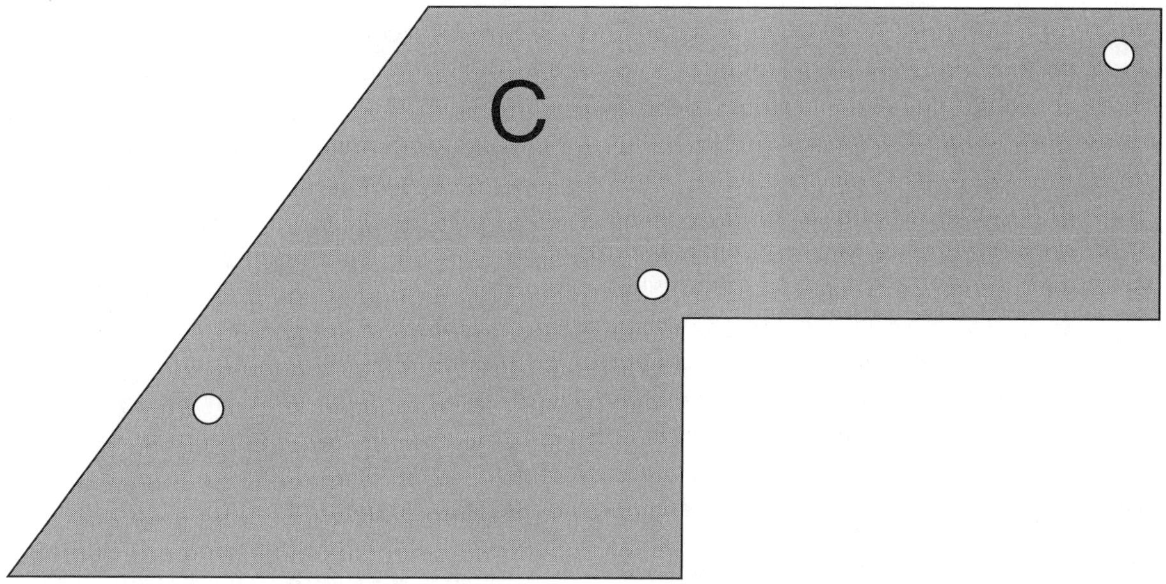

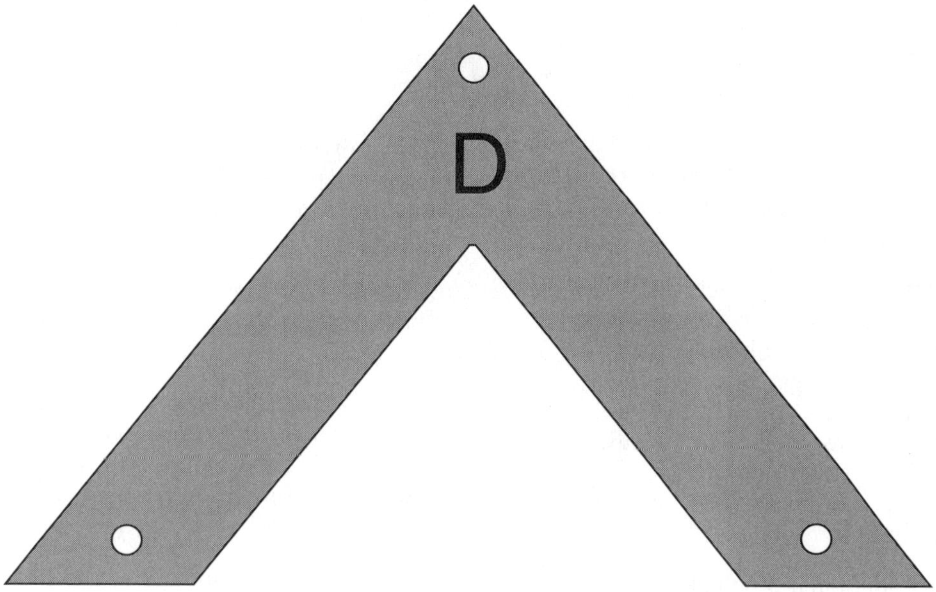

For use with *Physics In Action*, Chapter 1, Activity 3: Center of Mass

ACTIVITY 4
Defy Gravity

Background Information

It is suggested that you read the **Physics Talk** and **Example Analysis** sections in the student text for this activity before proceeding in this section. It is also suggested that you read the teacher background and student text for **Chapter 3, Activity 10**, "Energy in the Pole Vault."

Work, the product force x distance, is expressed in joules. Work is equivalent to energy and, indeed, is transformed into kinetic energy and gravitational potential energy in the vertical jump.

Research has shown that the location of the center of mass within the jumper's body varies only slightly for the body positions assumed during the process of the vertical jump.

The force which lifts and accelerates the body's center of mass during a vertical jump is provided by muscles of the leg, ankles, and feet. The method of analysis used for this activity assumes that the muscular force is constant as the body rises from "ready" to "launch" positions; this is not entirely accurate—in a real jump, the force varies—but is a reasonable approximation of reality.

Active-ating the Physics InfoMall

While "hang time" is discussed on the InfoMall, it is in the sense of how long a football stays in the air during a punt, and not how long a basketball player stays (or seems to stay) in the air.

Note that gravitational potential energy is mentioned in this activity, a topic we will encounter at the end of **Chapter 3**. And, like at the end of **Chapter 3**, you are encouraged to browse the textbooks. If you perform a search of the InfoMall for Work, Potential Energy, or Kinetic Energy, you will want to limit your search to only one or two stores at a time, or use additional keywords to restrict your search.

Should you desire additional problems for your students to work on, consult the Problems Place. For example, *Progressive Problems in Physics* has 16 problems on Work, and 25 on Energy.

Planning for the Activity

Time Requirements

- One class period.

Materials Needed

For the class:

- *Active Physics Sports* content video

(Segments: ice skater performing triple axel jump, basketball player "hanging" during slam dunk)

- VCR and TV monitor

For each group:

- *Sports* Content Video
- masking tape, 3/4" x 60 yds
- stick, meter, 100 cm, hardwood
- VCR & monitor

Advance Preparation and Setup

Reserve a VCR and TV monitor for showing segments of the *Active Physics Sports* video.

Teaching Notes

View the slow-motion sequences of the jumping figure skater and the jumping basketball player to examine "hang time" and to see if either athlete remains suspended at the peak of flight. These are clear cases of free fall; since the basketball player has high horizontal running speed at take-off, the top of the trajectory is quite flat, giving illusion, when viewed in "real time" that the player "hangs" in the air.

Students may be expected to need help when applying their own data to replicate the calculations presented as an example in **Physics Talk**.

If a sonic ranger is available, monitor a jump from above and analyze the graphs of distance, speed, and acceleration versus time with the entire class. Directions for using a sonic ranger are provided at the end of this activity.

Suggest to students who have access to VCRs with slow-motion playback capability that they could record jumps during athletic contests and perform analysis similar to those conducted using the *Active Physics Sports* video.

Activity 4 Defy Gravity

GOALS

In this activity you will:

- Measure changes in height of the body's center of mass during a vertical jump.
- Calculate changes in the gravitational potential energy of the body's center of mass during a vertical jump.
- Understand and apply the definition of work.
- Recognize that work is equivalent to energy.
- Understand and apply the joule as a unit of work and energy using equivalent forms of the joule.
- Apply conservation of work and energy to the analysis of a vertical jump, including weight, force, height, and time of flight.

What Do You Think?

No athlete can escape the pull of gravity.

- **Does the "hang time" of some athletes defy the above fact?**
- **Does a world-class skater defy gravity to remain in the air long enough to do a triple axel?**

Record your ideas about these questions in your *Active Physics* log. Be prepared to discuss your responses with your small group and the class.

For You To Do

1. Your teacher will show you a slow-motion video of a world-class figure skater doing a triple axel jump. The image of the skater will appear to "jerk," because a video camera completes one "frame," or one complete picture, every $\frac{1}{30}$ s. When the video is played at normal speed, you perceive the action as continuous; played at slow motion, the individual frames can be detected and counted. The duration of each frame is $\frac{1}{30}$ s.

Chapter 1

Activity Overview

In this activity students measure the positions of the C of M of a student during a vertical jump, and then analyze the amount of force and energy required by the student to perform the jump.

Student Objectives

Students will:

- Measure changes in height of the body's center of mass during a vertical jump.
- Calculate changes in the gravitational potential energy of the body's center of mass during a vertical jump.
- Understand and apply the definition of work, Work = *fd*.
- Recognize that work is equivalent to energy.
- Understand and apply the joule as a unit of work and energy using equivalent forms of the joule:

  $1\ J = 1\ Nm = 1\ (kg)m/s^2 \times (m) = 1\ (kg)m^2/s^2$
- Apply conservation of work and energy to analysis of a vertical jump, including weight, force, height, and time of flight.

ANSWERS FOR THE TEACHER ONLY

What Do You Think?

The answer to both questions is the same. There is no evidence that athletes are able to defy gravity.

ANSWERS

For You To Do

1. a) The skater is in the air for 15 frames.

   b) Time in air (s) = Number of frames × $\frac{1}{30}$ s

   $= 15 \times \frac{1}{30}$ s

   $= \frac{15}{30}$ s $= \frac{1}{2}$ s

   c) During the time frame as viewed on the video, the skater's position is constantly changing. There is no "hang" time.

2. a) The basketball player is in the air for 31 frames.

   Time in air (s) = Number of frames × $\frac{1}{30}$ s

   $= 31 \times \frac{1}{30}$ s

   $= \frac{31}{30}$ s $= 1\ \frac{1}{30}$ s

   b) During the time frame as viewed on the video, the basketball player's position is constantly changing. There is no "hang" time. (Since the ball is moving upward before the player leaves the ground, and since on the way down his arms are extending and lifting the ball into the net, the illusion of hanging in the air may be created.)

3. a) Students' answers will vary according to their weight in pounds.

   The following equations were presented on page 24.

   Weight in newtons = Weight in pounds × 4.38 N/lb.

   Weight (N) = Weight (lbs) × 4.38 N/1b

   Weight = mg

   Weight (N) = m(kg) × g(m/s^2)

   $$m\ (\text{kg}) = \frac{\text{Weight (N or kg·m/s}^2)}{g\ (\text{m/s}^2)}$$

a) Count and record in your log the number of frames during which the skater is in the air.

b) Calculate the skater's "hang time." (Show your calculation in your log.)

$$\text{Time in air (s)} = \text{Number of frames} \times \frac{1}{30}\ \text{s}$$

c) Did the skater "hang" in the air during any part of the jump, appearing to "defy gravity"? If necessary, view the slow-motion sequence again to make the observations necessary to answer this question in your log. If your observations indicate that hanging did occur, be sure to indicate the exact frames during which it happened.

2. Your teacher will show you a similar slow-motion video of a basketball player whose hang time is believed by many fans to clearly defy gravity.

a) Using the same method as above for the skater, show in your log the data and calculations used to determine the player's hang time during the "slam dunk."

b) Did the player hang? Cite evidence from the video in your answer.

3. How much force and energy does a person use to do a vertical jump? A person uses body muscles to "launch" the body into the air, and, primarily, it is leg muscles that provide the force. First, analyze only the part of jumping that happens before the feet leave the ground. Find your body mass, in kilograms, and your body weight, in newtons, for later calculations. If you wish not to use data for your own body, you may use the data for another person who is willing to share the information with you. (See **Activity 2, Physics To Go, Question 7**, for how to convert your body weight in pounds to weight in newtons and mass in kilograms.)

a) Record your weight, in newtons, and mass, in kilograms, in your log.

4. Recall the location of your body's center of mass from **Activity 3**. Place a patch of tape on either the right or left side of your clothing (above one hip) at the same level as your body's center of mass. Crouch as if you are ready to make a vertical jump. While crouched, have an assistant measure the vertical distance, in meters, from the floor to the level of your body's center of mass (C of M).
   a) In your log, record the distance, in meters, from the floor to your C of M in the ready position.

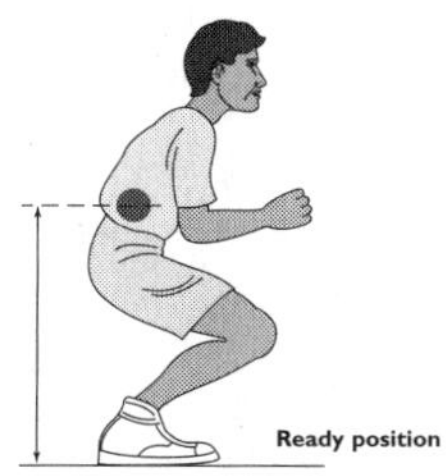

5. Straighten your body and rise to your tiptoes as if you are ready to leave the floor to launch your body into a vertical jump, but don't jump. Hold this launch position while an assistant measures the vertical distance from the floor to the level of your center of mass.
   a) In your log, record the distance, in meters, from the floor to your C of M in the launch position.
   b) By subtraction, calculate and record the vertical height through which you used your leg muscles to provide the force to lift your center of mass from the "ready" position to the "launch" position. Record this in your log as legwork height.

   Legwork height = Launch position – Ready position

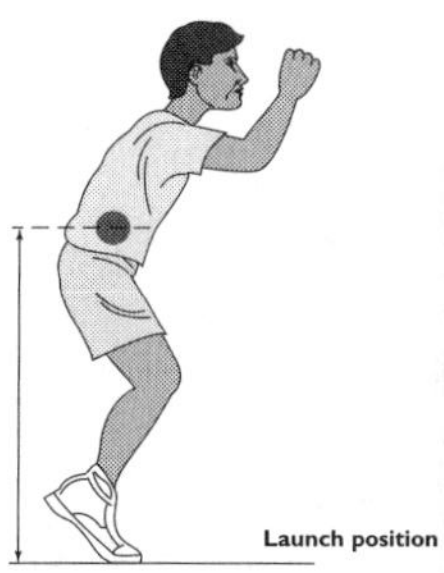

6. Now it's time to jump! Have an assistant ready to observe and measure the vertical height from the floor to the level of your center of mass at the peak of your jump. When your assistant is ready to observe, jump straight up as high as you can. (Can you hang at the peak of your jump for a while to make it easier for your assistant to observe the position of your center of mass? Try it, and see if your assistant thinks you are successful.)

Make sure the floor is dry and the area in which you are jumping is clear of obstructions.

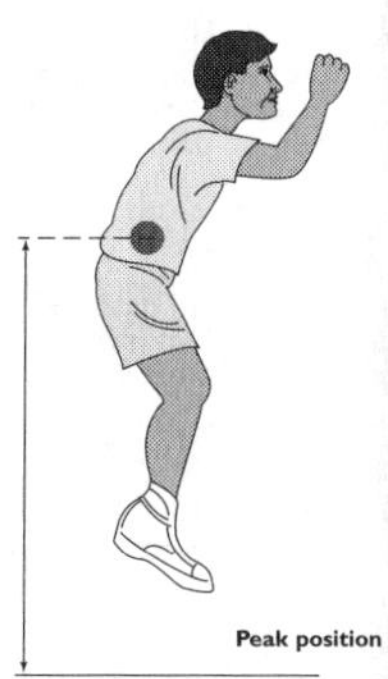

ANSWERS

For You To Do *(continued)*

4.-6. You may wish to provide the students with a copy of the Calculating Hang Time and Force during a Vertical Jump Worksheet provided after this activity. Expect to help students when applying their own data to replicate the calculations presented as an example in **Physics Talk**.

ANSWERS

For You To Do *(continued)*

6. a) Answers will vary.
   b) Answers will vary.
7. a) Answers will vary.
8. Student activity.

Physics Words

work: the product of the displacement and the force in the direction of the displacement; work is a scalar quantity.

potential energy: energy that is dependent on the position of the object.

kinetic energy: the energy an object possesses because of its motion.

a) In your log, record the distance from the floor to C of M at peak position.

b) By subtraction, calculate and record the vertical height through which your center of mass moved during the jump.

Jump height = Peak position – Launch position

7. The information needed to analyze the muscular force and energy used to accomplish your jump—and an example of how to use sample data from a student's jump to perform the analysis—is presented in **Physics Talk** and the Example Analysis.

a) Use the information presented in the **Physics Talk** and Example Analysis sections and the data collected during above **Steps 4** through **6** to calculate the hang time and the total force provided by *your* leg muscles during your vertical jump. Show as much detail in your log as is shown in the Example Analysis.

8. An ultrasonic ranging device coupled to a computer or graphing calculator, which can be used to monitor position, speed, acceleration, and time for moving objects, may be available at your school. If so, it could be used to monitor a person doing a vertical jump. This would provide interesting information to compare to the data and analysis that you already have for the vertical jump. Check with your teacher to see if this would be possible.

Activity 4 Defy Gravity

FOR YOU TO READ

Conservation of Energy

In this activity you jumped and measured your vertical leap. You went through a chain of energy conversions where the total energy remained the same, in the absence of air resistance. You began by lifting your body from the crouched "ready" position to the "launch" position. The **work** that you did was equal to the product of the applied force and the distance. The work done must have lifted you from the ready position to the launch position (an increase in **potential energy**) and also provided you with the speed to continue moving up (the **kinetic energy**). After you left the ground, your body's potential energy continued to increase, and the kinetic energy decreased. Finally, you reached the peak of your jump, where all of the energy became potential energy. On the way down, that potential energy began to decrease and the kinetic energy began to increase.

When you are in the ready position, you have elastic potential energy. As you move toward the launch position, you have exchanged your elastic potential energy for an increase in gravitational potential energy and an increase in kinetic energy. As you rise in the air, you lose the kinetic energy and gain more gravitational potential energy. You can show this in a table.

Energy→ / Position↓	Elastic potential energy	Gravitational potential energy = mgh	Kinetic energy = $\frac{1}{2}mv^2$
ready position	maximum	0	0
launch position	0	some	maximum
peak position	0	maximum	0

The energy of the three positions must be equal. In this first table, the sum of the energies in each row must be equal. The launch position has both gravitational potential energy and kinetic energy. Using the values in the activity, the total energy at each position is 410 J.

Energy→ / Position↓	Elastic potential energy	Gravitational potential energy = mgh	Kinetic energy = $\frac{1}{2}mv^2$
ready position	410 J	0	0
launch position	0	150 J	260 J
peak position	0	410 J	0

→

In the ready position, all 410 J is elastic potential energy. In the peak position, all 410 J is gravitational potential energy. In the launch position, the total energy is still 410 J but 150 J is gravitational potential energy and 260 J is kinetic energy.

Consider someone the same size, who can jump much higher. Since that person can jump much higher, the peak position is greater, and therefore the gravitational potential energy of the jumper is greater. In the example shown below, the gravitational potential energy is 600 J. Notice that this means the elastic potential energy of the jumper's legs must be 600 J. And when the jumper is in the launch position, the total energy (potential plus kinetic) is also 600 J.

Energy→ / Position↓	Elastic potential energy	Gravitational potential energy = mgh	Kinetic energy $= \frac{1}{2}mv^2$
ready position	600 J	0	0
launch position	0	150 J	450 J
peak position	0	600 J	0

A third person of the same size is not able to jump as high. What numbers should be placed in blank areas to preserve the principle of conservation of energy?

Total energy must be conserved. Therefore, in the launch position the kinetic energy of the jumper must be 50 J. In the peak position, all the energy is in potential energy and must be 200 J.

The conservation of energy is a unifying principle in all science. It is worthwhile to practice solving problems that will help you to see the variety of ways in which energy conservation appears.

Energy→ / Position↓	Elastic potential energy	Gravitational potential energy = mgh	Kinetic energy $= \frac{1}{2}mv^2$
ready position	200 J	0	0
launch position	0	150 J	50 J
peak position	0	200 J	0

A similar example to jumping from a hard floor into the air is jumping on a trampoline (or your bed, when you were younger). If you were to jump on the trampoline, the potential energy from the height you are jumping would provide kinetic energy when you landed on the trampoline. As you continued down, you would continue to gain speed because you would still be losing gravitational potential energy. The trampoline bends and/or the springs holding the trampoline stretch. Either way, the trampoline or springs gain elastic potential energy at the expense of the kinetic energy and the changes in potential energy.

Energy→ / Position↓	Elastic potential energy	Gravitational potential energy = *mgh*	Kinetic energy $= \frac{1}{2}mv^2$
High in the air position	0	2300 J	0
Landing on the trampoline position	0	500 J	1800 J
Lowest point on the trampoline position	2300 J	0	0

A pole-vaulter runs with the pole. The pole bends. The pole straightens and pushes the vaulter into the air. The vaulter gets to his highest point, goes over the bar, and then falls back to the ground, where he lands on a soft mattress. You can analyze the pole-vaulter's motion in terms of energy conservation. (Ignore air resistance.)

A pole-vaulter runs with the pole. (*The vaulter has kinetic energy.*) The pole bends. (*The vaulter loses kinetic energy, and the pole gains elastic potential energy as it bends.*) The pole unbends and pushes the vaulter into the air. (*The pole loses the elastic potential energy, and the vaulter gains kinetic energy and gravitational potential energy.*) The vaulter gets to his highest point (*the vaulter has almost all gravitational potential energy*) goes over the bar, and then falls back to the ground (*the gravitational potential energy becomes kinetic energy*), where he lands on a soft mattress (*the kinetic energy becomes the elastic potential energy of the mattress, which then turns to heat energy*). The height the pole-vaulter can reach is dependent on the total energy that he starts with. The faster he runs, the higher he can go.

The conservation of energy is one of the great discoveries of science. You can describe the energies in words (elastic potential energy, gravitational potential energy, kinetic energy, and heat energy). There is also sound energy, →

Physics in Action

light energy, chemical energy, electrical energy, and nuclear energy. The words do not give the complete picture. Each type of energy can be measured and calculated. In a closed system, the total of all the energies at any one time must equal the total of all the energies at any other time. That is what is meant by the conservation of energy.

If you choose to look at one object in the system, that one object can gain energy. For example, in the collision between a player's foot and a soccer ball, the soccer ball can gain kinetic energy and move faster. Whatever energy the ball gained, you can be sure that the foot lost an equal amount of energy. The ball gained energy, the foot lost energy, and the "ball and foot" total energy remained the same. The ball gained energy because work (force x distance) was done on it. The foot lost energy because work (force x distance) was done on it. The total system of "ball and foot" neither gained nor lost energy.

Physics provides you with the means to calculate energies. You may wish to practice some of these calculations now. Never lose sight of the fact that you can calculate the energies because the sum of all of the energies remains the same.

The equations for work, gravitational potential energy, and kinetic energy are given below.

The equation for work is:

$$W = F \cdot d$$

Work is done only when the force and displacement are (at least partially) in the same (or opposite) directions.

The equation for gravitational potential energy is:

$$PE_{gravitational} = mgh = wh$$

The w represents the weight of the object in newtons, where $w = mg$. On Earth's surface, when dealing with g in this course, consider it to be equal to 9.8 m/s^2. (Sometimes we use 10 m/s^2 for ease of calculations.)

The equation for kinetic energy is:

$$KE = \frac{1}{2}mv^2$$

Sample Problem

A trainer lifts a 5.0-kg equipment bag from the floor to the shelf of a locker. The locker is 1.6 m off the floor.

a) How much force will be required to lift the bag off the floor?

b) How much work will be done in lifting the bag to the shelf?

c) How much potential energy does the bag have as it sits on the shelf?

38

Active Physics

d) If the bag falls off the shelf, how fast will it be going when it hits the floor?

Strategy: This problem has several parts. It may look complicated, but if you follow it step by step, it should not be difficult to solve.

Part (a): Why does it take a force to lift the bag? It takes a force because the trainer must act against the pull of the gravitational field of the Earth. This force is called weight, and you can solve for it using Newton's Second Law.

Part (b): The information you need to find the work done on an object is the force exerted on it and the distance it travels. The distance was given and you calculated the force needed. Use the equation for work.

Part (c): The amount of potential energy depends on the mass of the object, the acceleration due to gravity, and the height of the object above what is designated as zero height (in this case, the floor). You have all the needed pieces of information, so you can apply the equation for potential energy.

Part (d): The bag has some potential energy. When it falls off the shelf, the potential energy becomes kinetic energy as it falls. When it strikes the ground in its fall, it has zero potential energy and all kinetic energy. You calculated the potential energy. Conservation of energy tells you that the kinetic energy will be equal to the potential energy. You know the mass of the bag so you can calculate the velocity with the kinetic energy formula.

Givens:

$m = 5.0$ kg
$h = 1.6$ m
$a = 9.8$ m/s^2

Solution:

a)

$$F = ma$$
$$= (5.0 \text{ kg})(9.8 \text{ m/s}^2)$$
$$= 49 \text{ kg} \cdot \text{m/s}^2 \text{ or } 49 \text{ N}$$

b)

$$W = F \cdot d$$
$$= (49 \text{ N})(1.6 \text{ m})$$
$$= 78.4 \text{ Nm or } 78 \text{ J (Nm = J)}$$

c)

$$PE_{gravitational} = mgh$$
$$= (5.0 \text{ kg})(9.8 \text{ m/s}^2)(1.6 \text{ m})$$
$$= 78 \text{ J}$$

Should you be surprised that this is the same answer as **Part (b)**? No, because you are familiar with energy conservation. You know that the work is what gave the bag the potential energy it has. So, in the absence of work that may be converted to heat because of friction, which you did not have in this case, the work equals the potential energy.

d)

$$KE = \frac{1}{2}mv^2$$
$$v^2 = \frac{KE}{\frac{1}{2}m}$$
$$= \frac{78 \text{ J}}{\frac{1}{2}(5.0 \text{ kg})}$$
$$= 31 \text{ m}^2/\text{s}^2$$
$$v = 5.6 \text{ m/s}$$

Physics in Action

PHYSICS TALK

Work

When you lifted your body from the ready (crouched) position to the launch (standing on tiptoes) position before takeoff during the vertical jump activity, you performed what physicists call work. In the context of physics, the word *work* is defined as:

The work done when a constant force is applied to move an object is equal to the amount of applied force multiplied by the distance through which the object moves in the direction of the force.

You used symbols to write the definition of work as:

$$W = F \cdot d$$

where F is the applied force in newtons, d is the distance the object moves in meters, and work is expressed in joules (symbol, J). At any time it is desired, the unit "joule" can be written in its equivalent form as force times distance, "(N)(m)."

The unit "newton" can be written in the equivalent form "(kg)m/s^2." Therefore, the unit joule also can be written in the equivalent form (kg)m^2/s^2. In summary, the units for expressing work are:

$$1\text{ J} = 1\text{ (N)(m)} = 1\text{ (kg)m}^2/\text{s}^2$$

As you read, it is very common in sports that work is transformed into kinetic energy, and then, in turn, the kinetic energy is transformed into gravitational potential energy. This chain of transformations can be written as:

$$\text{Work} = KE = PE$$

$$Fd = \frac{1}{2}mv^2 = mgh$$

These transformations are used in the analysis of data for a vertical jump.

Example:
Calculation of Hang Time and Force During Vertical Jump

DATA: Body Weight = 100 pounds = 440 N
Body Mass = 44 kg
Legwork Height = 0.35 m
Jump Height = 0.60 m

Analysis:
Work done to lift the center of mass from ready position to launch position without jumping ($W_{R\ to\ L}$):

$$W_{R\ to\ L} = Fd = (\text{Body Weight}) \times (\text{Legwork Height}) = 440\ \text{N} \times 0.35\ \text{m} = 150\ \text{J}$$

Gravitational Potential Energy gained from jumping from launch position to peak position (PE_J):

$$PE_J = mgh = (\text{Body Mass}) \times (g) \times (\text{Jump Height}) = 44\ \text{kg} \times 10\ \text{m/s}^2 \times 0.60\ \text{m} = 260\ (\text{kg})\text{m}^2/\text{s}^2 = 260\ (\text{N})(\text{m}) = 260\ \text{J}$$

The jumper's kinetic energy at takeoff was transformed to increase the potential energy of the jumper's center of mass by 260 J from launch position to peak position. Conservation of energy demands that the kinetic energy at launch be 260 J:

$$KE = \frac{1}{2}mv^2 = 260\ \text{J}$$

This allows calculation of the jumper's launch speed:

$$v = \sqrt{2(KE)/m} = \sqrt{2(260\ \text{J})/(44\ \text{kg})} = 3.4\ \text{m/s}$$

From the definition of acceleration, $a = \Delta v/\Delta t$, the jumper's time of flight "one way" during the jump was:

$$\Delta t = \Delta v/a = (3.4\ \text{m/s}) / (10\ \text{m/s}^2) = 0.34\ \text{s}$$

Therefore, the total time in the air (hang time) was $2 \times 0.34\ \text{s} = 0.68\ \text{s}$.

→

The total work done by the jumper's leg muscles before launch, W_T, was the work done to lift the center of mass from ready position to launch position without jumping, $W_{R\ to\ L}$ = 150 J, plus the amount of work done to provide the center of mass with 260 J of kinetic energy at launch, a total of 150 J + 260 J = 410 J. Rearranging the equation $W = F \cdot d$ into the form $F = W/d$, the total force provided by the jumper's leg muscles, F_T was:

$$F_T = \frac{W_T}{\text{(Legwork Height)}}$$

$$= 410 \text{ J} / 0.35 \text{ m}$$

$$= 1200 \text{ N}$$

Approximately one-third of the total force exerted by the jumper's leg muscles was used to lift the jumper's center of mass to the launch position, and approximately two-thirds of the force was used to accelerate the jumper's center of mass to the launch speed.

Reflecting on the Activity and the Challenge

Work, the force applied by an athlete to cause an object to move (including the athlete's own body as the object in some cases), multiplied by the distance the object moves while the athlete is applying the force explains many things in sports. For example, the vertical speed of any jumper's takeoff (which determines height and "hang time") is determined by the amount of work done against gravity by the jumper's muscles before takeoff. You will be able to find many other examples of work in action in sports videos, and now you will be able to explain them.

Physics To Go

1. How much work does a male figure skater do when lifting a 50-kg female skating partner's body a vertical distance of 1 m in a pairs competition?
2. Describe the energy transformations during a bobsled run, beginning with team members pushing to start the sled and ending when the brake is applied to stop the sled after crossing the finish line. Include work as one form of energy in your answer.
3. Suppose that a person who saw the video of the basketball player used in **For You To Do** said, "He really can hang in the air. I've seen him do it. Maybe he was just having a 'bad hang day' when the video was taken, or maybe the speed of the camera or VCR was 'off.' How do I know that the player in the video wasn't a 'look-alike' who can't hang?" Do you think these are legitimate statements and questions? Why or why not?
4. If someone claims that a law of physics can be defied or violated, should the person making the claim need to provide observable evidence that the claim is true, or should someone else need to prove that the claim is not true? Who do you think should have the burden of proof? Discuss this issue within your group and write your own personal opinion in your log.
5. Identify and discuss two ways in which an athlete can increase his or her maximum vertical jump height.
6. Calculate the amount of work, in joules, done when:
   a) a 1.0-N weight is lifted a vertical distance of 1.0 m.
   b) a 1.0-N weight is lifted a vertical distance of 10 m.
   c) a 10-N weight is lifted a vertical distance of 1.0 m.
   d) a 0.10-N weight is lifted a vertical distance of 100 m.
   e) a 100-N weight is lifted a distance of 0.10 m.
7. List how much gravitational potential energy, in joules, each of the weights in **Question 6** above would have when lifted to the height listed for it.

ANSWERS

Physics To Go

1. Work = $fd = (mg)d$
   $= 50 \text{ kg} \times 10 \text{ m/s}^2 \times 1 \text{ m} = 50$ j.
2. Team members do work while running and pushing the sled to give it and their bodies kinetic energy before jumping on the sled, $fd = 1/2mv^2$. After the team has jumped on the sled, the total energy of the team + sled is equal to the kinetic energy gained during the pushing phase plus the gravitational potential energy $= mgh$, where h is the vertical distance to the bottom of the hill. At the bottom of the hill, the kinetic energy of the sled should be equal to the kinetic energy gained during the pushing phase plus the loss in potential energy due to coming down the hill, $1/2mv^2 + mgh$. The brake must do enough work to cause the sled to lose all of its kinetic energy by exerting a force in the direction opposite the sled's motion.
3. It is apparent that the person wants to believe that the player can defy gravity and is attempting to justify that belief by rejecting scientific evidence. It could be said that the person is not reflecting open-mindedness, a desirable attribute in scientific pursuits.
4. The burden of proof rests with the person making the claim.
5. Increase the force the athlete is able to exert using muscles, lose weight without decreasing muscular force.
6. a) 1.0 N × 1.0 m = 1 J
   b) 1.0 N × 10 m = 10 J
   c) 10 N × 1.0 m = 10 J
   d) 0.10 N × 100 m = 10 J
   e) 100 N × 0.10 m = 10 J
7. All answers are the same as for #6 above.

Chapter 1

ANSWERS

Physics To Go
(continued)

8. All answers are the same as for #6 above.

9. $W = F \cdot d = (50.0\ \text{N})(43\ \text{m})$
$= 2150\ \text{J}\ (2200\ \text{J})$

10. $KE = 1/2\ mv^2 = 1/2\ (62\ \text{kg})$
$(8.2\ \text{m/s})^2 = 2084\ \text{J}\ (2100\ \text{J})$

11. a) $F = ma$
$a = F/m = 30.0\ \text{N}/5.0\ \text{kg}$
$= 6\ \text{m/s}^2$

b) $W = F \cdot d = (30.0\ \text{N})$
$(18.75\ \text{m}) = 563\ \text{J}$

12. a) $W = F \cdot d$
$d = W/F = 40{,}000\ \text{J}/3200\ \text{N}$
$= 12.5\ \text{m}\ (12\ \text{m})$

b) $F = ma$
$a = F/m = 3200\ \text{N}/1200\ \text{kg}$
$= 2.7\ \text{m/s}^2$

13. The work done is equal to the change in KE. The final KE is 0. The initial KE can be found.

$KE = 1/2\ mv^2 = 1/2\ (0.150\ \text{kg})$
$(40\ \text{m/s})^2 = 120\ \text{J}$

14. The change in KE is equal to the work done. Calculate the change in KE and then calculate the distance.

$KE = 1/2\ mv^2 = 1/2\ (64.0\ \text{kg})$
$(15.0\ \text{m/s})^2 = 7200\ \text{J}$

$W = F \cdot d$

$d = W/F = 7200\ \text{J}/417\ \text{N}$
$= 17.3\ \text{m}$

8. List how much kinetic energy, in joules, each of the weights in **Questions 6** and **7** would have at the instant before striking the ground if each weight were dropped from the height listed for it.
9. How much work is done on a go-cart if you push it with a force of 50.0 N and move it a distance of 43 m?
10. What is the kinetic energy of a 62-kg cyclist if she is moving on her bicycle at 8.2 m/s?
11. A net force of 30.00 N acts on a 5.00-kg wagon that is initially at rest.
    a) What is the acceleration of the wagon?
    b) If the wagon travels 18.75 m, what is the work done on the wagon?
12. Assume you do 40,000 J of work by applying a force of 3200 N to a 1200-kg car.
    a) How far will the car move?
    b) What is the acceleration of the car?
13. A baseball (m = 150.0 g) is traveling at 40.0 m/s. How much work must be done to stop the ball?
14. A boat exerts a force of 417 N pulling a water-skier (m = 64.0 kg) from rest. The skier's speed is now 15.0 m/s. Over what distance was this force exerted?

Activity 4 A

High-Tech Alternative for Monitoring Vertical Jump Height

FOR YOU TO DO

1. Place a computer motion sensor near the ceiling, pointing straight down.

2. Adjust the software so that the duration of the time axis is 5 s or less.

3. Activate the Distance versus Time graph.

4. Click the start button and, as soon as you hear the motion sensor clicking, jump. Try not to get closer than 50 cm to the sensor or you will get erroneous results.

5. Look at the resulting graph and try to find the following parts of the motion:
   - The initial bending of your knees in preparation for the jump.
   - The part of the motion when you were in the air.
   - The bending of your knees upon landing.

   a) Describe what each part of the jump looks like on the graph.

6. Use the software to zoom in on the part of the graph that contains the above-mentioned parts.

7. Switch to a Velocity versus Time graph.

   a) Describe your velocity while in the air.

8. Repeat the experiment for a higher and a lower jump. Compare and contrast your results.

For use with *Physics In Action*, Chapter 1, Activity 4: Defy Gravity

Calculating Hang Time and Force During a Vertical Jump

Use a calculator to complete the following analysis of a vertical jump:

DATA:

Calculate body weight.

Weight (N) = Weight (lb.) × 4.38 N/lb.

= ____________ × 4.38 N/lb.

= ____________

Body Weight = ____________

Calculate body mass.

Weight (N) = mg

$$m \text{ (kg)} = \frac{\text{weight (kg·m/s}^2)}{g(\text{m/s}^2)}$$

$$= \frac{\text{\_\_\_\_\_\_\_\_\_\_\_\_}}{10 \text{ m/s}^2}$$

= ____________

Body Mass = ____________

Calculate your legwork height.

Legwork Height = Launch position – Ready position

= ____________ – ____________

= ____________

Legwork Height = ____________

Calculate your jump height.

Jump Height = Peak position – Launch position

= ____________ – ____________

= ____________

Jump Height = ____________

Calculate the work done to lift the center of mass from ready position to launch position without jumping ($W_{R\ to\ L}$)

$W_{R\ to\ L} = fd$ = (Body Weight) × (Legwork Height)

= ____________ N × ____________ m

= ____________ N·m or J (joules)

Calculate the gravitational potential energy gained from jumping from launch position to peak position (PE_J).

$PE_J = mgh$ = (Body Mass) × (g) × (Jump Height)

= ________ kg × 10 m/s^2 × ________ m

= ____________ kg·m^2/s^2

= ____________ N·m or J (joules)

Conservation of energy demands that the kinetic energy at launch is equal to the gravitational potential energy at peak position.

$KE = PE$

= ____________ J

(Insert the figure you calculated for PE above.)

For use with *Physics In Action*, Chapter 1, Activity 4: Defy Gravity

Calculate the jumper's launch speed by writing the following equation in a different form.

$KE = 1/2mv^2$

$v = \sqrt{2(KE)/m}$

$= \sqrt{2\_\_\_\_\_\_\_\_\_\_ / \_\_\_\_\_\_\_\_\_\_}$

(Insert the KE from above) (Insert the figure for mass from data above)

= ____________________ m/s

Calculate the jumper's time of flight "one way" during the jump by using the following equation (acceleration is change in velocity divided by the time). The change in velocity is the jumper's final velocity subtracted from the jumper's launch velocity. In this case, the final velocity at the top of the jump will be zero, the velocity before the jumper begins to come down. The acceleration is the acceleration due to gravity ($a = g$).

$a = \Delta v/\Delta t$

$g = \Delta v/\Delta t$

This equation can be rearranged in the following form to find the jumper's time of flight one way.

$\Delta t = \Delta v/g$

Δt = ________________________ $/10\ m/s^2$

(Insert the value of jumper's launch speed)

= ________________________ s

Calculate the total time in the air (hang time) by multiplying by two, to account for the time going up and the time coming down.

Hang time = 2 × jumper's flight one way

= 2 × ____________________

(Insert the value of time for one-way trip.)

= ________________________

Calculate the total work done by the jumper's leg muscles before launch, W_T. This was the work done to lift the center of mass from ready position to launch position without jumping ,$W_{R\ to\ L}$, plus the amount of work done to provide the center of mass with kinetic energy (KE) at launch.

$W_T = W_{R\ to\ L}, + KE$

= ____________ + ____________

= ______________ J

Calculate the total force provided by the jumper's leg muscles, f_T by rearranging the following equation

$W = fd$

$f = W/d$,

$f_T = W_T$__________

(Legwork Height)

= ____________

= ____________

ACTIVITY 5
Run and Jump

Background Information

It is recommended that you read the section **Physics Talk:** Newton's Third Law of Motion in the student text for this activity before proceeding in this section.

The explanation of forces involved in walking given in the teacher's **Background Information** for **Activity 4** will serve to explain the forces involved with walking and running brought up in this activity. You may wish to review the **Background Information** for **Activity 4** before proceeding.

The pairs of equal and opposite forces identified during earlier activities to explain friction and walking are examples of Newton's Third Law of Motion, often stated as: "For every action there is an equal and opposite reaction." Another equal and opposite pair of forces arises during this activity when a student standing on a skateboard sets himself into motion by using a leg and foot to push off from the wall.

Inevitably, forces exist in equal and opposite pairs, and often the force which we identify as the force responsible for motion is not the correct one. For example, a person who says, "I pushed down on the trampoline with a mighty force, and my force launched me upward in a high jump," is mistaken; it was the equal and opposite reaction force provided by the trampoline that launched the person upward.

Active-ating the Physics InfoMall

In addition to looking for information on Newton's Third Law (look at problem 4.14 in *Schaum's 3000 Solved Problems in Physics*, in the Problems Place), perform a search using "force diagrams" as the keywords, and the first hit is a great one! It is, again, from Arons' *A Guide to Introductory Physics Teaching: Elementary Dynamics*, Chapter 3. Section 3.12 is on Newton's Third Law and Free Body Diagrams. Arons mentions common problems and suggests solutions, including suggestions of what not to do.

Arons also notes that "Students do not really begin to understand the concept of force until they become able to apply the third law correctly and draw proper, isolated force diagrams of interacting objects," in his article "Thinking, reasoning, and understanding in introductory physics courses," in *The Physics Teacher*, vol. 19, issue 3, 1981. Check out this article.

This same search produces the warning that "Introductory textbooks are liberally decorated with diagrams, but they fail to convey to students the essential role of diagrams in problem solving or, indeed, to distinguish the roles of different kinds of diagrams" from "Toward a modeling theory of physics instruction," in the *American Journal of Physics*, vol. 55, issue 5, 1987. It is clear that the practice and ability to draw force diagrams are important.

Stretching Exercise: Add the word "elevator" to the search above (so now it is "force diagrams" AND "elevator*") for some nice discussions related to the **Stretching Exercise.**

Planning for the Activity

Time Requirements

- One class period.

Materials Needed

For each group:

- chair with wheels
- elbow pads
- knee pads
- safety helmet
- skateboard
- stick, meter, 100 cm, mardwood
- washers, metal, 3/4"
- weight, 100 g slotted mass

Advance Preparation and Setup

If you do not have access to a skateboard and safety equipment ask the students if you may borrow theirs for use in class.

Teaching Notes

SAFETY PRECAUTIONS: Close supervision is needed to prevent injury or damage to equipment and surrounding items when students set themselves into motion by pushing off from the wall.

You may wish to emphasize that the reaction forces which we "feel" throughout a day are not limited to the reaction forces felt while running or walking, but include the reaction force which occurs whenever we touch something.

You may wish to do **Activity 5 A**, Using a Bathroom Scale to Measure Forces, following this activity in the Teacher's Edition. If you do, be sensitive to the fact that many students will not wish to step on scales in front of their peers. Be sure that the students are given the opportunity to volunteer for this activity. Also, forewarn students that the reading of the bathroom scale "goes crazy" when the applied force increases or decreases by great amounts in short time intervals because the scales inertia causes it to "overshoot" maximum or minimum reading. Therefore, the scale reading must be observed very soon after a dramatic change is made in the force applied to the scale. Also point out to students that a specific value of force is not expected to be read on the scale during the push-off phase of the vertical jump; only the nature of the change in force—e.g., greater, no change, less—needs to be observed.

NOTES

Activity 5 Run and Jump

GOALS

In this activity you will:

- Understand the definition of acceleration.
- Understand meters per second per second as the unit of acceleration.
- Use an accelerometer to detect acceleration.
- Use an accelerometer to make semiquantitative comparisons of accelerations.
- Distinguish between acceleration and deceleration.

What Do You Think?

The men's high jump record is over 8 feet.

- **Pretend that you have just met somebody who has never jumped before. What instructions could you provide to get the person to jump up (that is, which way do you apply the force)?**

Record your ideas about this question in your *Active Physics* log. Be prepared to discuss your responses with your small group and the class.

For You To Do

1. Carefully stand on a skateboard or sit on a wheeled chair near a wall. By touching only the wall, not the floor, cause yourself to move away from the wall to "coast" across the floor. Use words and diagrams to record answers to the following questions in your log:

a) When is your motion accelerated? For what distance does the accelerated motion last? In what direction do you accelerate?

ANSWERS

For You To Do

1. a) Your motion is accelerated when you push away from the wall. The acceleration lasts for a short distance after which you move at a constant velocity, and then slow down. The direction of acceleration is away from the wall.

Activity Overview

In this activity students analyze the forces involved in running, stopping, and jumping. First they push against a wall while standing on a skateboard. Then they perform a thought experiment about the forces involved.

Student Objectives

Students will:

- Understand the definition of acceleration.
- Understand meters per second per second as the unit of acceleration.
- Use an accelerometer to detect acceleration.
- Use an accelerometer to make semi-quantitative comparisons of accelerations.
- Distinguish between acceleration and deceleration.

ANSWERS FOR THE TEACHER ONLY

What Do You Think?

Jumping involves the downward force exerted by the feet on the jumping surface and the equal and opposite reaction force exerted by the jumping surface on the feet and, in turn, the body.

ANSWERS

For You To Do

(continued)

1. b) Motion is at a constant speed just after the initial acceleration, when the force is acting on you. If you neglected friction you would keep moving until another force acted on you to slow you down or stop you.

   c) The force is supplied by the wall. The force must be acting in the direction of motion, away from the wall.

   d) You push on the wall, in a direction towards the wall.

   e) The two forces are equal, but opposite in direction.

2. a)-b) You push your foot on the ground backwards from yourself. How much you can push your foot parallel to the surface of the sidewalk depends on how much frictional force can be sustained by the interaction of the sole of the shoe and the sidewalk's surface. If the shoe does not slip on the surface, the sidewalk surface's equal and opposite reaction to the rearward force of friction causes your body to accelerate forward.

   c) On the slippery ice surface, the force of friction is greatly reduced. You are unable to apply much of a backwards force on the ice surface because your shoe will slide on the surface. In turn the opposite reaction of the sidewalk will also be minimal with the result that you go nowhere, there is no force to push you forward.

3. a)-b) Students generate a force diagram similar to the one shown.

Physics Words

Newton's Third Law of Motion: forces come in pairs; the force of object A on object B is equal and opposite to the force of object B on object A.

b) When is your motion at constant speed? Neglecting the effects of friction, how far should you travel? (Remember Galileo's Principle of Inertia when answering this question.)

c) Newton's Second Law, $F = ma$, says that a force must be active when acceleration occurs. What is the source of the force, the push or pull, that causes you to accelerate in this case? Identify the object that does the pushing on your mass (body plus skateboard) to cause the acceleration. Also identify the direction of the push that causes you to accelerate.

d) Obviously, you do some pushing, too. On what object do you push? In what direction?

e) How do you think, on the basis of both amount and direction, the following two forces compare?
- The force exerted by you on the wall
- The force exerted by the wall on you

2. Do a "thought experiment" about the forces involved when you are running or walking on a horizontal surface. Use words and sketches to answer the following questions in your log:

a) With each step, you push the bottom surface of your shoe, the sole, horizontally backward. The force acts parallel to the surface of the ground, trying to scrape the ground in the direction opposite your motion. Usually, friction is enough to prevent your shoe from sliding across the ground surface.

b) Since you move forward, not rearward, there must be a force in the forward direction that causes you to accelerate. Identify where the forward force comes from, and compare its amount and direction to the rearward force exerted by your shoe with each step.

c) Would it be possible to walk or run on an extremely slippery skating rink when wearing ordinary shoes? Discuss why or why not in terms of forces.

3. Think about the vertical forces acting on you while you are standing on the floor.

a) Copy the diagram of a person at left in your log.

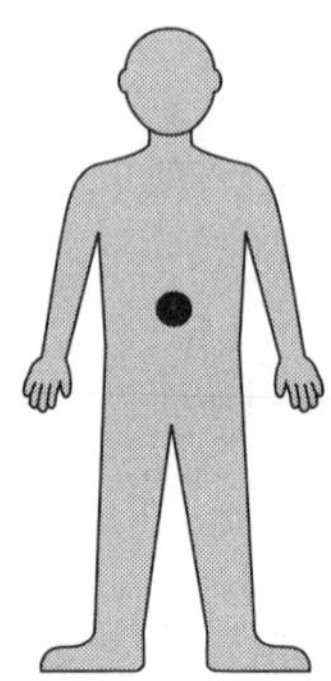

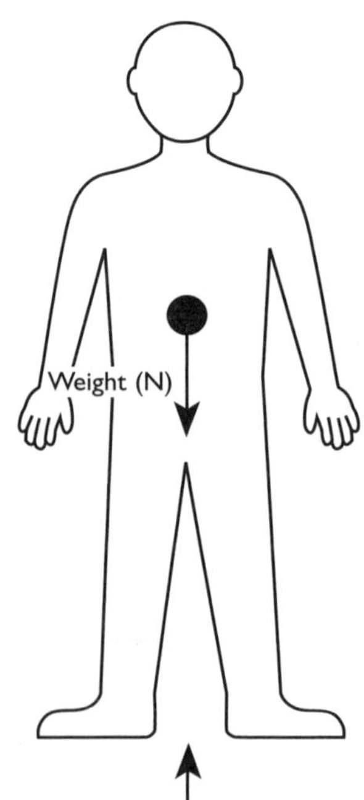

b) Identify all the vertical forces. Use an arrow to designate the size and direction of the force. Draw the forces from the dot.

c) How can you be sure that the force with which you push on the floor and the floor pushes on you are equal?

4. Set up a meter stick with a few books for support as shown.

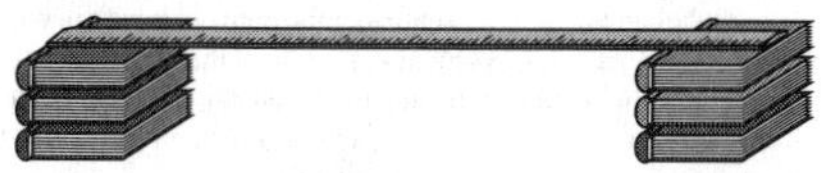

5. Place a washer in the center of the meter stick.

a) In your log, record what happens.

6. Remove the washer and replace it with 100 g (weight of 100 g = 1.0 N). Continue to place 1.0 N weights on the center of the meter stick. Note what happens as you place each weight on the stick.

Do not exceed 10 N of weight.

a) Measure the deflection of the meter stick for each 1.0 N of weight and record the values for these deflections.

b) How does the deflection of the meter stick compare to the weight it is supporting? In your log, sketch a graph to show this relationship.

c) Write a concluding statement concerning the washer and the deflection of the meter stick.

PHYSICS TALK

Newton's Third Law of Motion

Newton's **Third Law of Motion** can be stated as:

For every applied force, there is an equal and opposite force.

If you push or pull on something, that something pushes or pulls back on you with an equal amount of force in the opposite direction. This is an inescapable fact; it happens every time.

ANSWERS

For You To Do

(continued)

3. c) The forces must be equal because you are not moving.

4. Student activity.

5. a) Nothing happens.

6. a)-c) The more weight that is added to the meter stick, the greater the deflection of the stick. As the meter stick is deflected, the restoring forces in the wood build up until they exert an upward force equal to the downward force of the weight.

ANSWERS

Physics To Go

1. Yes, the forces are equal and opposite.
2. The restoring forces within the material from which the chair is made build up until the upward force exerted by the chair equals the downward force caused by your weight; if someone sits in your lap, the chair "bends" (or is otherwise deformed) more, resulting in a higher equal and opposite reaction force.
3. The forces on the ball and the bat are equal and opposite; sometimes the force exerted by the ball on the bat is enough to break the wood of the bat.
4. The forces on the players are equal and opposite, but the smaller player experiences a greater acceleration, which can have more harmful effects on the human body than a lesser acceleration.
5. The forces are equal and opposite; the hockey player is more likely than the boards to complain about the pain involved.
6. Gloves having padding which compresses and/or webbing which deforms when the ball hits the glove. The "softness" of a glove reduces the force which the glove exerts on the ball to a lower amount than a stationary hand would need to exert to stop the ball. The lower force causes the ball to decelerate at a lower rate, also reducing the reaction force which the ball exerts on the glove during stopping. A sure way to reduce the forces during a collision is to increase the amount of time that the objects exert forces on each other during the collision; that is one reason why airbags reduce injuries in automobile collisions.

Physics in Action

Reflecting on the Activity and the Challenge

According to Newton's Third Law, each time an athlete acts to exert a force on something, an equal and opposite force happens in return. Countless examples of this exist as possibilities to include in your video production. When you kick a soccer ball, the soccer ball exerts a force on your foot. When you push backward on the ground, the ground pushes forward on you (and you move). When a boxer's fist exerts a force on another boxer's body, the body exerts an equal force on the fist. Indeed, it should be rather easy to find a video sequence of a sport that illustrates all three of Newton's Laws of Motion.

Physics To Go

1. When an athlete is preparing to throw a shot put ball, does the ball exert a force on the athlete's hand equal and opposite to the force the hand exerts on the ball?
2. When you sit on a chair, the seat of the chair pushes up on your body with a force equal and opposite to your weight. How does the chair know exactly how hard to push up on you—are chairs intelligent?
3. For a hit in baseball, compare the force exerted by the bat on the ball to the force exerted by the ball on the bat. Why do bats sometimes break?
4. Compare the amount of force experienced by each football player when a big linebacker tackles a small running back.
5. Identify the forces active when a hockey player "hits the boards" at the side of the rink at high speed.
6. Newton's Second Law, $F = ma$, suggests that when catching a baseball in your hand, a great amount of force is required to stop a high-speed baseball in a very short time interval. The great amount of force is needed to provide the great amount of deceleration required. Use Newton's Third Law to explain why baseball players prefer to wear gloves for catching high-speed baseballs. Use a pair of forces in your explanation.

48

Active Physics

7. Write a sentence or two explaining the physics of an imaginary sports clip using Newton's Third Law. How can you make this description more exciting so that it can be used as part of your sports voice-over?

8. Write a sentence or two explaining the concept that a deflection of the ground can produce a force. How can you make this description more exciting so that it can be used as part of your sports voice-over?

Stretching Exercises

Ask the manager of a building that has an elevator for permission to use the elevator for a physics experiment. Your teacher may be able to help you make the necessary arrangements.

1. Stand on a bathroom scale in the elevator and record the force indicated by the scale while the elevator is:
   a) At rest.
   b) Beginning to move (accelerating) upward.
   c) Seeming to move upward at constant speed.
   d) Beginning to stop (decelerating) while moving upward.
   e) Beginning to move (accelerating) downward.
   f) Seeming to move downward at constant speed.
   g) Beginning to stop (decelerating) while moving downward.
2. For each of the above conditions of the elevator's motion, the Earth's downward force of gravity is the same. If you are accelerating up, the floor must be pushing up with a force larger than the acceleration due to gravity.
   a) Make a sketch that shows the vertical forces acting on your body.
   b) Use Newton's Laws of Motion to explain how the forces acting on your body are responsible for the kind of motion—at rest, constant speed, acceleration, or deceleration—that your body has.

ANSWERS

Physics To Go *(continued)*

7.–8. Students may wish to use parts of the voice-overs generated for these questions for their **Chapter Challenge**.

ANSWERS

Stretching Exercises

1. For the answers below, m is the person's mass, g is the acceleration due to gravity, mg is the person's weight, and a is the acceleration of the elevator.
   a) mg
   b) $mg + ma$
   c) mg
   d) $mg - ma$
   e) $mg - ma$
   f) mg
   g) $mg + ma$

Activity 5 A

Using a Bathroom Scale to Measure Forces

FOR YOU TO DO

1. Stand on a bathroom scale. Preferably, the scale should be calibrated in newtons; if is not, see page 16, of the Student Edition, **Activity 2, Physics to Go, Problem #5** for how to convert the calibration of the scale to newtons.

a) Reproduce the sketch of you standing on a bathroom scale in your log. Use a dot to show the approximate location of your center of mass.

b) Add an arrow pointing downward from your center of mass to show your weight, in newtons. Label the arrow: Weight = ___ N (enter the amount).

c) Since, when standing on the scale, you are not moving, the scale must be pushing upward on you with a force which balances your weight. Decide upon the amount and direction of the second force which must be acting on you. Starting from your center of mass in the sketch, draw the arrow to represent the force. Label the arrow: Force of Scale = ___ N (enter the amount).

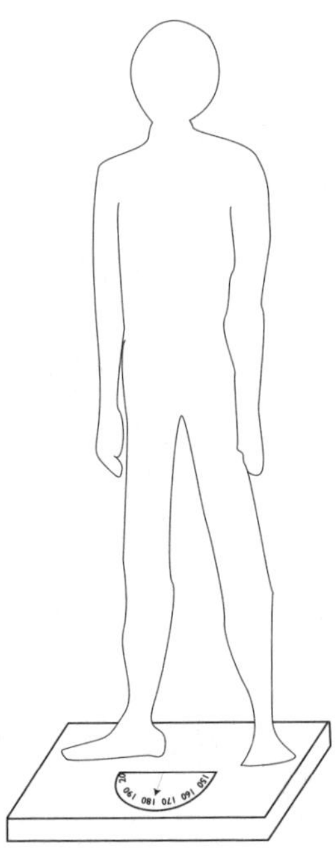

2. Stand on the bathroom scale again, but stand in a crouched position.

a) Reproduce the position of you crouching on a bathroom scale sketch in your journal.

b) Without moving any part of your body while in a crouched position are you able to "bear down" to increase the scale reading to more than when you are standing upright on the scale? Explain why or why not, and show the forces acting on your center of mass when crouched on the scale.

3. Crouch on the scale and rise to full standing position at very low, constant speed. Accelerate as little as possible as you move; you will need to accelerate somewhat to start moving upward, but, after that, try to move your center of mass at low, constant speed upward.

a) Describe in your log that you are moving your center of mass upward at constant speed from the positions used in the above steps.

b) Even though the scale reading may "wiggle" somewhat due to small, not-on-purpose accelerations as you move your center of mass upward at constant speed, how does the scale reading compare to when you were at rest in both crouched and standing positions? In your sketch, draw arrows to represent the forces acting on your center of mass as it moves upward at constant speed.

For use with *Physics In Action*, Chapter 1, Activity 5: Run and Jump

4. Explain how Newton's First Law of Motion applies to the above situations when you were:

a) At rest, standing upright on the scale.

b) At rest, crouched on the scale.

c) Standing on the scale while moving your center of mass upward at constant speed.

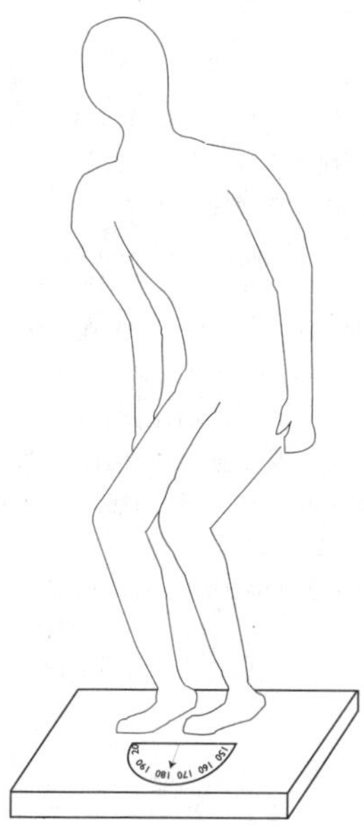

5. Again stand on the scale in crouched position. This time, you will accelerate your center of mass to jump upward. Design your jump to go slightly forward so that you land on the floor, not the scale, when you come back down – this is to prevent damage to the scale by landing on it.

Have an assistant ready to observe how the scale reading changes as you are accelerating your center of mass upward, but before you "launch" into the air. This observation will be "tricky" because bathroom scales cannot rapidly respond to changing forces. Also, inertia causes the scale to "overshoot" the maximum reading when the force on the scale is suddenly increased. Therefore, observation of how the scale reading changes will need to be made a fraction of a second after you begin rising from crouched position.

a) Sketch the situation in your log.

b) Jump as your assistant observes the scale. Compared to when your center of mass is at rest or moving with constant upward speed, is the downward force you exert on the scale when accelerating your center of mass upward more or less? Since the scale couldn't "keep up" with the action to allow reading a specific value of force, report in your journal whether the force was "going up" or "going down" as you were pushing on the scale before launching into your jump.

c) In your sketch, draw arrows to represent the forces acting on your center of mass as it accelerates upward before launch, including (i) the downward pull of gravity on your body (your weight) and (ii) the upward push of the scale. You do not have a value, in newtons, to determine how long to make the arrow representing the upward push of the scale, so indicate, according to how the scale reading was changing, that the amount of force is either greater or less than your weight.

d) Which of Newton's Laws of Motion best applies to this situation, the First Law or the Second Law? How does the law which you choose apply?

ACTIVITY 6
The Mu of the Shoe

Background Information

When it is desired to accelerate an object (change its speed, its direction, or both), a net force must act on the object. If you are standing still on a sidewalk and want to get moving, you must somehow cause a force to be exerted on your body's mass. You accomplish application of a force by pushing your foot in the backward direction parallel to the surface of the sidewalk. How much you can push your foot parallel to the surface of the sidewalk depends on how much frictional force can be sustained by the interaction of the sole of the shoe and the sidewalk's surface. If the shoe does not slip on the surface, the sidewalk surface's equal and opposite reaction to the rearward force of friction causes your body to accelerate forward; if there is ice on the sidewalk, the available force will be reduced, and your shoes may just slide on the surface with the result that you go nowhere.

The maximum frictional force, F, that can be generated between the surfaces of two materials in contact is expressed by the equation $F = \mu N$ where N is the "normal force" (the word "normal" in this context means "perpendicular") that is perpendicularly pushing the two surfaces together, and μ is the "coefficient of friction" for the pair of materials from which the surfaces are made. In the above example, the normal force, N, would be equal to your weight. The value of μ depends on the quality of the two materials in contact. For a given pair of materials, such as leather shoe soles on a concrete sidewalk surface, there are two kinds of μ, the "coefficient of static (starting) friction" and the "coefficient of sliding friction" (sometimes called the coefficient of "kinetic"—meaning "moving"—friction). The larger of the two, the "coefficient of static (stationary) friction" applies when the surfaces are at rest with respect to each other; the value of F resulting from calculations using the coefficient of static friction is the minimum force required to "tear the surfaces loose" to cause them to begin sliding across one another. The second kind of μ, the smaller of the two kinds, applies when the surfaces are moving with respect to each other in a sliding mode; the value of F resulting from calculations using the coefficient of sliding friction is the minimum force required to cause the surface to slide across one another at constant speed. It is the second kind, the coefficient of sliding friction, that is measured in this activity

The frictional force generated between the shoe and the sidewalk—due to pushing the foot rearward as gravity and the upward restoring force of the sidewalk squeeze the sole of the shoe and the surface of the sidewalk together—is answered by a corresponding equal forward push by the sidewalk on the foot. The latter force, the forward push by the sidewalk, is the push you "feel" and which causes you to accelerate forward.

It is not always the case that the normal force, N, is equal to the weight of the object. In the case of an object on a sloped surface, the normal force is less than the weight, equaling the component, or effectiveness, of the object's weight in the direction perpendicular to the sloped surface. Other examples of a cases where N is not equal to the weight of an object bearing on a surface would include the frictional force between belts riding on pulleys in machines; in such cases, tensioned springs usually are used to force the surfaces together to provide sufficient N to prevent sliding, or, intentionally as when stopping a machine, to reduce tension to cause N to be reduced to an amount where a belt will slide on a pulley.

Active-ating the Physics InfoMall

Discussions of friction can be found throughout the InfoMall. If you choose to do a search using the keyword "friction," you will need to limit your search to only a few stores at a time. If you look in the Articles and Abstracts Attic, one of the titles that may interest you is "Twas the class before Christmas," from *The Physics Teacher*, vol. 24, issue 9, 1986. At the very least, the problems involving friction can be amusing.

Try searching the Demo & Lab Shop with the keyword "friction." You will want to look at these yourself, so no examples are included here. Choose the demonstration that best suits your style and situation.

And you will not be surprised to find that there are many, many problems you can find involving friction in the Problems Place.

Planning for the Activity

Time Requirements

- One class period.

Materials Needed

For each group:

- rough horizontal surface, 6" x 36"
- shoe
- smooth surface (desk top, tile)
- scale, 0-20 newton range
- stack of books
- weight, 100 g slotted mass
- wood board, 4 ft

Advance Preparation and Setup

You will need a variety of athletic shoes and samples of floor materials. If enough students wear athletic shoes to school, use their shoes as samples; if not, arrange to have one shoe per group available. Floor materials may include your classroom floor, a table top to simulate a floor, samples of floor materials from a retail store, or, if they are different from the floor in your classroom, floors in other areas of your school such as the gymnasium. It is desired to have contrasting degrees of "roughness" represented in the samples. Surfaces which would be of most interest to your students would be best to use.

Teaching Notes

It is recommended that a brief discussion of the symbolism used in physics be conducted with the class after students have read the section "What is Mu?" and before beginning **For You To Do.** Students may feel it is "cool" to ask other students not taking physics about the "μ" of their athletic shoes.

Students may wish to measure μ for more kinds of shoes and floor materials called for in the instructions. The instructions include only the minimum number of samples needed to acquire meaningful data; encourage students to test more samples of shoes and/or floor materials if time allows.

Coefficients of friction for many pairs of materials are listed in the *Handbook of Chemistry and Physics* and in many traditional physics textbooks. You may wish to have listings available for students to examine.

In addition to athletic shoes which are specialized for particular sports, the variety of waxes used by skiers to control friction in various conditions may be of high interest to some students.

This activity can also be done with a smart pulley, computer, and weights hanging over the smart pulley.

You may wish to do the **Activity 6 A**: Alternative Activity for Measuring the Mu of the Shoe presented after this activity in the Teacher's Edition as a **Stretching Exercise** with the class, after you have completed the activity in the textbook. The explanation for this activity is also presented following the activity.

Activity Overview

In this activity students investigate the effect of different surfaces, and different weights on the coefficient of friction.

Student Objectives

Students will:

- Understand and apply the definition of the coefficient of sliding friction, μ.
- Measure the coefficient of sliding friction between the soles of athletic shoes and a variety of floor surface materials.
- Calculate the effects of frictional forces on the motion of objects.

ANSWERS FOR THE TEACHER ONLY

What Do You Think?

Vast amounts of engineering knowledge and research about friction are applied to the design of athletic footgear.

Activity 6 The Mu of the Shoe

GOALS

n this activity you will:

- Understand and apply the definition of the coefficient of sliding friction, μ.
- Measure the coefficient of sliding friction between the soles of athletic shoes and a variety of floor surface materials.
- Calculate the effects of frictional forces on the motion of objects.

What Do You Think?

A shoe store may sell as many as 100 different kinds of sport shoes.

- **Why do some sports require special shoes?**

Record your ideas about this question in your *Active Physics* log. Be prepared to discuss your responses with your small group and the class.

For You To Do

1. Take an athletic shoe. Use a spring scale to measure the weight of the shoe, in newtons.

   a) Record a description of the shoe (such as its brand) and the shoe's weight in your log.

ANSWERS

For You To Do

1. a) Answers will vary depending on the brand of shoe used and the size of the shoe.

2. Place the shoe on one of two horizontal surfaces (either rough or smooth) designated by your teacher to be used for testing. Attach the spring scale to the shoe as shown below so that the spring scale can be used to slide the shoe across the surface while, at the same time, the amount of force indicated by the scale can be read.

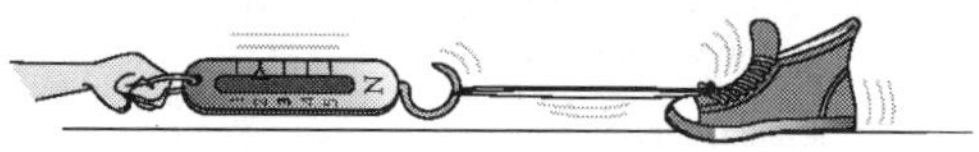

a) Record in your log a description of the surface on which the shoe is to slide.

b) Measure and record the amount of force, in newtons, needed to cause the shoe to slide on the surface at constant speed. Do not measure the force needed to start, or "tear the shoe loose," from rest. Measure the force needed, after the shoe has started moving, to keep it sliding at low, constant speed. Also, be careful to pull horizontally so that the applied force neither tends to lift the shoe nor pull downward on the shoe.

c) Use the data you have gathered to calculate μ, the coefficient of sliding friction for this particular kind of shoe on the particular kind of surface used. Show your calculations in your log.

The coefficient of sliding friction, symbolized by μ, is calculated using the following equation:

$$\mu = \frac{\text{force required to slide object on surface at constant speed}}{\text{perpendicular force exerted by the surface on the object}}$$

Example:

Brand X athletic shoe has a weight of 5 N. If 1.5 N of applied horizontal force is required to cause the shoe to slide with constant speed on a smooth concrete floor, what is the coefficient of sliding friction?

$$\mu_{x \text{ on concrete}} = \frac{1.5 \text{ N}}{5.0 \text{ N}} = 0.30$$

Chapter 1

ANSWERS

For You To Do *(continued)*

2. a)-c) Students' answers will vary depending on the surface and the shoe used.

ANSWERS

For You To Do

(continued)

3. a)-c) Taking into account possible errors in measurement, students should find that the value of μ is not affected by the weight of the shoe.

4. a) Student sketch (see bottom right).

b)-c) Students should recognize that the value of μ will be different for different surfaces. The "rougher" the surface, the greater the coefficient of friction. For example, the coefficient of friction for rubber on dry concrete is 140 times greater than rubber on ice.

d) The previous step in the activity indicates that the weight of the shoe should not make a difference in the coefficient of friction.

3. Add "filler" to the shoe to approximately double its weight and repeat the above procedure for measuring the μ of the shoe.
   - a) Calculate μ for this surface, showing your work in your log.
   - b) Taking into account possible errors of measurement, does the weight of the shoe seem to affect μ? Use data to answer the question in your log.
   - c) How do you think the weight of an athlete wearing the shoe would affect μ? Why?

4. Place the shoe on the second surface designated by your teacher and repeat the procedure.
   - a) Make another sketch to show the forces acting on the shoe.
   - b) Calculate μ.
   - c) How does the value of μ for this surface compare to μ for the first surface used? Try to explain any difference in μ.
   - d) Would it make any difference if you used the empty shoe or the shoe with the filler to calculate μ in this activity? Explain your answer.

Reflecting on the Activity and the Challenge

Many athletes seem more concerned about their shoes than most other items of equipment, and for good reason. Small differences in the shoes (or skates or skis) athletes wear can affect performance. As everyone knows, athletic shoes have become a major industry because people in all "walks" of life have discovered that athletic shoes are great to wear, not only on a track but, as well, just about anywhere. Now that you have studied friction, a major aspect of what makes shoes function well when need exists to be "sure-footed," you are prepared to do "physics commentary" on athletic footgear and other effects of friction in sports. Your sports commentary may discuss the μ of the shoe, the change in friction when a playing field gets wet, and the need for friction when running.

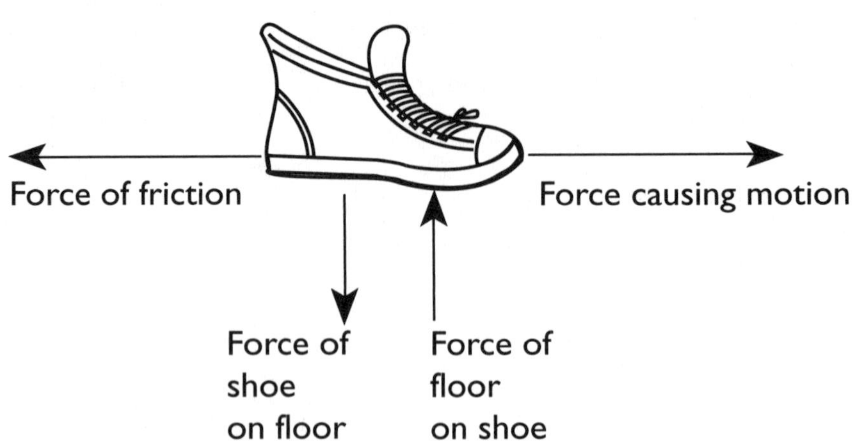

PHYSICS TALK

Coefficient of Sliding Friction, μ

There are not enough letters in the English alphabet to provide the number of symbols needed in physics, so letters from another alphabet, the Greek alphabet, also are used as symbols. The letter μ, pronounced "mu," traditionally is used in physics as the symbol for the "coefficient of sliding friction."

The coefficient of sliding friction, symbolized by μ, is defined as the ratio of two forces:

$$\mu = \frac{\text{force required to slide object on surface at constant speed}}{\text{perpendicular force exerted by the surface on the object}}$$

Facts about the coefficient of sliding friction:

- **μ does not have any units because it is a force divided by a force; it has no unit of measurement.**
- **μ usually is expressed in decimal form, such as 0.85 for rubber on dry concrete (0.60 on wet concrete).**
- **μ is valid only for the pair of surfaces in contact when the value is measured; any significant change in either of the surfaces (such as the kind of material, surface texture, moisture, or lubrication on a surface, etc.) may cause the value of μ to change.**
- **Only when sliding occurs on a horizontal surface, and the pulling force is horizontal, is the perpendicular force that the sliding object exerts on the surface equal to the weight of the object.**

Chapter 1

ANSWERS

Physics To Go

1. In football, players change the length of shoe cleats or sometimes wear shoes without cleats to improve footing in bad weather.
2. Downhill skiers use wax to reduce friction.
3. A common misconception is that the coefficient of friction—and, therefore, the force of friction—depends only on the shoe; it depends as much as on the nature of the surface beneath the shoe. No, the athlete cannot be assured that the same amount of frictional force will be present when the same shoe is used on a court having a different surface.
4. $F = 0.03 \times 600\text{ N} = 18\text{ N}$
5. Normal force, FN= Weight of vehicle = $mg = 1000\text{ kg} \times 10\text{ m/s}^2 = 10{,}000\text{ N}$

Frictional (stopping) force = $\mu N = 0.55 \times 10{,}000\text{ N} = 5500\text{ N}$

Work to stop vehicle = fd = $5500\text{ N} \times 100\text{ m} = 550{,}000$ joules

Work to stop vehicle = *KE* of vehicle before brakes were applied:

$550{,}000J = 1/2\ mv^2$

Therefore,

$v = \sqrt{2\ (550{,}000\ J)/m}$

$= \sqrt{1{,}100{,}000\ J/1000\text{ kg}}$

$= \sqrt{1100\text{ m}^2/\text{s}^2}$

$= 33\text{ m/s} = 75\text{ miles/hr}$

The driver has a problem because the laws of physics will prevail in court.

Physics To Go

1. Identify a sport and changing weather conditions that probably would cause an athlete to want to increase friction to have better footing. Name the sport, describe the change in conditions, and explain what the athlete might do to increase friction between the shoes and ground surface.
2. Identify a sport in which athletes desire to have frictional forces as small as possible and describe what the athletes do to reduce friction.
3. If a basketball player's shoes provide an amount of friction that is "just right" when she plays on her home court, can she be sure the same shoes will provide the same amount of friction when playing on another court? Explain why or why not.
4. A cross-country skier who weighs 600 N has chosen ski wax that provides $\mu = 0.03$. What is the minimum amount of horizontal force that would keep the skier moving at constant speed across level snow?
5. A race car having a mass of 1000 kg was traveling at high speed on a wet concrete road under foggy conditions. The tires on the vehicle later were measured to have $\mu = 0.55$ on that road surface. Before colliding with the guardrail, the driver locked the brakes and skidded 100 m, leaving visible marks on the road. The driver claimed not to have been exceeding 65 miles per hour (29 m/s). Use the equation:

   Work = Kinetic Energy

   to estimate the driver's speed upon hitting the brakes. (Hint: In this case, the force that did the work to stop the car was the frictional force; calculate the frictional force using the weight of the vehicle, in newtons, and use the frictional force as the force for calculating work.)

6. Identify at least three examples of sports in which air or water have limiting effects on motion similar to sliding friction. Do you think forces of "air resistance" and "water resistance" remain constant or do they change as the speeds of objects (such as athletes, bobsleds, or rowing sculls) moving through them change? Use examples from your own experience with these forms of resistance as a basis for your answer.
7. If there is a maximum frictional force between your shoe and the track, does that set a limit on how fast you can start (accelerate) in a sprint? Does that mean you cannot have more than a certain acceleration even if you have incredibly strong leg muscles? What is done to solve this problem?
8. How might an athletic shoe company use the results of your experiment to "sell" a shoe? Write copy for such an advertisement.
9. Explain why friction is important to running. Why are cleats used in football, soccer, and other sports?
10. Choose a sport and describe an event in which friction with the ground or the air plays a significant part. Create a voice-over or script that uses physics to explain the action.

Answers

Physics To Go
(continued)

6. Any sports which involve objects moving through air or water will involve fluid resistance.
7. Yes, the maximum frictional force between your shoe and the track does place a limit on acceleration. Sprinters use blocks when beginning a race, thereby providing a surface that is not horizontal, and does not depend on the weight of the runner.
8. See the **Assessment Rubric** following this activity in the Teacher's Edition.
9. Without friction it would be impossible to walk or run. Cleats increase the friction between the shoe and the ground by increasing the amount of surface in contact.
10. Students provide a voice-over related to friction.

Assessment Rubric: Physics To Go Question 8

How might an athletic shoe company use the results of your experiment to sell a shoe? Write a copy for an advertisement.

Descriptors	Levels of Attainment			
	poor	average	good	excellent
1. Physics concepts are accurately presented. • Newton's Third Law of Motion is explained in terms of running: For every applied force, there is an equal and opposite force. • The force required to cause the shoe to move on different surfaces is explained: *Coefficient of friction is expressed as a ratio of the force needed to move the shoe at a constant speed, by the force exerted by the surface of the shoe.*	1	2	3	4
2. Physics concepts explained in everyday language. • Examples are provided for each concept. • Presentation style does not talk down to the consumer.	1	2	3	4
3. The need for buying specialized sports shoes is established. • Examples are provided illustrating why different shoes are used for different sporting events. *Presentation identifies the amount of friction provided by different surfaces and the advantages of solid traction for different sports.*	1	2	3	4
4. Design and appeal. • Organization of information is short and snappy. • Message is clearly identified with a target audience. • Presentation designed for a specific media: i.e., visual images utilize color effectively and enhance message for television presentation.	1	2	3	4

For use with *Physics In Action*, Chapter 1, Activity 6: The Mu of the Shoe

NOTES

Activity 6 A

Alternative Activity for Measuring the Mu of the Shoe

FOR YOU TO DO

"Step right up. Have the Mu of your shoe measured right here! Takes only a minute." Can the coefficient of sliding friction be measured in a quicker, easier way than the method used in **For You To Do**. Well, it can.

1. For a quick-and-easy way to measure, all you need is a sample of "floor" material (such as a wooden board) which can be arranged into a ramp, a carpenter's square (if not available, a ruler can be substituted), a calculator, and, of course, a shoe. Use a stack of books (or some other method) to raise one end of the floor sample to form a sloped ramp and place the shoe on the ramp facing "downhill."

2. Adjust the slope of the ramp to find the slope at which the shoe, given only a "nudge" to start it sliding, continues sliding down the ramp "on its own" (without further pushing) at low, constant speed:

   - If the shoe does not slide down the ramp on its own after it is tapped, increase the slope of the ramp.
   - If the shoe accelerates while sliding down the ramp after it is tapped, decrease the slope of the ramp.

3. When you have found the proper slope for the ramp, use a carpenter's square (or a ruler) to measure the vertical rise and corresponding horizontal run of the ramp.

   a) Record the rise and the run in your log.

4. The coefficient of sliding friction is:

$$\mu = \frac{\text{Length of vertical rise of ramp}}{\text{Length of horizontal run of ramp}}$$

   a) Calculate the coefficient of sliding friction of the shoe.
   This method gives the same result as the method used in **For You To Do**. Your teacher may be able to help you understand why the two methods are equivalent.

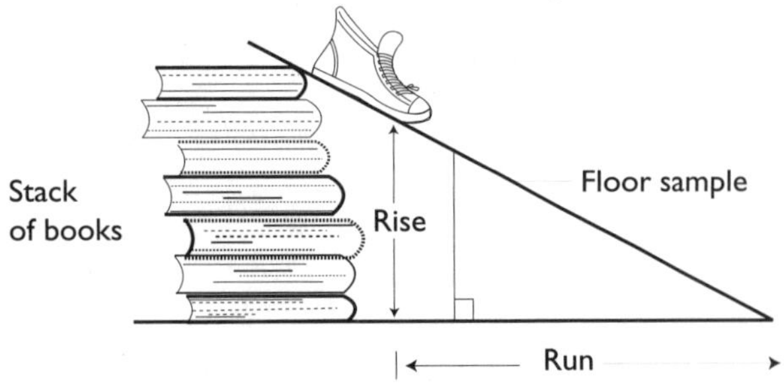

For use with *Physics In Action*, Chapter 1, Activity 6: The Mu of the Shoe

Background Information for Activity 6 A: Alternative Activity for Measuring the Mu of the Shoe

Below is the theoretical basis for the method of measuring μ presented in the **Alternative Activity 6 A.** Highly able students who have had experience with geometry may be able to understand the principles underlying the method. It is recommended that the procedure presented in the alternative activity be read carefully before proceeding to read the below explanation.

The purpose of this explanation is to prove that μ = (Rise) ÷ (Run) when the slope of the inclined plane made from a sample of floor material is adjusted so that the shoe, or other object represented by mass *m*, whose coefficient of friction with the floor material is desired to be measured—slides down the slope at low, constant speed when given a "tap" start. (A tap start is required to overcome the force of "static," or starting, friction which is greater than the force of sliding friction.) Referring to the below diagram, side lengths A and B of triangle ABC correspond, respectively, to the Rise and Run distances. Therefore, it is to be proven that μ = A/B.

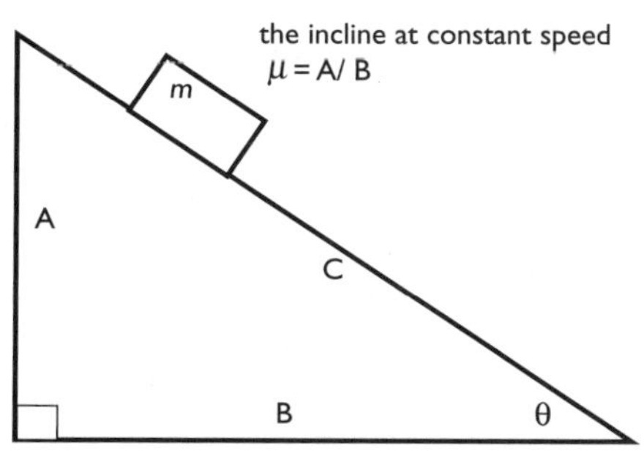

The weight, *W*, of mass *m* is a force which acts straight downward, as shown in the below diagram. The amount of force due to the object's weight, *W* = *mg*, is represented by the length of vector *W*.

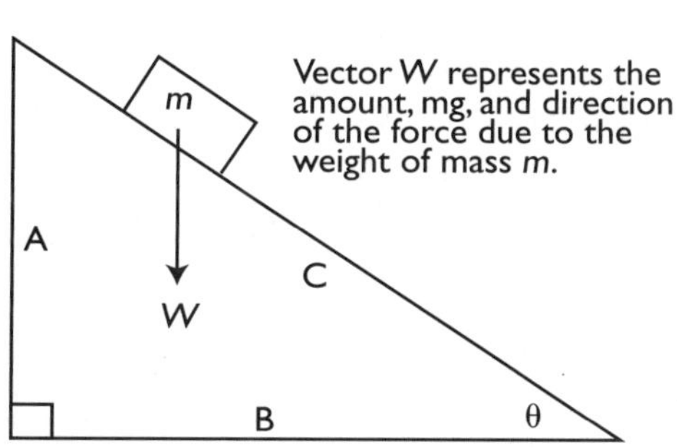

The weight vector, *W*, can be resolved into, or thought of as having the same effect as, two vectors called "components" of vector *W*: (1) a vector *R* which has a direction normal, or perpendicular, to the surface represented by side C of triangle ABC and (2) a vector *P* which has a direction parallel to side C. Vector *R* is the force which mass m exerts perpendicular to the surface on which sliding occurs, and vector *P* is the force which causes mass m to slide on the surface at constant speed.

The below diagram shows vectors *R* and *P* as components of vector *W*.

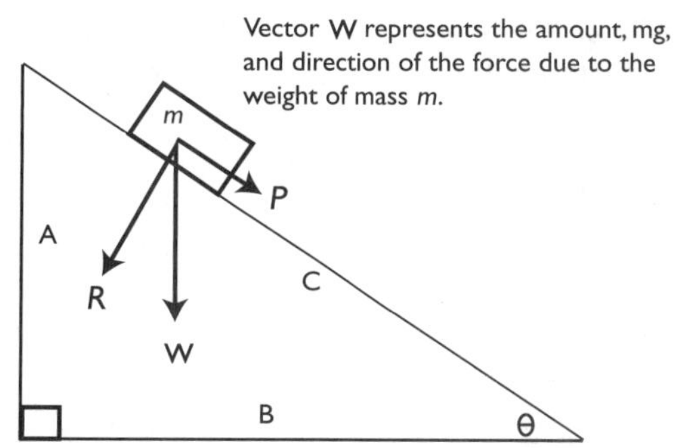

Vector mathematics requires that component vectors, in this case vectors *P* and *R*, must, when added together as vectors, equal the vector of which they are components, in this case vector *W*. On the right-hand side of the above diagram, vectors *P* and *R* are shown repositioned, but not altered in length or direction, for head-to-tail vector addition. As shown, when the component vectors are placed head-to-tail with one of the component vectors having its tail end corresponding to the tail end of vector *W*, the head end of the second component vector drawn in the head-to-tail vector addition process arrives at the head end of vector *W*. The fact that the component vectors "close" on the head end of *W* when placed head-to-tail shows, according to vector algebra, that the components add together to equal vector *W*. Since the directions of the component vectors *P* and *R* were specified as, respectively, parallel and normal to triangle side C, only the lengths of *P* and *R* shown in the diagram would satisfy the requirement of the vectors to close on the head end of vector *W*.

Referring to the above diagram, triangle ABC is similar to triangle *PRW* formed for the vector addition process. This is true by the theorem that two triangles which each contain a right angle and which have two mutually perpendicular sides are similar. Therefore, the sides of the two triangles exist in equal proportions, and:

A/B = *P*/*R*

Since, by definition, μ is the ratio of the force required to slide mass m at constant speed on the surface to the force which *m* exerts perpendicular to the surface:

$\mu = P/R = A/B$

Therefore, it is proven: the Rise divided by the Run, A/B, when mass *m* slides at constant speed is equal to μ.

For those familiar with trigonometry, A/B in triangle ABC is the tangent of angle θ.

Therefore, as an alternative to measuring the Rise and Run to determine μ, θ can be measured instead, and μ can be determined using a calculator or table by the relation:

$\mu = \tan\theta$

All of the forces active as mass *m* slides on the incline at constant speed are shown in the below diagram:

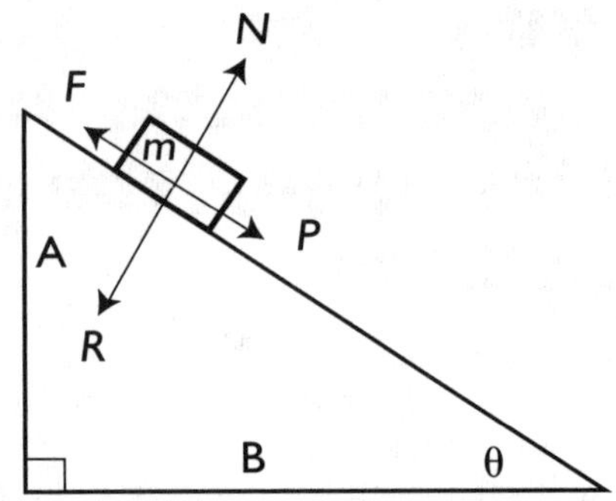

Where the new forces introduced to the diagram, *F* and *N*, are, respectively, the frictional force and the restoring force of the material represented by side C of triangle ABC. It is important to observe that, since mass *m* moves at constant speed, the sum—vector sum, that is—of the four forces acting on mass *m* is zero, or, in other words, all of the forces acting on mass *m* are balanced.

For students who wish to do the activity , samples of floor materials will need to be able to be arranged on an adjustable slope; boards 2 or 3 feet in length could be used as a base for samples to be arranged on a slope.

NOTES

ACTIVITY 7
Concentrating on Collisions

Background Information

This **Background Information** will serve for both this and the next activity because both involve the same topic, momentum. Momentum is introduced in this activity, and conservation of momentum is implied; the next activity, **Activity 8**, focuses on the Law of Conservation of Momentum.

Before proceeding in this section it is recommended that you read the following sections in this chapter of the student text: **For You To Read:** Mass, Velocity and Momentum in **Activity 7**, and **For You To Read:** Conservation of Momentum in **Activity 8**. Both selections are assumed as background for the below discussion of momentum.

A corollary to Newton's Third Law is that action/reaction pairs of forces always act for the same amount of time as each other. In any interaction, forces may vary in complicated ways during the time of interaction; nevertheless, the average force during the interaction often can be used to provide an accurate, but greatly simplified, analysis of the net result of an interaction such as a collision.

If F is used to represent the average force on one object during a collision and Δt is used to represent the duration of the collision, one can produce the following mathematical path to the quantity of ultimate interest for this discussion, the quantity called momentum. Newton's Second Law, written in rearranged form, may be applied to the object involved in the collision as:

$$F\Delta t = m\Delta v = m(v_f - v_i) = (mv_f - mv_i) = \Delta(mv)$$

$$F\Delta t = \Delta(mv)$$

where v_i is the speed with which the object entered into the collision and v_f is the speed of the object after the collision.

In the above equation $F\Delta t = \Delta(mv)$, the product of force and time, $F\Delta t$, is called the "impulse," and the quantity mv is called the "momentum." The equation can be read literally as, "When during a collision an object of mass m experiences an impulse $F\Delta t$, the object experiences a change in momentum $\Delta(mv)$. Indeed, it is the impulse that causes the object's change in momentum.

The force-time product, impulse, is expressed in the unit newton-seconds; dimensional analysis shows that Ns is equivalent to the unit of momentum, which, as a mass-velocity product, must be (kg)m/s:

Ns = [(kg)(m/s^2)] (s) = (kg)m/s (True, the unit of impulse equals the unit of momentum.)

But, according to Newton's Third Law, whatever happens to one of a pair of objects during a collision should, in terms of force, happen equally, but in the opposite direction, to the other object. Also, the objects involved in a collision touch each other for equal amounts of time, the duration of mutual contact. Thus, whatever change in momentum one object experiences, the other object must experience an equal and opposite change in momentum. For example, if one object gains momentum in a head-on collision, the other object must lose an equal amount of momentum. It must always be true that the combined momenta of the two objects before the collision are preserved, or conserved, after the collision. It is a far-reaching law of physics that momentum is conserved in all interactions, no matter how complex, convoluted, and intense the interactions may be.

This chapter considers only collisions that are "head-on," so the motions all are along the same line before and after the collision. However, the caution is made that when dealing with momentum in collision events, one must take care to keep track of the directions of travel of the objects both before and after the collision. For example, for a collision in which an object of mass m leaves the collision traveling at the same rate but in an opposite direction is not one for which the momentum change is zero. In this case, the momentum change is $2mv$ or $-2mv$, depending on how positive and negative signs are attributed to direction of travel.

Similarly, if two objects enter a collision with equal magnitudes of momentum, the opposite signs mean that the sum of momenta is zero. As a result, when the collision is completed, one possible outcome is to have both objects at rest at the point of impact. The only other possibility is that both objects will rebound with equal and opposite momenta, but not necessarily equal and opposite speeds because the masses may be unequal.

Depending on the nature of the objects involved in a collision, kinetic energy may be conserved to extents ranging from not at all (as when the objects stop upon colliding) to almost completely (as when extremely hard objects such as ball bearings or gas molecules collide and rebound). Whether or not kinetic energy is conserved, momentum is conserved, which makes the Law of Conservation of Momentum a very powerful tool.

Active-ating the Physics InfoMall

The physics of collisions is yet another of those topics discussed in almost every textbook around. To keep InfoMall searches interesting and related to this *Active Physical Science* book, search the Articles and Abstracts Attics using keywords "collisions" AND "sport*". Several articles result, including "Batting the ball," from the *American Journal of Physics*, vol. 31, issue 8, 1963. You will also find plenty on billiard balls, kicking footballs, and using tennis rackets.

You may wish to investigate the known misconceptions students have regarding momentum. Try a search with "momentum" AND "misconcept*". One of the hits is "Verification of fundamental principles of mechanics in the computerized student laboratory," in *American Journal of Physics,* vol. 58, issue 10, 1990. This article is also great for using computers for teaching.

Planning for the Activity

Time Requirements

- One class period.

Materials Needed

For each group:

- balls, bocci
- ball, golf
- ball, nerf,
- bBall, practice golf
- ball, tennis
- *Sports* Content Video
- starting ramp for bocci ball
- tape, masking, 3/4" x 60 yds.

Advance Preparation and Setup

A pair of (nearly) matched bocci balls is recommended for each group. Contacting bocci establishments in your area perhaps will result in donations of balls, and putting out a call to your school community—parents, faculty, etc.—also should be productive. If you simply cannot obtain bocci balls, hard balls, such as croquet balls, could be substituted.

You will also need a soccer ball, a golf ball, and a tennis ball for each group. You may be able to borrow the balls from your school's athletic department.

Teaching Notes

Discuss with the students what is happening in the humorous illustration on page 56. What will happen to the soccer ball? Students who enjoy drawing, may wish to draw the "next frame" in this action. You may wish to return to the illustration at the end of the activity to compare students responses to their original predictions.

Bocci balls, with their large inertias, are less sensitive to annoyances like small imperfections in the floor or a little grit in the path. They also have the advantage of a large diameter, so it is easier to create collisions that are nearly head on. They also demand a lot of attention. Their disadvantage is their large mass!

SAFETY PRECAUTION: Moving bocci balls can hurt people (especially toes and fingers) and property. Instruct students to use extreme care and provide close supervision when bocci balls are in use.

Contact sports are, of course, rich sources of collisions to be analyzed in terms of momentum.

Need to assign positive and negative values to velocity and momentum is implied in the discussion in **For You To Read.** The unit of momentum, (kg)m/s is new, but is sufficiently straightforward that it should not be a problem for students; somehow, the unit of momentum seems to have escaped being collected up under someone's name.

Chapter 1

Activity Overview

In this activity students investigate the principle of momentum by rolling balls down ramps and staging collisions. They will infer the relative masses of two balls by staging and observing collisions between them.

Student Objectives

Students will:

- Understand and apply the definition of momentum: Momentum = mv.
- Conduct semi-quantitative analyses of the momentum of pairs of objects involved in one-dimensional collisions.
- Infer the relative masses of two objects by staging and observing collisions between the objects.

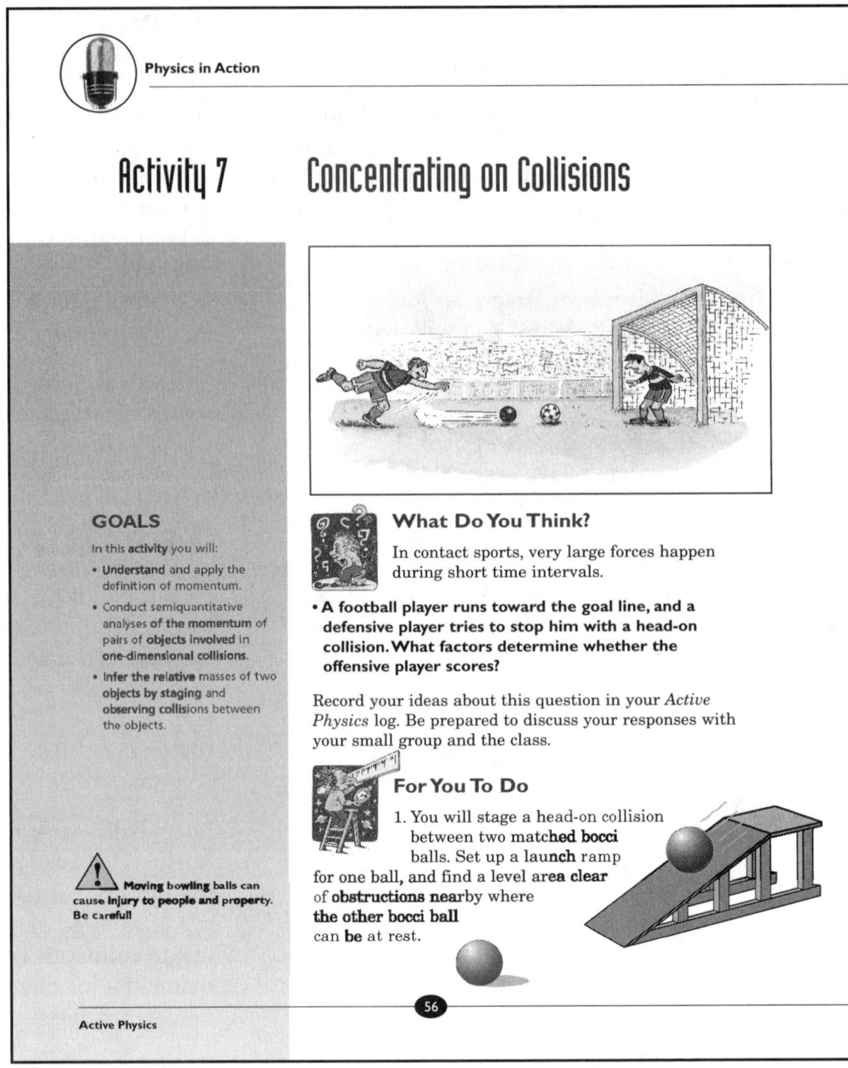

Physics in Action

Activity 7 Concentrating on Collisions

GOALS

In this activity you will:

- Understand and apply the definition of momentum.
- Conduct semiquantitative analyses of the momentum of pairs of objects involved in one-dimensional collisions.
- Infer the relative masses of two objects by staging and observing collisions between the objects.

What Do You Think?

In contact sports, very large forces happen during short time intervals.

- **A football player runs toward the goal line, and a defensive player tries to stop him with a head-on collision. What factors determine whether the offensive player scores?**

Record your ideas about this question in your *Active Physics* log. Be prepared to discuss your responses with your small group and the class.

For You To Do

1. You will stage a head-on collision between two matched bocci balls. Set up a launch ramp for one ball, and find a level area clear of obstructions nearby where the other bocci ball can be at rest.

Moving bowling balls can cause injury to people and property. Be careful!

Active Physics 56

ANSWERS FOR THE TEACHER ONLY

What Do You Think?

Both mass and speed have roles in collisions between two objects. The mass of each player and their speeds must be considered.

ANSWERS

For You To Do

1. Student activity.

Activity 7 Concentrating on Collisions

2. Temporarily remove the "target" bocci ball. Find a point of release within the first one-fourth of the ramp's total length that gives the ball a slow, steady speed across the floor. Mark the point of release on the ramp with a piece of tape.

3. Replace the target ball. Adjust the aim until a good approximation of a head-on collision is obtained. Stage the collision.
   a) Record the results in your log. Use a diagram and words to describe what happened to each ball.

4. Repeat the above type of collision, but this time move the release point up the ramp to at least double the ramp distance.
   a) Describe the results in your log.
   b) How did the results of the collision change from the first time?
   c) Identify a real-life situation that this collision could represent.

5. Arrange another head-on collision between the balls, but this time have both balls moving at equal speeds before the collision. Using a second, identical ramp, aim the second ramp so that the second ball's path is aligned with the first ball's path. Mark a release point on the second ramp at a height equal to the mark already made on the first ramp. This should ensure that the balls will have low, approximately equal speeds. On a signal, two persons should release the balls simultaneously from equal ramp heights.

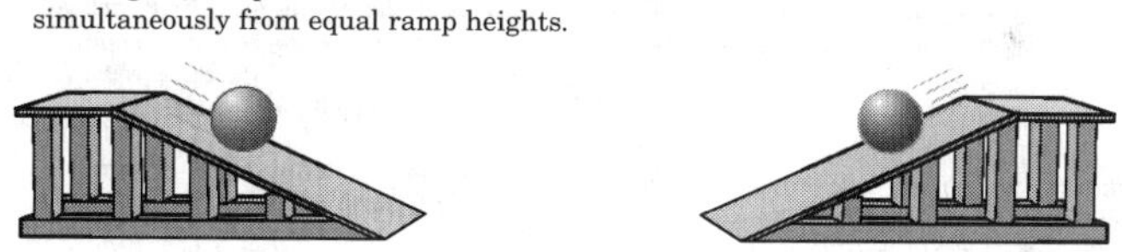

   a) Describe the results in your log.
   b) Identify a real-life situation that this collision could represent.

Answers

For You To Do

(continued)

2. Student activity.

3. a) Assume the bocci balls have reasonably well-matched masses. A moving ball striking a stationary ball of equal mass should result in the moving ball stopping upon colliding and the stationary ball moving away from the collision at about the same speed that the incoming ball had before the collision. bocci balls have a high coefficient of restitution, and the first ball can be expected to come nearly to a stop, while the other leaves at something close to the speed of the incident ball.

4. a)-b) As the speed of the released ball increases, the speed at which the stationary ball moves away will increase.

   c) This type collision may occur when a forward collides with a goaltender, or the head of a golf club collides with a golf ball.

5. a) Balls of equal mass and equal speed colliding head-on should result in both balls rebounding at speeds less than or equal to their speeds before the collision.

   b) This type of collision may occur between two football players. However, since people are less elastic than bocci balls, they won't bounce off each other so far. However, you can expect both players to bounce in such a manner that each finds himself on his back.

ANSWERS

For You To Do
(continued)

6. a) In collisions between a nerf ball and a bocci ball it is generally the case that the nerf ball "loses," or undergoes the greatest change in velocity. When a stationary nerf ball is hit head-on by a moving bocci ball, the nerf ball will leave the collision at a speed higher than the incoming speed of the bocci ball, and the bocci ball will keep moving in its original direction of motion after the collision, but with a slightly reduced speed.

7. a) When a stationary bocci ball is hit head-on by a moving nerf ball, it should occur, for ordinary incoming speeds of the nerf ball, that the bocci ball will move away from the collision at a relatively low speed and the nerf ball will rebound from the collision.

8. a) Students will stage a collision between a golf ball and a wiffle golf ball and determine from observing the balls after the collision that the golf ball is much more massive.

   b) Students determine mass of both balls. Their results from the previous method will be verified.

Physics Words

momentum: the product of the mass and the velocity of an object; momentum is a vector quantity.

6. Repeat **Steps 1, 2, and 3**, but replace the stationary bocci ball with a soccer ball.
   a) Be sure to write all responses, including identification of a similar situation in real life, in your log.
7. Repeat **Steps 1, 2,** and **3**, but in this case have the soccer ball roll down the ramp to strike a stationary bocci ball.
   a) Be sure to write all responses, including identification of a similar situation in real life, in your log.
8. Using your observations, determine the relative mass of a golf ball compared to a tennis ball by staging collisions between them.
   a) Which ball has the greater mass? How many times more massive is it than the other ball? Describe what you did to decide upon your answer.
   b) Use a scale or balance to check your result. Comment on how well observing collisions between the balls worked as a method of comparing their masses.

FOR YOU TO READ

Momentum

Taken alone, neither the masses nor the velocities of the objects were important in determining the collisions you observed in this activity. The crucial quantity is **momentum** (mass × velocity). A soccer ball has less mass than a bocci ball, but a soccer ball can have the same momentum as a bocci ball if the soccer ball is moving fast. A soccer ball moving very fast can affect a stationary bocci ball more than a soccer ball moving very slowly. This is similar to the damage small pieces of sand moving at very high speeds can cause (such as when a sand blaster is used to clean various surfaces).

Sportscasters often use the term *momentum* in a different way. When a team is doing well, or "on a roll," that team has momentum.

A team can gain or lose momentum, depending on how things are going. This momentum clearly does not refer to the mass of the entire team multiplied by the team's velocity.

Other times, sportscasters use the term momentum to mean exactly how it is defined in the activity (mass × velocity), when they say things such as, "Her momentum carried her out of bounds."

Reflecting on the Activity and the Challenge

You already have identified several real-life situations that involve collisions, and many such situations happen in sports. Some involve athletes colliding with one another as in hockey and football. Others cases include athletes colliding with objects, such as when kicking a ball. Still others include collisions between objects such as a golf club, bat, or racquet with a ball. Some spectacular collisions in sports provide fun opportunities for demonstrating your knowledge about collisions during voice-over commentaries. Use the concept of momentum when describing collisions in your sports video.

Physics To Go

1. Sports commentators often say that a team has momentum when things are going well for the team. Explain the difference between that meaning of the word momentum and its specific meaning in physics.
2. Suppose a running back collides with a defending linebacker who has just come to a stop. If both players have the same mass, what do you expect to see happen in the resulting collision?
3. Describe the collision of a running back and a linebacker of equal mass running toward each other at equal speeds.

ANSWERS

Physics To Go

1. To say that a team or a candidate for election has momentum is to say that things are going well, are "on the rise;" momentum in physics is defined as mass times velocity.
2. You would expect the defending linebacker to be set in motion, and the speed of the running back to be greatly reduced.
3. The two players will bounce back from each other, at least to the extent that they will end up on their backsides.

ANSWERS

Physics To Go
(continued)

4. a) The heavier bat has, for the same speed, more momentum than the lighter bat and will transfer more momentum to the ball, hitting the ball farther.

   b) For the same effect, the lighter bat would need to be swung at a speed 38/30 = 1.3 times faster than the heavy bat.

5. It is difficult to change the momentum of a massive person.

6. The relative speeds of the players determines who gets knocked backward; if the small player moves fast enough, the big player can get knocked backward.

7. 100 kg × 10 m/s =(0.10 kg) v, v = 10,000 m/s

8. Before the collision one puck is moving and the other is stationary; after the collision, the pucks have "traded" conditions, with the puck which originally was moving being stationary and the puck which originally was stationary moving at about the same speed which the other puck had before the collision.

4. Suppose that you have two baseball bats, a heavy (38-ounce) bat and a light (30-ounce) bat.
   a) If you were able to swing both bats at the same speed, which bat would allow you to hit the ball the farthest distance? Explain your answer.
   b) How fast would you need to swing the light bat to produce the same hitting effect as the heavy bat? Explain your answer.
5. Why do football teams prefer offensive and defensive linemen who weigh about 300 pounds?
6. What determines who will get knocked backward when a big hockey player checks a small player in a head-on collision?
7. A 100.0-kg athlete is running at 10.0 m/s. At what speed would a 0.10-kg ball need to travel in the same direction so that the momentum of the athlete and the momentum of the ball would be equal?
8. Use the words *mass*, *velocity*, and *momentum* to write a paragraph that gives a detailed "before and after" description of what happens when a moving shuffleboard puck hits a stationary puck of equal mass in a head-on collision.
9. Describe a collision in some sport by using the term *momentum*. Adapt this description to a 15-s dialogue that could be used as part of the voice-over for a video.

NOTES

ACTIVITY 8
Conservation of Momentum

Background Information

The **Background Information** for this activity is presented in the **Background Information** for **Activity** 7, Concentrating on Collisions, because the same topic, momentum, is involved in both activities.

Active-ating the Physics InfoMall

In addition to the references to **Activity** 7, you may wish to examine the Problems Place for even more exercises in momentum conservation. Remember, *Schaum's 3000 Solved Problems in Physics* has the problem and the solution. It can be a source for you, as well as a way to provide your students with solved problems for them to study!

Planning for the Activity

Time Requirements

- One class period.

Materials Needed

For each group:

- dynamics cart
- modeling clay
- spring scale, 0-10 newton range
- tape, masking, 3/4" x 60 yds
- ticker tape timer
- weight, 100 g slotted mass

Advance Preparation and Setup

Two matters need to be considered in advance: (1) the pairs of objects to be used by each group in the collisions, including how the masses will be varied and how the masses will be caused to stick together upon colliding, and (2) how the velocity will be measured before and after the collision.

Regarding the objects to be collided, two possibilities seem to exist, air track gliders or laboratory carts. At a minimum, three collisions are desired, involving mass ratios of 1:1, 2:1 and 1:2 (the moving mass is listed first in the ratios listed – see the data table in **For You To Do** for details). Masses of 1 and 2 kg are recommended, but certainly not required. Instead, one laboratory cart could collide with another identical laboratory cart to provide a 1:1 mass ratio, and a 2:1 (and 1:2) ratio could be obtained by loading one cart to double its mass. Similarly, air track gliders could be rigged to provide 1:1 and 2:1 mass ratios.

The colliding objects need to stick together upon colliding to move as a single object after the collision. Stick-on patches are very convenient for this purpose, but double-stick tape or modeling clay also works.

Whatever objects are used, have the students measure and use their masses, in kilograms, to call attention to and engage students in using the unit of momentum, (kg)m/s. If the masses are not to be 1 and 2 kg, have students change the values listed in the data table in **For You To Do** to the values to be used in your class.

The speed, in meters per second, must be measured before and after the collision. It is necessary to measure the speed of the incoming mass before the collision and the speed of the combined masses after the collision. One way to accomplish this would be to use a sonic ranging device to monitor the speed of Object 1 (the mass moving before the collision) before, during, and after the collision. Other possibilities for measuring the speeds include stop action video, strobe photography, a ticker-tape timer or a spark timer. The particular method to be used depends on the equipment available at your school. Whatever method of measuring speeds is used, students should record speeds in m/s.

For better data if friction is involved (as when using laboratory carts) it would be best to use the speed values which occur just before and just after the collision; this would help to avoid changes in speed (deceleration) as a source of error.

You may wish to provide additional mass ratios for students to use for staging additional collisions. A 3:1 mass ratio run "both ways," 3:1 and 1:3, gives particularly interesting results; if organized on bifilar supports to collide head-on as pendulums, hard wooden or metal spheres having a 3:1 mass ratio provide a cyclically repeating sequence of collisions.

"Nonsticky" collisions present problems for measuring speed because both masses move simultaneously at different speeds. This is difficult, but not impossible to overcome with ordinary equipment and may present an interesting challenge to interested students.

Teaching Notes

Monitor the class as students begin to stage collisions. They may need assistance choosing appropriate push-off speeds for the incoming mass and getting equipment to function.

Sample data is not provided because it cannot be anticipated what masses and speeds will be used in your class. However, conservation of momentum allows you to predict with ease what the velocity of the combined masses after the collision should be compared to the velocity of Object 1 before the collision:

Conservation of momentum applied to this kind of collision:

$$m_1 v_{before} = (m_1 + m_2) v_{after}$$

Solving the above equation for v_{after}:

$$v_{after} = v_{before}\, m_1 / (m_1 + m_2)$$

The final equation above will allow you to predict before class begins what the relative speeds before and after each collision should be. Another way of saying the same thing is that $v_{after} / v_{before} = m_1 / (m_1 + m_2)$.

Allowing for errors of measurement, students should find that the momenta before and after each collision are equal (in the Analysis table, the momentum of Object 2 will be zero, unless an additional collision is staged which is not included in the written procedure; this is pointed out only to avoid confusion).

It is important to discuss the collision analyzed as an example in **For You To Read** because it is of a general kind where both objects are moving before and aftera head-on collision. The same method of analysisis needed to solve some of the problems in **Physics To Go.**

Assigning positive and negative values to directions of velocities is very useful, if not necessary, for solving many collision problems.

NOTES

Activity 8 Conservation of Momentum

GOALS

In this activity you will:

- Understand and apply the Law of Conservation of Momentum.
- Measure the momentum before and after a moving mass strikes a stationary mass in a head-on, inelastic collision.

What Do You Think?

The outcome of a collision between two objects is predictable.

- **What determines the momentum of an object?**
- **What does it mean to "conserve" something?**

Record your ideas about these questions in your *Active Physics* log. Be prepared to discuss your responses with your small group and the class.

For You To Do

1. From the objects provided arrange to have a head-on collision between two objects of equal mass. Before the collision, have one object moving and the other object at rest. Arrange for the objects to stick together to move as a single object after the collision. Stage a head-on, sticky collision between equal masses. Measure the velocity, in meters per second, of the moving mass before the collision and the velocity of the combined masses after the collision.

Activity Overview

In this activity students stage collisions between objects to investigate the conservation of momentum.

Student Objectives

Students will:

- Understand and apply the Law of Conservation of Momentum.
- Measure the momentum before and after a moving mass strikes a stationary mass in a head-on, inelastic collision.

Answers for the Teacher Only

What Do You Think?

An object's mass and velocity determine its momentum, *mv*.

To conserve means to keep the same amount.

Answers

For You To Do

Sample data is not provided because it cannot be anticipated what masses and speeds will be used in your class. However, conservation of momentum allows you to predict with ease what the velocity of the combined masses after the collision should be compared to the velocity of Object 1 before the collision. (See **Teaching Notes.**)

Chapter 1

Physics in Action

a) Prepare a data table in your log similar to the one shown below. Provide enough horizontal rows in the table to enter data for at least four collisions.

Sticky Head-on Collisions: One Object Moving before Collision

Mass of Object 1 (kg)	Mass of Object 2 (kg)	Velocity of Object 1 before Collision (m/s)	Velocity of Object 2 before Collision (m/s)	Mass of Combined Objects after Collision (kg)	Velocity of Combined Objects after Collision (m/s)
1.0	1.0		0.0	2.0	
2.0	1.0		0.0	3.0	
1.0	2.0		0.0	3.0	
			0.0		

b) Record the measured values of the velocities in the first row of the data table.

2. Stage other sticky head-on collisions using the masses listed in the second and third rows of the data table. Then stage one or more additional collisions using other masses. Measure the velocities before and after each collision.

a) Enter the measured values in the data table.

3. Organize a table for recording the momentum of each object before and after each of the above collisions.

a) Prepare a table similar to the following example in your log:

Momentum of Object before and after Collisions
Momentum = Mass × Velocity

Before the Collision		After the Collision
Momentum of Object 1 kg (m/s)	Momentum of Object 2 kg (m/s)	Momentum of Combined Objects 1 and 2 kg (m/s)

62

Active Physics

ANSWERS

For You To Do *(continued)*

Sample data is not provided because it cannot be anticipated what masses and speeds will be used in your class. However, conservation of momentum allows you to predict with ease what the velocity of the combined masses after the collision should be compared to the velocity of Object 1 before the collision. (See **Teaching Notes.**)

Activity 8 Conservation of Momentum

b) Calculate the momentum of each object before and after each of the above collisions and enter each momentum value in the table.

c) Calculate and compare the total momentum before each collision to the total momentum after each collision.

d) Allowing for minor variations due to errors of measurement, write in your log a general conclusion about how the momentum before a collision compares to the momentum afterward.

FOR YOU TO READ

The Law of Conservation of Momentum

In this activity, you investigated another conservation principle that is a hallmark of physics—the conservation of momentum. If you sum all of the momenta before a collision or explosion, you know that the sum of all the momenta after the collision will be the same.

If the momentum before a collision is 500 kg·m/s, then the momentum after the collision is 500 kg·m/s. A football player stops to catch a pass. The player is not moving and therefore has momentum equal to zero. If an opponent that has a momentum of 500 kg·m/s then hits the player, both players move off with (a combined) 500 kg·m/s of momentum. Any time you see a collision in sports, you can explain that collision using the conservation of momentum.

Conservation of momentum is an experimental fact. Physicists have compared the momenta before and after a collision between pairs of objects ranging from railroad cars slamming together to subatomic particles impacting one another at near the speed of light. Never have any exceptions been found to the statement, "The total momentum before a collision is equal to the total momentum after the collision if no external forces act on the system." This statement is known as the Law of Conservation of Momentum. In all collisions between cars and trucks, between protons and protons, between planets and meteors, the momentum before the collision equals the momentum after.

ANSWERS

For You To Do *(continued)*

Sample data is not provided because it cannot be anticipated what masses and speeds will be used in your class. However, conservation of momentum allows you to predict with ease what the velocity of the combined masses after the collision should be compared to the velocity of Object 1 before the collision. (See **Teaching Notes**.)

A single cue ball hits a rack of 15 billiard balls and they all scatter. It would seem like everything has changed. Physicists have discovered that in this collision, as in all collisions and explosions, nature does keep at least one thing from changing—the total momentum. The sum of the momenta of all of the billiard balls immediately after the collision is equal to the momentum of the original cue ball. Nature loves momentum. Irrespective of the changes you visually note, the total momentum undergoes no change whatsoever. The objects may move in new directions and with new speeds, but the momentum stays the same. There aren't many of these conservation laws that are known.

Conservation of momentum can be shown to emerge from Newton's Laws. Newton's Third Law states that if object *A* and object *B* collide, the force of object *A* on *B* must be equal and opposite to the force of object *B* on *A*.

$$F_{A\text{ on }B} = -F_{B\text{ on }A}$$

The negative sign shows mathematically that the equally sized forces are in opposite directions.

Newton's Second Law states that $F = ma$. Also, acceleration, a, equals the change in velocity divided by the change in time ($a = \Delta v/\Delta t$):

$$m_B a_B = -m_A a_A$$

$$m_B \frac{\Delta v_B}{\Delta t} = -m_A \frac{\Delta v_A}{\Delta t}$$

$$m_B \frac{(v_f - v_i)_B}{\Delta t} = -m_A \frac{(v_f - v_i)_A}{\Delta t}$$

Since the change in time must be the same for both objects (*A* acts on *B* for as long as B acts on *A*), then Δt can be eliminated from both sides of the equation.

Combining the initial velocities (v_i) on one side of the equation and the final velocities (v_f) on the other side of the equation:

$$m_A v_{iA} + m_B v_{iB} = m_A v_{fA} + m_B v_{fB}$$

$$(m_A v_A)_{before} + (m_B v_B)_{before} = (m_A v_A)_{after} + (m_B v_B)_{after}$$

Newton's Laws have yielded the conservation of momentum. The momentum of object *A* *before* the collision plus the momentum of object *B* *before* the collision equals the momentum of object *A* *after* the collision plus the momentum of object *B* *after* the collision

This equation not only works in one-dimensional collisions, but works equally well in the extraordinarily complex two-dimensional collisions of billiard balls and three-dimensional collisions of bowling.

Solving conservation of momentum problems is easy. Calculate each object's momentum

Activity 8 Conservation of Momentum

before the collision. Calculate each object's momentum after the collision. The totals before the collision must equal the total after the collision.

There are a variety of collisions involving two objects. In each collision, momentum is conserved and the same equation is used. The equation gets simpler when one of the objects is at rest and has zero momentum. You may wish to draw two sketches for each collision—one showing each object before the collision and one showing each object after the collision. By writing the momenta you know directly on the sketch, the calculations become easier.

Collision Type 1: One moving object hits a stationary object and both stick together and move off at the same speed:

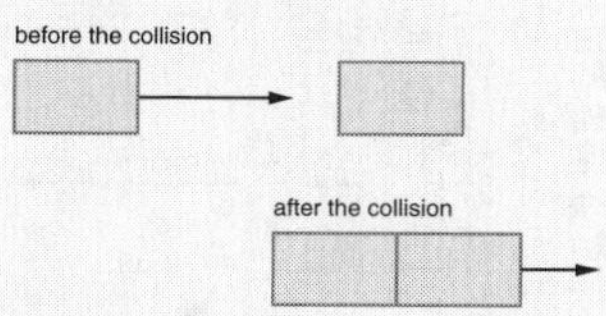

Collision Type 2: Two stationary objects explode and move off in opposite directions.

Collision Type 3: One moving object hits a stationary object. The first object stops, and the second object moves off.

Collision Type 4: One moving object hits a stationary object, and both move off at different speeds.

Collision Type 5: Two moving objects collide, and both objects move at different speeds after the collision.

Collision Type 6: Two moving objects collide, and both objects stick together and move off at the same speed.

Sample Problem 1

A 75.00-kg ice skater is moving to the east at 3.00 m/s toward his 50.00-kg partner, who is moving toward him (west) at 1.80 m/s. If he catches her up and they move away together, what is their final velocity?

Strategy: This is a problem involving the Law of Conservation of Momentum. The momentum of an isolated system before an interaction is equal to the momentum of the system after the interaction. As you are working through this problem, remember that the v in this expression is velocity and that it has direction as well as magnitude. Make east the positive direction, and then west will be negative.

Givens:

$m_b = 75.00$ kg
$m_g = 50.00$ kg
$v_b = 3.00$ m/s
$v_g = -1.80$ m/s

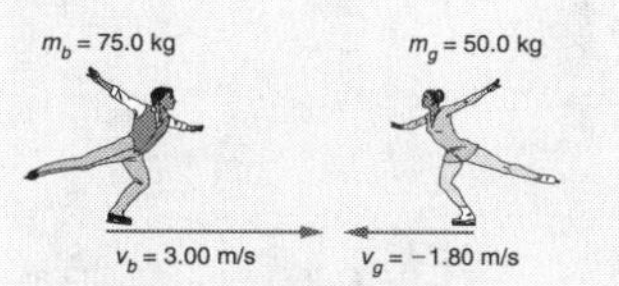

→

Physics in Action

Solution:

$(m_b v_b)_{before} + (m_g v_g)_{after} = [(m_b + m_g)v_{bg}]_{after}$

$(75.00\ kg)(3.00\ m/s) + (50.00\ kg)(-1.80\ m/s) = (75.00\ kg + 50.00\ kg)v_{bg}$

$$v_{bg} = \frac{225\ kg{\cdot}m/s - 90.0\ kg{\cdot}m/s}{125.00\ kg}$$

$$= 1.08\ m/s$$

Sample Problem 2

A steel ball with a mass of 2 kg is traveling at 3 m/s west. It collides with a stationary ball that has a mass of 1 kg. Upon collision, the smaller ball moves away to the west at 4 m/s. What is the velocity of the larger ball?

Strategy: Again, you will use the Law of Conservation of Momentum. Before the collision, only the larger ball has momentum. After the collision, the two balls move away at different velocities.

Givens:

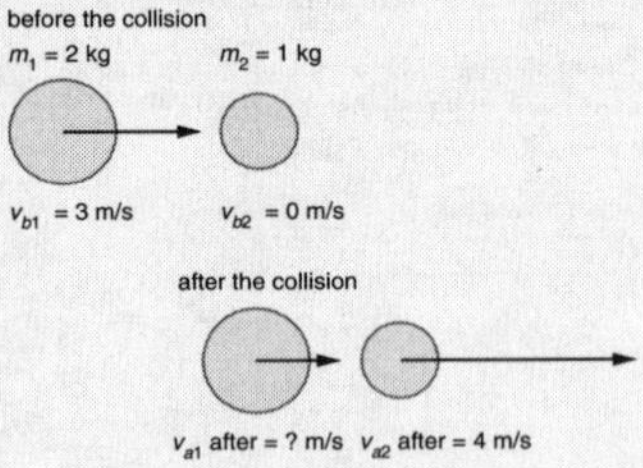

$m_1 = 2$ kg
$m_2 = 1$ kg
$v_{b1} = 3$ m/s
$v_{b2} = 0$ m/s
$v_{a2} = 4$ m/s

Solution:

$(m_1v_1)_b + (m_2v_2)_b = (m_1v_1)_a + (m_2v_2)_a$

$(2\ kg)(3\ m/s) + (1\ kg)(0\ m/s) = (2\ kg)v_{a1} + (1\ kg)(4\ m/s)$

$6\ kg{\cdot}m/s = (2v_{a1})\ kg + 4\ kg{\cdot}m/s$

$v_{a1} = 1\ m/s$

66

Active Physics

Reflecting on the Activity and the Challenge

The Law of Conservation of Momentum is a very powerful tool for explaining collisions in sports and other areas. The law works even when, as often happens in sports, one of the objects involved in a collision "bounces back," reversing the direction of its velocity and, therefore, its momentum, as a result of a collision. When describing a collision between a bat and ball, or a collision between two people, you can describe how the total momentum is conserved.

Physics To Go

1. A railroad car of 2000 kg coasting at 3.0 m/s overtakes and locks together with an identical car coasting on the same track in the same direction at 2.0 m/s. What is the speed of the cars after they lock together?
2. In a hockey game, an 80.0-kg player skating at 10.0 m/s overtakes and bumps from behind a 100.0-kg player who is moving in the same direction at 8.00 m/s. As a result of being bumped from behind, the 100.0-kg player's speed increases to 9.78 m/s. What is the 80.0-kg player's velocity (speed and direction) after the bump?
3. A 3-kg hard steel ball collides head-on with a 1-kg hard steel ball. The balls are moving at 2 m/s in opposite directions before they collide. Upon colliding, the 3-kg ball stops. What is the velocity of the 1-kg object after the collision? (Hint: Assign velocities in one direction as positive; then any velocities in the opposite direction are negative.)
4. A 45-kg female figure skater and her 75-kg male skating partner begin their ice dancing performance standing at rest in face-to-face position with the palms of their hands touching. Cued by the start of their dance music, both skaters "push off" with their hands to move backward. If the female skater moves at 2.0 m/s relative to the ice, what is the velocity of the male skater? (Hint: The momentum before the skaters push off is zero.)

Answers

Physics To Go

Chapter 1

1. $m(3.0 \text{ m/s}) + m(2.0 \text{ m/s}) = (2m)v$

   $(3.0 \text{ m/s} + 2.0 \text{ m/s}) = 2mv$

   $v = m(5.0 \text{ m/s}) / 2m$
   $= (5.0 \text{ m/s})/2 = 2.5 \text{ m/s}$

2. $(80.0 \text{ kg})(10.0 \text{ m/s}) + (100 \text{ kg})(8.00 \text{ m/s}) = (80 \text{ kg})v + (100 \text{ kg})(9.78 \text{ m/s})$

   $800 \text{ (kg)m/s} + 800 \text{ (kg)m/s} = (80 \text{ kg})v + 978 \text{ (kg)m/s}$

   $1600 \text{ (kg)m/s} = (80 \text{ kg})v + 978 \text{ (kg)m/s}$

   $622 \text{ (kg)m/s} = (80 \text{ kg})v$

   $v = [622 \text{ (kg)m/s}] / 80 \text{ kg} = 7.78 \text{ m/s}$

3. The direction of travel of the 3-kg ball before the collision is assigned as positive:

   $(3 \text{ kg})(2 \text{ m/s}) + (1 \text{ kg})(-2 \text{ m/s}) = (1 \text{ kg})v$

   $6 \text{ (kg)m/s} - 2 \text{ (kg)m/s} = (1 \text{ kg})v$

   $4 \text{ (kg)m/s} = (1 \text{ kg})v$

   $v = [4 \text{ (kg)m/s}] / (1 \text{ kg}) = 4 \text{ m/s}$

   The 1-kg ball bounces back after the collision at twice the speed it had coming into the collision.

4. The direction of the female skater after pushoff is assigned as positive:

   $0 = (45 \text{ kg})(2 \text{ m/s}) + (75 \text{ kg})v$

   $0 = 90 \text{ (kg)m/s} + (75 \text{ kg})v$

   $-90 \text{ (kg)m/s} = (75 \text{ kg})v$

   $v = [-90 \text{ (kg)m/s}] /(75 \text{ kg}) = -1.2 \text{ m/s}$

The male skater moves at 1.2 m/s in the direction opposite the female skater.

5. A 0.35-kg tennis racquet moving to the right at 20.0 m/s hits a 0.060-kg tennis ball that is moving to the left at 30.0 m/s. The racquet continues moving to the right after the collision, but at a reduced speed of 10.0 m/s. What is the velocity (speed and direction) of the tennis ball after it is hit by the racquet?

6. A stationary 3-kg hard steel ball is hit head-on by a 1-kg hard steel ball moving to the right at 4 m/s. After the collision, the 3-kg ball moves to the right at 2 m/s. What is the velocity (speed and direction) of the 1-kg ball after the collision? (Hint: Direction is important.)

7. A 90.00-kg hockey goalie, at rest in front of the goal, stops a puck (m = 0.16 kg) that is traveling at 30.00 m/s. At what speed do the goalie and puck travel after the save?

8. A 45.00-kg girl jumps from the side of a pool into a raft (m = 0.08 kg) floating on the surface of the water. She leaves the side at a speed of 1.10 m/s and lands on the raft. At what speed will the girl-raft system begin to travel across the pool?

9. Two cars collide head on. Initially, car A (m = 1700.0 kg) is traveling at 10.00 m/s north and car B is traveling at 25.00 m/s south. After the collision, car A reverses its direction and travels at 5.00 m/s while car B continues in its initial direction at a speed of 3.75 m/s. What is the mass of car B?

10. A proton ($m = 1.67 \times 10^{-27}$ kg) traveling at 2.50×10^5 m/s collides with an unknown particle initially at rest. After the collision the proton reverses direction and travels at 1.10×10^5 m/s. Determine the change in momentum of the unknown particle.

11. You shoot a 0.04-kg bullet moving at 200.0 m/s into a 20.00-kg block initially at rest on an icy pond.
    a) What is the velocity of the bullet-block combination?
    b) The coefficient of friction between the block and the ice is 0.15. How far would the block slide before coming to rest?

12. Write a 15- to 30-s voice-over that highlights the conservation of momentum in a sport of your choosing.

ANSWERS

Physics To Go *(continued)*

5. To the right is assigned as the positive direction:

$$(0.35\text{ kg})(20\text{ m/s}) + (0.060\text{kg})(30\text{ m/s}) = (0.35\text{ kg})(10\text{ m/s}) + (0.060\text{ kg})v$$
$$7.0\text{ (kg)m/s} + 1.8\text{ (kg)m/s} = 3.5\text{ (kg)m/s} + (0.060\text{ kg})v$$
$$5.3\text{ (kg)m/s} = (0.060\text{ kg})v$$
$$v = [5.3\text{ (kg)m/s}]\,/(0.060\text{ kg}) = 88\text{ m/s to the right}$$

6. To the right is assigned as the positive direction:

$$0 + (1\text{ kg})(4\text{ m/s}) = (3\text{ kg})(2\text{ m/s}) + (1\text{ kg})v$$
$$-2\text{ (kg)m/s} = (1\text{ kg})v$$
$$v = -2\text{ m/s}$$

The 1-kg ball rebounds from the collision, moving to the left at half the speed it had coming into the collision; its speed after the collision also is observed to be the equal and opposite of the 3-kg ball's speed after the collision. If after this collision each ball hit a bumper and the balls came back to collide again, this would take us back to the beginning of **Problem 3** above, and if they rebounded after that collision, we'd be back to this problem again, and so on...

ANSWERS

Physics To Go *(for Questions 7–12)*

7.
$$\text{Momentum}_{before} = \text{Momentum}_{after}$$
$$(m_{goalie}v_{goalie})_{before} + (m_{puck}v_{puck})_{before} = [(m_g + m_p)v_{gp}]_{after}$$
$$(90.00 \text{ kg})(0 \text{ m/s}) + (0.16 \text{ kg})(30.00 \text{ m/s}) = (90.16 \text{ kg})(v_{gp})$$
$$v = 0.05 \text{ m/s}$$

8.
$$\text{Momentum}_{before} = \text{Momentum}_{after}$$
$$(m_{girl}v_{girl})_{before} + (m_{raft}v_{raft})_{before} = [(m_g + m_r)v_{gr}]_{after}$$
$$(45.00 \text{ kg})(1.10 \text{ m/s}) + (0.08 \text{ kg})(0 \text{ m/s}) = (45.08 \text{ kg})\ v_{gr}$$
$$v_{gr} = 1.098 \text{ m/s } (1.1 \text{ m/s})$$

9.
$$\text{Momentum}_{before} = \text{Momentum}_{after}$$
$$(m_Av_A)_{before} + (m_Bv_B)_{before} = (m_Av_A)_{after} + (m_Bv_B)_{after}$$
$$(1700.0 \text{ kg})(10.0 \text{ m/s}) + m_B(-25 \text{ m/s}) = (1700.0 \text{ kg})(-5.00 \text{ m/s}) + m_B(-3.75 \text{ m/s})$$
$$m_B = (25{,}500 \text{ kg m/s})/(21.25 \text{ m/s})$$
$$m_B = 1200 \text{ kg}$$

10. The change in momentum of the proton must be equal and opposite to the change in momentum of the unknown particle.

Change in momentum = $m\Delta v$ = $(1.67 \times 10^{-27}$ kg$)(1.10 \times 10^5$ m/s$) - (-2.50 \times 10^5$ m/s$)$ = 6.0×10^{-22} kg m/s.

The unknown particle changed momentum by exactly this much but in the opposite direction.

11. a)
$$(m_Av_A)_{before} + (m_Bv_B)_{before} = (m_Av_A)_{after} + (m_Bv_B)_{after}$$
$$(0.04 \text{ kg})(200.0 \text{ m/s}) + (20.00 \text{ kg})(0 \text{ m/s}) = (20.04 \text{ kg})(v)$$
$$v = 0.4 \text{ m/s}$$

b) The change in KE of the block as it slides will be equal to the work done. This will allow the student to find the distance.

$\Delta KE = 1/2\ mv^2 = 1/2\ (20.04 \text{ kg})(0.4 \text{ m/s})^2 = 1.6 \text{ J}$

$W = F \cdot d$

$d = W/F = 1.6 \text{ J}/F$

Find the force using the coefficient of kinetic friction.

$F_f = \mu F_N = (0.15)(20.04 \text{ kg})(9.8 \text{ m/s}^2) = 29 \text{ N}$

$d = W/F = 1.6 \text{ J}/F = 1.6 \text{ J}/29 \text{ N} = 0.06 \text{ m}$

12. Student answers will vary. The voice-over should include a collision of a player with a player or a player with object. It should also include a statement of the conservation of momentum.

ACTIVITY 9
Circular Motion

Background Information

It already has been established that if an object is moving along a straight line at constant velocity (constant speed, always in the same direction, along the line of motion) all of the forces on the object are balanced; in other words, the net force on the object is zero. If a sudden, momentary force is applied to the same object in a direction exactly sideways, the speed of the object neither increases nor decreases because the sideways force has no effectiveness in either the same or opposite direction as the object's motion. The result of applying a sudden, momentary, sideways force to the object is to cause the object to turn in the direction of the applied force; after the force has been removed, the object would be found moving at the same constant speed as it was before application of the force, but it would be moving along a different direction line, its direction of motion having been changed by the force. Such a force is called a "deflecting force," and the result of a deflecting force—applied exactly sideways to the motion of an object—is to cause the object to change, or deflect, to a new direction of motion.

If instead of being applied suddenly and momentarily, a deflecting force is applied continuously, always in the same amount, and continuously adjusted in a direction to act exactly sideways to the object's motion, the object moves in a circular path at constant speed. In this case the force is called a "centripetal force" which is defined as "the force required to keep a mass moving in a circular path at constant speed." It can be experimentally determined that the relationship between the centripetal force, the object's mass and speed, and the radius of the object's circular path is:

$$F_c = mv^2/r$$

where F_c is the centripetal force in newtons, m is the object's mass in kilograms, v is the objects speed in meters/second, and r is the radius of the circular path in meters. A convenient alternate equation for centripetal force can be used when the period, T, of the objects circular motion (T is the time, in seconds, for the object to travel once around the circle) is known but the speed is not known:

$$F_c = m4\pi^2 R/T^2$$

According to Newton's Second Law of Motion, when there is an unbalanced force acting on an object, the object must be accelerating. This certainly must apply to an object moving in a circular path under the influence of an unbalanced, centripetal force. However, it is puzzling to contemplate that an object moving at constant speed, even on a circular path, somehow can be construed to be accelerating, because it is neither gaining nor losing speed. Consider the below "strobe" diagram showing an object moving at constant speed along a circular path at four instants equally separated in time:

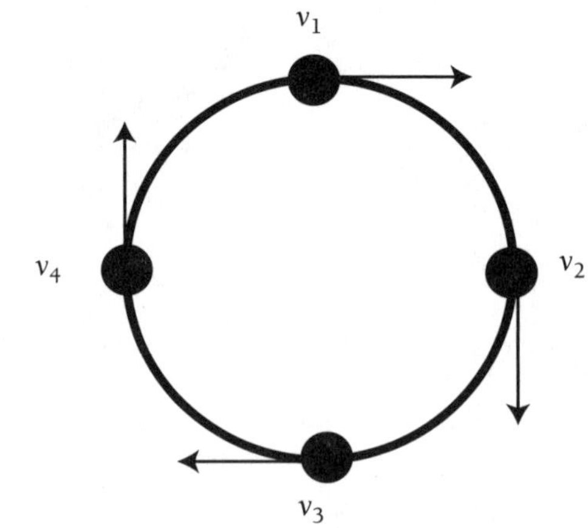

Diagram of object moving in a circular path.

The definition of acceleration, $a = \Delta v/\Delta t = (v_2 - v_1)/\Delta t$ will be applied to the above diagram to attempt to discover any basis for existence of acceleration. Since the instantaneous velocities v_1 and v_2 in the diagram are vectors which do not share a common direction, the quantity Δv in the defining equation for acceleration must be found by treating v_1 and v_2 as vectors during the subtraction process; that is, a vector subtraction method must be applied to find the difference $v_2 - v_1 = \Delta v$. To do so, the negative of vector v_1 (a vector having the same length but opposite direction as v_1) will be added to vector v_2 using the tip-to-tail method of adding vectors:

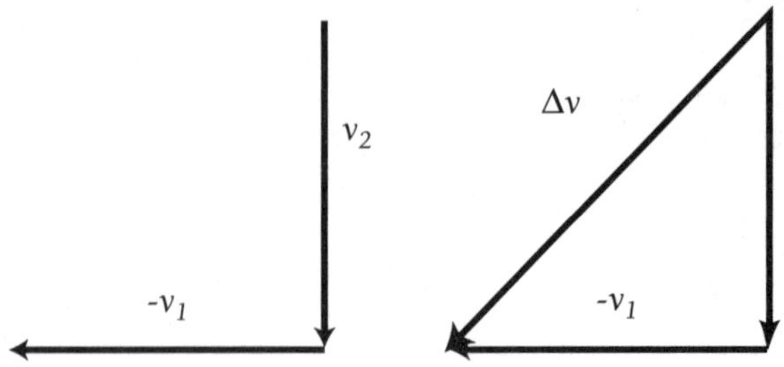

Clearly in the above diagram a change in velocity, Δv, occurred during the time interval, Δt, that the object moved at constant speed 1/4 of the distance around its circular path. If a numerical values were assigned to time and to the object's speed, a value for Δv could be determined by comparing the length of the Δv vector to the velocity vectors, and the average acceleration for the time interval could be calculated from $a = \Delta v/\Delta t$. Therefore, there is an acceleration associated with a centripetal force, but the nature of the acceleration is that it does not alter the object's speed, but does alter the direction of the object's velocity. It is informative to move the vector Δv in the above diagram into the original diagram to see Δv in relationship to the motion along the circular path. This is done in the below diagram. Notice that both the length and direction of the vector Δv are preserved and that Δv is positioned at the middle of the time interval from which it was derived:

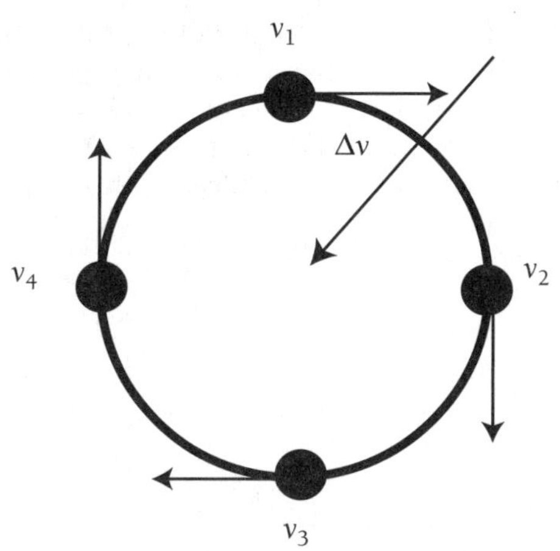

Object in circular motion with vector

Interestingly and not accidentally, Δv points toward the center of the circular path. Since the acceleration calculated from Δv using the defining equation $a = \Delta v/\Delta t$ would have the same direction as Δv, the acceleration also points toward the center of the circle. Indeed, it is a centripetal acceleration, caused, as one may suspect by the centripetal force, which also points toward the center of the circular path. Now it can be reasoned that the equations presented above for centripetal force are nothing more that Newton's Second Law, $f = ma$, applied to the special case of circular motion. The terms on the right-hand side of the equation other than the mass, m, are the centripetal acceleration, a_c:

$$F_c = ma_c = m(v^2/r) = m(4\pi^2 r/t^2)$$

In summary, the velocity, acceleration, and force vectors of an object moving at constant speed on a circular path constantly change in direction and remain constant in amount. Each vector rotates once during each trip of the object around the circular path.

Active-ating the Physics InfoMall

When teaching about circular motion, you will almost certainly have to dispel ideas about centrifugal forces. Perhaps you noticed, while performing some of the searches mentioned in the previous activities, the article "Centrifugal force: fact or fiction?," in *Physics Education*, vol. 24, issue 3, 1989. If not, you may want to check it out now.

Search the InfoMall using keywords "circular motion" AND "misconcept*" for a list of articles discussing known problems students have with this common concept. You may also find useful demonstrations by searching the Demo & Lab Shop using the keywords "circular motion."

Planning for the Activity

Time Requirements

- One class period.

Materials Needed

For each group:

- accelerometer
- accelerometer, cork
- accelerometer-to-cart
- balls, bocci
- ball, nerf
- rotating stool or chair
- ruler, flexible plastic, 30 cm
- ruler, metric, mm marking

Advance Preparation and Setup

A ball will be needed to serve as the subject of each group's inquiry. The ball needs to be sufficiently massive so that, when rolling at low speed across the floor, students can use a force meter to push sideways to the ball's motion to cause it to turn on a curve of observable, reasonable radius. The best choice of ball may be the bocci balls; if not available another kind of ball must be substituted. Try this yourself in advance of class to be sure that the ball's mass, the ball's speed, and the amount of force used, combine to produce a nice, observable curve of the ball's path.

If you have not done an earlier activity in *Sports*, yet, you will need to take the time to construct a cork accelerometer.

Teaching Notes

The cork accelerometer will indicate an acceleration toward the center of the circular path in which the accelerometer moves.

SAFETY PRECAUTION: If a rotating stool or chair is used to place a student holding an accelerometer in a state of circular motion, provide close supervision and maintain safe conditions; the effect can be observed without a rotating stool or chair if students simply twirl around while holding an accelerometer in the hands.

If students have not already done so, some time may be required to familiarize them with the use of the accelerometer.

Students can be expected to need practice at keeping alongside the ball while applying a constant force always sideways (at a right angle to) the ball's motion; in fact, to do so extremely well perhaps is nearly impossible, but "close" will do well enough for students to observe the tendency for the ball to move in a circular path.

Some students may raise the question, "How can there be an acceleration when the object moves at constant speed?" You may wish to see the explanation in the **Background Information** for the Teacher for this activity to decide how you will deal with that question.

Students also may ask about, or bring up, "centrifugal force." This is addressed in the **Background Information** for the Teacher. A first response to a question about centrifugal force would be to refer students to **Physics To Go, Question 2.**

Students may have an inclination to "run together," or treat as the same phenomenon, circular motion with spinning motion. They are separate, but related, phenomena. This activity applies to the former, an object whose center of mass is moving along a circular path. When a figure skater does an in-place spin, the skater's center of mass does not move; it is a different phenomenon. It is possible, however, to treat part of the skater, such as an extended foot at the end of the spin, as an object in circular motion to which the ideas in this activity could be applied.

Activity 9 Circular Motion

GOALS

In this activity you will:

- Understand that a centripetal force is required to keep a mass moving in a circular path at constant speed.
- Understand that a centripetal acceleration accompanies a centripetal force, and that, at any instant, both the acceleration and force are directed toward the center of the circular path.
- Apply the equation for circular motion.
- Understand that centrifugal force is the reaction to centripetal force.

⚠ To avoid becoming too dizzy, limit your spins while standing to about four.

What Do You Think?

Racecars can make turns at 150 mph.

- **What forces act on a racecar when it moves along a circular path at constant speed on a flat, horizontal surface?**

Record your ideas about this question in your *Active Physics* log. Be prepared to discuss your responses with your small group and the class.

For You To Do

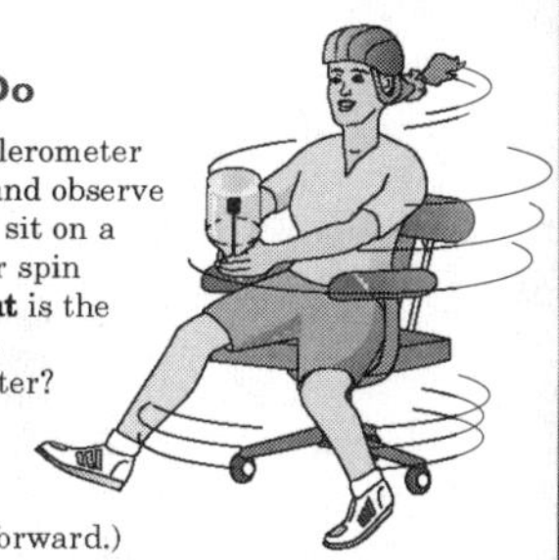

1. Hold an accelerometer in your hands and observe it as you either sit on a rotating stool or spin around while standing. What is the direction of the acceleration indicated by the accelerometer? (You can find out how the cork indicates acceleration by holding it and noting its behavior as you accelerate forward.)

Answers

For You To Do

1. a) Students provide a sketch similar to the one shown.

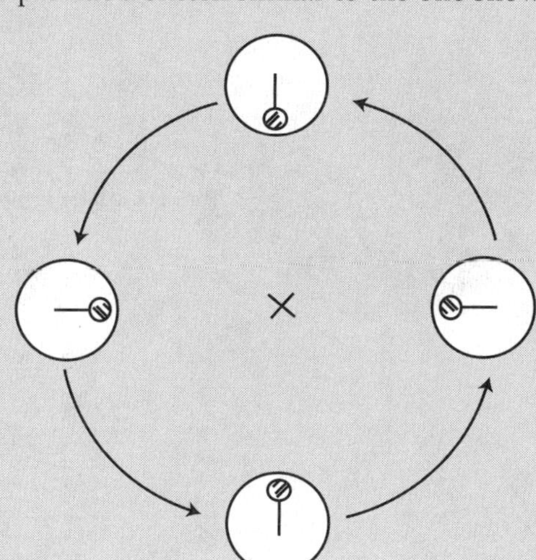

Activity Overview

In this activity students use an accelerometer to identify the direction of centripetal acceleration. They then use a force meter to provide the centripetal force needed to deflect an rolling ball moving in a straight line into a curved path.

Student Objectives

Students will:

- Understand that a centripetal force is required to keep a mass moving in a circular path at constant speed.
- Understand that a centripetal acceleration accompanies a centripetal force, and that, at any instant, both the acceleration and force are directed toward the center of the circular path.
- Apply the equation $F_c = ma_c = m(v^2/r)$ to calculations involving circular motion.
- Understand that centrifugal force is the reaction to centripetal force.

Answers for the teacher only

What Do You Think?

The forces acting on a race car include gravity downward, the force of the road up and centrifugal force.

ANSWERS

For You To Do
(continued)

2. a) Students should indicate that the greater the force applied to a mass, the greater the acceleration. The acceleration occurs in the direction of the unbalanced force.

3. a) Students provide a sketch similar to the one shown.

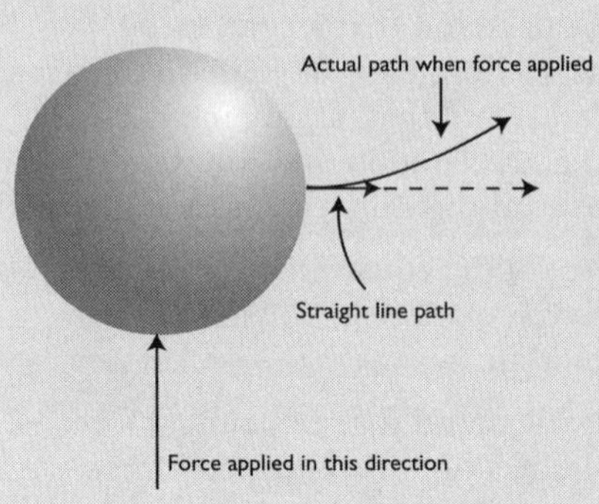

b) The speed of the ball does not change.

c) A constant force sideways needs to be applied to the ball.

d) If you stop pushing on the ball, the ball will follow a straight path. Students should provide a sketch similar to the one shown at right.

a) Make a sketch in your log to simulate a snapshot photo taken from above as the accelerometer was moving along a circular path. Show the circular path, the accelerometer "frozen" at one instant, the cork "frozen" in leaning position, and an arrow to represent the velocity of the accelerometer at the instant represented by your sketch.

2. Review in your textbook and your log how you used a force meter to apply a constant force to objects to cause the objects to accelerate in **Activity 2**.

a) Based on the results of **Activity 2**, write a brief statement in your log that summarizes how the amount and direction of acceleration of an object depends on amount and direction of the force acting on the object.

3. Start a ball rolling across the floor. While it is rolling, catch up with the ball and use the force meter to push exactly sideways, or perpendicular, to the motion of the ball with a fixed amount of force. Carefully follow alongside the ball and, as will be necessary, keep adjusting the direction of push so that it is always perpendicular to the motion of the ball.

a) Make a top view sketch in your log that shows:
- a line to represent the straight-line path of the ball before you began pushing sideways on the ball
- a dashed line to represent the straight-line path on which the ball would have continued moving if you had not pushed sideways on it
- a line of appropriate shape to show the path taken by the ball as you pushed perpendicular to the direction of the ball's motion with a constant amount of force.

b) When you pushed on the ball exactly sideways to its motion, did you cause the ball to move either faster or slower? Explain your answer.

c) Assuming that friction could be eliminated to allow the ball to continue moving at constant speed, describe what you would need to do to make the ball keep moving on a circular path.

d) If you stop pushing on the ball, how does the ball move? Try it, and use a sketch and words in your log to describe what happens.

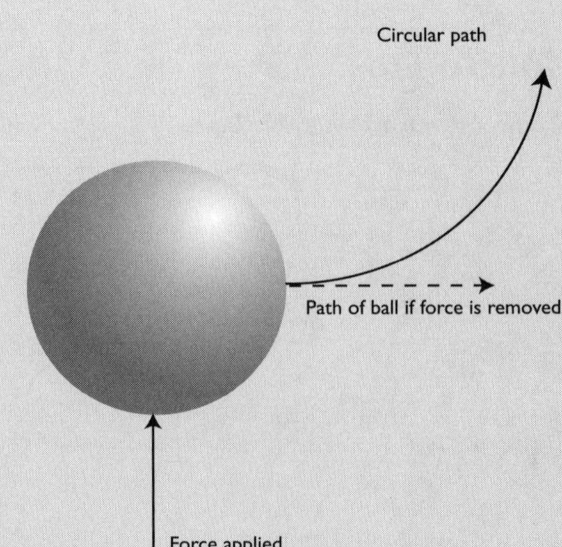

4. Review each of the items listed below. Copy each item into your log and write a statement to discuss how each item is related to an object moving along a circular path. If an item does not apply to circular motion, explain why.
   a) Galileo's Principle of Inertia
   b) Newton's First Law of Motion
   c) Newton's Second Law of Motion

Physics Words

centripetal acceleration: the inward radial acceleration of an object moving at a constant speed in a circle.

FOR YOU TO READ

The Unbalanced Force Required for Circular Motion

During the above activities you saw two things that are related by Newton's Second Law of Motion, $F = ma$. First, the accelerometer showed that when an object moves in a circular path there is an acceleration that at any instant is toward the center of the circle. This acceleration has a special name, **centripetal acceleration.** The word centripetal means "toward-the-center"; therefore, centripetal acceleration refers to acceleration toward the center of the circle when an object moves in a circular path.

You also saw that a centripetal force, a toward-the-center force, causes circular motion. When a centripetal force is applied to a moving object, the object's path curves; without the centripetal force, the object follows the tendency to move in a straight line. Therefore, a centripetal force, when applied, is an unbalanced force, meaning that it is not "balanced off" by another force.

Newton's Second Law seems to apply to circular motion just as well as it applies to accelerated motion along a straight line, but with a strange "twist." It is a clearly correct application of $F = ma$ to say that a centripetal force, F, causes a mass, m, to experience an acceleration, a. However, the strange part is that when an object moves along a circular path at constant speed, acceleration is happening with no change in the object's speed. The force changes the direction of the velocity.

Velocity describes both the amount of speed and the direction of motion of an object. Thinking about the velocity of an object moving with constant speed on a circular path, it is true that the velocity is changing from one instant to the next not in the amount of the velocity, but with respect to the direction of the velocity. The diagram shows an object moving at constant speed on a circular path. Arrows are used to represent the velocity of the object at several instants during one trip around the circle.

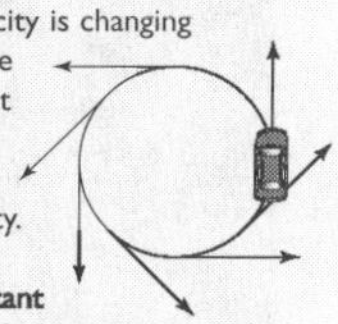

ANSWERS

For You To Do *(continued)*

4. a) Inertia is the natural tendency of an object to remain at rest or to remain moving with constant speed in a straight line.

   When there is no centripetal force acting on an object, it will move in a straight line.

   b) In the absence of an unbalanced force, an object at rest remains at rest and an object already in motion remains in motion with constant speed on a straight line path.

   In the absence of a sideways, towards-the-center force (centripetal force), the object going in a circle continues to move straight (at a tangent to the circle).

   c) The acceleration of an object is directly proportional to the unbalanced force acting on it and is inversely proportional to the object's mass. The direction of the acceleration is the same as the direction of the unbalanced force.

   When a centripetal force is applied to an object, it moves in a circular path. The acceleration is toward the center of the circle. It is definitely an application of $F = ma$ because the force acts (toward the center of the circle) on the mass giving it an acceleration (toward the center of the circle).

Physics in Action

Physicists have shown that a special form of Newton's Second Law governs circular motion:

$$F_C = ma_C = \frac{mv^2}{r}$$

where F_C is the centripetal force in newtons,
m is the mass of the object moving on the circular path in kilograms,
a_C is the centripetal acceleration in m/s^2,
v is the velocity in m/s, and
r is the radius of the circular path in meters.

Sample Problem

Find the centripetal force required to cause a 1000.0-kg automobile travelling at 27.0 m/s (60 miles/hour) to turn on an unbanked curve having a radius of 100.0 m.

Strategy: This problem requires you to find centripetal force. You can use the equation that uses Newton's Second Law to calculate F_C.

Givens:

$m = 1000.0$ kg

$v = 27.0$ m/s

$r = 100.0$ m/s

Solution:

$$F_C = \frac{mv^2}{r}$$

$$= \frac{(1000.0 \text{ kg } (27.0 \text{ m/s})^2}{100.0 \text{ m}}$$

$$= \frac{(1000.0 \text{ kg} \times 730 \text{ m}^2\text{/s}^2)}{100.0 \text{ m}}$$

$$= 7300 \text{ N}$$

If the force of friction is less than the above amount, the car will not follow the curve and will skid in the direction in which it is travelling at the instant the tires "break loose."

72

Active Physics

Reflecting on the Activity and the Challenge

Both circular motion and motion along curved paths that are not parts of perfect circles are involved in many sports. For example, both the discus and hammer throw events in track and field involve rapid circular motion before launching a projectile. Track, speed skating, and automobile races are done on curved paths. Whenever an object or athlete is observed to move along a curved path, you can be sure that a force is acting to cause the change in direction. Now you are prepared to provide voice-over explanations of examples of motion along curved paths in sports, and in many cases you perhaps can estimate the amount of force involved.

Physics To Go

1. For the car used as the example in the **For You To Read**, what is the minimum value of the coefficient of sliding friction between the car tires and the road surface that will allow the car to go around the curve without skidding? (Hint: First calculate the weight of the car, in newtons.)
2. If you twirl an object on the end of a string, you, of course, must maintain an inward, centripetal force to keep the object moving in a circular path. You feel a force that seems to be pulling outward along the string toward the object. But the outward force that you detect, called the "centrifugal force," is only the reaction to the centripetal force that you are applying to the string. Contrary to what many people believe, there is no outward force acting on an object moving in a circular path. Explain why this must be true in terms of what happens if the string breaks while you are twirling an object.
3. A 50.0-kg jet pilot in level flight at a constant speed of 270.0 m/s (600 miles per hour) feels the seat of the airplane pushing up on her with a force equal to her normal weight, $50.0 \text{ kg} \times 10 \text{ m/s}^2 = 500 \text{ N}$. If she rolls the airplane on its side and executes a tight circular turn that has a radius of 1000.0 m, with how much force will the seat of the airplane push on her? How many "*g*'s" (how many times her normal weight) will she experience?

ANSWERS

Physics To Go

1. μ = (frictional force)/(weight)
   = 7300 N / mg
   = 7300 N / (1,000 kg × 10 m/s^2)
   = 7300 N / 10,000 N = 0.73
2. If there were an outward force acting on the object, it would be expected to fly radially outward when the string breaks; when the string breaks, the object flies tangent to the circular path, indicating that there is no outward force.
3. $F_c = m(v^2/r)$
   = (50 kg) (270 m/s)^2/(1000 m)
   = (50 kg)(73,000 m^2/s^2)/(1000 m)
   = 3600 N

   The pilot's normal weight is $mg = 50 \text{ kg} \times 10 \text{ m/s}^2 = 500 \text{ N}$

   Therefore, the pilot "pulls" 3600 N / 500 N = 7.2 *g*'s during the turn; that is, she feels 7.2 times her normal weight. Assuming an inside turn with the top of the pilots head facing the center of the circle, the blood in her brain would tend to keep going straight ahead, tangent to the circle, draining from her brain and causing her to lose consciousness. However, her automatic pressure suit would inflate to squeeze against her legs, pushing blood upward to her brain to keep her from "blacking out."

ANSWERS

Physics To Go

(continued)

4. For the discus event in track and field, assume mass of disc = 1 kg, radius of twirling action before throw = 1 m, and speed during rotation before throw = 10 m/s:

   $F_c = m(v^2/r)$
   $= (1\ \text{kg})\ (10\ \text{m/s})^2/(1\ \text{m})$
   $= (1\ \text{kg})(100\ \text{m}^2/\text{s}^2)/(1\ \text{m}) = 100\ \text{N}$

   The athlete must "hold on" to the disc, using the throwing hand to provide an inward force of 100 N while twirling prior to release of the disc.

5. Viewed from a helicopter above the event, the passenger would be observed to keep going in a straight line as the car turns. The seat would slide under the passenger, and the door on the passenger side would hit the right shoulder of the passenger, providing the centripetal force thereafter to push the passenger into the same curve as the car.

6. The tilt-a-whirl ride is fun (but sometimes sickening) because two circular motions are "superimposed" on the body. One motion carries the body around in a large circular path corresponding to the circular path around which the chair moves; the second, superimposed, motion is provided as the chair spins as it revolves on the circular path, causing the body simultaneously to move in a circle of small radius. Sometimes the two centripetal forces add together, and sometimes they are in different directions. All of the effects on the body are too numerous to understand or explain, but one thing is for sure, the hot dogs in the stomach don't know which way to go!

7. a) They need a frictional force to turn. On a wet field, without friction, they continue moving in the same direction.

   b) Friction supplies the centripetal force. On a wet field, without friction, the players continue moving in the same direction, obeying Newton's First Law.

   c) Student work.

4. Imagine a video segment of an athlete or an item of sporting equipment moving on a circular path in a sporting event. Estimate the mass, speed, and radius of the circle. Use the estimated values to calculate centripetal force and identify the source of the force.

5. Below are alternate explanations of the same event given by a person who was not wearing a seat belt when a car went around a sharp curve:

   a) "I was sitting near the middle of the front seat when the car turned sharply to the left. The centrifugal force made my body slide across the seat toward the right, outward from the center of the curve, and then my right shoulder slammed against the door on the passenger side of the car."

   b) "I was sitting near the middle of the front seat when the car turned sharply to the left. My body kept going in a straight line while, at the same time due to insufficient friction, the seat slid to the left beneath me, until the door on the passenger side of the car had moved far enough to the left to exert a centripetal force against my right shoulder."

   Are both explanations correct, or is one right and one wrong? Explain your answer in terms of both explanations.

6. People seem to be fascinated with having their bodies put in a state of circular motion. Describe an amusement park ride based on circular motion that you think is fun, and describe what happens to your body during the ride.

7. a) Explain why football players fall on a wet field while changing directions during a play.

   b) Include the concepts centripetal force and Newton's Laws in a revised explanation.

   c) In a new revision, make the explanation exciting enough to include in your sports video voice-over.

PHYSICS AT WORK

Dean Bell

TELEVISION PRODUCER USES SPORTS TO TEACH MATH AND PHYSICS

Dean Bell is an award-winning filmmaker and television writer, director, and producer. His show *Sports Figures* is a highly acclaimed ESPN educational television series designed to teach the principles of physics and mathematics through sports. His approach has been to tell a story, pose a problem, and then follow through with its mathematical and scientific explanation. "But always," he says, "you must make it fun. It has to be both educational and entertaining."

Dean began his career as a filmmaker after college. He landed the apprentice film editor's position on a Woody Allen film. From there, he worked his way up in the field, from assistant editor, to editor and finally writer, director, and producer.

"I've always been a fan of educational TV," he states, "although I never thought that was where my career would take me. It's one of life's little ironies that I've ended up producing this type of show. You see, my father worked in scientific optics and was very science oriented. He was always delighted in finding out how things worked, and was even on the Mr. Wizard TV show a few times."

Dean writes the script for each segment, working together with top educational science consultants. "We spend a day coming up with ideas and then researching each subject thoroughly. Our job is to illustrate the relationship between a sports situation and the related mathematical or physics principles.

"At the end of the day," says Dean, "it really is nice to be working on a show that means something and that is so worthwhile. I'm still getting ahead in my career as a film and TV producer, but now I'm also an educator."

Chapter 1

Physics in Action

Chapter 1 Assessment

Your big day has arrived. You will be meeting with the local television station to audition for a job as a "physics of sports" commentator. Whether you will get the job will be decided on the quality of your voice-over.

With what you learned in this chapter, you are ready to do your science commentary on a short sports video. Choose a videotape from a sports event, either a school event or a professional event. Each of you will be responsible for producing your own commentary, whether or not you worked in cooperative groups during the activities. You are not expected to give a play-by-play description, but rather describe the rules of nature that govern the event. Your viewers should come away with a different perspective of both sports and physics. You may produce one of the following:

- **a written script**
- **a live narrative**
- **a video soundtrack**
- **an audiocassette**

Review the criteria by which your voice-over dialogue or script will be evaluated. Your voice-over should

- **use physics principles and terms correctly**
- **have entertainment value**

After reviewing the criteria, decide as a class the point value you will give to each of these criteria:

- **How important is the physics content? How many physics terms and principles should be illustrated to get the minimum credit? The maximum credit?**
- **What value would you place on the entertainment aspect? How do you fairly assess the excitement and interest of the broadcast?**

Physics You Learned

Galileo's Principle of Inertia

Newton's First Law of Motion

Newton's Second Law of Motion

Newton's Third Law of Motion

Weight

Center of mass; center of gravity

Friction between different surfaces

Momentum

Law of Conservation of Momentum

Centripetal acceleration

Centripetal force

Alternative Chapter Assessment

Multiple Choice: Select the letter of the choice that best answers the question or best completes the statement.

1. Which of the following best illustrates Newton's First Law of Motion?

   a) A collision between a running back and a linebacker in football.

   b) An ice hockey puck sliding along the ice after being hit by a player's stick.

   c) A bowling pin being struck by a bowling ball.

   d) A volleyball being "spiked" across the net.

2. A small ball rests on a circular turntable, rotating clockwise at a constant speed as illustrated in the below diagram. Which of the below pair of arrows best describes the direction of the acceleration and the net force acting on the ball at the point indicated in the diagram?

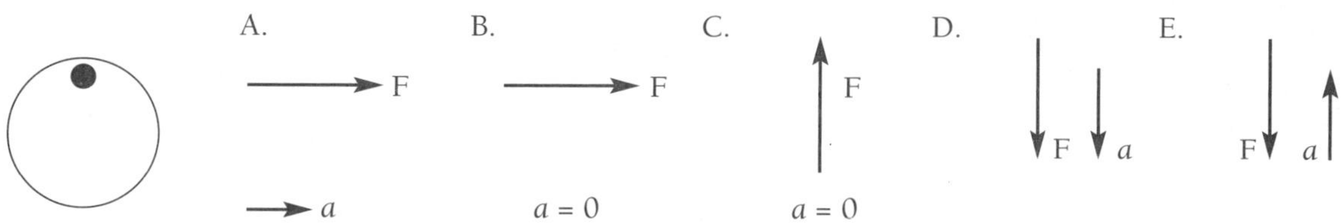

3. A person sliding into second base continues to slide past the base due to:

   a) inertia

   b) friction

   c) weight

   d) gravity

4. Newton's First Law of Motion states that an object at rest stays at rest unless acted upon by a:

   a) balanced force

   b) net force

   c) weak force

   d) strong force

5. In the absence of air, a penny and a feather dropped from the same height:

   a) fall at different rates

   b) float

   c) fall at equal rates

   d) do not have momentum

6. An object rolling across a level floor without any horizontal net force acting on it will:

a) slow down

b) speed up

c) keep moving forever

7. An object falling to Earth in the absence of air resistance:

a) falls with a constant speed of 9.8 m/s

b) falls with constant acceleration of 9.8m/s2

c) slows down

8. A constant net force acting on an object causes the object to move with constant:

a) speed

b) velocity

c) acceleration

d) momentum

9. Which of the following is NOT one of Newton's Laws of Motion?

a) An object in motion stays in motion unless acted upon by an unbalanced force.

b) A constant net force acting on an object produces a change in the object's motion.

c) For every action, there is an equal and opposite reaction.

d) Energy is neither created not destroyed; it simply changes form.

10. Newton's First Law is known as the law of:

a) impetus

b) inertia

c) acceleration

d) resistance

True or Replace False Word: Determine whether the word in bold print makes each statement true or false. If a statement is true, write "true" in the answer space. If a statement is false, write in the answer space a replacement for the word in bold print which would make the statement become true.

11. If a net force acts on an object, the object will change speed, direction or **neither**.

Answer: __________

12. **Gravity** is the tendency of an object to resist any change.

Answer: __________

13. A net force acting on an object causes the object to move with constant **velocity**.

Answer: __________

14. **Forces** that are equal in amount and opposite in direction are balanced forces.

Answer: __________

15. A bowling ball has more **inertia** than a tennis ball.

Answer: __________

Short Answer: Write a brief response to each item. Show your work for responses which require calculations.

16. Describe how Galileo's experiments with balls and ramps led to Newton's First Law of Motion.

17. Using Newton's First and Second Laws of Motion, explain why a ball thrown into the air follows a parabolic path.

18. On the diagrams below, draw arrows to show the vertical forces acting on a person jumping into the air. Use diagram A to show the forces acting during the push-off, and use diagram B to show the forces acting as the person lands on the ground. Write a paragraph comparing the forces in each situation, including a discussion on the relative sizes of the forces.

A. 

B. 

19. Describe the collision between a bat and a ball in terms of conservation of momentum.

20. Imagine a tug-of-war between two people. Draw a sketch indicating all of the forces acting in the tug-of-war. Using Newton's Laws and your picture, explain how and why one side will win.

21. For each of the following forces, identify the "reaction" force from Newton's Third Law:

a) Volleyball hitting the floor.

Answer: __________

b) Softball bat hitting a softball.

Answer: __________

c) Punter kicking a football.

Answer: __________

22. Explain why sprinters prefer to use longer spikes on their shoes than long distance runners, even though both run on same track surface.

23. A 75-kg ice hockey forward moving at 5.0 m/s collides with a stationary 85-kg defenseman and they become entangled. With what velocity will the pair move across the ice?

24. Explain how a soccer player can cause a ball to spin as a result of kicking it. What can the player do to change the direction of spin on the ball?

25. Two objects that have the same mass are dropped from the top of a 20-meter high building. One object is larger and flatter than the other object. Which hits the ground first? Use the terms gravity, acceleration, and air resistance correctly in your discussion.

Alternative Chapter Assessment Answers

Multiple Choice

1. b
2. d
3. a
4. b
5. c
6. c
7. b
8. c
9. d
10. b

True/Replace False Word

11. both
12. inertia
13. acceleration
14. true
15. true

Short Answer

16. Galileo used two ramps, initially in a V-shape, and demonstrated that a ball rolled from a specific height would roll up the opposite side until it reached the original height. If the angle of the second ramp was decreased, the ball still rolled to the same height but this ball traveled further on the ramp. He reasoned that the ball will roll until it reaches the height from which it was released, therefore, if the second ramp were horizontal the ball would continue to roll in the horizontal direction since it would not be able to reach this height. This supports the idea that the "natural" state of an object's motion is not necessary to be at rest.

17. When a ball is thrown into the air the primary force acting on it is its weight. Since weight is a force that acts "down", from Newton's Second Law the acceleration of the ball will be down, therefore the vertical velocity of the ball will be changing. If we assume no air resistance, horizontally the ball will continue in its original state of motion according to Newton's First Law, minimal air resistance would at least show very little change in horizontal motion. During equally spaced time intervals the horizontal displacement will be constant and the vertical displacement will be changing, resulting in a parabolic path for the ball.

18. In each diagram the vertical forces that need to be shown are the person's weight (acting down) and the normal force from the ground (acting up). In both cases the normal force will exceed the weight since both require an acceleration in the upward direction.

Chapter 1

19. When a bat collides with a ball the total momentum of the system will remain the same as long as no other forces acted. The bat will slow down upon collision, thus decreasing the momentum of the bat. The momentum that the bat lost will be gained by the ball, thus changing the momentum of the ball, since the bat was originally traveling in the opposite direction from the ball, the ball's motion will change such that its direction will change.

20. Consider the entire tug-of-war as one object. The forces acting are as follows:
"outside forces" — The weight of each person (down) and the normal force from the ground (up). Each side will have a friction force acting (horizontally) as a reaction to each person pushing sideways on the ground.
"internal forces" — Each side applies a force to the rope, the rope applies a force back on each side.
"Outside" forces change the motion of objects, the internal forces have no effect on the total motion of the object. The vertical forces (weights and normal forces) will add to zero, the horizontal friction forces will add to show a net force in one direction that will cause the entire tug-of-war to accelerate in that direction.

21. a) Floor applies a force on volleyball.

b) Softball applies a force on bat.

c) Football applies force on punter's foot.

22. Sprinters require larger accelerations since they wish to reach maximum velocity in a short period of time. The spikes allow for the sprinter to push harder on the ground so the ground can push pack with a larger force.

23. $(75*5.0) + (85*0) = (75+85)v$

$v = 2.3$ m/s

24. To cause a ball to spin a force must be applied off-center. To change the direction of the spin the player need apply the force at different points on the ball, e.g., to have the ball spin to the left, the force needs to be applied to the right of center.

25. The larger and flatter object will be influenced by the air as it falls. The force of air resistance will oppose the force of gravity acting on the object causing the net downward force to be less. Therefore, the downward acceleration of the object will be less than 9.8 m/s/s. It will take longer to fall the 20 m.

Chapter 2
SAFETY

Chapter 2-Safety

National Science Education Standards

Chapter Challenge

Dangers inherent in travel provide the context for this chapter. Students are challenged to design or build a safety device, or system, for protecting automobile, airplane, bicycle, motorcycle, or train passengers. New laws, increased awareness, and improved safety systems are explored as students work on this challenge. They are also encouraged to design improvements to existing systems and to find ways to minimize harm caused by accidents.

Chapter Summary

To meet this challenge, students engage in collaborative activities that explore motions and forces and the principles of design technology. These experiences engage students in the following content from the National Science Education Standards.

Content Standards

Unifying Concepts

- Systems, order and organization
- Evidence, models and explanations
- Constancy, change, and measurement

Science as Inquiry

- Identify questions and concepts that guide scientific investigations
- Use technology and mathematics to improve investigations
- Formulate and revise scientific explanations and models using logic and evidence
- Communicate and defend a scientific argument

History and Nature of Science

- Nature of scientific knowledge

Physical Science

- Motions and forces

Science in Personal and Social Perspectives

- Personal and community health
- Natural and human-induced hazards

Science and Technology

- Understandings about science and technology
- Ability to apply technology

Key Physics Concepts and Skills

Activity Summaries	Physics Principles
Activity 1: Response Time Using a response timer, students explore the time required for a driver to respond to a hazard. This activity introduces students to the process of beginning with their own ideas and predictions, then implementing an investigation that results in both qualitative and quantitative data.	• Series circuits • Switches • Response time
Activity 2: Speed and Following Distance Strobe, or multiple exposure photos of a moving vehicle are used to discuss speed and acceleration. Students then use a sonic ranger to measure how fast they walk and obtain a computer generated graph of their speed. Information about speed is then connected to response time with a discussion of tailgating.	• Average speed • Using data as basis for predictions • Speed, distance, and time relationships
Activity 3: Accidents Following an investigation crashing cars against barriers, students use advertisements and consumer reports to learn about safety devices on automobiles. Each is analyzed to determine the type of collision-related injuries it prevents, and to identify if the device could in fact increase injuries in a unique setting.	• Physical properties of matter • Effect of forces on motion
Activity 4: Life (and Death) before Seat Belts Using a lump of clay on a motion cart to represent a person in a car, students explore "objects in motion stay in motion." They then relate this to actual automobile collisions.	• Acceleration • Inertia
Activity 5: Life (and Fewer Deaths) after Seat Belts Students focus on the design and materials used in seat belt construction as they study force and pressure. They investigate how increasing surface area decreases the pressure exerted. They relate this to the challenge by finding ways to increase the area of impact in a collision.	• Inertia • Newton's Laws of Motion • Force and pressure • Newton as a unit of measure
Activity 6: Why Air Bags? A model of an air bag is used in an investigation of what happens on impact when objects of different mass are dropped from different heights. They observe the amount of damage in each case and relate this to the concept of "impulse" and how spreading out the time of the impulse reduces damage.	• Inertia • Force and pressure • Impulse
Activity 7: Automatic Triggering Devices In this inquiry investigation, students design a device that will trigger an air bag to inflate. These simulations allows them to apply concepts of inertia and impulse as they test ideas that help them address the chapter challenge.	• Inertia • Force and pressure • Impulse
Activity 8: The Rear-End Collision Students investigate the effect of rear-end collisions on passengers by using a model of the neck muscles and bones of the vertebral column. They then read to learn more about Newton's Second Law of Motion and consider how they can apply this information in designing a safety device that prevents movement of the head in a collision.	• Collisions • Newton's Second Law of Motion • Momentum
Activity 9: Cushioning Collisions (Computer Analysis) Using a force probe, students investigate the effectiveness of different types of systems designed to minimize the impact of collisions. The systems include sand canisters around bridge supports and padded car interiors. This investigation provides an opportunity to develop deeper understanding of the concepts of acceleration, velocity, and momentum.	• Inertia • Impulse • Momentum • Change in Momentum • Conservation of Momentum

GETTING STARTED WITH EQUIPMENT NEEDED TO CONDUCT THE ACTIVITIES.

Items needed—not supplied in Material Kits

Preparing the equipment needed for each activity in this chapter is an important procedure. There are some items, however, needed for the chapter that are not supplied in the It's About Time material kitpackage. Many of these items may already be in your school and would be an unnecessary expense to duplicate. Please read carefully the list of items to the right which are not found in the supplied kits and locate them before beginning activities.

Items needed—not supplied by It's About Time:

- **Electrical Clock Measuring 1/100 seconds**
- **Digital Photogates**
- **MBL or CBL With Sonic Ranger**
- **Concrete block or similar barrier**
- **Camcorder on tripod**
- **VCR having single frame advance mode**
- **Video monitor**
- **Baseball glove**
- **VCR & TV monitor**

Chapter 2

Equipment List For Chapter 2

PART	ITEM	QTY	ACTIVITY	TO SERVE
AC-6003-T2	Alligator Clip Leads	3	1, 7	Group
SS-6206-T2	Balance Platform	1	4A, 8, 9	Group
BH-6600-T2	Ball, Heavy	1	6	Group
WS-6731-T2	Bare Copper, #22 Wire, 1ft.	1	5	Group
BS-1596-T2	Battery, D-Cell	1	1, 7	Group
BS-6701-T2	Brochures on Auto Safety Features	1	3	Group
LS-6960-T2	Bulb Base for Miniature Screwbase	1	1, 7	Group
LS-6959-T2	Bulb, Miniature Light	3	1, 7	Group
MC-5418-T2	Clay, Modeling.	3	4, 4A, 5	Group
CB-1060-T2	Clear Bag, Inflatable	6	6	Group
MS-6726-T2	Cushioning Materials Set, Variety of	6	9	Group
DU-0026-T2	Duct Tape, Roll	1	8	Group
GC-0001	Dynamics Carts	2	4, 5, 6, 7, 8	Group
SS-6710-T2	Landing Surface Materials of 3 Hardnesses	1	6	Group
RS-6719-T2	Ribbon, Various Widths	18	5	Group
RS-2826	Ruler	1	1	Group
BH-6068-T2	Single Battery Holder, D-Cell	1	1, 7	Group
SS-2303-T2	Spring Scale, 0-10 Newton Range	1	8	Group
SR-6736-T2	Starting Ramp for Lab Cart	1	4, 5, 7, 8, 9	Group
SM-1676-T2	Stick, Meter, 100 cm, Hardwood	1	8	Group
SS-7778	Stopwatches	2	1	Group
SS-9019-T2	Switch, Spst.	2	1, 7	Group
TS-2662-T2	Tape, Masking, 3/4 x 36 yds.	1	8, 9	Group
WS-6732-T2	Wood Piece, 1" x 2" x 2"	1	8	Group
WS-6733-T2	Wood Piece, 1" x 3" x 10"	1	8	Group
WS-6734-T2	Wood Piece, 2" x 4" x 1'	1	8	Group
ITEMS NEEDED – NOT SUPPLIED BY IT'S ABOUT TIME				
	Electrical Clock Measuring 1/100 seconds	1	1A	
	Digital Photogates	2	1A	
	MBL or CBL With Sonic Ranger	1	2, 9	
	Concrete block or similar barrier		2, 7	
	Camcorder on tripod		4	
	VCR having single frame advance mode		4	
	Video monitor		4	
	Baseball glove		4A	
	VCR & TV monitor		3, 9	

Organizer for Materials Available in Teacher's Edition

Activity in Student Text	Additional Material	Alternative / Optional Activities
ACTIVITY 1: Response Time p. 80	Assessment: Group Work, p.156 Assessment: Scientific and Technological Thinking, p 157	Activity 1A: High-Tech Alternative, pgs. 158-159
ACTIVITY 2: Speed and Following Distance p. 86	Assessment: Graphing Skills, p.172	
ACTIVITY 3: Accidents p. 94	Assessment: Participation in Discussion, p.181	
ACTIVITY 4: Life (and Death) before Seat Belts p. 99		Activity 4 A: Dropping a Clay Ball to Investigate Inertia, pgs. 191-192 Activity 4 B: Low-Tech Alternative, p. 193
ACTIVITY 5: Life (and Fewer Deaths) after Seat Belts p. 105	Assessment, p. 204	
ACTIVITY 6: Why Air Bags? p. 111	Assessment, p. 215	
ACTIVITY 7: Automatic Triggering Devices p. 117	Assessments for Activity 7, p. 224 and for Scientific and Technological Thinking, p. 224	
ACTIVITY 8: The Rear-End Collision p. 122		
ACTIVITY 9: Cushioning Collisions (Computer Analysis) p. 129		

Scenario

Probably the most dangerous thing you will do today is travel to your destination. Transportation is necessary, but the need to get there in a hurry, and the large number of people and vehicles, have made transportation very risky. There is a greater chance of being killed or injured traveling than in any other common activity. Realizing this, people and governments have begun to take action to alter the statistics. New safety systems have been designed and put into use in automobiles and airplanes. New laws and a new awareness are working together with these systems to reduce the danger in traveling.

What are these new safety systems? You are probably familiar with many of them. In this chapter, you will become more familiar with most of these designs. Could you design or even build a better safety device for a car or a plane? Many students around the country have been doing just that, and with great success!

Challenge

Your design team will develop a safety system for protecting automobile, airplane, bicycle, motorcycle, or train passengers. As you study existing safety systems, you and your design team should be listing ideas for improving an existing system or designing a new system for preventing accidents. You may also consider a system that will minimize the harm caused by accidents.

Chapter and Challenge Overview

Divide the class into design teams consisting of three or four students. As there is a diversity of skills required – design, construction, writing, speaking, etc. – all students should be involved in the process. You may want to allow some time early in the week for group meetings for the purpose of brainstorming as well as an opportunity for you to check on the progress of the teams.

Encourage broad thinking. The projects do not have to be limited to the vehicle, but can include the roadway, traffic control, etc. Students may use their experiences with skateboarding, cycling, in-line skating or other athletic pursuits to help them get started.

Explain the two-day presentation format. On the first day a poster session is conducted during which students informally explain their projects to classmates and answer their questions. Meanwhile, the students are writing down questions about the projects, and placing them into envelopes, provided by you for each project. The formal presentations the next day may be quite brief, since most students will have seen the projects the day before. After a few sentences addressing the points listed in the student text, randomly draw a student question from the appropriate envelope for the team to answer. Use as many as time might allow.

Scoring the project might be based on assigning credit for each of the items listed in the student text with, perhaps, greater emphasis on how the physics concepts are utilized and explained. Discuss with the students the relative credit weighting of the project. Use as a starting point, the criteria mentioned in the student text, with the total points being 100. An example might be Part 1 – 30, Part 2 – 20, Part 3 – 20, with a teacher assigned (or peer assessment) of 30 points, adding to a total of 100.

The intention in this exercise is to motivate students into learning about the safety of various modes of transportation. As they develop their project, while studying this chapter, they should be revising and changing their safety system.

Your final product will be a working model or prototype of a safety system. On the day that you bring the final product to class, the teams will display them around the room while class members informally view them and discuss them with members of the design team. During this time, class members will ask questions about each others products. The questions will be placed in envelopes provided to each team by the teacher. The teacher will use some of these questions during the oral presentations on the next day.

The product will be judged according to the following three parts:

1. The quality of your safety feature enhancement and the working model or prototype.
2. The quality of a five-minute oral report that should include:
   - **the need for the system**
   - **the method used to develop the working model**
   - **the demonstration of the working model**
   - **the discussion of the physics concepts involved**
   - **the description of the next-generation version of the system**
   - **the answers to questions posed by the class**
3. The quality of a written and/or multimedia report including:
   - **the information from the oral report**
   - **the documentation of the sources of expert information**
   - **the discussion of consumer acceptance and market potential**
   - **the discussion of the physics concepts applied in the design of the safety system**

Criteria

You and your classmates will work with your teacher to define the criteria for determining grades. You will also be asked to evaluate your own work. Discuss as a class the performance task and the points that should be allocated for each part. A starting point for your discussions may be:

- **Part 1 = 40 points**
- **Part 2 = 30 points**
- **Part 3 = 30 points**

Since group work is made up of individual work, your teacher will assign some points to each individual's contribution to the project. If individual points total 30 points, then parts 1, 2 and 3 must be changed so that the total remains at 100.

Assessment Rubric for Challenge: Group Work in Designing Safety Feature Content

Total = 9 marks

1. Low level – indicates minimum effort or effectiveness.
2. Average – acceptable standard has been achieved, but the group could have worked more effectively if better organized.
3. Good – this rating indicates a superior effort. Although improvements might have been made, the group was on task all of the time and completed all parts of the activity.

Descriptor	Values		
1. The group worked cooperatively to design a safety feature. Comments:	1	2	3
2. The group was organized. Materials were collected and the problems were addressed by the entire group. Comments:	1	2	3
3. Data was collected and recorded in an organized fashion in data tables in their logs. Comments:	1	2	3

For use with *Safety*, Chapter 2

Assessment Rubric for Challenge: Safety Feature and Working Model

Descriptor	5	4	3	2	1
Skills Required for Working Model/Prototype					
understands the need to control variables					
has run at least three trials with safety feature					
demonstrates or explains why there is a need for several trials					
has rebuilt or modified safety feature as necessary					
uses appropriate materials in the construction of the safety feature					
care and attention has been given in assembling the working model					
working model functions appropriately during demonstration					
group has worked efficiently as a team in assembling the model					
Oral Report					
understands and explains the need for the system					
describes the method used to develop the working model					
demonstrates the working model					
discusses the physics concepts illustrated by the safety feature					
describes the next generation of the system					
answers questions posed by the class					
Written and/or Multimedia Report					
contains the points included in the oral report					
spelling, punctuation, grammar, and sentence structure are correctly used					
science vocabulary and symbols are used correctly					
documents sources of expert information					
discusses consumer acceptance and marketing potential					
data is presented in tables and graphs as appropriate					

For use with *Safety*, Chapter 2

What is in the Physics InfoMall for Chapter 2?

Chapter 2, *Safety* deals with the physics of safety systems in automobiles, airplanes, and bicycles. The Physics InfoMall CD-ROM contains an enormous amount of material related to the physics of many phenomena, and safety is one of them. At first, it seems like a good idea to see what the InfoMall has to say about "safety systems." So the first thing you may want to do is perform a search on the entire CD-ROM for "safety system*" (the asterisk is a wild character asking the search engine to look for any words that share the same beginning, such as "system", "systems", or even "systematic"). The only result from this search that is relevant to our needs is "The science of traffic safety," by Leonard Evans, and found in *The Physics Teacher*, volume 26, issue 7. Early in this article, Evans states "No one who lives in a motorized society can fail to be concerned about the enormous human cost of traffic crashes; as many young males are killed in traffic crashes as by all other causes combined. The United States Department of Transportation maintains a file containing information on all fatal traffic crashes in the United States since 1975. This data file now documents over half a million fatalities, and of course, injuries are enormously more numerous. Recent research, discussed below, should demonstrate that the study of phenomena related to traffic safety presents problems of intellectual challenge similar in character and difficulty to those encountered in physics. Traffic safety means the safety of the overall traffic system, as distinct from more specific properties of individual components, such as laboratory crash tests of vehicles." This passage lends support to the **Scenario** described at the beginning of this chapter. Following this article is a list of references that you may also find useful in teaching traffic safety. While this article is specific to traffic safety, it is a great beginning to this chapter.

ACTIVITY 1
Response Time

Background Information

Background Information for most activities is provided for the interest and insight of the teacher only. It is not intended to be part of the classroom instruction.

Reaction time can be understood by grouping physiological processes into three categories: input of sensory information, coordination by the central nervous system, and the response by motor nerves and their effectors, muscles and/or glands. The simplest reaction pathway is that of a reflex arc. Sensory receptors identify environmental stimuli causing a sensory nerve cell to become excited. The sensory nerve transmits an electrochemical impulse to the spinal cord. Here an intermediary nerve cell transmits the sensory impulse to a motor nerve cell. The impulse is carried by the motor nerve cell to a muscle (or in some cases a gland). The contraction of the muscle signals the response. A knee-jerk response provides an excellent example of this simple nerve pathway. The impulse is carried between three nerve cells: sensory nerve, interneuron, and motor nerve cell, toward the muscle. Surprisingly, no integration is required by the brain. These reactions occur without thinking.

Reactions that require integration by the central nervous system, such as those that occur when driving, take considerably longer to occur. A moose running in front of a vehicle is identified by visual receptors within the eye. Sensory impulses are carried toward the brain by the optic nerve. Here the information is accumulated and the driver is made aware of the problem. Multiple nerve connections carry the impulses toward the motor area of the brain. A conscious decision is made to lift the foot from the accelerator peddle and push down on the brake. Because the sensory nerves are connected with motor nerves through a maze of circuits within the brain, the reaction time is much longer than that of a reflex arch. Each time an impulse passes between connecting nerve cells, the speed of transmission is slowed.

Conscious decisions, such as braking for a moose, depend upon a number of variables. The time it takes to catch sight of the moose may well be the largest variable. Any distraction or driver fatigue will increase reaction times. Most impulses travel at approximately 100 m/sec along a nerve cell, but the time required for the impulse to travel between two different nerve cells varies greatly. Transmitter chemicals diffuse between connecting nerve cells. Because diffusion takes much more time than the movement of an impulse along a nerve cell, the connections between nerve cells slows reaction time. Not surprisingly, the complexity of integration of sensory impulses by the brain to create a visual image and the number of nerve cells involved also affects response time. The greater the number of interconnecting nerves, the slower is the processing time. Moving images require greater time to process and interpret than still images.

To accurately determine response times, we must consider how the reaction is measured. The removal of the foot from the driver's pedal takes considerably more time than just pushing down on the brake. The distance the leg moves, the amount of muscle required, and the health of the muscle also affect reaction rates.

As people age reaction rates are said to decline. The buildup of pigmented Nissl Bodies with nerve cells, slows the transmission of nerve impulses. In addition, the production of transmitter chemicals, the things that allow impulses to travel between nerves, decreases with age. Older people also tend to have less healthy muscles, further increasing the time it takes to respond to a stimulus. But age is not the major factor when considering reaction rates. The alertness of the driver is far more important.

In this activity students are asked to wire a series circuit. A series circuit has all of the current from the battery traveling through every part of the circuit. If either switch is open, the current is not able to traverse the entire circuit. In the reaction time circuit, one switch begins in the closed position and the other in the open position. One student is able to complete the circuit by closing the switch and lighting the bulb. The other student will then turn the light off by opening the other switch.

The reaction time graph is created by using the equation for free fall motion

$d = 1/2at^2$

where
a is the acceleration due to gravity (9.8 m/s^2),
t is the elapsed time and
d is the distance fallen. Since all objects fall at the same rate, there is no need to be concerned with the mass of the ruler.

Solving the above equation for time:

$t = \sqrt{2d/a}$

allows us to compute the reaction tome for any given distance. The students will be introduced to this equation later in the course. To provide the equation with no evidence of constant acceleration would not help their understanding at this point. If, on the other hand, they have studied acceleration previously, you may use this equation to provide a reinforcement of this concept.

Active-ating the Physics InfoMall

This activity is primarily about reaction times. There are several good items from the InfoMall that relate to this. If you choose to search the InfoMall CD-ROM for this activity, choose "reaction time" rather than "response time," as the latter will find mostly items that describe mechanical or chemical systems rather than people. All of the items found for **Activity 1**, and many for the following activities, were found searching for "reaction time" in all stores on the InfoMall at the same time (select stores to be searched using "compound search" and choose "select..." below "search in databases:").

The InfoMall has several methods for measuring a person's reaction time. The methods used in *Active Physics* can also be found on the InfoMall. Although not related to driving, the effect of reaction time in analyzing the Kennedy assassination can be found in the Articles and Abstracts Attic, *American Journal of Physics*, volume 44, issue 9, "A Physicist Examines the Kennedy Assassination Film."

Some good places to look for the effect on driving are the following:

For You To Do

Step 1: Testing the reaction time for the foot may be different than for the hand. This is discussed briefly in Articles and Abstracts Attic, *The Physics Teacher*, volume 8, issue 4, "Problems for Introductory Physics," problem 49. This can be found most easily by scrolling down to near the bottom of the article and then searching up, rather than down. Included are some questions to consider about the effect of reactions time on driving.

Step 2: The reaction time for visual stimuli can differ from the time for audible stimuli. This is discussed, along with methods for measuring the difference, in Articles and Abstracts Attic, *The Physics Teacher*, volume 28, issue 6, "Speed of Sound in a Parking Lot." Reaction time is discussed as a source for error in the measurement of the speed of sound for this particular activity, but the difference in audible versus visual stimuli is measured using a meter stick.

Steps 7 and 8: Alternate methods for measuring reaction times can also be found on the InfoMall.

A circuit using an oscilloscope can be found in the Demo and Lab Shop, *Laboratory Manual to Accompany Physics Including Human Applications*, by Fuller, Fuller, & Fuller, The Oscilloscope, Application I: Reaction Time Measurement. A graphic showing how the circuit should be set up is included.

Using a clock (if it uses a large sweep second hand and displays time in small increments) to measure reaction time is discussed in the Demo and Lab Shop, *Demonstration Handbook for Physics*, Mechanics, Kinematics, Reaction Time, and scroll down to Mb-1. Also, the use of a meter stick is discussed here, as it is in many other places.

Step 9: *Laboratory Manual to accompany Physics Including Human Applications*, by Fuller, Fuller, & Fuller; Human senses, Part III: reaction time. This shows how a meter stick can be used to measure reaction time.

Physics To Go

Methods mentioned in this activity include using a ruler or a dollar bill. These methods, including graphics are also discussed in the following places on the InfoMall:

Book Basement, *A Guidebook for Teaching Physics*, by Yurkewicz, Motion, topic IV: Uniform Acceleration. Activity #9 is about using a dollar bill, and Activity #10 is about using a meter stick.

The same methods are mentioned in the Demo and Lab Shop, *Physics Demonstrations and Experiments for High School*, Part II - Lab Experiments, #2 Using Acceleration of Gravity to Calculate Reaction Time," and uses a dollar bill and a meter stick.

The use of a dollar bill is also mentioned in Teacher Treasures, *Demonstration Guide for High School Physics*, and scroll down to "$ Bill & Reaction Time."

Planning for the Activity

Time Requirements

Allow about 40 minutes to construct the electrical circuit and complete **Steps 1** to **6**. An additional 20 minutes to complete the remaining steps and record data may be required.

Materials Needed

For each group:

- ruler
- alligator clip leads
- battery, d-cell
- bulb base for miniature screwbase
- bulb, miniature light
- ruler
- single battery holder, d-cell
- stopwatches
- switch, spst.

Advance Preparation and Setup

Search your school for a response timer formerly used in drivers' education. These units, about the size of an old movie projector, used to be quite common and included apparatus for testing peripheral vision and color blindness. It may well be in a closet somewhere in your building.

Should a response timer from a drivers' education class be unavailable, a usable circuit can be rigged if a clock measuring hundredths of a second is available. Set it up in series with its power source, a switch, and a normally closed foot switch. Response time is measured by having one student watch the clock, with a foot resting on the floor near the foot switch. When the student's partner starts the clock, the one being tested stops it by pressing the foot switch.

Teaching Notes

Active Physical Science uses a modified constructivist model. By confronting students' misconceptions and by having them do hands-on exploration of ideas, we seek to replace their misconceptions with correct perceptions of reality. In order to do this, a consistent scheme is integrated into the course activities to elicit the students' misconceptions early in any activity. Students' current mental models are sampled by one or more **What Do You Think?** questions. Students are not expected to know a "right" answer. These questions are supposed to elicit from students their beliefs regarding a very specific prediction or outcome, and students should commit to a written specific answer in their logs.

In **Activity 1** the term "response time" is used instead of the more common "reaction time" to differentiate the behavior from the reflex reaction.

Discuss the response-time circuit. The circuit is not complicated, but will provide the students with the experience of wiring a circuit. This is an opportunity to point out the characteristics of a simple series circuit. Do not begin an extensive lesson on circuit theory unless the class is really excited by the topic. Let students follow the direction in the text, and answer the questions. The sophistication of the circuit can be improved with the use of specialized normally on and normally off switches, if available.

You may be tempted to skip building the circuit and get right into measuring response time accurately. Do not skip this step. The qualitative estimates before building the circuit and using the circuit provide the foundation for understanding the short time intervals measured next.

Encourage students to work cooperatively by assigning tasks prior to beginning the activity. The following tasks are designed for groups of four students:

- **Organizer:** helps focus discussion and ensures that all members of the group contribute to the discussion. The organizer ensures that all of the equipment has been gathered and that the group completes all parts of the activity.
- **Recorder:** provides written procedures when required, diagrams where appropriate and records data. The recorder must work closely with the organizer to ensure that all group members contribute.
- **Researcher:** seeks written and electronic information to support the findings of the group. In addition, where appropriate, the researcher will develop and test prototypes. The researcher will also exchange information gathered among different groups.
- **Diverger:** seeks alternative explanations and approaches. The task of the diverger is keep the discussion open. "Are other explanations possible?"

The technique using two stopwatches in **Step 8** can become a popular game. Count your stopwatches before the students leave.

Students often believe that age is the largest factor when determining response time. Many will indicate that because of their much faster response times, they are better equipped to travel at higher speeds than older people. Variations due to age increase response time by as little a 0.01 sec.

The identification of sensory information has the greatest impact upon reaction time. Therefore, the alertness of the driver has the greatest impact upon stopping distances. The driver's experience may also play an important part in response time.

See the High Technology Alternative to **Activity 1:** Response Time. The use of photogates and 0:00 sec timers will be required if you choose to do this Alternative Activity.

Activity Overview

This activity addresses questions of response time in its relation to the problem of bringing a car to rest.

Student Objectives

Students will:

- Identify the parts of the process of stopping a car.
- Measure reaction time.
- Wire a series circuit.

Answers for the Teacher Only

What Do You Think?

Response times will vary. The most important factor is speed, and distance is proportional to the square of the speed. The activity will provide a basis for this relationship. The identification of sensory information has the greatest impact on reaction time. Therefore, the alertness of the driver has the greatest impact on stopping distances. The driver's experience may also play an important part in response time. **What Do You Think?** is designed to provoke a discussion of the effects of listening to loud music on response time, as well as the distraction of talking or the effect of fatigue, alcohol, or drugs.

Safety

Activity 1 Response Time

GOALS

In this activity you will:

- Identify the parts of the process of stopping a car.
- Measure reaction time.
- Wire a series circuit.

What Do You Think?

Many deaths that occur on the highway are drivers and passengers in vehicles that did not cause the accident. The driver was not able to respond in time to avoid becoming a statistic.

- **How long would it take you to respond to an emergency?**

Record your ideas about this question in your *Active Physics* log. Be prepared to discuss your responses with your small group and the class.

For You To Do

1. To stop a car, you must move your foot from the gas pedal to the brake pedal. Try moving your right foot between imaginary pedals.

Active Physics 80

a Estimate how long it takes to move your foot between the imaginary pedals. Record your estimate.

2. The first step in stopping a car happens even before you move your foot to the brake. It takes time to see or hear something that tells you to move your foot. Test this by having a friend stand behind you and clap. When you hear the sound, move your foot between imaginary pedals.

a) Estimate how long it took you to respond to the loud noise. Record your estimate.

3. Create a simple electric circuit to test your response time. Your group will need a battery in a clip, two switches, a flashlight bulb in a socket, and connecting wires. Connect the wires from one terminal of the battery to the first switch, then to the second switch, to the light bulb, and back to the battery.

Have your teacher approve your circuit before proceeding to Step 4.

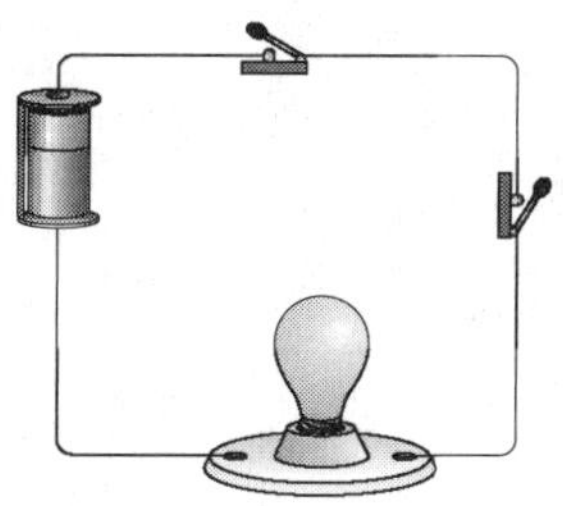

4. Close one switch wh:le the other :s open. Close the other switch. Take turns turning the light off and on with each person operating only one switch.

a) Record what happens in each case.

5. Try to keep the light on for exactly one second, then five seconds. You can estimate one second by saying "one thousand one."

a) How quickly do you think you can turn the light off after your partner turns it on? The time the bulb is lit is your response time. Record an estimate of your response time in your log.

ANSWERS

For You To Do

1. a) A reasonable estimate would be about half a second.

2. a) Estimates will vary. A reasonable estimate would be about half a second.

3. Students set up circuit.

4. a) The light bulb glows when both switches are closed.

5. a) Estimates will vary. Most students will be able to turn the light off in a quarter of a second or less. This is considerably less than the time it takes to move the foot from the gas pedal to the brake pedal.

Chapter 2

ANSWERS

For You To Do
(continued)

6. a) If the light bulb were replaced by a clock, the accuracy of the measurement would be improved. Students should also note that it takes a lot longer to move your foot and press a pedal than it takes to move a finger that is already posed for action. Students might suggest replacing the switch with a foot pedal.

   b) By averaging the results of a number of trials, the accuracy can be improved.

7. a) Response times will vary. Response times will probably be between 0.5 s and 0.8 s.

8. a) Expect the response times using this method to be slower than for previous trials.

9. a) Expect the distances to vary. If students are anticipating the release by closing their fingers periodically, and their fingers are close enough together, results may be as low as 2 cm. Suggest students average a number of trials to obtain a more reasonable result.

   b) If interpreting the graph of distance vs. time for a body falling from rest: $d = 1/2gt^2$ is difficult for your students, they could use the formula: $t = 0.45\sqrt{d}$.

Safety

response time. Record an estimate of your response time in your log.

6. Find your response time using the electric circuit.
   a) How could you improve the accuracy of the measurement?
   b) How would repeating the investigation improve the accuracy?

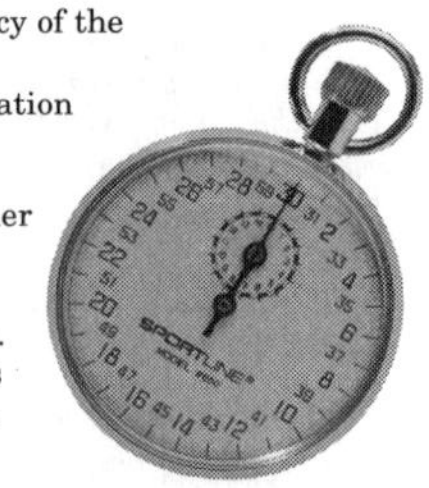

7. Test your response time with the other equipment set up in your classroom. Use a standard reaction time meter, such as one used in driver education. You will need to follow the directions for the model available in your class.
   a) Record your response time.

8. Use two stopwatches. One person starts both stopwatches at the same time, and hands one to her lab partner. When the first person stops her watch, the lab partner stops his. The difference in the two times is the response time.
   a) Record your response time.

9. Use a centimeter ruler. Hold the centimeter ruler at the top, between thumb and forefinger, with zero at the bottom. Your partner places thumb and forefinger at the lower end, but does not touch the ruler. Drop the ruler. Your partner must stop the ruler from falling by closing thumb and forefinger.

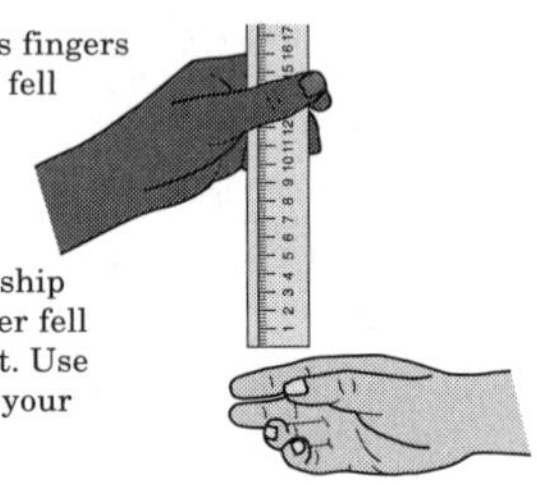

   a) The position of your partner's fingers marks the distance the ruler fell while her nervous system was responding. Record the distance in your log.
   b) The graph at the top of the next page shows the relationship between the distance the ruler fell and the time it took to stop it. Use the graph to find and record your response time.

82

Active Physics

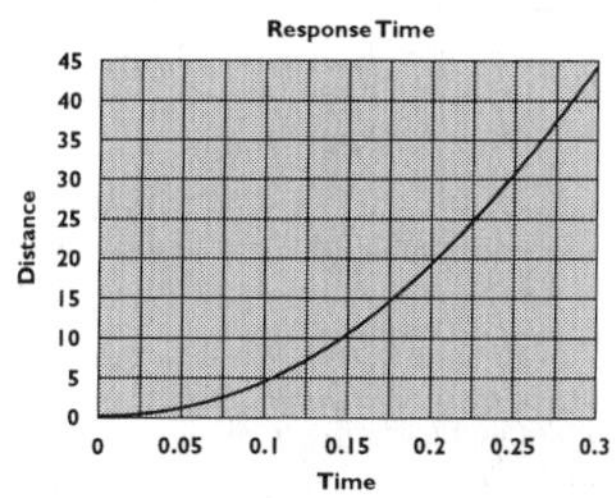

10. Compare the measures of your response obtained from each strategy.
    - a) Explain why they were not all the same.
    - b) What measure do you think best reports your response time? Why?
11. Compare the measures you obtained with those of other students.
    - a) Record the results for the fastest, slowest, and average response times.
    - b) Why do you think response times vary for people of the same age? Discuss this with your group and then record your answer.

Reflecting on the Activity and the Challenge

The amount of time people require before they can act has a direct impact on their driving. It takes time to notice a situation and more time to respond. A person who requires a second to respond to what he or she sees or hears is more likely to have an accident than someone who responds in half a second. Your **Chapter Challenge** is to design and build an improved safety device for a car. You may be able to design a car that helps drivers to stay alert and helps them become more aware of their surroundings. Anything that you can do to decrease a driver's response time will make the car safer.

ANSWERS

For You To Do

(continued)

10.a) Students may suggest that some of the methods permit them to be ready and poised to respond better than others. Estimating the point on the ruler where the fingers are located may cause errors.

b) Students may suggest that the standard reaction timer used in drivers' education might be the most accurate. Accept any reasonable explanation.

11.a) Students compare classroom data.

b) Students answers will vary. Students may consider: physiological difference in nervous systems, the amount and health of the muscles being used, the alertness of the subject, the time of day the test was conducted, use of medications such as cough syrups.

ANSWERS

Physics To Go

1. Encourage the students to continue the discussion of safe driving at home.
2. Students may find that response time for much older and very young family members may be slower. On the other hand, parents and siblings at home may be more motivated to produce excellent response times and therefore may be more alert than the subjects tested in class.
3. The length of a dollar bill is 15.7 cm. The free fall time is under 0.2 s, which makes it nearly impossible to catch the bill unless the hand is lowered or the release is anticipated. The grasping action with the thumb and forefinger is a much more familiar one than between the forefinger and middle finger. Refer the students to the graph, or provide them with a simple mathematical equation they can use to quantify their results.
4. The speed at which the race car driver is traveling requires a very quick response time. Although students will investigate the relationship between distance and time in the next activity, most will be able to answer this question from their previous experience. Encourage students to become aware of the distractions that the average driver faces, as well as the potential dangers. Compare the focused, alert race car driver encountering oil on the track, to a distracted student reacting to someone who has been pushed in front of the car "as a joke or dare."
5. Students should be assigned these questions routinely throughout the chapter. By answering this question, they have completed a part of their chapter challenge. Suggest that the students read the **Reflecting on the Activity and the Challenge** before they proceed with the answer. Answers will vary greatly. Some students who are familiar with video games may show "superhuman" responses.

Safety

Physics To Go

1. Test the response time of some of your friends and family with the centimeter ruler. Bring in the results from at least three people of various ages.
2. How do the values you found in **Question 1** compare with those you obtained in class? What do you think explains the difference, if any?
3. Take a dollar bill and fold it in half lengthwise. Have someone try to catch the dollar bill between his or her forefinger and middle finger. Most people will fail this task.
   a) Explain why it is so difficult to catch the dollar bill.
   b) Repeat the dollar bill test, letting them catch it with their thumb and forefinger.
   c) Explain why catching it with thumb and forefinger may have been easier. Try to include numbers in your answer such as the length of the dollar, the time for the dollar to fall, and average response time.
4. Does a racecar driver need a better response time than someone driving around a school? Explain your answer, giving examples of the dangers each person encounters.
5. Apply what you learned from this activity to describe how knowing your own response time can help you be a safer driver.

84

Active Physics

Activity 1 Response Time

Stretching Exercises

1. Build a device with a red light and a green light. If the red light turns on, you must press one button and measure the response time. If the green light turns on, you must press a second button and measure the response time. Have your teacher approve your design before proceeding. How do response times to this “decision” task compare with the response times measured earlier?

2. Use the graph for response time to construct a response-time ruler with the distance measurement converted to time. You can now read response times directly.

3. Do you think some groups of people have better or worse response times than others? Consider groups such as basketball players, video game players, taxi drivers, or older adults. Plan an investigation to collect data that will help you find an answer. Include in your plan the number of subjects, how you will test them, and how you will organize and interpret the data collected. Have your teacher approve your plan before you proceed.

ANSWERS

Stretching Exercises

1. This assignment may interest students who are enrolled in a technology program. It may be a way of getting a special reaction timer for use next year.

2. The students might want to mark a time scale on a strip of masking tape which could be affixed to a ruler or paint stirrer.

3. Students should be reminded to be courteous to their subjects.

Assessment: Group Work

The following rubric can be used to assess group work during the first six steps of the activity. Each member of the group can evaluate the manner in which the group worked to solve problems in the activity.

Maximum value = 12

1. Low level — indicates minimum effort or effectiveness.
2. Average — acceptable standard has been achieved, but the group could have worked more effectively if better organized.
3. Good — this rating indicates a superior effort. Although improvements might have been made, the group was on task all of the time and completed all parts of the activity.

Descriptor		Values	
1. The group worked cooperatively to design a circuit that would measure reaction times. Comments:	1	2	3
2. A plan was established before beginning and a light bulb was used to test the circuit. Comments:	1	2	3
3. The group was organized. Materials were collected and the problems were addressed by the entire group. Comments:	1	2	3
4. Data was collected and recorded in an organized fashion in data tables and in journals. Comments:	1	2	3

For use with *Safety*, Chapter 2, Activity 1: Response Time

Assessment: Scientific and Technological Thinking

Scientific and Technological Thinking can be assessed using the rubric below. Allow one mark for each check mark.

Maximum value = 10

Descriptor	Yes	No
1. A complete circuit is constructed.		
2. A light bulb is used to check for battery life and the functioning circuit.		
3. A switch is used to time reaction rates.		
4. Controls, such as the distance of the hand from the switch, are maintained throughout the experiment.		
5. Proper units are used to measure reaction times.		
6. A clock or timing device is integrated into the circuit to provide accuracy of measurement.		
7. Timing devices are tested and/or modified prior to collecting final data.		
8. Response time can be determined by using a distance vs time graph.		
9. Students can identify variables used in the experiment that would alter response time.		
10. Student is able to relate response times to the need for safe driving.		

For use with *Safety*, Chapter 2, Activity 1: Response Time

Activity 1 A

Response Time: High-Tech Alternative

FOR YOU TO DO

1. Photogates can be used along with an electronic timer to determine reaction times. Hide the first photogate within a cardboard box, so that observation and reaction times can be accurately monitored. Use the setup below. The subject observes the timer clock for the "GO" signal.

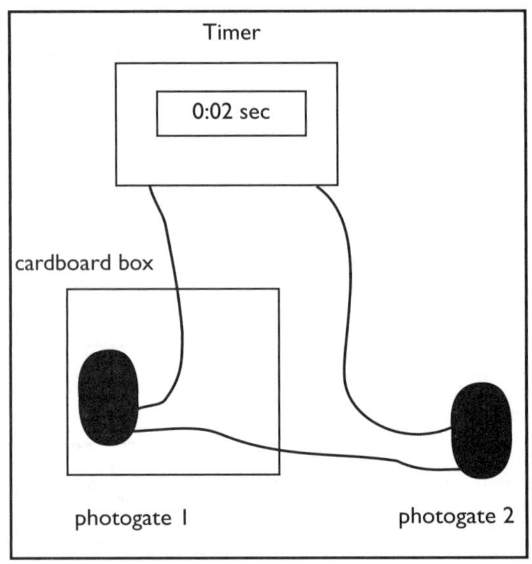

2. The tester moves his or her hand inside of the cardboard box and activates the first photocell.

3. The subject sees the clock begin to move and quickly moves his or her hand in front of the second photogate to stop the clock.

   a) Complete multiple trials and record your data. Calculate the mean time taken to stop the electronic timer.

   b) Explain why the mean time was determined.

   c) Repeat the procedure but this time use your non-dominant hand. Account for any differences in reaction rate.

   d) Why must the test subject's hand be held a specified distance from the timer?

   e) When braking, the right foot is moved from the gas to the brake. What advantage is gained from only using one foot to control both the accelerator pedal and the brake?

   f) A faster reaction time could be obtained by holding the left foot just millimeters above the brake pedal. As the left foot tramps on the brake, the right foot would leave the gas pedal. Explain why the practice of two-foot driving (using both the right and left foot) is discouraged.

For use with *Safety*, Chapter 2, Activity 1: Response Time

Activity 1 A

Response Time: High-Tech Alternative

Time Requirements

Approximately 40 minutes is required to complete the experiment.

Materials needed

- electrical clock (measuring 1/100 sec)
- 2 photogates

ANSWERS

For You To Do

3. a)-b) Small variations can be expected. Sometimes the subject may anticipate when the experimenter was about to throw the switch. Other times small distractions may have increased the reaction time. By taking an average, variables that increase and decrease reaction times can be eliminated.

c) In general, the dominant hand responds faster. The more neural circuits are used, the faster is the response time.

d) The further the hand is from the photogate, the greater the time it takes to reach the photocell.

e) You can't accelerate and brake at the same time. Not only would this increase wear on the brakes, it would tend to throw the car into a spin. By using one foot, the problem of simultaneously braking and accelerating is eliminated.

f) Simultaneous pushing down on the gas pedal and the brake would increase the braking distance. Because the engine would be pulling or pushing the car forward, while the brakes are applied — the effectiveness of the braking system would be reduced. The car is not able to do both at the same time.

ACTIVITY 2
Speed and Following Distance

Background Information

Kinematics is the study of motion. Every person will have experienced kinematics. From the moment we are able to crawl, we have a basic understanding of kinematics. As we grow and gain more experience, we are able to recognize objects as moving "fast" or "slow". We can make comparisons between the speed of a hare with the speed of a tortoise. In physics, we observe an object in motion, and then, using measurement and graphs, we are able to analyze the motion of that object.

To do this analysis, tools of measurement must be established which are appropriate for the object in motion and the speed at which it's moving. For example, a geologist who studies the movement of plates within the Earth's crust would measure the distances in inches (or cm) and the time in years or even thousands of years. When measuring the speed of an electron in a particle accelerator, we would use distances measured in meters or kilometers and times in millionths of seconds.

Understanding speed is critical in understanding motion and two tools used by scientists to achieve this understanding are mathematics, and graphical analysis. In this activity the students will be analyzing motion both mathematically and graphically.

A mathematical analysis is using the formula $v = d/t$.

Speed (symbol for speed is v) is the distance an object moves in a given time.

Average speed is the total distance traveled/total time. For example, you can travel from one city to the next in two hours, a total distance of 100 miles. Your average speed is (calculated mathematically)

$v =$ total distance/total time

$v =$ 100 miles/2 hours

$v =$ 50 mph

However, if on the return trip you had a flat tire, and spent 30 minutes fixing your tire, the total time has now changed to 2.5 h and the average speed is now 100 miles/2.5 hours or 40 mph. Instantaneous speed, on the other hand, is the speed that you are traveling at a given moment. For example on the return trip, even though your average speed was 40 mph, your instantaneous speed at a given time may have been 65 mph. Instantaneous speed is the speed at which you happen to be traveling when you look down at your speedometer.

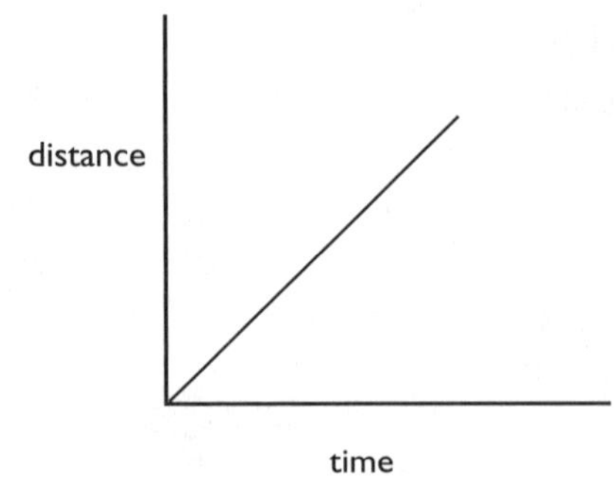

A graphical analysis of uniform motion (we will be using constant motion and uniform motion interchangeably) will involve collection of data, and then plotting that data onto a graph. Putting the information onto a distance-time graph will produce a straight line. (A straight line indicates uniform or constant motion.) The slope of that line ($\Delta d/\Delta t$) will give us the speed. (Δd (d_2 - d_1) = meters (m) (or miles) and Δt (t_2 - t_1) = seconds (s) (or hours), therefore the unit for the slope is m/s (or mph).) For distance, d_2 usually represents the final distance. In most situations d_1 is the starting point and is most often indicated by zero. Similarly, Δt represents change in time, where t_2 is the final or end time and t_1 is the initial time.

Although the following is beyond the intent of the activity, some extension is presented here for the teacher.

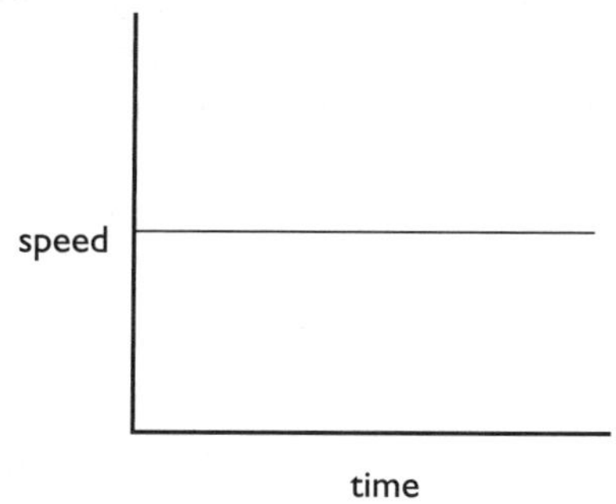

The speed-time graph of the same object can be obtained by taking the speed from the distance-time graph, at various times, and plotting them against time on a speed-time graph. The resulting line will be a horizontal straight line, which reinforces the concept of constant motion.

To find the distance an object travels while experiencing constant motion, we find the area under the graph. Area = length of side x the width of the side ($A = l \times w$). Therefore, $d = v \times t$.

Mathematically the formula $v = d/t$ will give the rearranged formula to solve for d as $d = v \times t$.

In later grades, students will move from the study of speed to the study of velocity. The concept is essentially the same, with the difference being speed is the only magnitude or how fast something is moving whereas velocity is the speed and the direction in which the object is moving.

Active-ating the Physics InfoMall

The effect of reaction time on following distance is discussed in Articles and Abstracts Attic, *The Physics Teacher,* volume 8, issue 4, "Problems for Introductory Physics," problem 49. This can be found most easily by scrolling down to near the bottom of the article and then searching up, rather than down. Included are some questions to consider about the effect of reactions time on driving.

Planning for the Activity

Time Requirements

During the class each group of students will require approximately 20 – 30 minutes to do the experiment and gather their data.

Materials Needed

For each group:

- MBL or CBL with sonic ranger

Advance Preparation and Setup

Become familiar with the sonic rangers and the software that controls their output. Instructions come with the equipment, but you should use it yourself before putting it into the classroom. It will expedite the experiment if you are aware of the near and far limits of range. Prepare a list of the keystrokes required by your software to erase one graph and get ready to gather new data. Some programs ask several questions before recording data. You should know the best responses so that there will be little delay. Check as well on an appropriate sampling rate for your machine, since too much detail will cloud the issues at the moment. The sampling rate should be relatively slow.

In preparing for the class, set up the ranger and computer ready to go. Put masking tape on the floor to mark the range and perhaps another tape to indicate the lane so that students will stay "on track."

Teaching Notes

After the students have read and reacted to the **What Do You Think?** question, perhaps enhanced with a personal story about being tailgated by a poor driver, get right to the sonic rangers to perform the exercises. If you do not have access to separate computer setups for each group of students, do this as a demonstration using student volunteers.

Make sure the computer equipment is in a secure and stable environment.

Even if you have only one setup available, it is likely that most students will remain engaged by the activity. It can be presented as a human-sized video game where the challenge is to create the straightest line, the smoothest curve, the steepest slope, etc. One student can be asked to create a graph while the others look away, after which another student can describe the motion that caused the graph. There are many variations, including having students copy graphs created by other students.

While it is fairly easy to obtain good speed-time graphs using the sonic ranger, an uneven gait or even a baggy sweatshirt can produce some irregularity. Teach the students to focus on the general trends. Students may accidentally collect data or construct graphs of displacement or acceleration instead of speed. You can use this error to your mutual advantage by asking them to explain what they see.

The graphs will quickly give students a secure understanding of the meaning of constant speed, and they can identify the different graphs almost at once. This provides preparation for algebraic analysis.

A common misconception held by students is that they go from zero to the speed they are walking or running instantly. In fact, there will always be some acceleration. Illustrate this with a set of data that has enough points to show that the first part of the graph is not straight.

Another common misconception of students that has been well documented is the confusion between distance and velocity. Look at the following pair of strobe pictures:

O   O   O   O   O
   X   X   X   X

Students may think that the X and O are traveling the same speed when they are aligned. They have the same speed when the distance between adjacent Xs is identical to the spacing between adjacent Os.

The sonic ranger has been shown to be extremely effective for students to gain an understanding of motion graphs. It is well worth the equipment investment.

The definition of velocity as the change in distance divided by the change in time can be written as an algebraic equation and can be solved for any of the variables.

$d = vt$

$v = d/t$

$t = d/v$

Some students deficient in algebra skills will need some help with this. You should emphasize the following points while students are learning this relationship:

The units of distance and time and velocity should always be presented with the numbers.

Any distance and any time can be used for velocity. (Cars move at miles/hour, km/h, or m/s; glaciers move at meters per year.)

To measure velocity, you need a ruler and a stopwatch.

Average velocity should be distinguished from initial velocity and final velocity. The average velocity for a trip does not give any indication of the initial and final velocity.

Before assigning the **Physics To Go Questions 4 to 7**, work a few examples with the students. Some students will need help in manipulating the speed equation. You may wish to provide them with the equations required to calculate distance and time.

NOTES

Activity Overview

This activity should provide the student with a feel and definition for the notion of speed.

Student Objectives

Students will:

- Define speed.
- Identify constant and changing speeds.
- Interpret distance-time and speed-time graphs.
- Contrast average and instantaneous speeds.
- Calculate the distance traveled at constant speed.

ANSWERS FOR THE TEACHER ONLY

What Do You Think?

The proper interval between your car and the vehicle in front is two seconds. (See the answer for **Physics To Go, Step 8** for more information.)

Activity 2 Speed and Following Distance

GOALS

In this activity you will:

- **Define** speed.
- **Identify** constant and changing speeds.
- **Interpret** distance-time and speed-time graphs.
- **Contrast** average and instantaneous speeds.
- **Calculate** the distance traveled at constant speed.

What Do You Think?

In a rear-end collision, the driver of the car in back is always found at fault.

- **What is a safe distance between your car and the car in front of you?**
- **How do you decide?**

Record your ideas about these questions in your *Active Physics* log. Be prepared to discuss your responses with your small group and the class.

For You To Do

1. A strobe photo is a multiple-exposure photo in which a moving object is photographed at regular time intervals. The strobe photo below shows a car traveling at 30 mph.

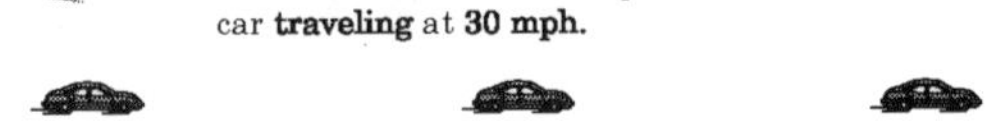

a) Copy the sketch in your log.

2. Think about the difference between the motion of a car traveling at 30 mph and one traveling at 45 mph.

a) Draw a sketch of a strobe photo, similar to the one above, of a car traveling at 45 mph.

b) Are the cars the same distance apart? Were they farther apart or closer together than at 30 mph?

c) Draw a sketch for a car traveling at 60 mph. Describe how you decided how far apart to place the cars.

3. The following sketch shows a car traveling at different speeds.

a) Copy the sketch in your log. Mark where the car is traveling fast, where it is traveling slowly, and where it is traveling at a constant speed. How did you know?

4. A sonic ranger connected to a computer will produce a graph that shows an object's motion. Use the sonic ranger setup to obtain the following graphs to print or sketch in your log.

Make sure the path of motion is clear of any hazards.

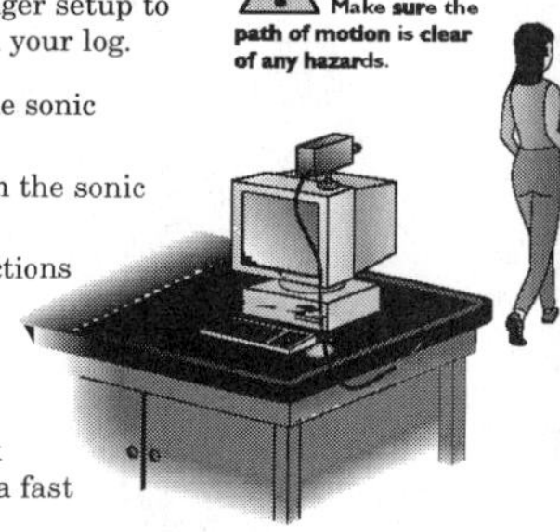

a) Sketch a graph of a person walking toward the sonic ranger at a normal speed.

b) Sketch a graph of a person walking away from the sonic ranger at a normal speed.

c) Sketch a graph of a person walking both directions at a very slow speed.

d) Sketch a graph of a person walking both directions at a fast speed.

5. Predict what the graph will look like if you walk toward the system at a slow speed and away at a fast speed. Test your prediction.

a) Record your prediction in your log.

b) Based on your measurements, how accurate was your prediction?

ANSWERS

For You To Do

1. a) If the students do not feel comfortable sketching cars, suggest that they use O or X symbols instead. Check to see that the students understand that the spaces between the cars are even.

2. a) The sketch should show the cars with larger, even gaps between them.

   b) The marks are a greater distance apart.

   c) The cars should be twice the distance apart as they were for 30 mph.

3. a) The car is traveling slowly for the first three intervals, then fast for the next two intervals, and slower and slower for the last two intervals.

4. a)-d) The slope will be positive for motion away from the detector and negative for motion towards the detector.

5. a)-b) The prediction of a steeper slope for high speed than for low speed will be confirmed.

ANSWERS

For You To Do
(continued)

6. a) For motion away from the sonic ranger, the total distance will be the maximum y-value of the graph. For motion towards the ranger the final y-value must be subtracted from the starting y-value.

   b) The total time is measured on the x-axis for the points corresponding to those for distance.

   c) Students divide the value they obtained in **6. a)** by the one in **6. b)** to obtain their average speed. Check that the units used are m/s.

6. Repeat any of the motions in **Steps 4** or **5** for a more thorough analysis.
   a) From your graph, determine the total distance you walked.
   b) How long did it take to walk that distance?
   c) Divide the distance you walked by the time it took. This is your average speed in meters per second (m/s).

Physics Words
speed: the change in distance per unit time; speed is a scalar, it has no direction.

PHYSICS TALK

Speed

The relationship between **speed**, distance, and time can be written as:

$$\text{Speed} = \frac{\text{Distance traveled}}{\text{Time elapsed}}$$

If your speed is changing, this gives your average speed. Using symbols, the same relationship can be written as:

$$v_{av} = \frac{\Delta d}{\Delta t}$$

where v_{av} is average speed
Δd is change in distance or displacement.
Δt is change in time or elapsed time.

Sample Problem 1

You drive 400 mi. in 8 h. What is your average speed?

Strategy: You can use the equation for average speed.

$$v_{av} = \frac{\Delta d}{\Delta t}$$

Activity 2 Speed and Following Distance

Givens:

$$\Delta d = 400 \text{ mi.}$$

$$v_{av} = \frac{\Delta d}{\Delta t} = \frac{400 \text{ mi.}}{8 \text{ h}} = 50 \text{ mph (miles per hour)}$$

Your average speed is 50 mph. This does not tell you the fastest or slowest speed that you traveled. This also does not tell you how fast you were going at any particular moment.

Sample Problem 2

Elisha would like to ride her bike to the beach. From car trips with her parents, she knows that the distance is 30 mi. She thinks she can keep up an average speed of about 15 mph. How long will it take her to ride to the beach?

Strategy: You can use the equation for average speed.

$$v_{av} = \frac{\Delta d}{\Delta t}$$

However, you will first need to rearrange the terms to solve for elapsed time.

$$\Delta t = \frac{\Delta d}{v_{av}}$$

Solution:

$$\Delta t = \frac{\Delta d}{v_{av}} = \frac{30 \text{ mi.}}{15 \text{ mph}} = 2 \text{ h}$$

Chapter 2

FOR YOU TO READ

Representing Motion

One way to show motion is with the use of strobe photos. A strobe photo is a multiple-exposure photo in which a moving object is photographed at regular time intervals. The sketches you used in **Steps 1, 2,** and **3** in **For You To Do** are similar to strobe photos. Here is a strobe photo of a car traveling at the average speed of 50 mph.

Another way to represent motion is with graphs. The graph below shows a car traveling at the average speed of 50 mph.

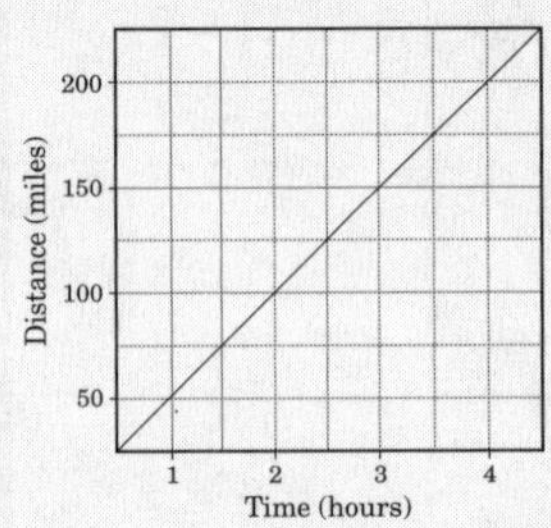

Kilometers and Miles

Highway signs and speed limits in the USA are given in miles per hour, or mph. Almost every other country in the world uses kilometers to measure distances. A kilometer is a little less than two-thirds of a mile. Kilometers per hour (km/h) is used to measure highway driving speed. Shorter distances, such as for track events and experiments in a science class, are measured in meters per second, m/s.

You will use mph when working with driving speeds, but meters per second for data you collect in class. The good news is that you do not need to change measures between systems. It is important to be able to understand and compare measures.

To help you relate the speeds with which you are comfortable to the data you collect in class, the chart below gives *approximate* comparisons.

School zone	25 mph	40 km/h	11 m/s
Residential street	35 mph	55 km/h	16 m/s
Suburban interstate	55 mph	90 km/h	25 m/s
Rural interstate	75 mph	120 km/h	34 m/s

Activity 2 Speed and Following Distance

Reflecting on the Activity and the Challenge

You now know how reaction time and speed affect the distance required to stop. You should be able to make a good argument about tailgating as part of the **Chapter Challenge**. If your car can be designed to limit tailgating or to alert drivers to the dangers of tailgating, it will add to improved safety.

Physics To Go

1. Describe the motion of each car moving to the right. The strobe pictures were taken every 3 s (seconds).

a) 

b) 

2. Sketch strobe pictures of the following:
   a) A car starting at rest and reaching a final constant speed.
   b) A car traveling at a constant speed then coming to a stop.

3. For each graph below, describe the motion of the car:

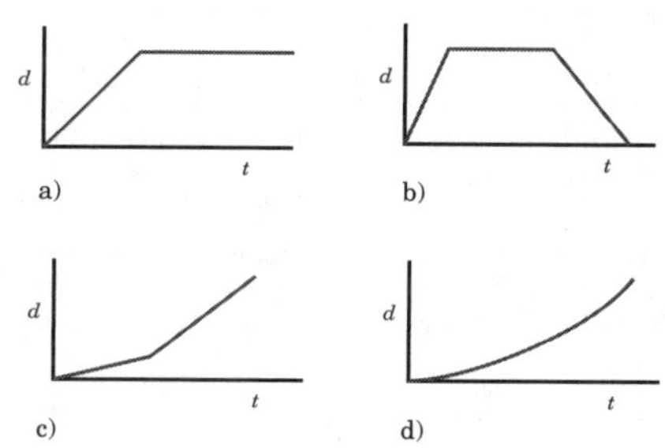

Answers

Physics To Go

1. a) The car is moving at a constant speed every three seconds.
   b) The car speeds up, slows down, speeds up, and then slows again.

2. a) Answers will vary, but there should be a measurable increase in the distance of the first few sketches, and then the same distance while the car travels at a constant speed.
   E.g.: ΔΔ Δ Δ Δ Δ Δ Δ Δ Δ
   b) The reverse of the above situation, moving at a constant speed then slowing down.
   E.g.: Δ Δ Δ Δ Δ Δ Δ Δ Δ ΔΔ

3. a) A car travels at a constant speed, and then stops.
   b) A car travels at a rapid, constant speed, stops, and then returns at a slower constant speed.
   c) A car travels at a slow constant speed, and then increases to travel at a faster constant speed.
   d) A car accelerates as it travels.

   Note: If students have difficulty interpreting this graph, suggest they return to the question after completing the next activity.

Chapter 2

ANSWERS

Physics To Go

(continued)

4. Using the formula $v = d/t$, therefore;
 $d = vt$, then;
 $d = 110 m/s \times 20 s$
 $d = 2200 m$ or 2.2 km
 The driver traveled 2200 m, or 2.2 km.

5. a) $v = d/t$ $v = 215$ miles/4.5 hours
 $v = 48$ mph. Her average speed was about 48 mph.

 b) This question cannot be answered with the data provided. We do not know her instantaneous speed at any one time. We can only assume that she was *probably* doing about 48 mph.

6. $v = d/t$ $v = 5$ miles/2 hours
 $v = 2.5$ mph You would need to keep up an average speed of 2.5 mph.

7. a) At 55 mph, the car will be moving at about 25 m/s, therefore to find the distance,
 $v = d/t$, then $d = vt$
 $d = 25$ m/s x response time (response times will vary according to student)

 For example, for a response time of 0.8 s: $d = vt$
 $d = 25$ m/s x 0.8 s $d = 20$ m

 b) Assume a response time of 0.8 s.At 40 mph, the car will be moving at about 19 m/s, therefore to find the distance,
 $v = d/t$, then $d = vt$
 $d = 19$ m/s x 0.8 s $d = 15$ m
 A greater speed produces a greater stopping distance with the same reaction time.

 c) You will travel double the distance.

8. Since distance traveled is directly proportional to time for constant speed, the two-second rule should work at any speed, and the distance can be described by the travel time. The premise for this rule is that the two cars have similar braking ability. The lead vehicle does not "stop on a dime." The two-second distance is not an adequate stopping distance, it is meant to cover the driver's response time and the time that the lead car has slowed before the driver's car begins to slow. Exceptions can be made for poor road, tire, or brake conditions.

9. Students' answers will vary depending on their response time. For a response time of 0.8s:
 a) 9 m b) 20 m c) 27 m

Safety

4. A racecar driver travels at 110 m/s (that's almost 250 mph) for 20 s. How far has the driver traveled?

5. A salesperson drove the 215 miles from New York City to Washington, DC, in $4\frac{1}{2}$ hours.
 a) What was her average speed?
 b) How fast was she going when she passed through Baltimore?

6. If you planned to walk to a park that was 5 miles away, what average speed would you have to keep up to arrive in 2 hours?

7. Use your average response time from **Activity 1** to answer the following:
 a) How far does your car travel in meters during your response time if you are moving at 55 mph (25 m/s)?
 b) How far does your car travel during your response time if you are moving at 35 mph (16 m/s)? How does the distance compare with the distance at 55 mph?
 c) Suppose you are very tired and your response time is doubled. How far would you travel at 55 mph during your response time?

8. According to traffic experts, the proper following distance you should leave between your car and the vehicle in front of you is two seconds. As the vehicle in front of you passes a fixed point, say to yourself "one thousand one, one thousand two." Your car should reach the point as you complete the phrase. How can the experts be sure? Isn't two seconds a measure of time? Will two seconds be safe on the interstate highway?

9. You calculated the distance your car would move during your response time. Use that information to determine a safe following distance at:
 a) 25 mph
 b) 55 mph
 c) 75 mph

Activity 2 Speed and Following Distance

10. Apply what you learned in this activity to write a convincing argument that describes why following a car too closely (tailgating) is dangerous. Include the factors you would use to decide how close counts as "tailgating."

Stretching Exercises

Measure a distance of about 100 m. You can use a football field or get a long tape or trundle wheel to measure a similar distance. You also need a watch capable of measuring seconds. Determine your average speed traveling that distance for each of the following:

a) a slow walk
b) a fast walk
c) running
d) on a bicycle
e) another method of your choice

ANSWERS

Physics To Go

(continued)

10. Students answers will vary. Tailgating increases the potential hazards of driving. During the time from when you see the brake light of the lead car flash on until you hit the brake and begin to slow, your car is still traveling at top speed while the lead car has been slowing down from the time the light first flashed. If the two cars have equivalent braking ability, there had better be an extra distance equal to the product of your speed and response time between the two cars. Since safe driving means minimizing risks, following distances must be commensurate with speed and response time, in addition to road conditions.

Student answers could also indicate that road conditions will increase the stopping time, physical conditions of the driver will increase the response time, distractions in and outside the car will increase the response time, etc. Also, the condition of the car in front, which is not in the control of the person tailgating, must be taken into consideration as an intangible. This could be that the car in front has faulty tail lights, or brakes that are very good, or that the driver in front is slowing down using the engine (gearing down) rather than brakes.

Assessment: Graphing Skills

The following assessment rubric provides insight into the attainment of graphing skills and monitors communication by way of mathematical expression. Place a check mark (√) in the appropriate box. Two check marks will be required for one point. A sample conversion scale is provided below the chart.

Descriptor	Yes	No
Analysis and Communication • data is recorded in an organized data table • manipulated and responding variables are identified in data table • appropriate units are recorded for distance (meters), time (seconds) and speed (m/sec) • multiple trials are used and averages are calculated • average speed (*v*) is calculated from formula *d/t* • distance or time can be calculated from v = *d/t* • distance can be calculated by pacing • student is able to explain why increasing the speed will require greater distance between cars to allow for reaction time		
Graphing Skills • graph has a title • the *x*-axis and *y*-axis are clearly labeled • units of measurement are provided for distance and time • manipulated variable (time) is plotted along the *x*-axis • responding variable (distance) is plotted along the *y*-axis • *x*-axis and *y*-axis are drawn to proper scale • student is able to plot distance/time coordinates • a best fit line is used to connect coordinates for a line graph • distance and time relationships taken from the sonic ranger can be interpreted from computer-generated graph • distance traveled can be determined from graphs where a constant velocity is provided • time of travel can be determined from graphs where a constant velocity is provided • average speed can be determined from distance/time graph by calculating the slope of the line		

Conversion

20 check marks = 10 points

18 + check marks = 9 points

16 + check marks = 8 points, etc.

For use with *Safety*, Chapter 2, Activity 2: Speed and Following Distance

NOTES

ACTIVITY 3
Accidents

Background Information

Most of the physics involved with this chapter on *Safety*, involves an understanding of Inertia and Momentum.

Symbols

v = speed	m = mass (kg)
d = distance	F_{net} = acceleration force
t = time	F_A = applied force
a = acceleration	F_f = force of friction
F = force (N)	

Newton's First Law of Motion states (Inertia) that an object in motion or at rest will remain in motion or at rest unless acted upon by an outside unbalanced force. An object at rest staying at rest is fairly obvious for students to understand. Even understanding that an object in motion remaining in motion should be easily understood once an understanding that friction is acting on things on Earth in one form or another.

Therefore, an object in motion, say an automobile, will continue in motion. However, students will observe that the automobile slows down (decelerates). Enter into a discussion on what slows the automobile. Friction (in the engine, axles, wheels, transmission, and tires on the road) is the outside unbalanced force acting on the automobile which stops it from moving at a constant speed. Therefore in order to keep it moving at a constant speed, you must be applying a force to it.
F_{net} (the accelerating force)
= F_A (the force applied by the engine) +
F_f (force of friction always acting opposite to the direction of motion).

$$F_{net} = F_A + F_f$$

When there is no acceleration (therefore, constant motion), there is no net force. This means that the force applied is the same as the force of friction; only opposite in direction. As long as the force applied and the force of friction are the same magnitude but opposite in direction, the object will continue to move at a constant speed.

Safety, then is a discussion on how inertia affects the movement of a body in an automobile. While the automobile is moving, everything and everyone in the automobile are moving at the same speed as the automobile. When the automobile stops, everything that is attached to the automobile (bumper, seats, steering wheel, etc.) are stopping or experiencing the deceleration as well. However, anything not attached to the automobile, (people, dogs, tape cases, hockey sticks, etc.) will continue to move according to Newton's First Law. In this chapter, we will be doing an analysis of the inertia of the objects inside a vehicle and how to prevent injury.

Active-ating the Physics InfoMall

The articles mentioned above are good for this activity. One of the safety devices mentioned in this activity is the air bag. A quick search for "air bags" found several interesting hits, including "Resource letter PE-1: Physics and the environment, *American Journal of Physics*, vol. 42, 267-273 (1974).

Planning for the Activity

Time Requirements

This activity is centered primarily on reading and answering the questions in **For You To Do**. After the students have answered the questions, there should be a class discussion, to investigate the understanding students have now about accidents and to help the students to open their minds to the seriousness of accidents. As the majority of students are entering the most dangerous time of their lives, in terms of learning to drive, and being involved in accidents, the discussion should be serious. Allow more time if a film and a discussion are added to the lesson.

Materials Needed

For each group:

- brochures on auto safety features

Teaching Notes

If students have a tough time with the discussion from **What Do You Think?** you may try using brainstorming on these activities. See **Assessment** for a possible rubric for this activity.

While discussing accidents, be aware that some students may have already been in a serious accident, or know of someone close to them, who has been injured or killed in an accident. Be sensitive to the student who sits quietly and doesn't want to participate in the discussion.

Activity Overview

This activity centers around the students own experiences with safety in transportation. The students will be exploring their own ideas, misconceptions and evaluating quantitatively their ideas in the "test".

Student Objectives

Students will:

- Evaluate their own understandings of safety.
- Evaluate the safety features on selected vehicles.
- Compare and contrast the safety features on selected vehicles.
- Identify safety features in selected vehicles.
- Identify safety features required for other modes of transportation (in-line skates, skate boards, cycling, etc.).

Activity 3 Accidents

GOALS

In this activity you will:

- Evaluate your own understandings of safety.
- Evaluate the safety features on selected vehicles.
- Compare and contrast the safety features on selected vehicles.
- Identify safety features in selected vehicles.
- Identify safety features required for other modes of transportation (in-line skates, skateboards, cycling, etc.).

What Do You Think?

Chances are you will not be able to avoid being in an accident at some time in the future.

- **How can you protect yourself from serious injury, or even death, should an accident occur?**
- **What do you think is the greatest danger to you or the people in an accident?**

Record your ideas about these questions in your *Active Physics* log. Be prepared to discuss your responses with your small group and the class.

ANSWERS FOR THE TEACHER ONLY

What Do You Think?

Students answers will vary. Most students will come up with the obvious – seat belts, air bags, roll cages, etc. However, some students may come up with ideas, which will convey their misconceptions of accidents, such as, bend over, and get into a ball, hang on to the dashboard.

Again, students will relay the most obvious, flying through the windshield, automobile rolling over you if you get thrown out, etc.; be prepared for answers such as being burned in the automobile, the automobile explodes on impact; look for sensible answers, that will steer the students toward an understanding of the physics, i.e., inertia, forces, momentum, without having to talk about the details of the "physics". Some of the answers which students may not think of might include: other people in the automobile flying around during the collision, being knocked unconscious as you enter a lake or slough, objects not tied or secured flying around inside the automobile.

Try not to limit the discussion to only cars and trucks. Expand if appropriate to motorcycles, snowmobiles, motorized tricycles and quads. Also, if it is brought up by the students, a discussion about the safety of bicycles could develop.

For You To Do

1. Many people think that they know the risks involved with day-to-day transportation. The "test" below will check your knowledge of automobile accidents. The statements are organized in a true and false format. Record a T in your log for each statement you believe is true and an F if you believe the statement is false. Your teacher will supply the correct answers for discussion at the end of the activity.

a) More people die because of cancer than automobile accidents.

b) Your chances of surviving a collision improve if you are thrown from the car.

c) The fatality rate of motorcycle accidents is less than that of cars.

d) A large number of people who are belted into their cars are killed in a burning or submerged car.

e) If you don't have a child restraint seat, you should place the child in your seat belt with you.

f) You can react fast enough during an accident to brace yourself in the car seat.

g) Most people die in traffic accidents during long trips.

h) A person not wearing a seat belt in your car poses a hazard to you.

i) Traffic accidents occur most often on Monday mornings.

j) Male drivers between the ages of 16 and 19 are most likely to be involved in traffic accidents.

k) Casualty collisions are most frequent during the winter months due to snow and ice.

l) More pedestrians than drivers are killed by cars.

m) The greatest number of roadway fatalities can be attributed to poor driving conditions.

n) The greatest number of females involved in traffic accidents are between the ages of 16 and 20.

o) Unrestrained occupant casualties are more likely to be young adults between the ages of 16 and 19.

Answers

For You To Do

1. a) False, more people die because of automobile accidents.

b) False, 30% of the people who die in car crashes die because they are thrown. One report indicates that at least 50% of these fatalities need not have occurred.

c) False, motorcycles are less common, but they are much more dangerous.

d) False, very few people are killed because of burning or drowning. One study indicates that it is less than 0.1%.

e) False, the child can be sandwiched by your mass. The abdomen of small children is particularly vulnerable to impact.

f) False, you don't have enough time to react even at 50 km/h.

g) False, most people die within 40 kilometers of home.

h) True, the person becomes a projectile once the car stops.

i) False, most traffic accidents occur on Friday afternoon or evening traffic.

j) True, that accounts for the higher insurance rates.

k) False, the months May, June, and August have the greatest number of casualty collisions.

l) False, pedestrians only account for 6.1% of traffic fatalities.

m) False, more accidents take place when road conditions are dry.

n) True, young female drivers, like their male counterparts, are responsible for the greatest number of accidents. Inexperience may be the major contributing factor.

o) True.

2. Calculate your score. Give yourself two points for a correct answer, and subtract one point for an incorrect answer. You might want to match your score against the descriptors given below.

21 to 30 points: Expert Analyst

14 to 20 points: Assistant Analyst

9 to 13 points: Novice Analyst

8 points and below: Myth Believer

a) Record your score in your log. Were you surprised about the extent of your knowledge? Some of the reasons behind these facts will be better understood as you continue to travel through this chapter.

Obtain permission from the cars' owners before proceeding.

3. Survey at least three different cars for safety features. The list on the next page will allow you to evaluate the safety features of each of the cars. Place a check mark in the appropriate square.

Number 1 indicates very poor or nonexistent, 2 is minimum standard, 3 is average, 4 is good, and 5 is very good.

For example, when rating air bags: a car with no air bags could be given a 1 rating, a car with only a driver-side air bag a 2, a car with driver and passenger side air bags a 3, a car with slow release driver and passenger-side air bags a 4, and a car which includes side-door air bags to the previous list a 5. You may add additional safety features not identified in the chart. Many additional features can be added!

a) Copy and complete the table in your log.

b) Which car would you evaluate as being safest?

ANSWERS

For You To Do *(continued)*

2. a) Students scores will vary. A short discussion can follow with the expectation that the students will be able to better understand the physics of the accidents after the next few chapters.

3. a) Students should be able to come up with at least five more safety features. These might include side impact beams in doors, shoulder belts for all outboard seats, laminated windshield glass, ABS brakes, tempered shatterproof glass, etc.

b) Students answers will vary. Look for reasonable arguments for their evaluation. They should include factors such as speed, forces on the body, analysis of the effect on the body from different kinds of collisions.

Car Tested: Make and Model ______	Year ______				
Safety Feature	Rating				
Padded front seats	1	2	3	4	5
Padded roof frame	1	2	3	4	5
Head rests	1	2	3	4	5
Knee protection	1	2	3	4	5
Anti-daze rear-view mirror that brakes on impact	1	2	3	4	5
Child proof safety locks on rear doors	1	2	3	4	5
Padded console	1	2	3	4	5
Padded sun visor	1	2	3	4	5
Padded doors and arm rests	1	2	3	4	5
Steering wheel with padded rim and hub	1	2	3	4	5
Padded gear level					
Padded door pillars					
Air bags					

Reflecting on the Activity and the Challenge

Serious injuries in an automobile accident have many causes. If there are no restraints or safety devices in a vehicle, or if the vehicle is not constructed to absorb any of the energy of the collision, even a minor collision can cause serious injury. Until the early 1960s, automobile design and construction did not even consider passenger safety. The general belief was that a heavy car was a safe car. While there is some truth to that statement, today's lighter cars are far safer than the "tanks" of the past.

The safety survey may have provided ideas for constructing a prototype of a safety system used for transportation. If it has, write down ideas in your log that have been generated from this activity.

Chapter 2

ANSWERS

Physics To Go

1. Looking at the chart from page 97, these are possible answers-
   - Padded front seats F, R
   - Padded roof frame T
   - Head rests F, R
   - Knee protection F, S
   - Anti-daze rear-view mirror F, S
   - Child-proof safety locks
   - Padded console F, R, S, T
   - Padded sun visor F, S, T
   - Padded doors and arm rests S
   - Steering wheel-padded F
   - Padded gear level F, S
   - Padded door pillars S, T
   - Air bags F, S, T

2. Safety features for cycling could be: padded handle bars, better brakes, padded seat and top bar, lots of lighting, and reflectors, safety flags for better visibility, safety training, like defensive driving, helmet, padding for knees, hands and elbows and shoulders.

3. Many of the same for bikes, but emphasis on padding and helmets.

4. As above, with an emphasis on padding again.

5. Students answers will vary. Look for a thorough evaluation, using the previous list, and emphasize to the students that the discussion with the owner of the vehicle is as important as the evaluation.

Safety

Physics To Go

1. Review and list all the safety features found in today's new cars. As you compile your list, write next to each safety feature one or more of the following designations:

   F: effective in a front-end collision.

   R: effective in a rear-end collision.

   S: effective in a collision where the car is struck on the side.

   T: effective when the car rolls over or turns over onto its roof.
2. Make a list of safety features that could be used for cycling.
3. Make a list of safety features that could be used for in-line skating.
4. Make a list of safety features that could be used for skateboarding.
5. Ask family members or friends if you may evaluate their car. Discuss and explain your evaluation to the car owners. Record your evaluation and their response in your log.

Stretching Exercises

1. Read a consumer report on car safety. Are any cars on the road particularly unsafe?
2. Collect brochures from various automobile dealers. What new safety features are presented in the brochures? How much of the advertising is devoted to safety?

98

ANSWERS

Stretching Exercises

1. There may be many cars on the road that have elements which are unsafe. Have the students evaluate which ones constitute true safety breaches, and which are more along cosmetic lines. For example, electronic fuel filters which may ignite, or dashboards which spontaneously start on fire, or brakes which fail would constitute major safety hazards. Radios that fail or brakes that squeal when stopping, or cars that prematurely rust, are more cosmetic, and don't constitute a safety hazard.

2. Students answers will vary. Students will likely come up with the observation that a great deal of advertising has to do with driver safety.

Assessment: Participation in Discussion

The following is an **Assessment Rubric**, designed for informal feedback to the students on their discussions related to their personal experiences with safety in a vehicle. While they are discussing this, the students may want to brainstorm on some of the safety features with which they are familiar.

Descriptor	most of the time	some of the time	almost never	comments
shows interest				
stays on task				
asks questions related to topic				
listens to other students' ideas				
shows cooperation in group brainstorming				
provides leadership in group activity				
demonstrates tolerance of others' viewpoints				

For use with *Safety*, Chapter 2, Activity 3: Accidents

ACTIVITY 4
Life (and Death) before Seat Belts

Background Information

Newton's First Law of Motion (Inertia) gives us an understanding of constant motion (or rest). Therefore, when an object such as a car is stopping, there is a change in velocity (previously introduced as speed). In other words there is an acceleration. Analyzing a collision involves examining the changes in velocity of a car. The crumple zone refers to the stopping distance as the car is pushed in, while stopping. Cars designed today are built with crumple zones. This is to increase the distance a car takes to stop, therefore reducing the acceleration and ultimately the force being applied to your body.

Increasing the crumple zone will affect the acceleration of the people inside the car, initially using the average velocity ($v_{ave} = d/t$). Average velocity is the sum of the two velocities divided by 2

$$v_{ave} = (v_1 + v_2)/2.$$

$$v_{ave} = (40 \text{ m/s} + 0 \text{ m/s})/2$$

$$v_{ave} = 20 \text{ m/s}$$

Find the time required to stop (crumple zone of the vehicle),

$$t = d/\ v_{ave}$$

$$t = 0.50 \text{ s}/20 \text{ m/s}$$

$$t = 0.025 \text{ s}$$

and substitute t into the equation for acceleration

$$a = \Delta v/\Delta t$$

$$a = 40 \text{ m/s} / 0.025 \text{ s}$$

$$a = 1600 \text{ m/s}^2$$

This is clearly enough to rip out the aorta.

Another way to emphasize the impact of this force, is that a 10 m/s² acceleration is roughly equivalent to holding 1 kg (2.2 pounds) mass (you need 10 N of force to lift 1 kg of mass). Therefore, if we look at the above acceleration, you could picture a 160 kg (352 pound) person being held up by the aorta...ouch that would hurt!

This model should help students realize that the external safety devices we use help, but there is not very much we can do to improve on the safety devices that are built into our own bodies.

Factors which may enter into the discussion, regarding race car drivers traveling at 200 mph, may be the fact that most race car drivers are in very good physical shape, and are generally younger.

One other way to help the students understand that the internal organs undergo acceleration, is how they feel while driving in a car over a hilly road. The feeling often expressed is that of your stomach rising and falling at the crests and valleys.

In the **FYTD** activity, the students will be releasing the carts at different heights on the ramp. If the carts experience an almost frictionless surface, the speed at which the cart hits the barrier can be determined using the height from which the cart is released.

Conservation of Energy Theory (First Law of Thermodynamics) states that energy can be neither created nor destroyed, only transferred from one form to another. Therefore, the energy at the top of the ramp (gravitational energy) $E_p = mgh$ (where m is the mass of the cart and clay, g is gravity (9.81 m/s²) and h is the height from which the cart is released) will be the same as the kinetic energy ($E_k = 1/2\ mv^2$ where m is the mass, v is the velocity) at the bottom.

$$E_p = E_k$$

$$mgh = 1/2mv^2$$

masses cancel out

$$gh = 1/2v^2$$

$$v = \sqrt{2gh}$$

For example: the cart is released at a height of 0.50 m. Therefore the speed of the cart as it reaches the bottom of the ramp would be

$$v = \sqrt{2gh}$$

$$v = \sqrt{2 \times 9.81 \text{ m/s}^2 \times 0.50 \text{ m}}$$

$$v = 3.1 \text{ m/s}$$

Active-ating the Physics InfoMall

The title of this activity leads naturally to a search for "seat belt*". The very first result from this search is from *The Fascination of Physics*, in the Textbook Trove, which says "Automobile accidents involve two collisions. The first occurs when the automobile strikes an object, such as a telephone pole. The pole provides the force needed to change the car's momentum, eventually bringing it to rest. A second collision, which occurs shortly after the first, involves the passengers. If they are not in some way attached to the car, the passengers do not experience the force exerted by the telephone pole. The car may stop, but the passengers continue moving forward at a constant velocity. According to Newton's First Law, their forward motion will continue until they experience a force. Unfortunately, this force is usually exerted by the dashboard or windshield, and serious injuries result." This can be compared to the **For You To Read** passage, which goes a little further by looking at three collisions. You should do this search and read the results.

The second hit from this search is the epilogue to *The Fascination of Physics*, and mentions that "The benefits of not using seat belts are the saving of a few second in buckling and unbuckling, a slight increase in the freedom of movement inside a car, and some psychological or emotional benefits that are difficult to define. The benefits and the belief that the probability of an accident is small convince most people to sit on top of their seat belts." This applies directly to **Question 2** of the **Stretching Exercise**.

A little further down the list from this "seat belt*" search is an article that you might not otherwise find - "The car, the soft drink can, and the brick wall," from *The Physics Teacher*, vol. 13, (1975). This article was written by a co-PI for the InfoMall. It describes an experiment very similar to this activity, but with some variations.

A big concept in this activity is the concept of force. Students' understanding of this concept has been studied extensively. An InfoMall search using "force" AND "misconception*" in only the Articles and Abstracts Attic produced many great references. The first such hit is the article containing the Force Concept Inventory. The second is "Common sense concepts about motion," *American Journal of Physics*, vol. 53, issue 11, 1985 in which it is mentioned that "(a) On the pretest (post-test), 47% (20%) of the students showed, at least once, a belief that under no net force, an object slows down. However, only 1% (0%) maintained that belief across similar tasks. (b) About 66% (54%) of the students held, at least once, the belief that under a constant force an object moves at constant speed. However, only 2% (1%) held that belief consistently." More results are reported in this article.

The third hit in this search is "Physics that textbook writers usually get wrong," in *The Physics Teacher*, vol. 30, issue 7, 1992. This article is good reading for any introductory physics teacher. The list of hits from this search is long. In fact, it had to be limited to just the Articles and Abstracts Attic to prevent the "Too many hits" warning. If you search the rest of the CD-ROM, you will find many other great hits, such as this quote from Chapter 3 of Arons' *A Guide to Introductory Physics Teaching: Elementary Dynamics*: "In the study of physics, the Law of Inertia and the concept of force have, historically, been two of the most formidable stumbling blocks for students, and, as of the present time, more cognitive research has been done in this area than in any other."

Speed is one of the first concepts introduced in introductory physics. It is also one that causes students problems. If you search the InfoMall for "student difficult*" OR "student understand*" you will find several articles that deal with research into how students learn fundamental concepts in physics. Some of these are "Investigation of student understanding of the concept of velocity in one dimension," *American Journal of Physics*, vol. 48, issue 12 (see "Diagnosis and remediation of an alternative conception of velocity using a microcomputer program," *American Journal of Physics*, vol. 53, issue 7 for a discussion of this); "Research and computer-based instruction: Opportunity for interaction," *American Journal of Physics*, vol. 58, issue 5 (if you look at other articles in this volume, you will find "Learning motion concepts using real-time microcomputer-based laboratory tools," *American Journal of Physics*, vol. 58, issue 9); and more.

These last two articles make specific mention of the use of computers in teaching physics. This is a trend that is gaining strength and shows great promise. Look for more such articles in journals that are newer than the CD-ROM. As a starting point, you can always look in the annual indices of the physics journals and look under the names of the authors of the articles mentioned above.

Note that **Physics To Go Step 1** mentions curved motion. This topic was discussed elswhere in this book, where we found that you can give students a feel for curved motion with *The Physics Teacher*, volume 21, issue 3, "People Demos." Curved motion often brings up "centrifugal force," which is discussed in *Physics Education* (in the Articles and Abstracts Attic), issue 3, "Centrifugal force: fact or fiction," by Michael D. Savage and Julian S. Williams. See also Robert P. Bauman, "What is centrifugal force?," *The Physics Teacher*, vol. 18, number 7. Another good reference for centripetal and centrifugal forces is Arnold Arons' book (found in the Book Basement) *A Guide to Introductory Physics Teaching*, Motion in Two Dimensions, sections 4.9 to 4.11. These can all be found with simple searches.

Planning for the Activity

Time Requirements

- At least one class period (40 – 50 minutes). Allow extra time for variations on their molded clay figures.

Materials Needed

For each group:

- camcorder on tripod
- clay, modeling, 1/2 lb.
- dynamics carts
- starting ramp for lab cart
- VCR having single frame advance mode
- video monitor

Advance Preparation and Setup

This experiment requires students to crash a loaded cart into a wall, or other suitable barrier. Test out different barriers (walls, desks, bricks, homemade structures) prior to laboratory day.

Teaching Notes

You might want to show the *Active Physics Transportation* Content Video showing collisions in which a dummy is thrown forward.

The students begin by forming a clay figure of a body with a relatively large head. With the figure seated on the lab cart, allow the cart to be released from various heights or various angles on the ramp. This will simulate the vehicle crashing into a barrier at various speeds. At high speeds, the figure should crash into the barrier head-first. At low speeds, the figure should topple head first, smashing its head into the "dashboard."

Some students may wish to analyze different scenarios. Have the students submit their plans for teacher approval, then allow them to carry out their plans. An example may be to release the cart backwards to demonstrate the rear-end collisions (to be studied later).

Be aware that the ramp needs to be secure on a desk, or the floor as the cart may fall onto the floor or on someone's toes! Remind the students to place down newspaper or some drop cloths while creating their figure.

Students may have the notion that if they were in an accident if they got their arms up in time they would be able to protect themselves. Review with them reaction times, and have them try to bring their hands up to the dashboard level in that time. To further emphasize the danger in this thinking refer to the **Additional Activities A** following this activity in the Teacher's Edition.

Activity 4 Life (and Death) before Seat Belts

Activity 4 Life (and Death) before Seat Belts

GOALS

In this activity you will:

- **Understand Newton's First Law of Motion.**
- **Understand** the role of safety belts.
- **Identify** the three collisions in every accident.

Perform the activity outside of traffic areas. Do not obstruct paths to exits. Do not leave carts lying on the floor.

What Do You Think?

Throughout most of the country, the law requires automobile passengers to wear seat belts.

- **Should wearing a seat belt be a personal choice?**
- **What are two reasons why there should be seat belt laws and two reasons why there should not?**

Record your ideas about these questions in your *Active Physics* log. Be prepared to discuss your responses with your small group and the class.

For You To Do

1. In this activity, you will investigate car crashes where the driver or passenger does not wear seat belts. Your model car is a laboratory cart. Your model passenger is molded from a lump of soft clay. With the "passenger" in place, send the "car" at a low speed into a wall.

99

Active Physical Science

Activity Overview

This activity is an investigation into the inertia of an automobile passenger and will lead directly toward the necessity of a restraint system.

Student Objectives

Students will:

- Understand Newton's First Law of Motion.
- Understand the role of safety belts.
- Identify the three collisions in every accident.

Answers for the Teacher Only

What Do You Think?

Some students will think it should and some will think it not. It is an excellent way to open discussion

Some reasons for wearing might include: safety, decrease the risk of death, decrease the risk of serious injury, lower the total cost of health insurance due to decrease in serious accidents.

Some of the reasons for not wearing a seat belt might include: the right to choose to wear or not, might get trapped in the car, the seat belt will do more damage than the accident.

Discussion should try to center around the effects of the accident rather than a debate on the issue of personal choice. Try to bring in statistics from your local police or automobile association to emphasize the risk of injury or death will always decrease over many different accidents. There are always exceptions to the rule, but the facts are that wearing seat belts will decrease the likelihood of serious injury or death.

Chapter 2

a) Describe, in your log, what happens to the "passenger."

2. Repeat the collision at a high speed. Compare and contrast this collision with the previous one.

   a) Compare and contrast requires you to find and record at least one similarity and one difference. A better response includes more similarities and differences.

3. You can conduct a more analytical experiment by having the cart hit the wall at varying speeds. Set up a ramp on which the car can travel. Release the car on the ramp and observe as it crashes into the wall. Repeat the collision for at least two ramp heights.

   a) Record the heights of the ramp and describe the results of the collision. Describe the collision by noting the damage to the "passenger."

Physics Words

Newton's First Law of Motion: an object at rest stays at rest and an object in motion stays in motion unless acted upon by an unbalanced, external force.

inertia: the natural tendency of an object to remain at rest or to remain moving with constant speed in a straight line.

PHYSICS TALK

Newton's First Law of Motion

Newton's First Law of Motion (also called the Law of **Inertia**) is one of the foundations of physics. It states:

An object at rest stays at rest, and an object in motion stays in motion unless acted upon by a net external force.

There are three distinct parts to Newton's First Law.

Part 1 says that objects at rest stay at rest. This hardly seems surprising.

Part 2 says that objects in motion stay in motion. This may seem strange indeed. After looking at the collisions of this activity, this should seem clearer.

Part 3 says that Parts 1 and 2 are only true when no force is present.

ANSWERS

For You To Do *(continued)*

2. a) The students should note that at a slower speed, the clay model will fall head-first into the front of the cart, or hit the wall with its head. Difference at a high speed would be that the figure will continue moving at approximately the same speed as the cart was moving before the accident. The difference as to why the figure will not continue at exactly the same speed, or that at higher speeds it is more pronounced, is that there is a certain amount of friction between the figure, and the cart. It should be noted also that the greater the speed, the greater damage there is to the figure.

3. a) The higher the ramp, the faster the speed of the cart down the ramp, and the faster the figure will crash into the wall. Therefore, there should continue to be greater damage as the ramp gets higher and higher.

Activity 4 Life (and Death) before Seat Belts

FOR YOU TO READ

Three Collisions in One Accident!

Arthur C. Damask analyzes automobile accidents and deaths for insurance companies and police reports. This is how Professor Damask describes an accident:

Consider the occupants of a conveyance moving at some speed. If the conveyance strikes an object, it will rapidly decelerate to some lower speed or stop entirely; this is called the first collision. But the occupants have been moving at the same speed, and will continue to do so until they are stopped by striking the interior parts of the car (if not ejected); this is the second collision. The brain and body organs have also been moving at the same speed and will continue to do so until they are stopped by colliding with the shell of the body, i.e., the interior of the skull, the thoracic cavity, and the abdominal wall. This is called the third collision.

Newton's First Law of Motion explains the three collisions:

- First collision: the car strikes the pole; the pole exerts the force that brings the car to rest.
- Second collision: when the car stops, the body keeps moving; the structure of the car exerts the force that brings the body to rest.
- Third collision: the body stops, but the heart and brain keep moving; the body wall exerts the force that brings the heart and brain to rest.

Even with all the safety features in our automobiles, some deaths cannot be prevented. In one accident, only a single car was involved, with only the driver inside. The car failed to follow the road around a turn, and it struck a telephone pole. The seat belt and the air bag prevented any serious injuries apart from a few bruises, but the driver died. An autopsy showed that the driver's aorta had burst, at the point where it leaves the heart.

Chapter 2

ANSWERS

Physics To Go

1. • *You step on the brakes to stop your car.*

   You and the car are moving forward. The brakes apply a force to the tires and stop them from rotating. Newton's law states that an object in motion will remain in motion unless a force acts upon it. In this case, the force is friction between the ground and the tires. You remain in motion since the force that stopped the car did not stop you.

   • *You step on the accelerator to get going.*

   You and the car are stopped. The engine provides a force to turn the wheels, which in turn causes the car to move forward. Inertia will keep you still unless a force acts upon you. This force is provided by the seat back which pushes you forward at the same rate as the car.)

   • *You turn the wheel to go around a curve.*

   You are moving forward with the car. Your force causes the wheels to turn, the friction of the road on the tires produces the centripetal force necessary to turn the vehicle. Inertia causes you to remain moving in a straight line, where the car (doors, seat belt, seat, etc.) produce the centripetal force to allow you to stay in the car, and move in the curved path.

   • *You step on the brakes, and an object in the back of the car comes flying forward.*

   The object was moving with the same velocity as the car. The force which caused the car to stop, was not acting on the object. Inertia of the object kept the object moving in a straight path (until it hits the driver or the windshield).

Safety

Reflecting on the Activity and the Challenge

In this activity you discovered that an object in motion continues in motion until a force stops it. A car will stop when it hits a pole but the passenger will keep on moving. If the car and passenger have a large speed, then the passenger will continue moving with this large speed. The passenger at the large speed will experience more damage from the fast-moving cart.

Have you ever heard someone say that they can prevent an injury by bracing themselves against the crash? They can't! Restraining devices help provide support. Without a restraining system, the force of impact is either absorbed by the rib, skull, or brain.

Use Newton's First Law of Motion to describe your design. How will your safety system protect passengers from low speed and higher speed collisions?

Physics To Go

1. Describe how Newton's First Law applies to the following situations:
   • You step on the brakes to stop your car.

   (Sample answer: You and the car are moving forward. The brakes apply a force to the tires and stop them from rotating. Newton's law states that an object in motion will remain in motion unless a force acts upon it. In this case, the force is friction between the ground and the tires. You remain in motion since the force that stopped the car did not stop you.)

   • You step on the accelerator to get going.

   • You turn the wheel to go around a curve. (Hint: You keep moving in a straight line.)

   • You step on the brakes, and an object in the back of the car comes flying forward.

102

Active Physics

2. Give two more examples of how Newton's First Law applies to vehicles or people in motion.
3. According to Newton's First Law, objects in motion will continue in motion unless acted upon by a force. Using Newton's First Law, explain why a cart that rolls down a ramp eventually comes to rest.
4. The skateboard, shown in the picture to the right, strikes the curb. Draw a diagram indicating the direction in which the person moves. Use Newton's First Law to explain the direction of movement.

5. Explain, in your own words, the three collisions during a single crash as described by Professor Damask in **For You To Read**.
6. Use the diagrams below to compare the second and third collisions described by Professor Damask with the impact of a punch during a boxing match.

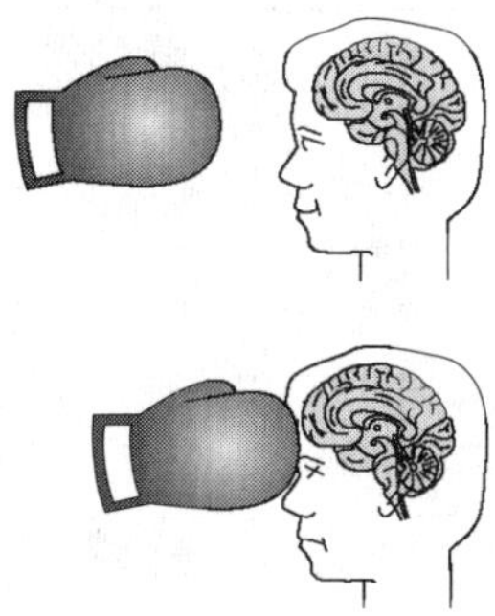

7. When was the law instituted requiring drivers to wear seat belts?

ANSWERS

Physics To Go
(continued)

2. Students' answers will vary. Look for sound physics when describing the situation. Some might include, the sensation of getting "heavier" or "lighter" as an elevator starts up or down, the problem of stopping or rounding a curve on ice, being slammed into the seat of a bus, as it accelerates from the stop.
3. Friction between the cart and the wheels, wheels and ground, and to a very small extent the air friction.
4. Inertia states that the person will continue to move in the same direction as the skateboard, thus indicating the need for helmet, knee pads, and gloves!
5. Answers will vary; first: car hitting tree; second: body hitting car; third: internal organs hitting the internal wall of thoracic cavity, or brain hitting internal skull.
6. First collision: glove hitting head; second collision: head moves backward, and the brain collides with the interior of the skull; third collision: elasticity of the brain attached to the brain stem, causes the brain to move toward the back of the brain colliding with the internal back of the skull.
7. Answers will vary, depending on the state. Forty-nine states and the District of Columbia have mandatory seat-belt laws. In some states, seat-belt laws date back to 1985.

Stretching Exercises

1. Determine what opinions people in your community hold about the wearing of seat belts. Compare the opinions of the 60+ years old and 25 to 59 year old groups with that of the 15 to 24 year old group. Survey at least five people in each age group: Group A = 15 to 24 years, Group B = 25 to 59 years, and Group C = 60 years and older. (Survey the same number of individuals in each age group.) Ask each individual to fill out a survey card.

A sample questionnaire is provided below. You may wish to eliminate any question that you feel is not relevant. You are encouraged to develop questions of your own that help you understand what attitudes people in your community hold about wearing seat belts. The answers have been divided into three categories: 1 = agree; 2 = will accept, but do not hold a strong opinion; and 3 = disagree. Try to keep your survey to between five and ten questions.

Age group:	Date of Survey:		
Statement	Agree	No strong opinion	Disagree
1. I believe people should be fined for not wearing seat belts.	1	2	3
2. I wouldn't wear a seat belt if I didn't have to.	1	2	3
3. People who don't wear seat belts pose a threat to me when they ride in my car.	1	2	3
4. I believe that seat belts save lives.	1	2	3
5. Seat belts wrinkle my clothes and fit poorly so I don't wear them.	1	2	3

2. Make a list of reasons why people refuse to wear seat belts. Can you challenge these opinions using what you have learned about Newton's First Law of Motion?

Activity 4 A

Dropping a Clay Ball to Investigate Inertia

FOR YOU TO DO

1. This could be a messy activity. Place newspaper or other drop cloth on the desks before building the clay balls. Form a ball using 1.0 kg of clay. Using a balance, measure the exact mass of the ball.

a) Record the mass of the ball in your log.

2. Drop the ball from a variety of heights into a hand wearing a baseball glove. Avoid trying to "cushion" the catch, which will ruin the effect of the force of impact. Look away while catching the ball, so you are better able to describe the force qualitatively.

a) In your log record the distance from which the ball was dropped.

b) Describe how the ball felt as it hit the glove when dropped from different heights.

3. The formula for finding the velocity of an object that is falling is:

$$v = \sqrt{2gh}$$

where $g = 9.81\ \text{m/s}^2$,

h is the height from which the ball is released to the glove.

From this formula you can determine the momentum (m x Δv) of the object in order to show the increase in the change momentum. Determine the time by averaging the velocities $[(v_2 + v_1)/2]$, then using that v to find the time in $t = d/v$. Now you can determine the force ($F = m\Delta v/\Delta t$) that is being exerted on their hands as the object is dropped from different heights.

a) Determine the force of the ball for the different heights.

Sample Calculation:

Mass of the ball = 1.0 kg; distance from glove to the desk = 1.5 m

- Velocity where $h = 1.5$ m

  $v = \sqrt{2gh}$

  $v = 5.42$ m/s

- Change in Momentum (as the ball goes from the final speed of 5.42 m/s to 0 m/s in the glove)

  $\Delta p = m\Delta v$

  $\Delta p = 5.42$ m/s

- Average velocity

  = (5.42 m/s+ 0 m/s)/2

  = 2.71 m/s

For use with *Safety*, Chapter 2, Activity 4: Life (and Death) before Seat Belts

- Time to stop the ball in 0.10 m

  $t = d/v$

  $t = 0.10\ \text{m}/\ 2.71\ \text{m/s}$

  $t = 0.037\ \text{s}$

- Force acting on the glove by the ball

  $F = m\Delta v/\Delta t$

  $F = (1.0\ \text{kg}\ 5.42\ \text{m/s})/0.037\ \text{s}$

  $F = 146\ \text{N}$

Or about 15 kg (31 lb.) of mass (Every 10 N of force is approximately the equivalent of 1 kg of mass.)

b) What was the speed of the clay ball when dropped from 2.0 m?

c) Estimate what the force might be if the clay ball were dropped from the gym roof (approximate height of 10 m).

If the possibility exists, take the clay ball outside and drop it into a cloth held in a way similar to the firefighters net.

a) The force of an accident can be compared to the forces you felt when you were dropping the ball of clay. Relate your experiences of catching the ball, and how they might compare with the forces that the car and the tree exert on each other during their collision.

4. Make clay balls of various sizes (exact masses are not necessary).

5. Place the balls on the ends of chopsticks so that only the end of the chopsticks is showing, as shown.

6. With the bottom of the chopstick in hand, slam your hand down on the desk. Observe what happens.

a) Why did the clay slide down the chopstick?

b) What stopped the clay once it was moving?

c) What could you do to stop the clay from moving down the chopstick when you hit it on the desktop?

Activity 4 B

Life (and Death) before Seat Belts Part B: Low-Tech Alternative.

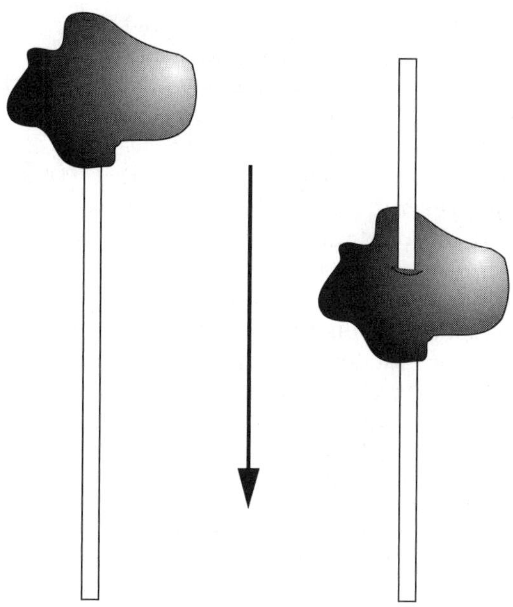

This is a low-tech activity to illustrate Newton's First Law. It is intended to give the students a firsthand look at inertia.

Materials needed

- quantity of play dough or clay
- chopsticks (or similar sticks)

Planning for the Activity

- Students will make clay balls of various sizes (exact masses are not necessary).
- Place the balls on the ends of the chopsticks so that only the end of the chopsticks is showing.
- With the bottom of the chopstick in hand, slam your hand down on the desk.
- Observe what happens

Time Requirements

Approximately 30 minutes to complete this activity.

Teaching Notes

As the mass of the ball increases, the students will want to think that the acceleration increases. While the ball appears to accelerate more as it increases its mass, the increases in the mass actually means there is an increase of the force (due to gravity) Fg.

Classroom Management Tips

This could be a messy activity. Remind the students to place newspaper or other drop cloth on the desks before building the clay balls.

Questions:

1. Why did the clay slide down the chopstick?
   A: Inertia of the clay kept the clay moving.
2. What stopped the clay once it was moving?
   A: The friction between the chopstick and the clay?
3. What could you do to stop the clay from moving down the chopstick, when you hit it on the desktop?
   A. Answers will vary. Look for something like seat belts or stops of some kind on the chopstick. Some may say change the friction between the clay and the chopstick (fire it in an oven to harden it), but we are looking for an application to be able to relate it to a moving car.

Chapter 2

ACTIVITY 5
Life (and Fewer Deaths) after Seat Belts

Background Information

Inertia is the tendency for an object to stay in motion. In order to stop that motion, there must be a force applied to that object in order to accelerate the object or change its velocity to 0 m/s ($a = \Delta v/\Delta t = v_2 - v_1/\Delta t$). For example, for an object moving at 3.0 m/s, is brought to 0 m/s in 1.0 s

$$a = \Delta v/\Delta t$$

$$a = v_2 - v_1/\Delta t$$

$$a = (0 \text{ m/s} - 3.0 \text{ m/s}) / 1.0 \text{ s}$$

$$a = -3.0 \text{ m/s}^2$$

Note that this is a negative acceleration, which we often refer to as deceleration. For the purposes of this course, we will refer to slowing down and speeding up as acceleration. There will be references to negative and positive acceleration.

Newton's Second Law of Motion gives a mathematical understanding of forces and acceleration – i.e.: changing velocity.

Symbols
v = velocity (m/s)
d = distance (m)
t = time (s)
a = acceleration (m/s)
F = force (N)
m = mass (kg)
Δ = change

Newton's Second Law states that the acceleration of an object is proportional to the force in the same direction, and inversely proportional to the mass.

$$a \quad F$$

$$a \quad 1/m$$

Therefore we can state this proportion as $a = F/m$, or as it is commonly known as $F = ma$.

If you have a wooden crate with nothing in it, it is very easy to push (apply a force to) the box. The empty crate is also very easy (or requires a smaller force) to stop. However, increasing the mass in the box, it becomes increasing difficult to push the box. This is caused by the increase in the inertia of the box. As well, it is very difficult (or requires a larger force) to stop the box.

Newton, after showing that forces cause objects to move (accelerate) asked himself where do these forces come from. This led to his third law, where forces come from other objects. Your hand pushing on a desk exerts a force on the desk. You can see, also that there must be a force acting on your hand because you can see the deformation of the "dent". Also, by applying more force, you realize that it hurts. Therefore, the only conclusion can be that the desk must be exerting a force on your hand. Newton's Third Law of Motion states that whenever an objects exerts a force on another object, the second object exerts a force equal in magnitude, but opposite in direction, on the first object. These are dealt with as action-reaction pairs. There can never be one force on an object without an equal but opposite force on the other

Therefore, your inertia will keep you moving in a straight line unless a force stops your motion. The force comes when your body comes in contact with the seat belt. This net force is enough to accelerate you to a stop.

In the previous activity the students examined the acceleration of a vehicle while stopping in a collision. In this activity, the students are manipulating the restraining device to give different pressures, without changing the forces. The physics involved with this activity is the understanding of how the pressure changes as you increase the area of the restraining device. The pressure is the force per unit area of the contact with the body. They will discover that the greater the area of contact, the more the force is spread out over the body.

$$P = F/A.$$

Therefore, even if the force is great (due to a large acceleration), the pressure on the individual will be much less.

Active-ating the Physics InfoMall

This activity has much in common with **Activity 4**. So perform searches that will find something new, and perhaps interesting. For example, see what the InfoMall says about crash dummies. One method is to search for "crash"" AND "dumm*", where we are using AND to mean the logical operation (returns hits that contains both search keywords) and the asterisk (wild character). The first hit is "Forensic physics of vehicle accidents," *Physics Today*, vol. 40, issue 3, (1987). You may wish to look this one up yourself.

Another suggestion is to search for "bed of nails". Try it out.

And if you want more problems for your students, don't forget the Problems Place! You can easily find many, many problems about pressure.

Planning for the Activity

Time Requirements

Allow approximately 40 minutes to complete this activity and record the data.

Materials Needed

For each group:

- clay, modeling, 1/2 lb.
- copper, bare #22 wire, 1ft.
- dynamics carts
- ribbon, various widths
- starting ramp for lab cart

For the class:

- *Active Physics Transportation* Content Video (Segment: Crash Dummy)
- VCR & TV monitor

Advance Preparation and Setup

Prepare as you did for the previous activity. Have several different types of material, as well as large quantities of each available.

Teaching Notes

Students find that the wire will cut more deeply, and that more, thicker supports will do the least damage. In debriefing, emphasis should still be on the fact that even at high speeds, three-point seat belts may still not save lives. There is lots of discussion that can come from this, especially when referring back to the **Chapter Challenge**. Again, the emphasis will be on decreasing the force that is exerted on the body by spreading out the force which occurs while the vehicle is stopping. The same force will occur if the body is stopped. The idea of having fatter seat belts is that the greater surface area of the belt allows for a smaller pressure.

As with any activity, a review of proper decorum in the classroom should be warranted. Students may use the clay to throw around the room. If the students are to complete the survey (**Question** #7 in **Physics To Go**) they need to have permission to survey other students, or parents. This is a good activity to have the students survey more than immediately around the school. Give them the assignment of surveying some members of their community.

Please note that students may still have the misconception that the forces exerted on the seat belts are not that significant.

NOTES

Activity 5 Life (and Fewer Deaths) after Seat Belts

GOALS

In this activity you will:

- Understand the role of safety belts.
- Compare the effectiveness of various wide and narrow belts.
- Relate pressure, force and area.

What Do You Think?

In a collision, you cannot brace yourself and prevent injuries. Your arms and legs are not strong enough to overcome the inertia during even a minor collision. Instead of thinking about stopping yourself when the car is going 30 mph, think about catching 10 bowling balls hurtling towards you at 30 mph. The two situations are equivalent.

- **Suppose you had to design seat belts for a racecar that can go 200 mph. How would they be different from the ones available on passenger cars?**

Record your ideas about this question in your *Active Physics* log. Be prepared to discuss your responses with your small group and the class.

Activity Overview

In this activity the students will be investigating the role and requirements of an effective restraint system. They will be simulating different styles of safety belts, and submitting their model to crash simulations as in **Activity 4**.

Student Objectives

Students will:

- Understand the role of safety belts.
- Compare the effectiveness of various wide and narrow belts.
- Relate pressure, force and area.

Answers for The Teacher Only

What Do You Think?

If you can, go to see the film *Speedway*. The film talks about how the race cars have changed in the past few decades. Because the speeds are greater than they were, they can no longer build the cars like tiny missiles, which do not crush. They have found, as was mentioned above, that the crush zone allows some of the energy of the collision to be absorbed by the car, rather than the driver. Even in well-protected cars, equipped with shoulder belts, air bags, and padded interiors, if the body of the driver is stopped at certain accelerations, the damage to the interior of the body may be fatal.

Most of the answers the students will come up with will be centered on more of what is already in the vehicle. However, the answers should reflect the idea that the faster you go, the greater protection you will need. Also in the discussion, you can emphasize that it is not only more padding, but rather the reduction of the forces, which will be talked about in this activity. In other words, the stopping from high speeds, into a tree, or the front of your dashboard can be equally dangerous with a tight seat belt on, to the interior organs of the body.

Safety

For You To Do

1. In this activity you will test different materials for their suitability for use as seat belts. Your model car is, once again, a laboratory cart; your model passenger is molded from a lump of soft clay. Give your passenger a seat belt by stretching a thin piece of wire across the **front** of the passenger, and attaching it on the sides or **rear** of the car.

Perform the activity outside of traffic areas. Do not obstruct paths to exits. Do not leave carts lying on the floor.

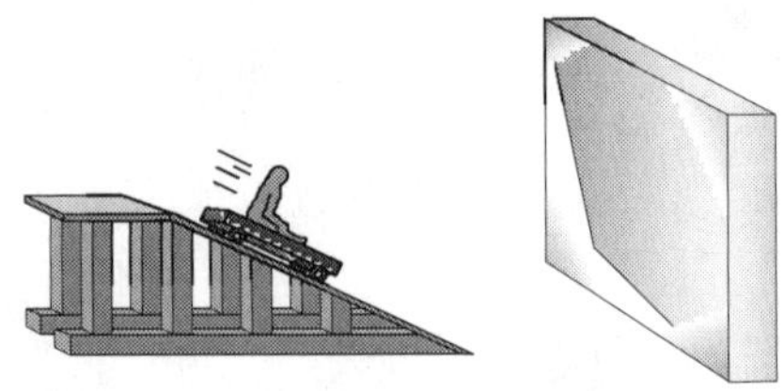

Physics Words

force: a push or a pull that is able to accelerate an object; force is measured in newtons; force is a vector quantity.

pressure: force per surface area where the force is normal (perpendicular) to the surface; measured in pascals.

2. Make a collision by sending the car down a ramp. Start with small angles of incline and increase the height of the ramp until you see significant injury to the clay passenger.
   a) In your log, note the height of the ramp at which significant injury occurs.

3. Use at least two other kinds of seat belts (ribbons, cloth, etc.). Begin by using the same incline of ramp and release height as in **Step 2**.
   a) In your log, record the ramp height at which significant injury occurs to the "passenger" using the other kinds of seat belt material.

4. Crash dummies cost $50,000! Watch the video presentation of a car in a collision, with a dummy in the driver's seat. You may have to observe it more than once to answer the following questions:
   a) In the collision, the car stops abruptly. What happens to the driver?
   b) What parts of the driver's body are in the greatest danger? Explain what you saw in terms of the law of inertia (Newton's First Law of Motion).

106

Active Physics

ANSWERS

For You To Do

1. Student activity.

2.a) Students' responses, but as the height of the ramp is increased, there should be an increase in the significant damage.

3.a) Students' responses.

4.a) The driver continues to move in the direction of the car (inertia) until stopped by the seat belt, (or dashboard).

b) The head is probably in the most danger as it is not secured by anything. In terms of the Law of Inertia, as the car stops, the body continues to move. As the body stops, by the seat belt, the head continues to move. This is where damage can occur and usually happens.

Activity 5 Life (and Fewer Deaths) after Seat Belts

FOR YOU TO READ

Force and Pressure

When you repeated this experiment accurately each time, the **force** that each belt exerted on the clay was the same each time that the car was started at the same ramp height. Yet different materials have different effects; for example, a wire cuts far more deeply into the clay than a broader material does.

The force that each of the belts exerts on the clay is the same. When a thin wire is used, all the force is concentrated onto a small area. By replacing the wire with a broader material, you spread the force out over a much larger area of contact.

The force per unit area, which is called **pressure**, is much smaller with a ribbon, for example, than with a wire. It is the pressure, not the force, that determines how much damage the seat belt does to the body. A force applied to a single rib might be enough to break a rib. If the same force is spread out over many ribs, the force on each rib can be made too small to do any damage. While the total force does not change, the pressure becomes much smaller.

PHYSICS TALK

Pressure is the force per unit area:

$$P = \frac{F}{A}$$

where F is force in newtons (N)

A is area in meters squared (m^2)

and P is pressure in newtons per meter squared (N/m^2).

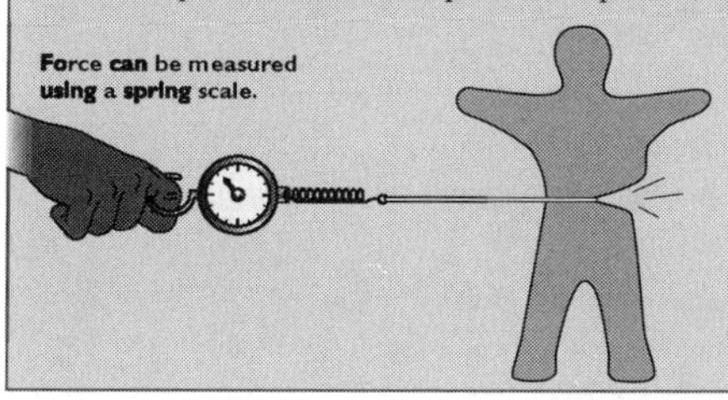

Force can be measured using a spring scale.

Sample Problem

Two brothers have the same mass and apply a constant force of 450 N while standing in the snow. Brother A is wearing snow shoes that have a base area of 2.0 m². Brother B, without snowshoes, has a base area of 0.1 m². Why does the brother without snowshoes sink into the snow?

Strategy: This problem involves the pressure that is exerted on the snow surface by each brother. You can use the equation for pressure to compare the pressure exerted by each brother.

Givens:

$F = 450$ N

$A_1 = 2.0\ m^2$

$A_2 = 0.1\ m^2$

Solution:

Brother A	Brother B
$P = \frac{F}{A}$	$P = \frac{F}{A}$
$= \frac{450\ N}{2.0\ m^2}$	$= \frac{450\ N}{0.1\ m^2}$
$= 225\ N/m^2$ or $230\ N/m^2$	$= 4500\ N/m^2$

The pressure that Brother B exerts on the snow is much greater.

Activity 5 Life (and Fewer Deaths) after Seat Belts

Reflecting on the Activity and the Challenge

In this activity you gathered data to provide evidence on the effectiveness of seat belts as restraint systems. The material used for the seat belt and the width of the restraint affected the distortion of the clay figure. By applying the force over a greater area, the pressure exerted by the seat belt during the collision can be reduced.

It is important to note that not every safety restraint system will be a seat belt or harness, but that all restraints attempt to reduce the pressure exerted on an object by increasing the area over which a force is applied.

How will your design team account for decreasing pressure by increasing the area of impact? Think about ways that you could test your design prototype for the pressure created during impact. Your presentation of the design will be much more convincing if you have quantitative data to support your claims. Simply stating that a safety system works well is not as convincing as being able to show how it reduces pressure during a collision.

Physics To Go

1. Use Newton's First Law to describe a collision with the passenger wearing a seat belt during a collision.
2. What is the pressure exerted when a force of 10 N is applied to an object that has an area of:
   a) 1.0 m^2?
   b) 0.2 m^2?
   c) 15 m^2?
   d) 400 cm^2?
3. A person who weighs approximately 155 lb. exerts 700 N of force on the ground while standing. If his shoes cover an area of 400 cm^2 (0.0400 m^2), calculate:
   a) the average pressure his shoes exert on the ground
   b) the pressure he would exert by standing on one foot

Answers

Physics To Go

1. The car is moving in a given direction. As the car stops, the driver or passenger continues to move forward until the seat belt stops them—a force acts on the body to change the motion from moving relative to the car to stopping with the car. Parts of the body, not directly attached to the seat belt, will continue also to move in the given direction. This is where significant damage to the body, mostly the head, can occur.

2. a) 1.0 m^2

   Using formula $P = F/A$,
   $P = 10\ N/ 1.0\ m^2$ therefore
   $P = 10\ N/m^2$)

   b) 0.2 m^2

   $P = 50\ N/m^2$

   c) 15 m^2

   $P = 0.67\ N/m^2$

   d) 400 cm^2

   $P = 0.025\ N/cm^2$ or $250\ N/cm^2$

3. a) 17,500 N/cm^2

   b) Double the pressure or 35,000 N/cm^2

ANSWERS

Physics To Go
(continued)

4. a) Using approximate values, of weight of 500 Newtons, and area of high heels about 1 cm^2, the $P = F/A$, = 500 N/ 0.0001 m^2, P = 5,000,000 N/m^2

   b) Using approximate values, of weight of 500 N, and area of your hands of about 900 cm^2, the $P = F/A$, = 500 N/ 0.09 m^2, P = 5555 N/m^2

   c) If your body was about 500 N, and there were 5000 nails, then there is approximately 500 N / 5000 nails, or 0.1 N per nail of pressure. In reality, that is probably less the weight of one nail, and if you were to support one nail on your finger it would not penetrate.

5. The force is identical, but on a smaller surface area. Therefore the pressure exerted on your body is greater. It would be enough to cut your body in two pieces.

6. Students answers will vary. Answers might include people who don't use seat belts may end up having greater injuries, therefore increasing the cost of health care, they may lose control of their vehicle more easily, therefore causing more damage to their own vehicle or others. Have the students explore the concept of social responsibility. What constitutes the need for any law? How does the use or lack of use of safety restraints affect us as a society? How do economics affect the passing of legislation in this area?

7. Have students try to survey more people, and not simply their immediate family.

4. For comparison purposes, calculate the pressure you exert in the situations described below. Divide your weight in newtons, by the area of your shoes. (To find your weight in newtons multiply your weight in pounds by 4.5 N/lb. You can approximate the area of your shoes by tracing the soles on a sheet of centimeter squared paper.)
   a) How much pressure would you exert if you were standing in high heels?
   b) How much pressure would you exert while standing on your hands?
   c) If a bed of nails contains 5000 nails per square meter, how much force would one nail support if you were to lie on the bed? With this calculation you can now explain how people are able to lie on a bed of nails. It's just physics!
5. Describe why a wire seat belt would not be effective even though the force exerted on you by the wire seat belt is identical to that of a cloth seat belt.
6. Do you think there ought to be seat belt laws? How does not using seat belts affect the society as a whole?
7. Conduct a survey of 10 people. Ask each person what percentage of the time they wear a seat belt while in a car. Be prepared to share your data with the class.

Stretching Exercises

The pressure exerted on your clay model by a thin wire can be estimated quite easily. Loop the wire around the "passenger," and connect the wire to a force meter.

a) Pull the force meter hard enough to make the wire sink into the model just about as far as it did in the collision.
b) Record the force as shown on the force meter (in newtons).
c) Estimate the frontal area of the wire—its diameter times the length of the wire that contacts the passenger. Record this value in centimeters squared (cm^2).
d) Divide the force by the area. This is the pressure in newtons per centimeter squared (N/cm^2).

NOTES

Assessment: Activity 5

Place a check mark in the appropriate box. Two check marks will be required for one point.
A sample conversion scale is provided below the chart.
Total marks = 10.

Descriptor	Yes	No
Lab Skills		
understands the need to control variables		
knows the manipulated variable		
is able to construct reasonable seat belts		
uses more than two types of materials in constructing seat belts		
uses appropriate materials in constructing seat belts		
runs at least three trials with each model of seat belt		
demonstrates or explains why there is a need for several trials		
rebuilds model as necessary		
Understanding Concepts		
demonstrates understanding of forces		
demonstrates understanding of pressure		
knows which parts of the body are most vulnerable in a collision		
demonstrates an understanding for seat belt laws		
is able to describe a collision of a person with and without a seat belt		
understands the societal impact of using a seat belt		
understands the societal impact of not using a seat belt		
Mathematical Skills		
can calculate the force per unit area (pressure)		
calculates the pressure of the wire		
calculates the pressure of their own weight on their shoes		

Conversion

20 check marks = 10 points
18 check marks = 9 points
16 check marks = 8 points, etc.

For use with *Safety*, Chapter 2, Activity 5: Life (and Fewer Deaths) after Seat Belts

NOTES

ACTIVITY 6
Why Air Bags?

Background Information

Symbols
v = velocity (m/s)
d = distance (m)
t = time (s)
a = acceleration (m/s)
F = force (N)
m = mass (kg)
Δ = change

Now that we have dealt with forces, how does this have anything to do with accidents? In order to examine accidents, we must first look at momentum. Momentum (p) is the product of the mass (m) of an object and the velocity (v) of that object ($p = mv$). Ask the students which is harder to stop, the 175 pound (80 kg) quarterback running at 4 m/s, or the fullback at 255 pounds (116 kg) running at 4 m/s? Most students will say that the quarterback would be easier to stop. We would say that the momentum of the quarterback is less (m x v) than the fullback (m x v)

momentum of quarterback

$$p_{qb} = m \times v$$

$$p_{qb} = (80 \text{ kg}) \times (4 \text{ m/s})$$

$$p_{qb} = 320 \text{ kg}\cdot\text{m/s}$$

momentum of fullback

$$p_{fb} = m \times v$$

$$p_{fb} = (116 \text{ kg}) \times (4 \text{ m/s})$$

$$p_{fb} = 464 \text{ kg}\cdot\text{m/s}$$

We can summarize momentum by stating that the greater the momentum of an object, the harder it will be to stop.

According to Newton's First Law – Inertia, an object in motion will remain in motion unless acted upon by an outside, unbalanced force. Therefore, to stop a moving object (the fullback), we need a force. Newton's Second Law

$$F = ma$$

states that the acceleration (or change in velocity) will be proportional to the force. Since acceleration is the change in velocity divided by the change in time, we can restate Newton's equation as

$$F = m\ \Delta v/\Delta t.$$

Remove Δt by multiplying both sides by Δt and we now have

$$F\Delta t = m\Delta v.$$

called the impulse (measured in Ns). Impulse is the product of the force applied and the change in time over which the force acted. In other words the force times the time interval is proportional to the change in velocity times the mass.

Therefore, (where $p = mv$)

$$\Delta p = m\Delta v,\ \ [\Delta p = m\ (v_2 - v_1)]$$

or the change in momentum is equal to the mass times the change in velocity.

Substituting Δp for $m\Delta v$ in $F\Delta t = m\Delta v$, we get

$$F\Delta t = \Delta p$$

Therefore we get impulse is equal to the change in momentum. As we analyze the equation, we see that in order to stop an object (change its momentum to zero), there must be a force applied to the object over a particular time. Increasing the time decreases the force and decreasing the time increases the force required to change the same momentum. For example: A bowling ball (m = 16 pounds or 7.3 kg) moving at 5.0 m/s. What is the impulse to stop the bowling ball?

$$\text{impulse} = \Delta p = m\Delta v$$

$$\Delta p = 7.3 \text{ kg} \times 5.0 \text{ m/s}$$

$$\Delta p = 36.5 \text{ kg}\cdot\text{m/s}$$

Therefore, what is the force required to bring the bowling ball to rest in 1.0 s?

Impulse = Δp, therefore $F\Delta t = \Delta p$, therefore

$$F = \Delta p/\Delta t$$

$$F = 36.5 \text{ kg}\cdot\text{m/s} / 1.0\text{s}$$

$$F = 36.5 \text{ kg}\cdot\text{m/s}^2 \text{ or } 36.5 \text{ N}$$

How much force is then needed to bring the ball to rest in 3.0 s?

$$F = 36.5 \text{ kg}\cdot\text{m/s} / 3.0 \text{ s}$$

$$F = 12.2 \text{ N}$$

How much force is then needed to bring the ball to rest in 0.10 s?

F = 36.5 kg•m/s / 0.10 s

F = 365 N

You can see as you increase the time needed to stop an object, you decrease the force necessary. As the time decreases, the force increases.

The physics of the air bag is to increase the time required for the object (the driver) to stop, and therefore reduce the force on the driver. The other aspect for which the air bag is designed is that the force, even though decreased, is spread over a larger area. The face slamming into the steering wheel can cause a lot of damage, as compared with the face slamming into an air bag.

Active-ating the Physics InfoMall

Search for "air bag*" and you will not be disappointed. There are several good comments made in various places on the InfoMall, including "Resource letter PE-1: Physics and the environment," *American Journal of Physics*, vol. 42, 267-273 (1974), and "How Things Work" in the Book Basement has a section on air bags (in the 1985 section).

The concept of impulse is introduced in this activity. Searching for "impulse" on the InfoMall provides many great references, including problems your you or your students to work. The textbook *Physics Including Human Applications* has a chapter devoted to Momentum and Impulse.

Planning for the Activity

Time Requirements

Allow at least one period for the students to design and run the tests on the air bag. If a video camera is used the students may want to take it home to analyze. If there are not enough video cameras to go around, then rotate through the groups, with the other groups analyzing the data they have collected, while the other students may be watching a video, or redesigning their air bags.

Materials Needed

For each group:

- ball, heavy
- clear bag, inflatable
- landing surface materials of 3 hardnesses

Advance Preparation and Setup

Depending on the material available, you may need to make the equivalent of several different substances to test against the air bag model.

Check proper functioning of the recording device. The VCR must be able to give a clear picture on single frame advance mode. If the students are to use the tape times, they may need additional instructions if they have not used them before. If you are going to use computer software, which digitizes the video, and then allows you to analyze the picture, prior use or training with the software would be an asset, both for the students and the teacher. If there is not enough software, and hardware for all students to use, then a demo would be beneficial to set up for the students in order for them to have a standard to shoot for.

Teaching Notes

Air bags are becoming more and more common in vehicles. Poll the students as to whether their family cars have air bags. Ask if they are willing to pay the extra money for them. Have them write the answers in their log books.

Remind the students how to make measurements using the videotape system. Encourage experimentation that includes various impact speeds and different stopping surfaces. If tape timers are to be used review the time interval specifications. Measurements of distances are not necessary; only counting the intervals between when the object touches the bag and when the object stops in order to get the stopping time. When using the tape timer, it may be difficult at first to get good results. Be prepared to use a lot of tape in getting accurate results.

There may be video tapes available which show slow motion air bags actually working. These may be available from your local automobile association, or from a car dealership.

Students should work in groups of three or four. Students will be dropping heavy objects onto surfaces which may cause the object to bounce in different and unpredictable directions, and therefore, should take appropriate precautions to prevent injury to damage to property.

Some students may have heard that air bags have caused suffocation once they inflated. This is an incorrect understanding of how the air bag works. The air bag inflates in about 1/32 of a second, and once inflated, it deflates after about one to two seconds. Other fallacies about air bags are that they impair your ability to see, and they inhibit you from exiting the vehicle. (For more information, visit any car dealer and ask for their information on air bags.) While there is some smoke and dust from the CO_2 cartridge, and some heat associated with the inflation of the airbag, it will not impair your ability to get out of the vehicle. The effectiveness of the air bag increases in conjunction with the seat and shoulder belts, but is not very effective in side impacts.

Local automotive dealers will have safety videos about air bags and ABS brakes. Many dealers will lend them, or may even give a copy to you.

You may also use the local Automobile Association affiliate in your area to give or lend safety videos. Some will even send instructors to give safety talks.

Some local driving companies (taxi, trucking, courier services) may also have safety supervisors who would be able to come to the classroom and talk about safety.

The *Active Physics Transportation* Content Video has excellent footage showing air bags inflating.

Activity 6 Why Air Bags?

GOALS

In this activity you will:

- Model an automobile air bag.
- Relate pressure to force and area.
- Demonstrate that the force of an impact can be reduced by spreading it out over a longer time.

What Do You Think?

Air bags do not take the place of seat belts. Air bags are an additional protection. They are intended to be used with seat belts to increase safety.

- **Why are air bags effective?**
- **How does the air bag protect you?**

Record your ideas about these questions in your *Active Physics* log. Be prepared to discuss your responses with your small group and the class.

Activity Overview

In this lesson the students will study the use of an air bag and how it will spread the force of the impact over both space and time.

Student Objectives

Students will:

- Model an automobile air bag.
- Relate pressure to force and area.
- Demonstrate that the force of an impact can be reduced by spreading it out over a longer time.

ANSWERS FOR THE TEACHER ONLY

What Do You Think?

Students' answers will vary. Some may think of the air bag as cushioning the stop, or making the stop softer. What they are actually saying, is that the air bag will increase the time in which you crash, therefore decreasing the force acting on your body (or more particularly the head).

Again, students' answers will vary. The air bag protects you by increasing the time and the space over which the force is acting on your body. See the **Background Information** for details and the mathematics behind the air bag.

For You To Do

1. You will use a large plastic bag or a partially inflated beach ball as a model for an air bag. Impact is provided by a heavy steel ball, or just a good-sized rock, dropped from a height of a couple of meters.

   Gather the equipment you will need for this activity. Your problem is to find out how long it takes the object to come to rest. What is the total time duration from when the object first touches the air bag until it bounces back?

2. With a camcorder, videotape the object striking the air bag from a given height, such as 1.5 m.

   a) Record the exact height, from which you dropped the object.

3. Play the sequence back, one frame at a time. Count the number of frames during the time the object is moving into the air bag—from the moment it first touches the bag until it comes to rest, before bouncing. Each frame stands for $\frac{1}{30}$ s. (Check your manual.)

   a) In your log, record the number of frames and calculate how long it takes for the object to come to rest.

   If a camcorder is not available, the experiment may be performed, although less effectively, by attaching a ticker-tape timer to the falling object.

   After the object is dropped, with the object still attached, stretch the tape from the release position to the air bag. Mark the dot on the tape that was made just as the object touched the air bag.

   Now push the object into the air bag, about as far as it went just before it bounced. Mark the tape at the dot that was made as the object came to rest. The dots should be close together for a short interval at this point.

   Now count the time that passed between the two marks you made. (You must know how rapidly dots are produced by your timer.)

Set up the activity in an area clear of obstruction. Arrange for containment of the dropped object.

ANSWERS

For You To Do

1. Student activity.

2. a) Students' response.

3. a) This may vary, with the camcorder. Most are 1/30 s. Students will be noticing, that the greater the time to stop, the less potential damage will be done.

4. Repeat **Steps 2** and **3**, but this time drop the ball against a hard surface, such as the floor. Keep the height from which the object is dropped constant.
   a) Record how long it takes for the object to come to rest on a floor.
5. Choose two other surfaces and repeat **Steps 2** and **3**.
   a) Record how long it takes the object to come to rest each time.
   b) In your log, list all the surfaces you tested in the order in which you expect the most damage to be done to a falling object, to the least damage.
   c) Is there a relationship between the time it takes for the object to come to rest and the potential damage to the object landing on the surface? Explain this relationship in your log.

PHYSICS TALK

Force and Impulse

Newton's First Law states that an object in motion will remain in motion unless acted upon by a net external force. In this activity you were able to stop an object with a force. In all cases the object was traveling at the same speed before impact. Stopping the object was done quickly or gradually. The amount of damage is related to the time during which the force stopped the object. The air bag was able to stop the object with little damage by taking a long time. The hard surface stopped the object with more damage by taking a short time.

Physicists have a useful way to describe these observations. An **impulse** is needed to stop an object. That impulse is defined as the product (multiplication) of the force applied and the time that the force is applied.

Physics Words

impulse: the product of force and the interval of time during which the force acts; impulse results in a change in momentum.

ANSWERS

For You To Do *(continued)*

4. a) Students' response. Again the students should be noticing that the greater the time to stop, the less damage will be done.

5. a) Students' response. The harder the surface, the less time, and potentially the greater damage done.

   b) Students' response. This will vary with the types of materials each group uses.

   c) Students' response. Again, the longer it takes the object to stop, the less damage will be done to the object.

Safety

Impulse = $F\Delta t$

where F is force in newtons (N);

Δt is the time interval during which the force is applied in seconds (s).

Impulse is calculated in newton seconds (Ns).

An object of a specific mass and a specific speed will need a definite impulse to stop. Any forces acting for enough time can provide that impulse.

If the impulse required to stop is 60 Ns, a force of 60 N acting for 1 s has the required impulse. A force of 10 N acting for 6 s also has the required impulse.

Force F	Time Interval Δt	Impulse $F\Delta t$
60 N	1 s	60 Ns
10 N	6 s	60 Ns
6000 N	0.01 s	60 Ns

The greater the force and the smaller the time interval, the greater the damage that is done.

Sample Problem

A person requires an impulse of 1500 Ns to stop. What force must be applied to the person to stop in 0.05 s?

Strategy: You can use the equation for impulse and rearrange the terms to solve for the force required.

$$\text{Impulse} = F\Delta t$$

$$F = \frac{\text{Impulse}}{\Delta t}$$

Active Physics 114

Givens:

Impulse = 1500 Ns

$\Delta t = 0.05$ s

Solution:

$$F = \frac{\text{Impulse}}{\Delta t}$$

$$= \frac{1500 \text{ Ns}}{0.05 \text{ s}}$$

$$= 30{,}000 \text{ N}$$

Reflecting on the Activity and the Challenge

People once believed that the heavier the automobile, the greater the protection it offered passengers. Although a heavy, rigid car may not bend as easily as an automobile with a lighter frame, it doesn't always offer more protection.

In this activity, you found that air bags are able to protect you by extending the time it takes to stop you. Without the air bags, you will hit something and stop in a brief time. This will require a large force, large enough to injure you. With the air bag, the time to stop is longer and the force required is therefore smaller.

Force and impulse must be considered in designing your safety system. Stopping an object gradually reduces damage. The harder a surface, the shorter the stopping distance and the greater the damage. In part this provides a clue to the use of padded dashboards and sun visors in newer cars. Understanding impulse allows designers to reduce damage both to cars and passengers.

ANSWER

Physics To Go

1. Any combination of F (e.g., 30 N) multiplied by Δt (e.g., 2 s) which give a result of 60 Ns.
2. $\Delta p = F\Delta t$, therefore, $F = \Delta p/\Delta t$,
   a) $F = 1000$ Ns / 0.01 s, $F = 100{,}000$ N;
   b) $F = 1000$ Ns / 0.1 s, $F = 10{,}000$ N;
   c) $F = 1000$ Ns / 1.0 s, $F = 1000$ N;
3. When a car stops, there is an impulse (change in the momentum). This impulse ($F\Delta t$) will be transferred to the body in the car as the body's change in momentum is the same. Therefore, the air bag changes the force being applied to the body by increasing the time that the body is slowing down because the impulse does not change.
4. Hitting a brick wall will cause greater damage, as the time that the car stops is smaller, therefore the force is greater than the car hitting a snow bank.
5. a) The bumper might be mounted on a piston or a spring which will compress when the bumper strikes an object. Even without a spring or piston, the bumper can dent before the rigid frame of the car strikes the object.
   b) The collapsible steering wheel breaks upon impact. Rather than impaling the driver's body on the rigid shaft, the steering column telescopes in, lengthening the time of the impact.
   c) The crush or crumple zones allow the car to compress like an accordion. Previously, the sheet metal parts of the car were welded continuously along the seams, conveying the impact from object to occupants. Now the crumpling of the car increases the time after the front of the car hits until the rest of the car stops.
   d) The padding is not rigid, so as a body hits it, it crushes, lengthening the time for the impulse to be delivered.
6. a) When you catch a hard ball, you use a mitt and you do not hold your arms rigid. The mitt is cushioned and the leather webbing stretches. Your flexed arms ride with the ball. All of this lengthens the time of the impulse, reducing the required force.
   b) When you land on the ground you allow your knees to bend, lengthening the time for your body to come to rest, reducing the required force that will stop your movement.
   c) The bungee cord exerts a force stopping your fall as it stretches. By the time it is fully extended, your speed should be reduced enough so that the jerk of the stop is neither sudden nor severe. The impulse received from the bungee acts over a long period of time.
   d) The net stretches, allowing the victim to stop gradually. The impulse is the same as if the victim had landed on the ground, but much less destruction since the force is less, (due to increased time).
7. Due to the higher speed, the design of a seat belt for the airplane might be similar to the four-point belt that is used by fighter pilots, and racecar drivers.
8. Air bags might be a very effective, but expensive, safety feature in a passenger plane. This is especially so because of the absence of a shoulder strap. One could argue that the seat in front of the passenger is not rigid, and acts as a modified air bag.

Safety

Physics To Go

1. If an impulse of 60 Ns is required to stop an object, list in your log three force and time combinations (other than those given in the **Physics Talk**) that can stop an object.
2. A person weighing 130 lb. (60 kg) traveling at 40 mph (18 m/s) requires an impulse of approximately 1000 Ns to stop. Calculate the force on the person if the time to stop is:
   a) 0.01 s
   b) 0.10 s
   c) 1.00 s

3. Explain in your log why an air bag is effective. Use the terms force, impulse, and time in your response.
4. Explain in your log why a car hitting a brick wall will suffer more damage than a car hitting a snow bank.
5. There are several other safety designs that employ the concept of spreading out the time interval of a force. Describe in your log how the ones listed below perform this task:
   a) the bumper
   b) a collapsible steering wheel
   c) frontal "crush" zones
   d) padding on the dashboard
6. There are many other situations in which the force of an impulse is reduced by spreading it out over a longer time. Explain in your log how each of the actions below effectively reduces the force by increasing the time. Use the terms force, impulse, and time in your response.
   a) catching a hard ball
   b) jumping to the ground from a height
   c) bungee jumping
   d) a fireman's net
7. The speed of airplanes is considerably higher than the speed of automobiles. How might the design of a seat belt for an airplane reflect the fact that a greater impulse is exerted on a plane when it stops?
8. Airplanes have seat belts. Should they also have air bags?

116

Assessment: Activity 6

Place a check mark in the appropriate box. Two check marks will be required for one point. A sample conversion scale is provided below the chart.

Maximum = 10

Descriptor	Yes	No
Lab Skills		
student understands the need to control variables		
student is able to identify the manipulated variable		
student is able to identify the responding variable		
student is able to accurately record the time required to stop		
student uses more than 2 types of materials in constructing air bags		
student uses appropriate materials in constructing air bags		
student runs at least 3 trials with each model of air bag		
student demonstrates or explains why there is a need for several trials		
student tested at least 3 different surfaces		
Understanding Concepts		
student demonstrates understanding of forces and effects of air bags		
student demonstrates understanding of pressure		
student understands how the air bag inflates		
student relates inertia with the forces an air bag exerts on the body		
student understands need for seat belt use with the air bag		
student relates increased time of stopping with decreased force on body		
student understands impulse and the air bag		
student understands how impulse is related to damage		
Mathematical Skills		
student can calculate impulse given force and different times		
given an impulse and time, student can calculate force		

Conversion

19 check marks = 10 points

17+ check marks = 9 points

15+ check marks = 8 points

For use with *Safety*, Chapter 2, Activity 6: Why Air Bags?

ACTIVITY 7
Automatic Triggering Devices

Background Information

Air bags and seat belts do save your life, and prevent serious injuries if used properly. If you pull on your seat belt while you are moving at normal speed, it seems as though there is no way the seat belt could stop you from moving forward. Yet, you have probably slammed on your brakes at some point and the seat belt did in fact tighten, and held you in place. However, the air bag didn't inflate. The safety devices in your car cannot be working all the time (you wouldn't be able to see past the air bag), so you must have a triggering device to make sure that the air bag inflates only when it is needed and when it is it will prevent further damage.

In this activity the students will be using their own imagination, ingenuity, and any materials you are able to bring into the classroom. They will attempt to design a device which will operate like the trigger device in a car.

One such design that is used in cars is a cylinder with a steel ball attached at one end to a magnet, and a open circuit at the other end (See diagram). The ball is held in place by the magnet and only dislodges from the magnet at certain accelerations. When it leaves one end and hits the other end, the circuit is then closed, and the air bag inflates inside the car. This acceleration must be such that it would only happen in a collision, where the speed of the car is greater than about 18 km/h (11 mph).

Active-ating the Physics InfoMall

Again, you may want to find what is known about student difficulties with a concept in this activity. One possible search is "student difficult*" AND "circuit*".

Planning for the Activity

Time Requirements

Allow at least 40 minutes for this activity, with the possibility of the students taking their project home to work on. For the presentation of their design, allow approximately 10 minutes for each group to explain their design, and show how it works.

Materials Needed

For each group:

- alligator clip leads
- battery, d-cell
- bulb base for miniature screwbase
- bulb, miniature light
- dynamics carts
- single battery holder, d-cell
- starting ramp for lab cart
- switch, spst.

For the class:

- Collection of components such as tape, rubber bands, string, wire, paper clips, metal foil

Teaching Notes

Explore the students thoughts on how a triggering device might operate. One type of the device is shown in the **Background Information**. Elicit the conditions under which the device should trigger. The test criterion is purposely vague: that the device must trigger only when the car collides with a large speed and not a slow speed.

Reliability is a key issue. The device that inflates the air bag is a good example of a spin-off from the space program. It is based on a device that demanded 100% reliability: the release mechanism from the lunar launch system.

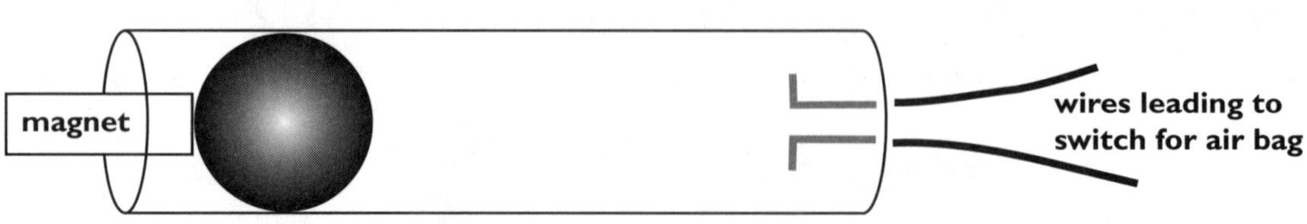

For a sophisticated class, you may wish to help the class determine the best release points for testing the car by using some kinematics, or through energy conservation. These areas have been explored. A sonic ranger or tape timer may be used in conjunction with your tests.

This activity can lead to increasing amounts of noise and mayhem. Keep students on task with constant monitoring and subtle encouragement while looking at each group. Use groups of four. In groups of three or less, there are generally fewer ideas, and groups greater than five give some an opportunity to blend into the woodwork.

Encourage students to work cooperatively by assigning tasks prior to beginning the activity. The following tasks are designed for groups of four students:

- **Organizer:** helps focus discussion and ensures that all members of the group contribute to the discussion. The organizer ensures that all of the equipment has been gathered and that the group completes all parts of the activity.
- **Recorder:** provides written procedures when required, diagrams where appropriate and records data. The recorder must work closely with the organizer to ensure that all group members contribute.
- **Researcher:** seeks written and electronic information to support the findings of the group. In addition, where appropriate, the researcher will develop and test prototypes. The researcher will also exchange information gathered among different groups.
- **Diverger:** seeks alternative explanations and approaches. The task of the diverger is keep the discussion open. "Are other explanations possible?"

Students will be working with many different electrical devices. Ensure that there are no wires connected to common household circuits. Ensure that the students are only using common dry cell or household batteries.

Some students may think that the triggering device for such safety devices are simple electronic switches. This activity will help them to understand that in order for a switch to be functional for a particular purpose (such as the *g*-forces that are enough to cause damage to the body, and hence need to have some way to reduce the overall force to the body), that the actual trigger must be linked to the cause of the damage. This is why seat belts are based on a rocker/pendulum switch, which catches the roll that the seat belt is rolled upon, when the forces are high enough—much like the blinds that when pulling down slowly will slowly come down, but to stop them rolling back up a sharp tug is needed to catch the roll.

NOTES

Activity 7 Automatic Triggering Devices

GOALS

In this activity you will:

- Design a device that is capable of transmitting a digital electrical signal when it is accelerated in a collision.

What Do You Think?

An air bag must inflate in a sudden crash, but must not inflate under normal stopping conditions.

- **How does the air bag "know" whether to inflate?**

Record your ideas about this question in your *Active Physics* log. Be prepared to discuss your responses with your small group and the class.

For You To Do

Inquiry Investigation

1. Form engineering teams of three to five students. Meet with your engineering team to design an automatic air bag triggering device using a knife switch, rubber bands, string, wires, and a flashlight bulb. Other materials may also be supplied by you or your teacher.

Activity Overview

In this activity the students will use imagination and creativity to design their own triggering device. The emphasis on this activity should be that when "inventing" something, there should be no limitations, within reason.

Student Objectives

Students will:

- Design a device that is capable of transmitting a digital electrical signal when it is accelerated in a collision.

ANSWERS FOR THE TEACHER ONLY

What Do You Think?

Students answers will vary. One common way in which the air bag is triggered, is the method described in the **Background Information**.

Chapter 2

ANSWERS

For You To Do

All the answers for this section involve students observations and their design of the triggering device.

Safety

Be sure to receive your teacher's approval before using any material.

2. The design parameters are as follows:
   - The device must turn a flashlight bulb on, or turn it off. This will be interpreted as the trigger signal.
   - The device must not trigger if the car is brought to a sudden stop from a slow speed.
   - The device must trigger if the sudden stop is from a high speed.
   - The car containing the device must be released down a ramp. The car will then strike a wall at the bottom of the ramp.
   - The battery and bulb must be attached to the car along with the triggering device. The bulb does not have to remain in the final on or off state, but it must at least flash to show triggering.
3. Follow your teacher's guidelines for using time, space, and materials as you design your triggering device.
4. Demonstrate your design team's trigger for the class.

FOR YOU TO READ

Impulse and Changes in Momentum

It takes an unbalanced, opposing force to stop a moving car. **Newton's Second Law** lets you find out how much force is required to stop any car of any mass and any speed.

The overall idea can be shown using a concept map.

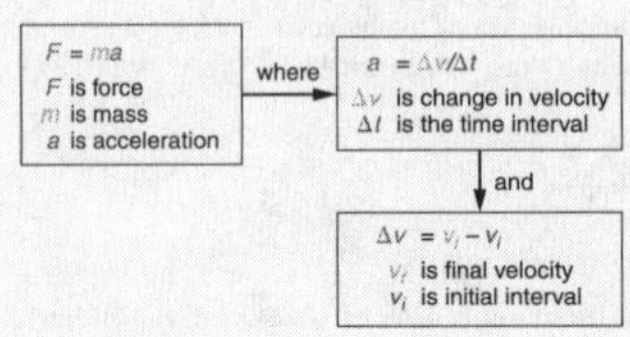

If you know the mass and can determine the acceleration, you can calculate the force using Newton's Second Law:

$$F = ma$$

A moving car has a forward **velocity** of 15 m/s. Stopping the car gives it a final velocity of 0 m/s.

The change in velocity $= v_{final} - v_{initial}$
$= 0 - 15$ m/s
$= -15$ m/s.

Any change in velocity is defined as **acceleration**.

In this case the change in velocity is −15 m/s. If the change in velocity occurs in 3 s, the acceleration is −15 m/s in 3 s, or −5 m/s every second, or −5 m/s^2. You can look at this as an equation:

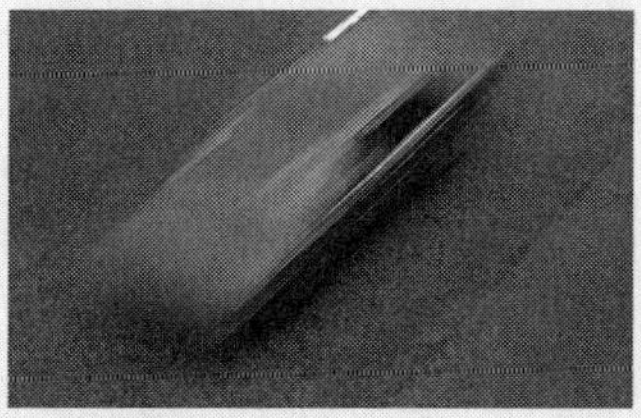

$$a = \frac{\Delta v}{\Delta t} = \frac{v_f - v_o}{\Delta t} = \frac{0 \text{ m/s}-15 \text{ m/s}}{3\text{s}} = -5 \text{ m/s}^2$$

If the car had stopped in 0.5 s, the change in speed is identical, but the acceleration is now:

$$a = \frac{\Delta v}{\Delta t} = \frac{v_f - v_o}{\Delta t} = \frac{0 \text{ m/s}-15 \text{ m/s}}{0.5 \text{ s}} = -30 \text{ m/s}^2$$

Newton's Second Law informs you that unbalanced outside forces cause all accelerations. The force stopping the car may have been the frictional force of the brakes and tires on the road, or the force of a tree, or the force of another car. Once you know the acceleration, you can calculate the force using Newton's Second Law. If the car has a mass of 1000 kg, the unbalanced force for the acceleration of –5 m/s *every* second would be –5000 N. The force for the larger acceleration of –30 m/s *every* second would be –30,000 N. The negative sign tells you that the unbalanced force was opposite in direction to the velocity.

The change in velocity, acceleration, and force give a complete picture.

There is another, equivalent picture that describes the same collision in terms of **momentum** and impulse.

Any moving car has momentum. Momentum is defined as the mass of the car multiplied by its velocity $p = mv$.

The impulse/momentum equation tells you that the momentum of the car can be changed by applying a force for a given amount of time.

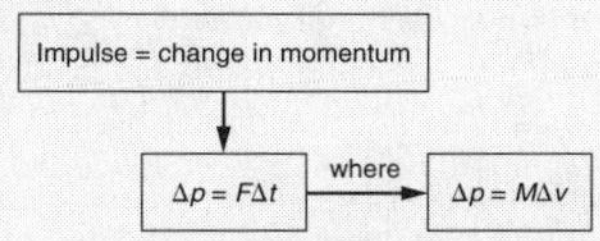

Using the impulse/momentum approach explains something different about a collision. Consider this question, "why do you prefer to land on soft grass rather than on hard concrete?" Soft grass is preferred because the force on your legs is less when you land on soft grass. Let's find out why using Newton's Second Law and then by using the impulse/momentum relation.

→

Chapter 2

Safety

Newton's Second Law explanation: Whether you land on concrete or soft grass, your change in velocity will be identical. Your velocity may decrease from 3 m/s to 0 m/s. On concrete, this change occurs very fast, while on soft grass this change occurs in a longer period of time. Your acceleration on soft grass is smaller because the change in velocity occurred in a longer period of time.

$$a = \frac{\Delta v}{\Delta t}$$

When the change in the period of time gets larger, the denominator of the fraction gets larger and the value of the acceleration gets smaller.

When landing on grass, Newton's Second Law then tells you that the force must be smaller because the acceleration is smaller for an identical mass. $F = ma$. Smaller acceleration on grass requires a smaller force. Smaller forces are easier on your legs and you prefer to land on soft grass.

Momentum/impulse explanation: Whether you land on concrete or soft grass, your change in momentum will be identical. Your velocity will decrease from 3 m/s to 0 m/s on either concrete or grass.

$$F\Delta t = \Delta p$$

You can get this change in momentum with a large force over a short time or a small force over a longer time.

If your mass is 50 kg, the amount of your change in momentum may be 150 kg m/s, when you decrease your velocity from 3 m/s to 0 m/s. There are many forces and associated times that can give this change in the value of the momentum.

If you could land on a surface that requires 3 s to stop, it will only require 50 N. A more realistic time of 1 s to stop will require a larger force of 150 N. A hard surface that brings you to a stop in 0.01 s requires a much larger force of 15,000 N.

On concrete, this change in the value of the momentum occurs very fast (a short time) and requires a large force. It hurts. On soft grass this change in the value of the momentum occurs in a longer time and requires a small force: it is less painful and is preferred.

Change in value of momentum	Force	Change in Time Δt	$F\Delta t$
150 kg m/s	50 N	3 s	150 kg m/s
150 kg m/s	75 N	2 s	150 kg m/s
150 kg m/s	150 N	1 s	150 kg m/s
150 kg m/s	1500 N	0.1 s	150 kg m/s
150 kg m/s	15,000 N	0.01 s	150 kg m/s

Physics To Go

1. How do impulse and Newton's First Law (the Law of Inertia) play a role in your air bag trigger design?
2. Imagine a device where a weight is hung from a string within a metal can. If the weight hits the side of the can, a circuit is completed. How do impulse and the Law of Inertia work in this device?
3. In cars built before 1970, the dashboard was made of hard metal. After 1970, the cars were installed with padded dashboards like you find in cars today. In designing a safe car, it is better to have a passenger hit a cushioned dashboard than a hard metal dashboard.
   a) Explain why the padded dashboard is better using Newton's Second Law.
   b) Explain why the padded dashboard is better using impulse and momentum.
4. Why would you prefer to hit an air bag during a collision than the steering wheel?
   a) Explain why the air bag is better using Newton's Second Law.
   b) Explain why the air bag is better using impulse and momentum.
5. Explain why you bend your knees when you jump to the ground.
6. Catching a fast ball stings your hand. Why does wearing a padded glove help?
7. When a soccer ball hits your chest, you can stiffen your body and the ball will bounce away from you. In contrast, you can "soften" your body and the ball drops to your feet. Explain how the force on the ball is different during each play.

Physics Words

Newton's Second Law of Motion: if a body is acted upon by an external force, it will accelerate in the direction of the unbalanced force with an acceleration proportional to the force and inversely proportational to the mass.

velocity: speed in a given direction; displacement divided by the time interval; velocity is a vector quantity, it has magnitude and direction.

acceleration: the change in velocity per unit time.

momentum: the product of the mass and the velocity of an object; momentum is a vector quantity.

Stretching Exercises

1. How does a seat belt "know" to hold you firmly in a crash, but allow you to lean forward or adjust it without locking? Write your response in your log.
2. Go to a local auto repair shop, junk yard, or parts supply store. Ask if they can show you a seat belt locking mechanism. How does it work? Construct a poster to describe what you have learned.

Answers

Physics To Go

1. Students' will have individual responses. Impulse is the change in momentum. As the car comes to a stop, the occupants and the triggering device will continue to move in a straight line (Newton's First Law of Motion—Inertia).
2. When in motion at a constant speed the metal pendulum hangs straight down. When the car stops, the pendulum continues to move forward until an impulse, the force of the string and the wall of the can, stop it.
3. a) Newton's Second Law states that the force is proportional to the acceleration. A padded dashboard will require more time for the passenger to stop. More time stopping implies a smaller acceleration. If the acceleration is smaller, the force is smaller ($F = ma$).

   b) $F\Delta t = \Delta mv$
   The change in momentum of the passenger will be identical whether stopped by a hard or soft dashboard. The padded dashboard will require more time to stop. The impulse momentum equation requires that the force be smaller.
4. a) Newton's Second Law states that the force is proportional to the acceleration. An air bag will require more time for the passenger to stop. More time stopping implies a smaller acceleration. If the acceleration is smaller, the force is smaller ($F = ma$).

   b) $F\Delta t = \Delta mv$
   The change in momentum of the passenger will be identical whether stopped by the steering column or the air bag. The air bag will require more time to stop. The impulse momentum equation requires that the force be smaller.
5. Jumping down and stopping requires you to change your momentum. This change in momentum will occur with a large force over a short time or a small force over a large time. You bend your knees in order to increase the time to stop, thereby decreasing the force on your legs.
6. Catching a baseball requires you to change the momentum of the ball. This change in momentum will occur with a large force over a short time or a small force over a large time. You use a padded glove in order to increase the time to stop, thereby decreasing the force on your hands.
7. A soccer ball hitting a hard surface like a stiff body will reverse its velocity and bounce away. A soccer ball hitting a soft surface like a "softened" body, will have an inelastic collision and the two objects will move off together after the collision with a small, almost zero velocity.

Assessment: Activity 7

The following rubric can be used to assess group work during **Activity 7**: Automatic Triggering Devices.

Each member of the group can evaluate the manner in which the group worked to solve problems in the activity.

Total = 12 marks

1. Low level – indicates minimum effort or effectiveness.
2. Average – acceptable standard has been achieved, but that the group could have worked more effectively if better organized.
3. Good – this rating indicates a superior effort. Although improvements might have been made, the group was on task all of the time and completed all parts of the activity.

Descriptor	Values		
1. The group worked cooperatively to engineer an automatic triggering device.	1	2	3
Comments:			
2. A plan was established before beginning and all tasks were shared equally.	1	2	3
Comments:			
3. The group was organized. Materials were collected and the problems were addressed by the entire group.	1	2	3
Comments:			
4. Data was collected and recorded in an organized fashion in data tables in journals.	1	2	3
Comments:			

Assessment: Scientific and Technological Thinking

Scientific and technological thinking can be assessed using the rubric below. Allow one mark for each check mark.

Maximum value = 10

Descriptor	Yes	No
1. The device (either light on or light off) does not trigger at low speeds.		
2. The device (either light on or light off) does trigger at high speeds.		
3. The device shows innovation and imagination.		
4. Controls, such as release height on ramp, are maintained throughout.		
5. The device is reliable (at least 75%).		
6. Students recognize the need for several trials.		
7. Triggering devices are tested and/or modified prior to final demonstration.		
8. Students identify the role of inertia and Newton's laws in this activity.		
9. Students can identify variables that would trigger the device.		
10. Students are able to explain their device to the class.		

NOTES

ACTIVITY 8
The Rear-End Collision

Background Information

Whiplash is the mechanism of injury, not an injury itself. However, most people associate whiplash with the injury itself. In this activity, whiplash will be referred to as the mechanism of injury, and the injury will be referred to as whiplash effect.

In a head-on collision, the head will continue to move in a forward direction while the body is strapped in, until the chin hits the chest. As there is a sudden movement, there will be minor soft tissue injury, not dissimilar to a mild sprain, but not as in the rear-end collisions.

In the rear-end collision, the head is snapped backward, and then as the collision stops, snapped forward, in a lashing motion (hence whiplash). However, if there is nothing that will stop the head from moving backward (such as a head rest), the head will keep moving as far as possible, very quickly, and doing tremendous damage to the soft tissue. The whipping back and forth will keep doing the damage, can leave the neck muscles very stretched and sore, in minor cases, or torn and not able to function properly, as well as the possibility of fracture cervical vertebrae in severe cases. Again, this will depend on the severity of the accident.

The physics involved with this accident, is again Newton's First Law – an object (head) in motion will remain in motion unless acted upon by an outside unbalanced force (the elasticity of the muscles in the neck, and the physical motion of the neck). Newton's Second Law, more commonly referred to in the formula $F = ma$, the force involved is proportional to the acceleration of the head. Therefore, if there is a larger vehicle crashing into a smaller vehicle, there will be a larger acceleration. This larger acceleration causes a larger force, which will cause the head to whip backwards with greater force, causing greater injury.

In this activity, car 1 will be released and will collide at the bottom of the ramp with car 2 with the passenger. For this collision we will assume conservation of momentum. Therefore, if the cars are the same mass, there will be conservation of momentum, and the second car will move off with a velocity the same as the first before the collision.

Momentum before the crash will equal momentum after the crash

$$m_1 v_1 + m_2 v_2 = m_1 v_1' + m_2 v_2'$$

because the first car is stationary before, $m_1 v_1 = 0$, and because the second car is stationary after $m_2 v_2' = 0$, then

$$m_2 v_2 = m_1 v_1'$$

Now, the passenger in the car, according to Newton's First Law of Motion, will remain stationary unless an outside, unbalanced force acts on it. Therefore, the body as a result of the seat pushing on the back of the individual, will move forward. However, the head remains stationary, and will only go forward after the elasticity of the neck muscles cause the head to snap forward. The result is the whiplash effect and injury.

The extent of the injury will have to do with Newton's Second Law of Motion, more commonly referred to in the formula $F = ma$ (F is the force in newtons, m is the mass in kg, and a is the acceleration in m/s^2). The force is proportional to the mass, and is also proportional to the acceleration of the object. In the case of the collision from behind, the greater the change in momentum, the greater the force acting on the car, and hence the body and head. Therefore, being hit from behind by a motorcycle will cause less damage than being hit from behind by a semi.

Active-ating the Physics InfoMall

Search for "whiplash" and you will find a nice discussion in *The Fascination of Physics*, from the Textbook Trove. This discussion includes some nice graphics.

Newton's Second Law is discussed in virtually every physics textbook in existence, not to mention the InfoMall. Depending on the level at which you wish to present this Law, you may wish to examine the conceptual-level texts, the algebra-based texts, or even the calculus-based textbooks on the InfoMall.

If you want more exercises to give to your students, searching the InfoMall is a bad idea - there are too many problems on the CD-ROM. Searching with keywords "force" AND "acceleration" AND "mass" in the Problems Place alone produces "Too many hits." However, you will find more than enough by simply going to the Problems Place and browsing a few of the resources you will find there. For example, *Schaum's 3000 Solved Problems in Physics* has a section on Newton's Laws of Motion. You will surely find enough problems there to keep any student busy for some time!

For You To Do Step 6 mentions ratios. This is one of those areas in which students are known to have problems. Perform a search using "student difficult*" AND "ratio*" for more information. Alternately, try "misconcept*" AND "ratio*". If any of these searches causes "Too many hits" simply reduce the number of stores you search in, or require that the words occur in the same paragraph.

Try similar searches to find student difficulties with the concept of acceleration. You should find, for example, "Investigation of student understanding of the concept of acceleration in one dimension," *American Journal of Physics*, vol. 49, issue 3.

Physics To Go, Step 6, asks students to predict the direction a cork will move. If you were to search the InfoMall to find more about the importance of predictions in learning, you would find that you need to limit your search. For example, a search for "prediction*" AND "inertia" resulted in several hits; the first hit is from *A Guide to Introductory Physics Teaching: Elementary Dynamics*, Arnold B. Arons' Book Basement entry. Here is a quote from that book: "Because of the obvious conceptual importance of the subject matter, the preconceptions students bring with them when starting the study of dynamics, and the difficulties they encounter with the Law of Inertia and the concept of force, have attracted extensive investigation and generated a substantial literature. A sampling of useful papers, giving far more extensive detail than can be incorporated here, is cited in the bibliography [Champagne, Klopfer, and Anderson (1980); Clement (1982); di Sessa (1982); Gunstone, Champagne, and Klopfer (1981); Halloun and Hestenes (1985); McCloskey, Camarazza, and Green (1980); McCloskey (1983); McDermott (1984); Minstrell (1982); Viennot (1979); White (1983), (1984)]." Note that students' preconceptions can have a large effect on how they learn something. It is important that they are forced to consciously acknowledge their preconceptions by making predictions.

Planning for the Activity

Time Requirements

Approximately 40 minutes are required to complete the procedure. Longer if the students would like to try to rig up different devices to prevent the injury.

Materials Needed

For each group:

- balance platform
- duct tape, roll
- dynamics carts
- spring scale, 0-10 newton range
- starting ramp for lab cart
- stick, meter, 100 cm, hardwood
- tape, masking, 3/4" x 36 yds.
- wood piece, 1" x 2" x 2"
- wood piece, 1" x 3" x 10"
- wood piece, 2" x 4" x 1'

Advance Preparation and Setup

Show the students your model of the passenger.

Teaching Notes

This assignment provides an application of the scientific principles of momentum, and Newton's First and Second Laws.

Students will probably be able to articulate the action or anticipated actions. However, it is important that the physics be explained, so as to help them realize that this demonstration can, in fact, be a realistic model.

Give a short explanation as to conservation of momentum. Most students will have a realistic view (from playing billiards, playing marbles, football, or hockey, or other contact sports).

Activity Overview

The students will be comparing the collision of their model "crash test dummy" with real life accidents, in order to analyze the effect of rear-end collisions on the neck muscles.

Student Objectives

Students will:

- Evaluate from simulated collisions, the effect of rear-end collisions on the neck muscles.
- Understand the causes of whiplash injuries.
- Understand Newton's Second Law of Motion.
- Understand the role of safety devices in preventing whiplash injury.

ANSWERS FOR THE TEACHER ONLY

What Do You Think?

Students will likely say that whiplash is an injury to the neck, when hit from behind. See **Background Information** for explanation of whiplash injury.

Students will likely say that when you go forward, your body will also move forward, so there is less force acting on the neck. However, when hit from behind, the head stays at rest due to inertia, as the body is accelerated forward by the seat. This is true to a certain extent, but the primary reason is due to the lashing effect of the head being whipped back and forth, with nothing to support the head in the backwards direction. In the front the chin will hit the chest, thus stopping the head's forward motion.

Safety

Activity 8 The Rear-End Collision

GOALS

In this activity you will:

- Evaluate from simulated collisions, the effect of rear-end collisions on the neck muscles.
- Understand the causes of whiplash injuries.
- Understand Newton's Second Law of Motion.
- Understand the role of safety devices in preventing whiplash injury.

What Do You Think?

Whiplash is a serious injury that is caused by a rear-end collision. It is the focus of many lawsuits, loss of ability to work, and discomfort.

- **What is whiplash?**
- **Why is it more prominent in rear-end collisions?**

Record your ideas about these questions in your *Active Physics* log. Be prepared to discuss your responses with your small group and the class.

Active Physics 122

Activity 8 The Rear-End Collision

For You To Do

1. You will use two pieces of wood to represent the torso (the trunk of the body) and the head of a passenger. Attach a small piece of wood (about 1" x 2" x 2") to a larger piece of wood (about 1" x 3" x 10") with some duct tape acting like a hinge between the two pieces.
   a) Make a sketch to show your passenger. Label what each part of the model passenger represents.

2. Set up a ramp against a stack of books about 40 cm high, as shown in the diagram below. Place the wooden model passenger at the front of a collision cart positioned about 50 cm from the end of the ramp. Release a second cart from a few centimeters up the ramp.
   a) In your log record what happens to the head and torso of the wooden model.

Perform the activity outside of traffic areas. Do not obstruct paths to exits. Do not leave carts lying on the floor.

3. With the first cart still positioned about 50 cm from the end of the ramp, release the second cart from the top of the ramp.
   a) Describe what happens to the head of the model passenger in this collision.
   b) Use Newton's First Law of Motion to explain your observations.

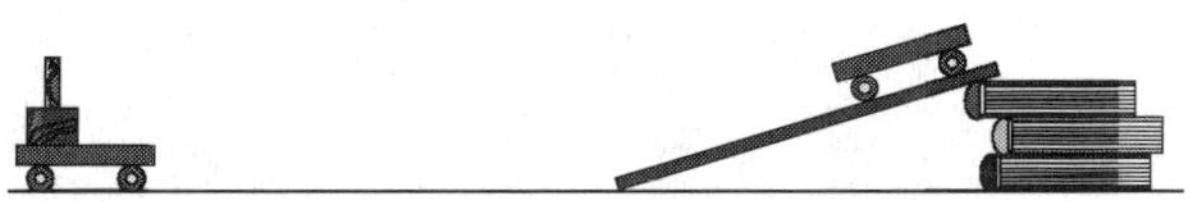

4. The duct tape represents the neck muscles and bones of the vertebral column. How large a force do the neck muscles exert to keep the head from flying off the body, and to return the head to the upright position? To answer this question, begin by estimating the mass of an average head.
   a) Estimate and record in your log the mass of an average human head. The mass would be close to the mass of a filled water container of the same size.

Answers

For You To Do

1. a) Students' sketches.

2. a) Students' response. They should note that the "head" may snap backward, and the torso may move backward, but without the snapping motion. Because of the low speed, there may not be a lot of movement.

3. a) The head will snap backward, or may even fall off.
   b) An object (the head) at rest will remain at rest unless acted upon by an outside unbalanced force (the first cart crashing into it).

4. a) One way to estimate the mass, is to find the approximate volume of the head, and multiply it by the mass of 1 mL of water. The average head of a human is about 14 pounds (estimate).

5. Mark off a distance about 30-cm long on the lab table or the floor. Obtain a piece of wood and attach it to a spring scale. Pull the wooden mass with the spring scale over the distance you marked.
   - a) In your log record the force required to pull the mass and the time it took to cover the distance.
   - b) Repeat the step, but vary the time required to pull the mass over the distance. Record the forces and the times in your log.
   - c) Use your observations to complete the following statement:
     The shorter the time (that is, the greater the acceleration) the ______ the force required.

6. The ratio of the mass of the wood to the estimated mass of the head is the same as the ratio of the forces required to pull them.
   - a) Use the following ratio to calculate how large a force the neck muscles exert to keep the head from flying off the body, and to return the head to the upright position under different accelerations.

$$\frac{\text{mass of head}}{\text{mass of wood}} = \frac{\text{force to move head}}{\text{force to move wood}}$$

7. Whiplash is a serious injury that can be caused by a rear-end collision. The back of the car seat pushes forward on the torso of the driver and the passengers and their bodies lunge forward. The heads remain still for a very short time. The body moving forward and the head remaining still causes the head to snap backwards. The neck muscles and bones of the vertebral column become damaged. The same muscles must then snap the head back to its place atop the shoulders.
   - a) What type of safety devices can reduce the delay between body and head movement to help prevent injury?
   - b) What additional devices have been placed in cars to help reduce the impact of rear-end collisions?

Answers

For You To Do *(continued)*

5. a) Students' response.
   b) Students' response.
   c) Students should note that the shorter the time period, the greater the force needed to accelerate the block of wood.

6. a) Students' response.

7. a) Head rests.
   b) Rear bumpers also have the collapsing bumpers, which absorb some of the energy of the collision, as well as the crumple zone in the rear of the car. Cars also are equipped with more visible brake lights, to enable the driver behind to see the braking vehicle in front.

Activity 8 The Rear-End Collision

FOR YOU TO READ

Newton's Second Law of Motion

Newton's First Law of Motion is limited since it only tells you what happens to objects if net force acts upon them. Knowing that objects at rest have a tendency to remain at rest and that objects in motion will continue in motion does not provide enough information to analyze collisions. Newton's Second Law allows you to make predictions about what happens when an external force is applied to an object. If you were to place a collision cart on a level surface, it would not move. However, if you begin to push the cart, it will begin to move.

Newton's Second Law states:

If a body is acted on by a force, it will accelerate in the direction of the unbalanced force. The acceleration will be larger for smaller masses. The acceleration can be an increase in speed, a decrease in speed, or a change in direction.

Newton's Second Law of Motion indicates that the change in motion is determined by the force acting on the object, and the mass of the object itself.

Analyzing the Rear-End Collision

This activity demonstrated the effects of a rear-end collision. Newton's First Law and Newton's Second Law can help explain the "whiplash" injury that passengers suffer during this kind of collision.

Imagine looking at the rear-end collision in slow motion. Think about all that happens.

1. A car is stopped at a red light. This is the car in which the driver is going to be injured with whiplash. The driver is at rest within the car.
2. The stopped car gets hit from the rear.
3. The car begins to move. The back of the seat pushes the driver forward and his torso moves with the car. The driver's head is not supported and stays back where it is.
4. The neck muscles hold the head to the torso as the body moves forward. The muscles then "whip" the head forward. The head keeps moving until it gets ahead of the torso. The head is stopped by the neck muscles. The muscles pull the head back to its usual position. Ouch!

Let's repeat the description of the collision and insert all of the places where Newton's First Law applies. Newton's First Law states that *an object at rest stays at rest and an object in motion stays in motion unless acted upon by an unbalanced, outside force.*

1. A car is stopped at a red light. This is the car in which the driver is going to be injured with whiplash. The driver is at rest within the car. *Newton's First Law: an object (the driver) at rest stays at rest.*

→

Chapter 2

Safety

2. The stopped car gets hit from the rear.
3. The car begins to move. The back of the seat pushes the driver forward and his torso moves with the car. *Newton's First Law: an object (the driver's torso) at rest stays at rest unless acted upon by an unbalanced, outside force.* The driver's head is not supported and stays back where it is. *Newton's First Law: an object (the driver's head) at rest stays at rest.*
4. The neck muscles hold the head to the body as the body moves forward. The muscles then "whip" the head forward. *Newton's First Law: an object (the head) at rest stays at rest unless acted upon by an unbalanced, outside force.* The head keeps moving until it gets ahead of the torso. *Newton's First Law: an object (the head) in motion stays in motion.* The head is stopped by the neck muscles. *Newton's First Law: an object (the head) in motion stays in motion unless acted upon by an unbalanced, outside force.* The muscles pull the head back to its usual position. *Newton's First Law: an object at rest stays at rest unless acted upon by an unbalanced, outside force.* Ouch!

Let's repeat the description of the collision and insert all of the places where Newton's Second Law applies. Newton's Second Law states that *all accelerations are caused by unbalanced, outside forces, F = ma.* An acceleration is any change in speed.

1. A car is stopped at a red light. This is the car in which the driver is going to be injured with whiplash. The driver is at rest within the car.
2. The stopped car gets hit from the rear.
3. The car begins to move. *Newton's Second Law: the car accelerates because of the unbalanced, outside force from the rear; F = ma.* The back of the seat pushes the driver forward and his torso moves with the car. *Newton's Second Law: the torso accelerates because of the unbalanced, outside force from the back of the seat; F = ma.* The driver's head is not supported and stays back where it is.
4. The neck muscles hold the head to the torso as the body moves forward. The muscles then "whip" the head forward. *Newton's Second Law: the head accelerates because of the unbalanced force of the muscles; F = ma.* The head keeps moving until it gets ahead of the torso. The head is stopped by the neck muscles. *Newton's Second Law: the head accelerates (slows down) because of the unbalanced force from the neck muscles; F = ma.* The muscles pull the head back to its usual position. *Newton's Second Law: the head accelerates because of the unbalanced force from the rear; F = ma.* Ouch!

Newton's Second Law informs you that all accelerations are caused by *unbalanced, outside* forces. It does not say that all forces cause accelerations. An object at rest may have many forces acting upon it. When you hold a book in

126

Active Physics

your hand, the book is at rest. There is a force of gravity pulling the book down. There is a force of your hand pushing the book up. These forces are equal and opposite. The "net" force on the book is zero because the two forces balance each other. There is no acceleration because there is no "net" force.

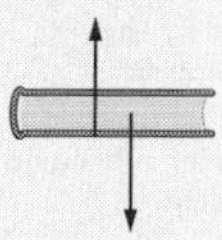

Both forces act through the center of the book. They are shifted a bit in the diagram to emphasize that the upward force of the hand acts on the bottom of the book and the downward force of gravity acts on the middle of the book.

As a car moves down the highway at a constant speed, there are forces acting on the car but there is no acceleration. This indicates that the net force must be zero. The force of the engine on the tires and road moving the car forward must be equal and opposite to the force of the air pushing the car backward. These forces balance each other in this case, where the speed is not changing. There is no net force and there is no acceleration. The car stays in motion at a constant speed. A similar situation occurs when you push a book across a table at constant speed. The push is to the right and the friction is to the left, opposing motion. If the forces are equal in size, there is no net force on the book and the book does not accelerate—it moves with a constant speed.

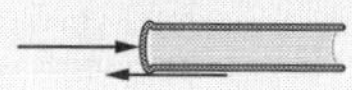

Reflecting on the Activity and the Challenge

The vertebral column becomes thinner and the bones become smaller as the column attaches to the skull. The attachment bones are supported by the least amount of muscle. Unfortunately, the smaller bones, with less muscle support, make this area particularly susceptible to injury. One of the greatest dangers following whiplash is the damage to the brainstem. The brainstem is particularly vital to life support because it regulates blood pressure and breathing movements. Consider how your safety device will help prevent whiplash following a collision. What part of the restraining device prevents the movement of the head?

Chapter 2

ANSWERS

Physics To Go

1. The huge forces that are associated with the rear-end collision, in conjunction with little or no support for the neck from behind.
2. Inertia. An object will continue to move in a straight path until an outside force acts on it.
3. The passengers on the bus are not moving. They have a tendency to remain at rest even after the bus begins moving.
4. Because the motorcycle has less mass it will move more quickly if struck by a car.
5. The headrest would be most beneficial if you were in a rear-end collision. The passengers and driver are forced backward during the collision.
6. a) The cork will appear to move in the opposite direction to the push. Emphasize to students, that the cork is not moving, but "trying to stay motionless" (inertia).

   b) Opposite to the original push.

Safety

Physics To Go

1. Why are neck injuries common after rear-end collisions?
2. Explain why the packages in the back move forward if a truck comes to a quick stop.
3. As a bus accelerates, the passengers on the bus are jolted toward the back of the bus. Indicate what causes the passengers to be pushed backward.
4. Why would the rear-end collision demonstrated by the laboratory experiment be most dangerous for someone driving a motorcycle?
5. Would headrests serve the greatest benefit during a head-on collision or a rear-end collision? Explain your answer.
6. A cork is attached to a string and placed in a jar of water as shown by the diagram to the right. Later, the jar is inverted.
   a) If the glass jar is pushed along the surface of a table, predict the direction in which the cork will move.
   b) If you place your left hand about 50 cm in front of the jar and push it with your right hand until it strikes your left hand, predict the direction in which the cork will move.

Be sure the outside of the jar is dry so it does not slip out of your hands.

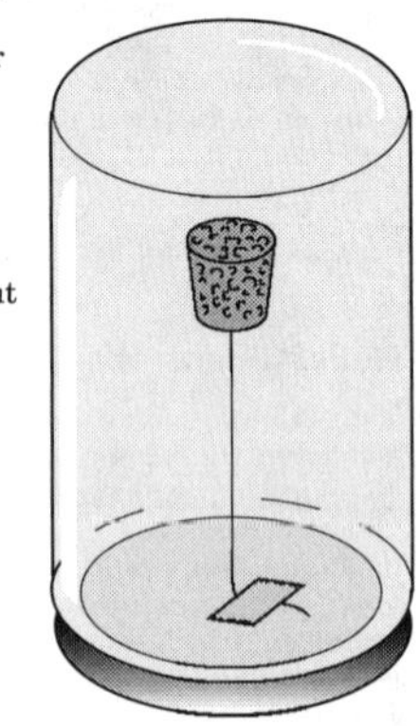

128

ACTIVITY 9
Cushioning Collisions (Computer Analysis)

Background Information

In this activity the students will be putting into practice the concepts that they have learned in the past few activities. Their task is to design a cushioning device which will reduce the damage that can be done in a collision. They should be looking for cushioning material that will reduce the force as much as possible. The change in momentum will be the same with each velocity, but the force that brings the car to a rest should be as small as possible.

One other concept that may help in understanding collisions is the Conservation of Energy.

Collisions can be either elastic (energy is conserved) or inelastic (energy is not conserved). The collisions that we study are collisions (in the laboratory) where there is always a conservation of momentum. While there is conservation of momentum (p), there may or may not be conservation of energy. For example, there can be conservation of energy with two billiard balls moving towards each other, and colliding and moving away with the exact opposite speed and direction. This is elastic. However, in the case of most collisions, there is no conservation of energy

In this activity, there will not be a conservation of energy. Therefore, the students will need to design a cushion which will be able to transform as much kinetic energy (energy of motion) as possible to another substance(s). For example, when you have a large ball and you drop it onto a solid floor (such as concrete), it will bounce back. Therefore, there is energy from kinetic energy falling, transferred to potential in the elastic potential energy of the ball in deformation, and subsequently transferred back into kinetic energy to move it upwards. (Assume an elastic collision, and that the ball will bounce back to its original height.)

However, if you were to drop that same ball into a box of sand, then the kinetic energy is transferred to each particle of sand moving in all directions. There is a conservation of momentum in both cases, but not a conservation of energy. In almost all cases of collisions, in reality, there is not a conservation of energy. Energy is transferred to sound, but mostly to heat due to friction of the parts heating up in the collision, and to the deformation of the car parts.

Active-ating the Physics InfoMall

In addition to the information we found previously (on momentum and graphs, for example), you may wish to examine the Problems Place for even more exercises in momentum conservation. Remember, *Schaum's 3000 Solved Problems in Physics* has the problem and the solution. It can be a source for you, as well as a way to provide your students with solved problems for them to study!

Planning for the Activity

Time Requirements

The time for this activity will vary depending on the materials at hand. Allow time to investigate as many different kinds of materials as available, and encourage bringing materials from home. This activity can be introduced (10 minutes) one day and then allow one class (approximately 40 minutes) for the design and measuring. Allow extra time for analyzing the data and graphing the data as necessary.

Materials Needed

For each group:

- balance platform
- cushioning materials set, variety of
- MBL or CBL with sonic ranger
- starting ramp for lab car
- VCR & TV monitor
- tape, masking, 3/4" x 36 yds.

Advance Preparation and Setup

If your materials are limited, use a large-screen monitor to display the data to the class. Prepare a kit for each team consisting of a toy car and a different cushioning material. Each team can prepare one car and run it as a demonstration station.

Chapter 2

Try a few sample runs in advance to make sure that the results fall within limits of the available probes. Adjust the height of the ramp accordingly. Demonstrate the operation of the sensors and the software to the class if necessary.

Providing materials produces a certain regularity and predictability to the activity. You may want to present this activity as a challenge a day in advance. The students are sure to bring some unusual cushioning material from home.

Teaching Notes

Replaying the video of auto crashes is one way to engage students in this activity. Ask students to explain and give examples of the three types of collisions that result from a car crash. Previous activities focused on the secondary collision—the occupants with the interior of the car. The tertiary collision—the internal organs of the occupants is not easily tested. Here the focus is on the primary collision—the vehicle with another vehicle or obstacle. How can the effects of this collision be minimized? Students should be encouraged to think of other systems besides sand canisters that are designed to cushion primary collisions such as crumple zones in cars, energy-absorbing bumpers, standardized height bumpers, break-away signs and light poles, types of plantings near roads, buried guard rail ends, shape of curbs or barriers (so-called "New Jersey barriers" are shaped to rub against the tire to slow the car), etc.

In order to help students with the **WDYT**, demonstrate the operation of the force probe and focus on the impulse graph (force vs. time). Demonstrate how the measurement of the impact force and velocity prior to impact changes when the ramp angle is changed. See if the students understand the operation of the apparatus, make a measurement of the velocity of the toy car before the collision and the impulse during the collision when no cushioning is used. This baseline information (the graphs) should be provided to each lab group or left on the chalkboard during the investigation.

As each group completes the design on the cushioning system, they should test the effectiveness of their system by using the probes at the demonstration work station. Once the students have a print-out of the F vs. t graphs, they can return to their regular work stations to analyze the data and figure out ways of increasing the effectiveness of their cushioning system.

The balance is needed to get the mass of each cushioned car in order to calculate its momentum.

Students can work in groups of four, using similar criteria to the previous assessment criteria for group work. This activity will inevitably lead to noise, but busy noise can still be productive noise. Keep students on task by asking them questions such as: Why are you using that material? What are the independent variables? Why are you keeping the height of the ramp the same for this activity? Why are you changing the material?

Refer to the safety notes in previous activities regarding a secure ramp, and objects falling on students' toes.

Students will need to be guided to realize, again, that the increase in the time of the collision is the most important factor here. The change in momentum will always be the same, as the mass of the object never changes, and the velocity should be constant if they are releasing the car at the same height each time they do the trial.

Activity 9 Cushioning Collisions [Computer Analysis]

GOALS

In this activity you will:

- Apply the concept of impulse in the analysis of automobile collisions.
- Use a computer's motion probe (sonic ranger) to determine the velocity of moving vehicles.
- Use a computer's force probe to determine the force exerted during a collision.
- Compare the momentum of a model vehicle before the collision with the impulse applied during the collision.
- Explore ways of using cushions to increase the time that a force acts during a primary collision.

What Do You Think?

The use of sand canisters around bridge supports and crush zones in cars are examples of technological systems that are designed to minimize the impact of collisions between a car and a stationary object or another car.

- **How do these technological systems reduce the impact of the primary collision?**

Record your ideas about this question in your *Active Physics* log. Be prepared to discuss your responses with your small group and the class.

For You To Do

1. In this investigation you will be using a force probe that is attached to a computer to determine the effectiveness of different types of cushions for a toy vehicle. Release a toy car at

ANSWERS

For You to Do

1. Student activity.

Activity Overview

The students will be analyzing how different types of materials will affect the impact of collisions. They will relate these designs with real-life materials, such as the bumper design on cars. The design of these materials will teach about impulse and momentum.

Student Objectives

Students will:

- Apply the concept of impulse in the analysis of automobile collisions.
- Use a computer's motion probe (sonic ranger) to determine the velocity of moving vehicles.
- Use a computer's force probe to determine the force exerted during a collision.
- Compare the momentum of a model vehicle before the collision with the impulse applied during the collision.
- Explore ways of using cushions to increase the time that a force acts during a primary collision.

ANSWERS FOR THE TEACHER ONLY

What Do You Think?

Looking back to the understanding of impulse and forces, we can say that the impulse is the same in any collision from the same speed—the car traveling on the highway at 55 mph is brought to rest. However, the car which crashes directly into the bridge support experiences tremendously large forces, as the time of the impulse is very small. The car that crashes into the sand canisters, increases the time of the collision, sometimes many times longer than the first car, which allows the force on the car to be decreased by the same factor. This is similar to the example of the ball falling into the sand.

ANSWERS

For You To Do
(continued)

2.–3. Data will vary.

4. a) Data will vary.

5. a) Data will vary.

Safety

Perform the activity outside of traffic areas. Do not obstruct paths to exits. Do not leave carts lying on the floor.

the top of a ramp and measure the force of impact as it strikes a barrier at the bottom. A sonic ranger can be mounted on the ramp to measure the speed of the toy car prior to the collision. Open the appropriate computer files to prepare the sonic ranger to graph velocity vs. time and the force probe to graph force vs. time.

2. Mount the sonic ranger at the bottom of a ramp and place the force probe against a barrier about 10 cm from the bottom of the ramp, as shown in the diagram. Attach an index card to the back of the car, to obtain better reflection of the sound wave and improve the readings of the sonic ranger.

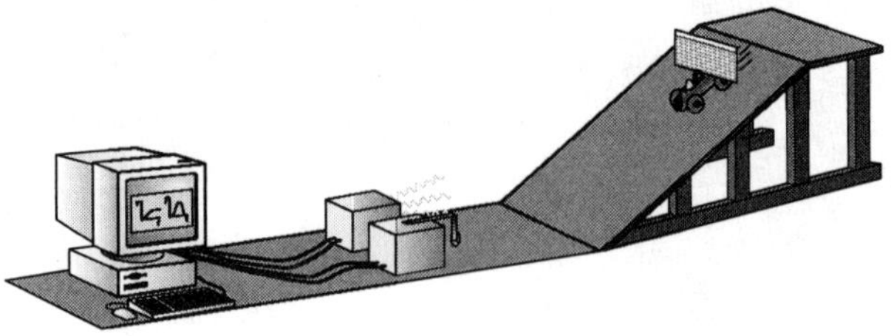

3. Conduct a few runs of the car against the force probe to ensure that the data collection equipment is working properly.
4. Attach a cushioning material to the front of the car. Conduct a number of runs with the same type of cushioning. Make sure that the car is coasting down the same slope from the same position each time.
   a) Make copies of the velocity vs. time and force vs. time graphs that are displayed on the computer.
5. Repeat **Step 4** using other types of cushioning materials.
   a) Record your observations in your log.

130

Active Physics

Activity 9 Cushioning Collisions (Computer Analysis)

6. Use the graphical information you obtained in this activity to answer the following:
   a) Compare the force vs. time graphs for the cushioned cars with those for the cars without cushioning.
   b) Compare the areas under the force vs. time graphs for all of the experimental trials.
   c) Compute the momentum of the car (the product of the mass and the velocity) prior to the collision and compare it with the area under the force vs. time graphs.
   d) Summarize your comparisons in a chart.
   e) How can impulse be used to explain the effectiveness of cushioning systems?
   f) Describe the relationship between impulse ($F\Delta t$) and the change in momentum ($m\Delta v$).

PHYSICS TALK

Change in Momentum and Impulse

Momentum is the product of the mass and the velocity of an object.

$$p = mv$$

where p is the momentum,
m is the mass,
and v is the velocity.

Change in momentum is the product of mass and the change in velocity.

$$\Delta p = m\Delta v$$

Impulse = change in momentum

$$F\Delta t = m\Delta v$$

ANSWERS

For You To Do
(continued)

6. a) Data will vary. Generally, students should notice that the force is greater on the collision without the cushioning.

   b) Data will vary. If the speeds before the collisions are the same, then the areas under the graphs will be the same.

   c) Mass x velocity should equal the area under the force vs. time graphs. The relevant velocity is the velocity at impact, which should be the highest value recorded by the sonic ranger. The area under the graph should equal the change in momentum of the car, since the final momentum is zero. Student answers may vary due to calculation and measurement errors.

   d) Data will vary.

   e) Impulse is equal to the change in momentum, and is therefore related to the mass and the velocity of the car. Depending on the cushioning, there will be an increase in the time, impulse remaining constant; thus the force acting on the car and individuals inside will be less.

   f) Impulse equals the change in momentum.

Sample Problem

A vehicle has a mass of 1500 kg. It is traveling at 15.0 m/s. Calculate the change in momentum required to slow the vehicle down to 5.0 m/s.

Strategy: You can use the equation for calculating the change in momentum.

$$\Delta p = m\Delta v$$

Recall that the Δ symbol means "the change in." If you know the final and initial velocities you can write this equation as:

$$\Delta p = m(v_f - v_i)$$

where v_f is the final velocity and

v_i is the initial velocity.

Givens:

$m = 1500$ kg

$v_f = 5.0$ m/s

$v_i = 15.0$ m/s

Solution:

$$\Delta p = m\,(v_f - v_i)$$
$$= 1500 \text{ kg } (5.0 \text{ m/s} - 15.0 \text{ m/s})$$
$$= 1500 \text{ kg } (-10.0 \text{ m/s})$$
$$= -15{,}000 \text{ kg} \cdot \text{m/s}$$

Activity 9 Cushioning Collisions (Computer Analysis)

Reflecting on the Activity and the Challenge

What you learned in this activity better prepares you to defend the design of your safety system. The principles of momentum and impulse must be used to justify your design. Previously, you discovered objects with greater mass are more difficult to stop than smaller ones. You determined that increasing the velocity of objects also makes them more difficult to stop. Objects that have a greater mass or greater velocity have greater momentum.

Linking the two ideas together allows you to begin examining the relationship between momentum and impulse. For a large momentum change in a short time, a large force is required. A crushed rib cage or broken leg bones often result. The change in the momentum can be defined by the impulse on the object.

What device will you use to increase the stopping time for the **Chapter Challenge** activity? Make sure that you include impulse and change in momentum in your report. Your design features must be supported by the principles of physics.

Physics To Go

1. Helmets are designed to protect cyclists. How would the designer of helmets make use of the concept of impulse to improve their effectiveness?
2. The Congress of the United States periodically reviews federal legislation that relates to the design of safer cars. For many years, one regulation was that car bumpers must be able to withstand a 5 mph collision. What was the intent of this regulation? The speed was later lowered to 3 mph. Why? Should it be changed again?
3. If a car has a mass of 1200 kg and an initial velocity of 10 m/s (about 20 mph) calculate the change in momentum required to:

   a) bring it to rest

   b) slow it to 5 m/s (approximately 10 mph)

ANSWERS

Physics To Go

1. Helmets use cushioning materials that lengthen the time of impact and thus decrease the force of impact. Note that helmets are designed to break and should not be used again after an accident. A rigid, unbreakable helmet would be useless.
2. The ability of bumpers to withstand collisions at various speeds is directly related to the ability of the bumpers to minimize the force of the impact via cushioning. The intent of the regulation is to protect the consumer from expensive repairs after every fender-bender or parking lot bump. The industry was able to get the regulation changed by arguing that the initial cost of effective bumpers was too high and the added weight contributed to higher fuel consumption. Arguments for another change can be based on consumer preference, cost of injuries, and the development of new materials and technologies that would reduce initial costs.
3. a) $\Delta p = m\Delta v$;

   $\Delta p = m \text{ x } (v_2 - v_1)$;

   $\Delta p = 1200 \text{ kg x } (0 - 10 \text{ m/s})$;

   $\Delta p = 12{,}000 \text{ kg}\cdot\text{m/s}$

   b) $\Delta p = m\Delta v$;

   $\Delta p = m \text{ x } (v_2 - v_1)$;

   $\Delta p = 1200 \text{ kg x } (5 \text{ m/s} - 10 \text{ m/s})$;

   $\Delta p = 6000 \text{ kg}\cdot\text{m/s}$

ANSWERS

Physics To Go
(continued)

4. Impulse $\Delta p = F\Delta t$

   Δp = 10,000 N x 1.2 s

   Δp = 12,000 kg•m/s

   change in velocity is Δv therefore:

   $\Delta p = m\Delta v$

   $\Delta v = \Delta p / m$

   Δv = 12,000 k.gm/s / 1200 kg

   Δv = 10 m/s

5. $\Delta p = F\Delta t$

   $m\Delta v = F\Delta t$

   $F = m\Delta v/\Delta t$

   F = 1500 kg x 5 m/s / 0.1 s

   F = 75,000 N

6. Increasing the time from 0.1 s to 2.8 seconds decreases the force acting on the car (and the driver).

   $\Delta p = F\Delta t$

   $m\Delta v = F\Delta t$

   $F = m\Delta v/\Delta t$

   F = 1500 kg x 5 m/s / 2.8 s

   F = 2700 N

7. Students' answers will vary.

8. The steering wheel increases the time of contact, and will decrease the force acting on the body. Similar to previous answer.

9. The first graph has a greater force over a shorter time period, and the second graph has a smaller force over a longer period of time. If you measure the area under the graph, you will find the impulse.

Safety

4. If the braking force for a car is 10,000 N, calculate the impulse if the brake is applied for 1.2 s. If the car has a mass of 1200 kg, what is the change in velocity of the car over this 1.2 s time interval?
5. A 1500-kg car, traveling at 5.0 m/s after braking, strikes a power pole and comes to a full stop in 0.1 s. Calculate the force exerted by the power pole and brakes required to stop the car.
6. For the car described in **Question 5**, explain why a breakaway pole that brings the car to rest after 2.8 s is safer than the conventional power pole.
7. Write a short essay relating your explanation for the operation of the cushioning systems to the explanation of the operation of the air bags.
8. Explain why a collapsible steering wheel is able to help prevent injuries during a car crash.
9. Compare and contrast the two force vs. time graphs shown.

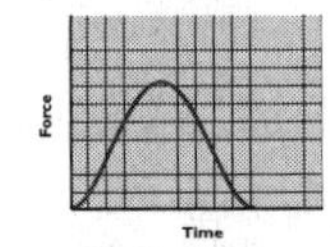

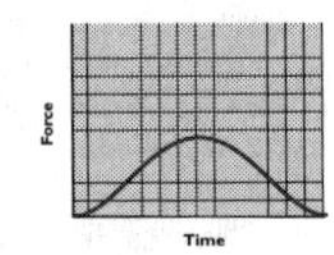

Stretching Exercises

Package an egg in a small container so that the egg will not break upon impact. Your teacher will provide the limitations in the construction of your package. You may be limited to two pieces of paper and some tape. You may be limited to a certain size package or a package of a certain weight. Bring your package to class so that it can be compared in a crash test with the other packages.

(Hint: Place each egg in a plastic bag before packaging to help avoid a messy cleanup.)

134

PHYSICS AT WORK

Mohan Thomas

DESIGNING AUTOMOBILES THAT SAVE LIVES

Mo is a Senior Project Engineer at General Motors North American Operation's (NAO) Safety Center and his responsibilities include making sure that different General Motors vehicles meet national safety requirements. Several of the design features that Mo has helped to develop have been implemented into vehicles that are now out on the road.

"This is how it works," he explains. "An engineer for a vehicle comes to us here at the Safety Center and requests technical assistance with design features to help them meet the side impact crash regulations required by the government. You have to analyze the physical forces of an event, which involves one car hitting another car on the side and then the door smashing into the driver," he continues. "We'll study the velocity, acceleration, momentum, and inertia in an event, as well as the materials used in the vehicle itself."

"The initial energy of an impact from one vehicle on another," states Mo, "has to be managed by the vehicle that's getting hit. Our goal is to manage the energy in such a way that the occupant in the vehicle being hit is protected. You take the forces that are coming into the vehicle and you redirect them into areas around the occupant. The framework of the car, therefore, is very important to the design, as well as energy-absorbing materials used in the vehicle."

Mo grew up in Chicago, Illinois, and has always enjoyed math and science, but he was also interested in creative writing. He wanted to combine math and science with creative work and has found that combination in the design work of engineering. "The nice part of being at the Safety Center," states Mo, "is that you know that you are contributing to something meaningful. The bottom line is that the formulas and problems that we are working on are meant to save people's lives."

Chapter 2 Assessment

Your design team will develop a safety system for protecting automobile, airplane, bicycle, motorcycle or train passengers. As you study existing safety systems, you and your design team should be listing ideas for improving an existing system or designing a new system for preventing accidents. You may also consider a system that will minimize the harm caused by accidents.

Your final product will be a working model or prototype of a safety system. On the day that you bring the final product to class, the teams will display them around the room while class members informally view them and discuss them with members of the design team. At this time, class members will generate questions about each others' products. The questions will be placed in envelopes provided to each team by the teacher. The teacher will use some of these questions during the oral presentations on the next day. The product will be judged according to the following:

1. The quality of your safety feature enhancement and the working model or prototype.
2. The quality of a 5-minute oral report that should include:
   - **the need for the system**
   - **the method used to develop the working model**
   - **demonstration of the working model**
   - **discussion of the physics concepts involved**
   - **description of the next-generation version of the system**
   - **answers to questions posed by the class**
3. The quality of a written and/or multimedia report including:
   - **the information from the oral report**
   - **documentation of the sources of expert information**
   - **discussion of consumer acceptance and market potential**
   - **discussion of the physics concepts applied in the design of the safety system**

Criteria

Review the criteria that were agreed to at the beginning of the chapter. If they require modification, come to an agreement with the teacher and the class.

Your project should be judged by you and your design team according to the criteria before you display and share it with your class. Being able to judge the quality of your own work before you submit it is one of the skills that will make you a "treasured employee"!

137

Active Physical Science

Chapter 2

Physics You Learned

Newton's First Law of Motion (inertia)

Pressure (N/m^2)

Pressure = $\frac{\text{force}}{\text{area}}$ $(P = \frac{F}{a})$

Distance vs. time relationships

Time interval =
time$_{(final)}$ − time$_{(initial)}$ $(\Delta t = t_f - t_i)$

Impulse (N × *time*),

Impulse =
force × time interval (Impulse = $F \times \Delta t$)

Stopping distance

Newton's Second Law of Motion, constant acceleration, net force
Force = mass × acceleration
$F = ma$

Momentum = mass × velocity ($p = mv$)

Conservation of momentum

Change in momentum is affected by mass and change in velocity

Change in momentum =
mass × change in velocity ($\Delta p = m\,\Delta v$)

Impulse =
change in momentum ($F\Delta t = m\Delta v$)

NOTES

Chapter 2

Alternative Chapter Assessment

Part A: Multiple Choice:

Choose the best answer and place on your answer sheet.

1. Safety belts

a) always protect you from death in an accident

b) seldom protect you from death in an accident

c) increase the chances of you suffering severe injuries or death

d) decrease the chances of you suffering severe injuries or death

2. Seat belts are more effective

a) when used alone

b) when used with shoulder belts

c) when used with air bags

d) when used with shoulder belts and air bags

3. Which of the following are safety features, that have been added to most vehicles in the last ten years?

a) shoulder belts in back seats

b) air bags

c) ABS brakes

d) all of these

4. A seat belt is a safety device used in a vehicle because in a low-speed accident

a) they slow down the vehicle

b) they slow down the occupants in the vehicle

c) they prevent the occupants from flying around inside the vehicle

d) they prevent the occupants from dying

5. In a collision, there is generally more than one collision. Which of the following would describe the collisions of an accident?

a) when the car hits a stationary object, and the occupant hits the dashboard

b) when the car hits a stationary object, the occupant hits the dashboard, and the brain collides with the inside of the skull

c) when the occupant hits the dashboard, and the brain collides with the inside of the skull

d) when the car hits a stationary object, the occupant flies out of the vehicle, and hits the ground and stops

6. Newton's First Law is

a) inertia

b) impulse

c) momentum

d) force

7. In order to have a change in inertia, you must have

a) no force

b) balanced forces

c) a force opposite to the object's motion

d) an accelerating force

8. Momentum is the product of

a) mass and acceleration

b) mass and velocity

c) velocity and acceleration

d) force and velocity

9. The reason for having a wide seat belt as opposed to having a thinner seat belt is

a) it is more comfortable when driving

b) it reduces the force on your body when stopping

c) it increases the impulse on your body when stopping

d) it decreases the pressure of the seat belt on your body while stopping

10. Impulse is the same as

a) change in velocity

b) change in impulse

c) change in momentum

d) change in acceleration

11. By wearing a seat belt, and a shoulder belt, you are potentially reducing the damage to your body because

a) you are decreasing the velocity of the car

b) you are increasing the impulse of the car

c) you are decreasing the change in momentum of the car

d) you are increasing the time of the impulse of the car

12. A car was travelling at 35 m/s. The car hit a tree, and the crush zone in the car was only 0.40 m. What was the acceleration of the driver of the car?

a) 1521 m/s^2

b) 3182 m/s^2

c) 7 m/s^2

d) 14 m/s^2

13. The average speed of a car that is slowing down from 25 m/s to 0 m/s in 3.0 s is

a) 25 m/s

b) 12.5 m/s

c) 8.3 m/s

d) 0 m/s

14. The acceleration of an airplane travelling at 65 m/s that stops on a runway 200 m long in 10 s is

a) 6.5 m/s^2

b) 650 m/s^2

c) 0.325 m/s^2

d) 3.1 m/s^2

15. The force to stop a vehicle is supplied by

a) the friction caused by the road against the tires

b) the friction caused by the motor slowing down

c) the friction caused by the brake pads against the moving wheels

d) the inertia of the car

16. The bungee cord works to reduce the force on your body by

a) decreasing the time you are falling

b) increasing the time you are falling

c) increasing the time the elastic part of the cord is slowing you down

d) decreasing the time the elastic part of the cord is slowing you down

17. What is the impulse required to bring a 1000-kg vehicle moving at 25 m/s to a stop in 5.0 s?

a) 25,000 kgm/s

b) 125,000 kgm/s

c) 40 kgm/s

d) 5000 kgm/s

18. What is the momentum of a 75-kg tennis player moving at 6.5 m/s?

a) 488 kgm/s

b) 736 kgm/s

c) 11.5 kgm/s

d) 81.5 kgm/s

Use the following information to answer questions 19 - 21.

A cyclist (mass of cyclist and bicycle is 90 kg) is travelling along a road at 58 km/h (16 m/s).

19. What force is necessary to bring him to rest in 10 s?

a) 56 N

b) 144 N

c) 522 N

d) 14,400 N

20. Where does the force necessary to bring him to a stop come from?

a) no force is necessary, as he will glide to a stop

b) the friction between the road and the tires

c) the friction between the wheels and the brakes

d) friction from wind resistance and the road

21. If time were to change, which time would most likely do the most damage to the cyclist and his bike?

a) 1 s

b) 5 s

c) 15 s

d) 20 s

22. Which of the following would be very important when designing an automatic triggering device?

a) the device must trigger 100 % of the time

b) the device must trigger at high speeds when brought to a slow stop

c) the device must trigger at low speeds when brought to a sudden stop

d) none of these

23. The sand or water canisters you see around bridge supports are similar to air bags in that they

a) are large and protect something hard (cement or dashboard)

b) reduce the impulse of the collision

c) increase the force by decreasing the time of the collision

d) decrease the force by increasing the time of the collision

Part B: Matching

Match the words or phrases on the left with the matching word or phrase on the right.

__ 1. average speed	A. increase the time of collision
__ 2. acceleration	B. final velocity + initial velocity /2
__ 3. impulse	C. change in momentum
__ 4. momentum	D. part of the car to reduce injuries in collisions
__ 5. decrease force	E. reduce energy and therefore velocity in bungee cord
__ 6. friction	F. change velocity divided by change in time
__ 7. crush zone	G. product of mass and velocity
__ 8. inertia	H. Newton's First Law of Motion

Part C: Written Response

1. Describe why seat belts are an important safety device. Be sure to include how physics plays a role.

   A: Answers will vary. Look for seat belts keep the occupant in the car; seat belts slow the occupant down in a more controlled fashion; they spread out the force over a large area of the body, rather than concentrate it in one spot (head on dashboard); they decrease the extent of injury, thus keeping the cost of health care lower.

2. Describe what type of material would make the best kind of seat belt. Include in your answer such qualities as comfort, safety, pressure, cost.

   A: Answers will vary. The ideal type of material, would be inexpensive, have high tensile strength, washable, would decrease the force per unit area (pressure) to the body to an acceptable level, easy to install, readily available materials, be soft to touch so as to be more comfortable, be relatively light.

3. Describe the role of inertia in the three collisions that occur in an accident.

   A: Inertia is the tendency for an object to stay in motion, therefore, when a vehicle stops abruptly (in a collision), the first collision is when the car hits the wall, and the force of the wall causes the vehicle to come to a rest. Because the passenger is in motion, the second collision comes when the force of the vehicle, at rest now, exerts the force against the passenger causing it to come to a rest. The third collision is the force exerted by the inside of the skull, now at rest, against the moving brain.

4. Describe how a parachute works in helping prevent injury to the body. (Be sure to include impulse and force into your answer.)

   A: A parachute will slow down your descent, where the velocity is much decreased. This slower velocity gives a lower momentum (m x v), which therefore lowers the impulse of the parachutist landing on the ground. $F\Delta t = m\Delta v$.

5. Explain why a helmet is designed to break upon impact.

   A: A helmet is designed to increase the time of the collision, therefore decreasing the force of the impact. If the helmet was rigid, then the total force of the collision would be transferred to the head, and serious damage would result.

6. Your baseball coach has been reminding you to stay in contact with the ball and follow through when you hit. Explain why, in terms of the physics you learned about in this chapter, this is important.

   A: The longer (Δt) you can keep in contact (F) with the ball the greater the impulse ($F\Delta t$). Therefore, the greater the impulse, the greater the change in momentum ($F\Delta t = m\Delta v$). (Mass stays constant).

7. Explain how an air bag works. Include the terms pressure, force, impulse, change in momentum.

   A: Answers will vary. An air bag essentially changes the force that is applied to the head during a collision because the time of the collision is larger, and the impulse or change in momentum is the same. Because the air bag has a large surface area, the force is spread out over a large area, causing the pressure on the head to be smaller.

8. You have recently dislocated your shoulder, but it doesn't hurt that badly and you are sitting in the seat next to the emergency exit on the plane. Your friend says that you should change seats with him so that he is next to the seat. Explain why your friend thinks it is a good idea.

   A: Because your need to be able to operate the emergency exit, efficiently, and with a dislocated shoulder, even though it doesn't hurt, may hinder or slow your actions and response time in an emergency.

Alternative Chapter Assessment Answers

Part A: Multiple Choice:

1. d	13. b
2. d	14. a
3. d	15. a
4. c	16. c
5. b	17. a
6. a	18. a
7. d	19. b
8. b	20. b
9. d	21. a
10. c	22. a
11. d	23. d
12. c	

Part B: Matching:

1. B
2. F
3. C
4. G
5. A
6. E
7. D
8. H.

NOTES

Chapter 3

THE TRACK & FIELD CHAMPIONSHIP

Chapter 3- The Track & Field Championship

National Science Education Standards

Chapter Challenge

The **Scenario** and **Challenge** for this chapter is for the physics class to write a physics manual about track and field training for your high school team to help improve its performance.

Chapter Summary

To gain knowledge and understanding of physics principles necessary to meet this challenge, students work collaboratively on activities in which they apply concepts of potential and kinetic energy as they collect and analyze data collected in investigations of speed, acceleration, velocity, projectile motion, and gravity. These experiences engage students in the following content identified in the National Science Education Standards.

Content Standards

Unifying Concepts

- Evidence, models and explanations
- Constancy, change and measurement

Science as Inquiry

- Identify questions and concepts that guide scientific investigations
- Use technology and mathematics to improve investigations
- Formulate and revise scientific explanations and models using logic and evidence

Physical Science

- Motions and Forces: Objects change their motion when a net force is applied
- Laws of Motion are used to calculate the effects of forces in motion

Key Physics Concepts and Skills

Activity Summaries	Physics Principles
Activity 1: Running the Race Students time classmates to calculate the average speed as they run set distances. They then compare and analyze differences in speed of one runner at different distances and among different runners for the same distance.	• Relationship of speed, distance and time • Kinetic energy and motion
Activity 2: Analysis of Trends Students examine graphs representing results from track meets to explore trends in average speed. From this, they learn about extrapolation of data and the use of data in making predictions.	• Average Speed • Using data as basis for predictions
Activity 3: Who Wins the Race? Students use a ticker-tape timer to investigate speed of cars on sloped tracks. Analysis of the tapes introduces the concept that the average speed is not the same as actual speed at all points.	• Acceleration • Average Speed • Friction
Activity 4: Understanding the Sprint Students create and analyze graphs of split time vs. distance as an introduction to instantaneous speed and how a runner changes speed during a race.	• Acceleration • Instantaneous Speed • Average Speed
Activity 5: Acceleration Using a simple accelerometer to monitor changes in motion, students investigate acceleration and deceleration while walking, then with physics lab carts.	• Acceleration • Instantaneous Speed • Average Speed
Activity 6: Measurement Students use direct and indirect measurements to find the area of a penny. Comparing results emphasizes the value of probability in developing theories. They then read about the role of probability and indirect measurement in the discovery of atomic nuclei.	• Atomic Particles • Nucleus of atoms • Nature of scientific discoveries
Activity 7: Increasing Top Speed Students measure their own stride length as an introduction to wavelengths. They then investigate the relationship of speed in a race to stride length and stride frequency of the runner.	• Wavelength • Velocity = Frequency × Wavelength
Activity 8: Projectile Motion To develop understanding of the shot put, students explore the differences between the motion and landing position of objects dropped straight down to those with projected motion.	• Projectile Motion • Gravity • Free Fall
Activity 9: The Shot Put Students compare mathematical and physical models of projectile motion to that of a shot put. They apply this to describe the vertical and horizontal motion of the projected object, and predict its trajectory.	• Projectile Motion • Gravity • Trajectories
Activity 10: Energy in the Pole Vault Students use a penny launched from a ruler to model motion during the pole vault. They connect their observations to the concept of energy conservation.	• Gravitational Potential Energy • Transfer of Mechanical Energy • Conservation of Energy

Chapter 3

GETTING STARTED WITH EQUIPMENT NEEDED TO CONDUCT THE ACTIVITIES.

Items needed—not supplied in Material Kits

Preparing the equipment needed for each activity in this chapter is an important procedure. There are some items, however, needed for the chapter that are not supplied in the It's About Time material kitpackage. Many of these items may already be in your school and would be an unnecessary expense to duplicate. Please read carefully the list of items to the right which are not found in the supplied kits and locate them before beginning activities.

Items needed—not supplied by It's About Time:

- **Graphing Calculator (TI89)**
- **Computer**
- **Excel Program**
- **5cm cardboard**
- **Bubble level**
- **Clear tape**
- **Smart pulley**
- **VCR & Monitor**
- **Sports content video**
- **Chair with wheels**
- **Chalkboard**
- **Chalk**
- **Ladder**

Equipment List For Chapter 3

PART	ITEM	QTY	ACTIVITY	TO SERVE
AH-9251	Accelerometer	1	5,5B	Group
AH-9252	Accelerometer, Cork	1	5,5B	Group
AH-9250	Accelerometer-To-Cart	1	5,5B	Group
BS-72ST	Ball, Steel, 1"	1	10	Group
BS-0790	Ball, Tennis	1	8, 9	Group
CM-1108	Calculator, Basic	1	1,2,3,4,6,7,9	Group
CS-5011	Cards,Unlined Index	100	9,10	Group
CS-3246	Clamp, C-Clamp	1	10	Group
PH-6814	Digital Photogate	1	3A, 9	Class
GC-0001	Dynamics Carts	2	3, 5,5B	Group
RS-2723	Flexible Ruler	1	10	Group
GS-7706	Graph Paper Pad of 50 Sheets	1	2,4, 6	Group
IT-0001	Incline Track-1Meter Long	1	3	Group
MS-1425	Marker, Felt Tip	1	1,7,10	Group
MH-2621	Mass Hanger	1	5,5B	Group
PF-0001	Picket Fence	1	9	Class
NS-1210	Post-It-Note	100	9	Group
PL-0001	Projectile Launcher	1	8	Group
TS-6120	Roll of Adding Machine Tape	1	3	Group
RB-6108	Rubberbands	4	10	Group
RS-2826	Ruler	1	10	Group
SR-0020	Starting Ramp 20 degrees	1	10	Group
PH-3070	Steel Pulley On Mount	1	5,5B	Group
SM-1676	Stick, Meter, 100 cm, Hardwood	4	1,3, 6, 7,8,9	Group
SS-7778	Stopwatches	4	1, 7	Group
SS-7722	String, Ball Of, 225 ft.	1	5,5B, 9	Group
TA-0025	Tape Measure, Windup	1	1, 7	Group
TS-2662	Tape, Masking, 3/4" x 60 yds.	1	1,3, 7, 10	Group
TT-6100	Ticker Tape Timer, AC Timer	1	3,9	Group
TH-2143	Trajectory Model	1	9	Class
WS-6910	Washer, Metal, 3/4"	2	8,10	Group
WH-9012	Weight Set, Slotted	1	5,5B	Group
FW-0002	Weight, Small Fishing	10	9	Class
ITEMS NEEDED – NOT SUPPLIED BY IT'S ABOUT TIME				
	Graphing Calculator (TI89)	1	2	
	Computer	1	2A	
	Excel Program	1	2A	
	5cm cardboard	1	3A	
	Bubble level	1	5A	
	Clear tape	1	5A	
	Smart pulley	1	5B	
	VCR & Monitor	1	7	
	Sports content video	1	7	
	Chair with wheels	1	8	
	Chalkboard	1	9	
	Chalk	1	9	
	Ladder	1	9	

Organizer for Materials Available in Teacher's Edition

Activity in Student Text	Additional Material	Alternative / Optional Activities
ACTIVITY 1: Running the Race p. 142	Sample Data of Records From High School Track Meets pgs. 278-279	
ACTIVITY 2: Analysis of Trends p. 147	Speed vs. Year, Men's Olympic 400-m Dash p. 287 Speed vs. Distance, Men and Women, Penn Relays p. 288 Performance-Based Assessment Rubric: Reading and Interpreting Graphs p. 291	ACTIVITY 2 A: Spreadsheet Games–Analysis of Long-Range Trends pgs. 289-290
ACTIVITY 3: Who Wins the Race? p. 151	One-Meter Sprints: Compiling the Results and Making Predictions p. 301	ACTIVITY 3 A: Who Wins the Race? (Using a Photogate) p. 300
ACTIVITY 4: Understanding the Sprint p. 156	Carl Lewis's World Record 100-m Dash Average Speed, 10-m Intervals p. 308	
ACTIVITY 5: Acceleration p. 160	Constructing a Cork Accelerometer p. 319	ACTIVITY 5 A: Investigating Different Types of Accelerometers pgs. 320-321 ACTIVITY 5 B: Constant Acceleration p. 322
ACTIVITY 6: Measurement, p. 167	Assessment: Measuring and Performance-Based, p. 329	
ACTIVITY 7: Increasing Top Speed p. 171	Alternative Solutions to Physics To Go p. 337	
ACTIVITY 8: Projectile Motion p. 176	Assessment Rubrics for Two Falling Coins and the Rolling Chair pgs. 348-349	
ACTIVITY 9: The Shot Put p. 184	Measuring the Acceleration Due to Gravity p. 364 Four-meter Tack Strip for Trajectory Model p. 366 2.4-Meter Stick and String Model of Projectile Trajectory p. 367	
ACTIVITY 10: Energy in the Pole Vault p. 194	Olympic Pole Vault: Actual Heights and Heights Predicted from Average Speed in Men's 100-m Dash, 1896 to 1992 p. 383	

Scenario

High school students across the country compete in track and field events locally. Some of the best competitors go on to compete at the state or national levels. The world's largest such event is the Penn Relays, which includes competitors at the high school, college, and professional levels. Your school has been invited to send a team to the Penn Relays (or some other big competition). Everyone is excited and nervous about the event. The team has read that some professional athletes hire physicists and other science professionals to help them improve their performances. Your school's track team cannot afford professional physicists but hopes that students studying *Active Physics* may be able to provide some assistance. Can the *Active Physics* students use their understanding to help the track team improve its performance?

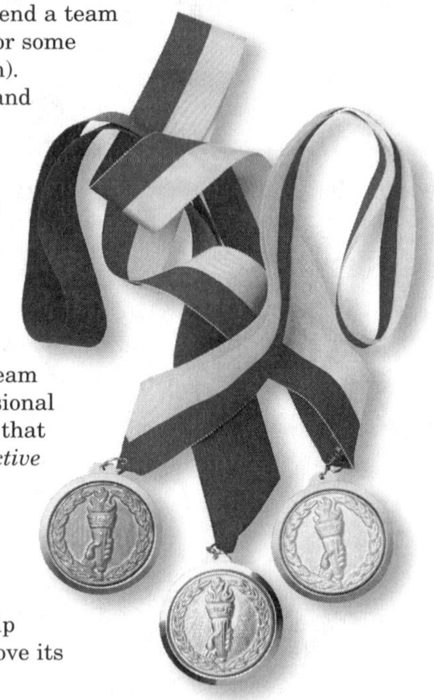

140

Chapter and Challenge Overview

In this chapter students investigate simple kinematics using an invitation for the school's track team to participate at the Penn Relays as the driving scenario. We chose track and field, since almost every student can run, jump, or throw, while not all engage in team sports. We chose the Penn Relays, as opposed to the Olympics or a local school league championship, for two reasons. We have available the winning times and distances for male and female high school students at the Relays for many years. Second, it is unlikely that the students in your school can compete at an Olympic level, while the Relays are very possible.

If the Penn Relays are not well known, your own state or district track and field championships might be substituted, but then you will have to find some of the performance records and substitute them for the Relays' records.

It would be very useful to make contact with the school track coach before this chapter is started, possibly giving him/her a copy of the text before students start asking the coach questions about the material in it. The coach might even drop in on an early class and say a few words about the Penn Relays, the track team, and how he/she is looking forward to working with the class if they have any problems or ideas. You will need some data from the coach about your own school records and present team's records for certain events. For example, it would be useful to obtain data for the 100-m sprint, 1500-m and 3000-m run, 4 × 100-m relay, high jump, long jump, and shot put.

There are a number of places in this and subsequent chapters where it might be useful and interesting to have the entire class perform some athletic event and have it timed or measured and then have the results analyzed. We did not include this type of activity in the text since some students may feel embarrassed about their athletic ability. You are in the best position to judge if this type of activity is suitable for your students. There are usually a few students who are very willing to run, jump, and throw, and they can be your volunteers.

As you review the **Chapter Challenge** assignments, reassure the students that while they may feel incompetent now, by the end of the chapter they will have the necessary skills and vocabulary to respond adequately.

On the following pages of the Teacher's Edition there are suggestions on how to evaluate students on this material. It is very important at this time that the students be made aware of the method you are going to use and how you will evaluate their work. Have the students actively participate in deciding the criteria for evaluation.

The **Physics To Go** at the end of each section often contains more questions than should ever be assigned for homework. This section has been written in such a way as to give you a choice as to how much work, and the nature of the work the students will be expected to do each day out of class.

As you begin *Active Physical Science*, be aware that the same physics concepts appear repeatedly in different contexts. It is not necessary for the students to achieve total understanding the first time that they study speed, speed vs. time graphs, acceleration, trajectory motion, or potential and kinetic energy.

Challenge

Your challenge is to write a physics manual about track and field training for your high-school team to help improve its performance. The manual should:

- **help students compare themselves with their competitors**
- **include a description of physics principles as they relate to track events**
- **provide specific techniques to improve performance**

Criteria

How will the manual be graded? What qualities should a good manual have? Discuss these issues in small groups and with your class. You may decide that some or all of the following qualities should be graded in your track and field manual:

- **physics principles**
- **inclusion of charts**
- **past records**
- **relevant equations**
- **definitions**
- **specific techniques**

Any advice you give should be understandable to athletes who have not studied physics. You can describe any activities you have done to explain how you know that the technique works, but you should not tell so much about each activity that the reader becomes bored.

Once you list the criteria for judging the track and field manuals, you should also decide how many points should be given for each criterion. If the group is going to hand in one manual, you must also agree on the way in which individuals will receive their grade. One method may be that the individual contributions toward the project receive 75% of the individual grade, and the group grade provides the remaining 25% of the individual grade. You should discuss different strategies and choose the one that is best suited to your school.

Assessment Rubric: Training Manual

Meets the standard of excellence. **5**	• Significant information is presented in a consistently appropriate manner. • Details are specific and consistently effective. • Concepts within the chapter are integrated in appropriate places. • Additional research, beyond basic concepts presented in the chapter, is evident. • The writing holds the readers interest.
Approaches the standard of excellence. **4**	• Significant information is most often presented in an appropriate manner. • Details are specific and effective. • Concepts within the chapter are integrated in appropriate places. • The writing holds the readers interest.
Meets an acceptable standard. **3**	• Sufficient information is presented in an appropriate manner. • Details are general but effective. • Some of the concepts, presented within the chapter, are integrated in appropriate places. • The writing generally holds the readers interest.
Below acceptable standard and requires remedial help. **2**	• Limited information presented in an inappropriate manner. • Details are limited and do not always pertain to the topic. • Few of the concepts, presented within the chapter, are integrated in appropriate places. • The writing generally does not hold the readers interest.
Basic level that requires remedial help or demonstrates a lack of effort. **1**	• Essential information is missing. • Details are limited and do not always pertain to the topic. • Few of the concepts, presented within the chapter, are integrated within the manual. • The writing is very difficult to follow. • Much of the manual remains incomplete.

Assessment Rubric: Scientific Language

Meets the standard of excellence. **5**	• Scientific vocabulary is used consistently and precisely. • Sentence structure is consistently controlled. • Spelling, punctuation, and grammar are consistently used in an effective manner. • Scientific symbols for units of measurement are used appropriately in all cases. • Where appropriate, data is organized into tables or presented by graphs.
Approaches the standard of excellence. **4**	• Scientific vocabulary is used appropriately in most situations. • Sentence structure is usually consistently controlled. • Spelling, punctuation, and grammar are generally used in an effective manner. • Scientific symbols for units of measurement are used appropriately in most cases. • Where appropriate, most of the data is organized into tables or presented by graphs.
Meets an acceptable standard. **3**	• Some evidence that the student has used scientific vocabulary although usage is not consistent or precise. • Sentence structure is generally controlled. • Spelling, punctuation, and grammar do not impede the meaning. • Some scientific symbols for units of measurement are used. Generally, the usage is appropriate. • Limited presentation of data by tables or graphs.
Below acceptable standard and requires remedial help. **2**	• Limited evidence that the student has used scientific vocabulary. Generally, the usage is not consistent or precise. • Sentence structure is poorly controlled. • Spelling, punctuation, and grammar impedes the meaning. • Some scientific symbols for units of measurement are used, but most often, the usage is inappropriate. • No presentation of data by tables or graphs.
Basic level that requires remedial help or demonstrates a lack of effort. **1**	• Limited evidence that the student has used scientific vocabulary and usage is not consistent or precise. • Sentence structure is poorly controlled. • Spelling, punctuation, and grammar impedes the meaning. • No attention to using scientific symbols for units of measurement. • No presentation of data by tables or graphs.

Maximum = 10 Points

What is in the Physics InfoMall for Chapter 3?

If you have had much experience with the Physics InfoMall CD-ROM, you have probably done a few searches, and no doubt some of the searches have resulted in "Too many hits." Surprisingly, searching the entire CD-ROM with the keyword "sport*" does not give "too many" hits, but provides some interesting hits. Note that the asterisk is a wild character; this searches for any word beginning with "sport."

If you do the search just mentioned, the first hit is a resource letter ("Resource letter PS-1: Physics of sports," *American Journal of Physics,* vol. 54, issue 7) that discusses the published discussions on the physics of sports. According to this letter, "there is surprisingly little published information about the basic physics underlying most sports, even though the relevant physics is all classical." Included is a list of places you might find such information, including journals and books. The letter contains a list of specific references grouped by sport, such as these two for Track and Field: "Behavior of the discus in flight," J. A. Taylor, Athletic J. 12(4), 9, 45 (1932), and "Bad physics in athletic measurement," P. Kirkpatrick, Am. J. Phys. 12, 7 (1944). Also in this list of hits are several articles on the physics of specific sports, such as basketball (Physics of basketball, *American Journal of Physics*, vol. 49, issue 4). Another interesting article is "Students do not think physics is 'relevant.' What can we do about it?," in the *American Journal of Physics*, vol. 36, issue 12.

Given that the physics in sports is classical, you might search for student difficulties learning classical physics in general. One article you might find is "Factors influencing the learning of classical mechanics," *American Journal of Physics,* vol. 48, issue 12. Knowledge of such factors affecting learning can be a valuable tool. Perform other searches that meet your needs, and the InfoMall is very likely to provide good information. And we have not even opened the Textbook Trove yet!

As mentioned above, there is a resource letter that contains references to published information about the physics of various sports. Track and Field, the topic of **Chapter 3**, is one of the sports mentioned. Of course, each sport uses several different concepts in physics, and these are analyzed in the following activities. Each of these concepts can always be found in the textbooks in the Textbook Trove, although we might not make such specific references here.

ACTIVITY 1
Running the Race

Background Information

The definition of average speed is introduced in Activity 1:

Average Speed = Distance ÷ Time

where the terms on the right-hand side of the equation mean the total distance travelled by an object and the total time required to travel that distance.

The standard units of measurement for distance and time are, respectively, the meter and the second. Therefore, the standard unit of average speed is meters per second. In some cases, subdivisions and multiples of the meter are used for measuring distance (e.g., centimeters and kilometers). Other units of measurement are used in cases where common practice involves measuring distance in feet, yards, or miles and measuring time in hours, resulting in units for average speed such as cm/s, km/hr, ft/sec, mi./hr.

It is important to note that the above definition of average speed does not take into account any variations in speed which an object may have as it travels along its path; the average speed is, simply, the total distance travelled divided by the total time taken to travel the distance. Therefore, an object's average speed may be more or less representative, or descriptive, of what really happens during particular cases of motion.

For example:

Suppose that a driver operates a vehicle on a superhighway for an hour with only minor variations in speed due to traffic. If the odometer shows that 65 miles were travelled during the hour, the average speed, 65 miles per hour, probably is quite representative of the vehicle's speed at most instants during the hour of travel.

Suppose that the same driver stopped for $^1/_2$ hour to fix a flat tire, causing the 65 mile trip to require $1^1/_2$ hours. According to the definition, the average speed for this trip is (65 mi.) / ($1^1/_2$ hr) = 43 mi./hr. True, the driver "averaged" 43 mi./hr, but the speedometer probably showed 43 mi./hr at only a few instants during the trip. This average speed is less representative of how the car moved (or, at some times, did not move) during the trip.

The second of the above two examples raises the caution that "average speed" should not be confused with "instantaneous speed;" the latter is an object's speed at a particular instant, which will be introduced later in this chapter.

This activity, and many others in this chapter, involves track events in which runners start from rest (zero speed) and "build up" (accelerate) to a "top speed" which is maintained at a reasonably constant level for the remainder of the race. The "rest start" usually causes the average speed for the entire race to be less than the runner's "top speed." The discrepancy between a runner's average speed and top speed becomes less for races of greater distance because the period of acceleration occupies a less significant part of a long race than a short race.

Active-ating the Physics InfoMall

Speed is one of the first concepts introduced in introductory physics. It is also one that causes students problems. If you search the InfoMall for "student difficult*" OR "student understand*" you will find several articles that deal with research into how students learn fundamental concepts in physics. Some of these are "Investigation of student understanding of the concept of velocity in one dimension," *American Journal of Physics* vol.48, issue 12 (see "Diagnosis and remediation of an alternative conception of velocity using a microcomputer program," *American Journal of Physics*, vol. 53, issue 7 for a discussion of this); "Research and computer-based instruction: Opportunity for interaction," *American Journal of Physics*, vol. 58, issue 5 (if you look at other articles in this volume, you will find "Learning motion concepts using real-time microcomputer-based laboratory tools," *American Journal of Physics*, vol. 58, issue 9); and more. These last two articles make specific mention of the use of computers in teaching physics. This is a trend that is gaining strength and shows great promise. Look for more such articles in journals that are newer than the CD-ROM. As a starting point, you can always look in the annual indices of the physics journals and look under the names of the authors of the articles mentioned above.

One of the strengths of the InfoMall is the ability to search the entire database quickly. Don't think that searching is the only, or even the best, method for finding what you want. For example, suppose you want a demonstration of lab activity that uses the concept of speed. Go to the Demo & Lab Shop and choose almost any of the selections. If you choose Physics Lab Experiments and Computer Aids, you may notice that choice number 8 is Straight-Line Motion at Constant Speed (the next choice involves acceleration, which may also be useful). This shows (with a nice graphic) how to use a toy bulldozer in the lab. Choice 7 shows how you can use strobe photographs in a lab as well.

If you want to find additional problems for your students, you may browse through Teachers Treasures and find that there is a section on Speed Problems. Or you can look in the Problems Place, where you will find *Schaum's 3000 Solved Problems in Physics* (among others). You can find plenty of speed problems (with their solutions!).

Planning for the Activity

Time Requirements

- One class period.

Materials Needed

For each group:

- calculator, basic
- marker, felt tip
- stick meter, 100 cm, hardwood
- stopwatches
- tape, masking, 3/4" x 60 yds.
- tape measure, windup

Advance Preparation and Setup

An area will be needed to serve as a straight running "track." The area needs to be a minimum of 30 m (about 100 feet) in length, allowing 20 m for running and a stopping distance of at least 10 m. Tape (or some other marking material) must be able to be placed on the running surface at 5-m intervals. The track must be wide enough for one person at a time to run while other persons serving as timers stand alongside the track and the entire class observes. Noise can be expected. An outdoor area such as a sidewalk or parking lot may be preferred to accommodate noise and to avert scheduling problems. Indoors, a gymnasium or hallway could be used.

If you do not have an adequate supply of stopwatches (see **Materials Needed**), you may wish to check in advance to see if class members or others have wrist watches having a stopwatch function which could be used instead of stopwatches.

Teaching Notes

Active Physical Science uses a modified constructivist model. By confronting students' misconceptions and by having them do hands-on exploration of ideas, we seek to replace their misconceptions with correct perceptions of reality. In order to do this, a consistent scheme is integrated into the course activities to elicit the students' misconceptions early in any activity.

Students' current mental models are sampled by one or more **What Do You Think?** questions. Students are not expected to know a "right" answer. Answers are provided for your use only. These questions are supposed to elicit from students their beliefs regarding a very specific prediction or outcome, and students should commit to a written specific answer in their logs.

Some students in the class may be aware that the average speed of a runner is determined by dividing the total distance travelled by the total elapsed time. Most runners change speed while running. In a short dash starting from rest, a runner does not reach top speed instantly; therefore, more time usually is required to run the first half of the total distance than the last half. When running a long distance, a runner may tire and run a lower speed during the final part of the distance, taking more time to travel the second half of the total distance than the first half.

Stage one or more practice runs wherein all students serving as timers simultaneously measure the time required for one runner to travel 10 m. Use the variation in times measured by different individuals to establish consensus on the accuracy to which times should be reported – probably to the nearest 1/10 s. Establish that distance can be measured more accurately than time in this activity, and that the accuracy of the time measurements, in turn, places a limitation on the accuracy of speeds calculated using the time measurements. Be aware that students tend to believe that if a stopwatch used in this activity shows, for example, a time of 3.37 s for a student to run 15 m, the time is known to the nearest 1/100 of a second when, in reality, student reaction times probably limit the accuracy to less than 1/10 s.

This is not an appropriate time to get bogged down in discussion of precision, accuracy and significant figures, but it also would be unwise to allow students to believe that they know the value of average speed to greater accuracy than they really do. A good way to put this is to say that the accuracy to which the speed is known can be no better than the least accurate measurement involved in calculating the speed; in this case, the least accurate measurement is time. During each 5-m interval we know the time to only one significant figure; therefore, the speed during each interval should be reported to only one significant figure.

You may wish to point out that the unit of measurement for average speed, m/s, is a "derived" unit, meaning that it is made up from a combination of "base," or fundamentally defined, units, the meter and the second.

You may also wish to show students the mathematical operations involved in transforming the equation for average speed to solve for any of the three variables, pointing out that this offers the advantage of needing to remember only one form of theequation.

The average speed of record holders at the Penn Relays varies from 9.6 m/s for the mens' 100-m dash to 4.9 m/s for the womens' 5000-meter race. Speeds generally decrease as the distance of race events increases.

NOTES

Activity Overview

In this activity students measure the time it takes a runner to reach each 5-m mark along a 20-m run. They then use their data to calculate the average speed for each interval. Students then compare the average speed for different segments as well as different runners.

Student Objectives

Students will:

- Measure distance in meters along a straight line.
- Use a stopwatch to measure the amount of time for a running person to travel a measured distance.
- Use measurements of distance and time to calculate the average speed of a running person.
- Compare the average speeds of a running person during segments of a run.
- Compare the average speeds of different persons running along the same path.
- Infer factors which affect the average speed of a running person.

The Track and Field Championship

Activity 1 Running the Race

What Do You Think?

Very few people in the world can run 100 meters in less than 10 seconds.

- **How can you measure a runner's speed?**
- **Does running twice as far take twice as much time?**

Record your ideas about these questions in your *Active Physics* log. Be prepared to discuss your responses with your small group and the class.

GOALS

In this activity you will:

- Measure distance in meters along a straight line.
- Use a stopwatch to measure the amount of time for a running person to travel a measured distance.
- Use measurements of distance and time to calculate the average speed of a running person.
- Compare the average speeds of a running person during segments of a run.
- Compare the average speeds of different persons running along the same path.
- Infer factors which affect the average speed of a running person.

For You To Do

1. Your teacher will indicate the location of a "track." Place a mark on the track at the 0-, 5-, 15-, and 10-m positions.
2. Place students with stopwatches at the 5-, 10-, 15-, and 20-m marks to serve as timers. Each timer should measure the amount of time for each runner to reach the timer's assigned mark.
3. Have someone serve as starter, saying "Ready, Set, Go!" for each runner. All watches should be started on the "Go" signal, and each watch should be stopped as the runner passes the timer's assigned mark. Measure the time for three runners.

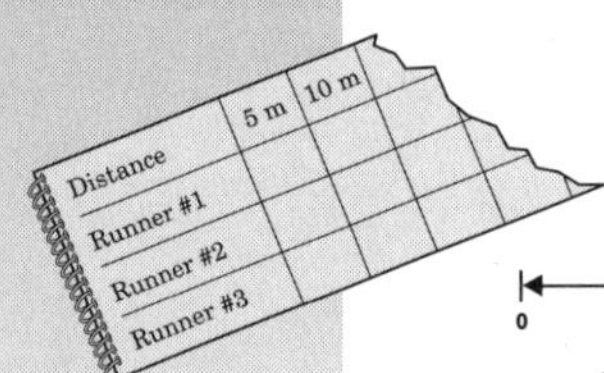

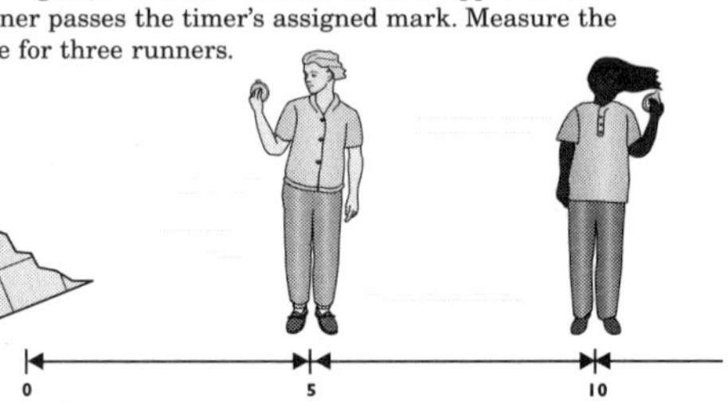

Active Physics 142

ANSWERS FOR THE TEACHER ONLY

What Do You Think?

The average speed of a runner is determined by dividing the total distance travelled by the total elapsed time.

Most runners change speed while running. In a short dash starting from rest, a runner does not reach top speed instantly; therefore, more time usually is required to run the first half of the total distance than the last half. When running a long distance, a runner may tire and run at a lower speed during the final part of the distance, taking more time to travel the second half of the total distance than the first half.

ANSWERS

For You To Do

1.–2. Student activity.

3. a) Sample data for one runner:

Distance	5 m	10 m	15 m	20 m
Runner #1 Time	**1.6 s**	**2.5 s**	**3.3 s**	**4.0 s**

a) In your log, record the time it takes each runner to reach the 5-, 10-, 15-, and 20-m positions.

4. Calculate the amount of time taken to run each 5-m interval. To calculate the time taken to run from the 5- to 10-m mark, you will need to subtract the time at the 5-m mark from the time at the 10-m mark.

a) Use a table similar to the following to write the results in your log.

5. Does running twice as far take twice as much time?

Distance	0-5 m	5-10 m	10-15 m	15-20 m
Runner #1 Time				
Speed				
Runner #2 Time				
Speed				
Runner #3 Time				
Speed				

a) Use data from the 20-m dashes to explain your answer in your log.

6. Calculate the average speed during each 5-m interval. Use the equation:

$$\text{Average speed} = \frac{\text{Distance traveled}}{\text{Time taken}}$$

a) Record the average speed during each 5-m interval in the table in your log.

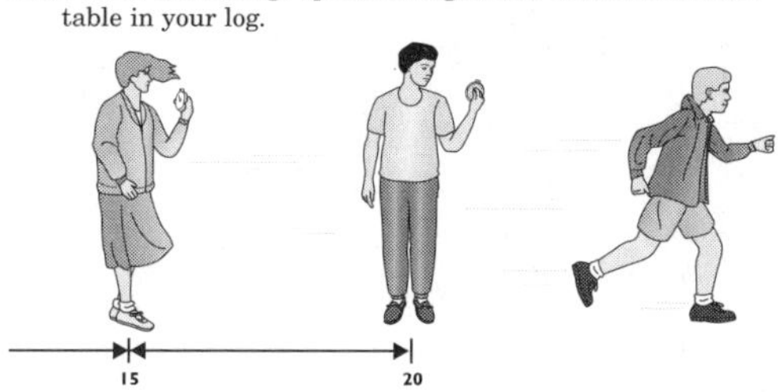

ANSWERS

For You To Do *(continued)*

4. a) Sample data for one runner:

Distance	5 m	10 m	15 m	20 m
Runner #1 Time	**1.6 s**	**0.9 s**	**0.8 s**	**0.7 s**
Speed	**3 m/s**	**6 m/s**	**6 m/s**	**7 m/s**

5. a) Running twice as far does not necessarily take twice as much time. Students may note that in the case of a short-distance run, such as 20 m, it takes less than twice the time to run 20 m than to run 10 m. Students may also suggest that a great deal will depend on the distance run, that in longer races the runner may fatigue.

6. a) Using the sample data: Average speed = 20 m/4.0 s = 5.0 m/s.

Chapter 3

7. Use your data to answer the following questions:
   a) In which distance interval did each runner have the greatest average speed? Circle the interval in your log for each runner.
   b) Was the interval of greatest speed the same or different for different runners?
   c) Which runner holds the record for the fastest 5-m interval?
   d) Describe each runner's total dash in terms of speed during distance intervals.
   e) Estimate the amount of time taken for each runner to reach maximum speed.

8. Calculate the average speed of each runner for the entire dash.
   a) Record the average speed in your log.
   b) If the three dashes had been an Olympic time trial, which runner would have won? What was the winning speed?

9. Use data from each runner's dash.
   a) Write suggestions in your log for the runners to improve their performances.

Physics Words

speed: the change in distance per unit time; speed is a scalar, it has no direction.

PHYSICS TALK

Average Speed

The relationship between average **speed**, distance, and time can be written as:

$$\text{Average speed} = \frac{\text{Distance}}{\text{Time}}$$

This relationship can be rearranged as follows:

$$\text{Distance} = \text{Average Speed} \times \text{Time}$$

$$\text{Time} = \frac{\text{Distance}}{\text{Average Speed}}$$

ANSWERS

For You To Do *(continued)*

7\. a)-e) Students will probably find that the greatest average speed was during the final distance intervals. The initial time interval will be greater because the student is accelerating from rest. Student average speeds will likely vary.

8\. a)-b) Answers will vary depending on experimental results.

9\. a) Students could suggest that the runner improve his/her start. Also, students could observe if a runner slows down for the approach of the finish line.

Activity 1 Running the Race

FOR YOU TO READ

The **Scenario** for this chapter mentioned that the Penn Relays is the world's largest track and field event. It includes athletes at the high school, college, and professional levels. Examine the record times of male and female athletes in running events at the Penn Relays in the table to the right.

Many students in your class may be able to run 100 m in less than 15 s. Some members of your school's track and field team might be able to do the same in under 12 s. Do a calculation to show yourself that even a slow person who could run 100 m in 15 s could, theoretically, break the Penn Relays record for the 1500-m race if it is true that "running 15 times farther takes 15 times as much time." It will be necessary to convert the Penn Relays record given in minutes and seconds into seconds. (Example: A time of 4:08 for the mile equals 248 s.)

Obviously, not only speed but also stamina is involved in winning a race. An athlete needs to be able to run at high speed and have the stamina, or strength, to keep the speed as high as possible for the entire race. Both speed and stamina can be improved through training and knowledge. Athletes who hold records are those who run both fast and smart.

Penn Relays Record Times

Distance (meters)	Time–Men (minutes:seconds)	Time–Women (minutes:seconds)
100	0:10.47	0:11.44
200	0:21.07	0:23.66
400	0:45.49	0:52.33
800	1:48.8	2:05.4
1500	3:49.67	4:24.0
mile	4:08.7	4:49.2
3000	8:05.8	9:15.3
5000	15:09.36	16:59.5

ANSWERS

For You To Read

A person who can run 100 m/15.0 s would have an average speed of 6.67 m/s. If that speed were maintained for a 1500-m race, the time would be Time = Distance/Speed = 1500-m/6.67 m/s = 225 s = 3:45 minutes.
This would break the record time for men (3:49.67) in the Penn Relays 1500-m race, and it would break the record for women in that race (4:24.0).

ANSWERS

Physics To Go

1. a) 1500 m/229.7 s = 6.53 m/s
   b) There might be students who can reach the same speed.
   c) A student probably could not maintain such a speed for a distance of 1500 m.
2. 100 m : 8.74 m/s
   200 m : 8.45 m/s
   400 m : 7.64 m/s
   1500 m : 5.68 m/s

   The average speeds of runners decrease as the distance of races increases.
3. Compare record times for school track team to record times for Penn Relays; also compare average speeds by event. If you are unable to obtain records from your school's track team, sample data is provided in **Additional Materials**.
4. Expect students to indicate that it probably would not be fair or meaningful to compare data for high school running events to world records due to differences in athletes' ages, levels of training, and other factors.
5. More time would be required during the first half of the run because the runner is starting from rest and building up to top speed during the first half, therefore having a lower average speed than during the last half.
6. Split times provide ability to calculate speeds during parts of a race, allowing runners to have detailed information which could be used to improve performance.

The Track and Field Championship

Reflecting on the Activity and the Challenge

You know from measurements of your classmates' running that the person who travels the entire distance of a race in the least time wins. The winner also has the highest average speed for the overall race. You also know that the speed of each runner in your class changed during the run.

Search for patterns in the distances and times of men and women who hold records in the Penn Relays. Do the speeds vary with the distance of the events? What information could you give your school's track and field team about the Penn Relays right now? What further information do you need?

Physics To Go

1. a) Calculate the average speed of the male who holds the Penn Relays record for the 1500-m run.
   b) From the data you gathered, are there students in your class who can reach the same speed as the male 1500-m record holder?
   c) Do you think the fastest student in your class could run the 1500 m in record time?
2. Calculate the average speeds of women who hold Penn Relays records in the 100-, 200-, 400-, and 1500-m runs. What is the pattern of speeds?
3. Find some record times for your school's track team in running events and compare them with the Penn Relays record times. Also compare the average speeds for races of equal distance for your school with those for the Penn Relays.
4. Is it fair to compare data for high school running events with data for world records? Why or why not?
5. How do the amounts of time taken by a champion 100-m runner to travel the first and last 50 m of a sprint compare? What is the basis for your answer?
6. Track coaches often measure "split times" for runners during races. For example, a runner's time might be measured every 100 m during a 400-m race. How could "splits" be useful for helping runners improve both speed and stamina?

146

NOTES

Sample Data of Records from High School Track Meets

Girls' Records for Track Events

Track Event	Name	Time	Year
100 Meter Dash	Carol C.	13.0	1989
200 Meter Dash	Antoinette C.	27.1	1986
400 Meter Dash	Meg K.	62.2	1984
800 Meter Run	Amy C.	2:19.3	1983
1500 Meter Run	Julie S.	4:47.2	1983
3000 Meter Run	Dawn E.	10:24.4	1985
100 Meter Hurdles	Holly F.	16.2	1987

Boys' Records for Track Events

Track Event	Name	Time	Year
100 Meter Dash	Bill C.	10.5	1985
200 Meter Dash	Chris B.	21.2	1984
400 Meter Dash	Jerome W.	46.9	1988
800 Meter Run	Mike S.	1:50.5	1983
1500 Meter Run	Mike S.	3:46.9	1982
3000 Meter Run	Mike S.	9:01.7	1983
100 Meter Hurdles	Art M.	13.5	1976
400 Meter Hurdles	Sherwin S.	52.7	1986

Girls' Records for Field Events

Field Event	Name	Distance	Year
Shot Put	Judy B.	33'1/2"	1998
Discus	Naomi J.	108'10"	1991
Javelin	Naomi J.	126'0"	1992
Long Jump	Tonya J.	17'1/2"	1974
Triple Jump	Roxanne R.	31'4 3/4"	1991

Boys' Records for Field Events

Field Event	Name	Distance	Year
Shot Put	C.J. H.	64'5 1/4"	1986
Discus	Phil C.	183'3"	1997
Javelin	Andy S.	225'0"	1986
Long Jump	Matt S.	24'9 1/4"	1990
Triple Jump	Sanya O.	15'8"	1996

For use with *The Track and Field Championship*, Chapter 3, Activity 1: Running the Race

Sample Data of Records from High School Track Meets

Girls' Records for Relays

Event	Times	Total Time	Year
400 Meter Relay	14.2	51.6	1987
	13.3		
	11.0		
	13.1		
800 Meter Relay	27.9	1:49.7	1997
	27.2		
	27.0		
	27.7		
1600 Meter Relay	63.3	4:14.9	1985
	63.7		
	62.7		
	65.3		
3200 Meter Relay	2:21.8	9:22.55	1985
	2:20		
	2:21.2		
	2:19.0		

Boys' Records for Relays

Event	Times	Total Time	Year
400 Meter Relay	10.7	42.4	1994
	10.6		
	10.6		
	10.5		
800 Meter Relay	21.8	1:25.4	1987
	21.5		
	21.3		
	20.8		
1600 Meter Relay	49.8	3:12.7	1966
	48.0		
	47.4		
	47.5		
3200 Meter Relay	1:56.2	7:45.8	1995
	1:56.8		
	1:57.1		
	1:55.7		

For use with *The Track and Field Championship*, Chapter 3, Activity 1: Running the Race

ACTIVITY 2
Analysis of Trends

Background Information

The concept of average speed is extended from **Activity 1**. Students are asked to use graphing techniques (curve fitting and extrapolation) to explore how the performance of athletes has improved over a century of time, and how the average speed of runners varies with the distance of a race.

It is of value for students to estimate the best fit of a curve to data points on a graph and to extrapolate the curve of a graph "by hand," as directed in the activity. However, it also is of great value for students to experience the power of technology-assisted graphing techniques. Routines are available for both personal computers and graphing calculators which essentially automate graphing procedures; only the ordered pairs of data and the range of values for the axes of the graph are required to be entered. Once entered, the "best fit" curve is displayed, and both interpolation and extrapolation can be accomplished by simple commands.

If students are to perform extensive graphical analysis of data from your school's track team, doing so manually would be discouragingly tedious. This would be an excellent example of using a personal computer or graphing calculator as a tool.

If you are not familiar with technological tools for graphing, it is possible that a math, science, or computer science teacher at your school already has the knowledge and tools needed to accomplish this. Seek their help. If you don't have the time to learn how to use the devices, a highly able student may be very willing to learn to use the tools and to teach you and the class how they work.

If the technological tools are not available at your school, they can be found in most science catalogs, in advertisements in professional journals, and on display at science education conferences. These technological tools are the "state of the art," exist in great variety, and are becoming quite inexpensive. If you and your students are not using them, perhaps it's time you did. (See also *Active*-ating the Physics InfoMall.)

Active-ating the Physics InfoMall

For You To Do Step 3 suggests using computers or graphing calculators to examine trends. In addition to the articles mentioned in **Activity 1** on the use of computers, you can find "Student difficulties with graphical representations of negative values of velocity," in *The Physics Teacher*, vol. 27, issue 4 on the InfoMall. This was one of the articles found in the search using "student difficult*" OR "student understand*". Also found in that search is "Student difficulties in connecting graphs and physics: Example from kinematics," in the *American Journal of Physics*, vol. 55, issue 6. Don't let these titles make you think that graphs are a bad thing! It is good to be aware that students do not always understand graphs, a valuable tool in the study of physics. This step also suggests extrapolation. For discussions on extrapolation, you can perform a search using the keyword "extrapolation". One of the hits is Chapter 11 of the textbook *Physics for the Inquiring Mind.* This warns that "both interpolation and extrapolation are easier if the graph is a straight line. Even then they are not equally safe as sources of information, and asks "Which of the two, interpolation and extrapolation, would you as a scientist value more?"

Planning for the Activity

Time Requirements

- One class period.

Materials Needed

For each group:

- calculator, basic
- graph paper pad of 50 sheets
- graphing calculator (TI89)

Advance Preparation and Setup

Allow time for reproduction of the blackline masters "Speed versus Year, Men's Olympic 400-meter Dash" and "Speed versus Distance, Men and Women, Penn Relays" found on the pages following this activity.

Obtain and reproduce records of the performances of members of your school's track team. Information about running events is needed now, and later it would be useful to have information about all track and field events. Your school's track coach should be able to provide this information. If possible, the coach could be invited to present the information to your class; this would require additional time beyond the one class period needed to complete this activity. If your school does not have a track team, perhaps you can obtain data from another school. Sample data are also available at the end of the previous activity.

Teaching Notes

There is no "right" trend line or curve to sketch through the points on either graph; some estimates of the trends shown by the data would be better than others, depending on the methods used to establish "best fits" to the data. Students should be encouraged to sketch a straight line or a curve through the array of data points on each graph which makes the most sense to them; students should be discouraged, however, from simply connecting consecutive data points with straight line segments to produce a "saw tooth" curve because doing so involves no interpretation of the data. It is desired that students take the "risk" to sketch lines or curves which show their impressions of the trends of the data.

When students have completed their work with the graphs, you may wish to share with them the examples shown of best fit lines generated for the same data using a computer program. The computer-generated graphs use sophisticated routines to estimate "best fits" to data and to accomplish extrapolations, but should not be taken as "absolute truths."

It is indicated in **Reflecting on the Activity and the Challenge** that you, the teacher, will help students obtain data on the performances of members of your school's track team in various events; students will need such data in order to proceed. If your school does not have a track team, try to get data from a neighboring school. Sample data is also provided in this Teacher's Edition.

If you have ready access to a computer and a spreadsheet program such as MS Excel, you may wish to do the alternative **Activity 2** provided on the pages following this activity.

Assessment

See the Performance-based **Assessment Rubric**: Reading and Interpreting Graphs following this activity.

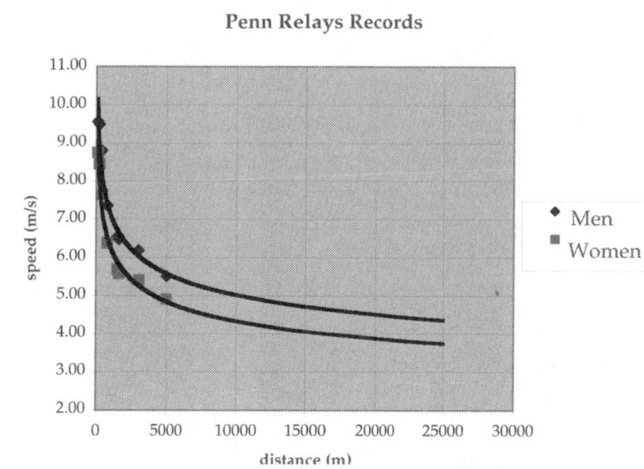

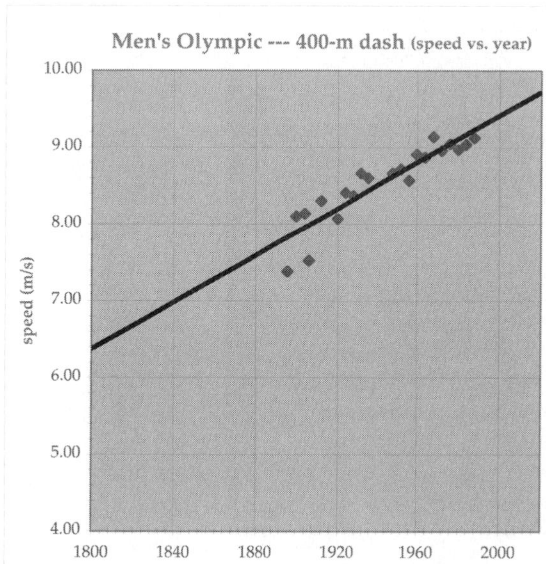

Chapter 3

NOTES

Activity 2 Analysis of Trends

Activity 2 Analysis of Trends

What Do You Think?

Current trends indicate that women will start outrunning men in 65 years.

- **Is it useful to compare track records over many years of time?**
- **Can future track records be predicted based on past performances?**

Record your ideas about these questions in your *Active Physics* log. Be prepared to discuss your responses with your small group and the class.

For You To Do

1. Look at the graph "Speed Versus Year: Men's Olympic 400-m Dash." The average speed of runners is shown on the vertical axis, and the year in which the race was run is shown on horizontal axis. Take a moment to be sure you understand that the plotted points show a 100-year history of the speeds of male athletes in the Olympic 400-m dash.

Speed Versus Year:
Men's Olympic—400-m Dash

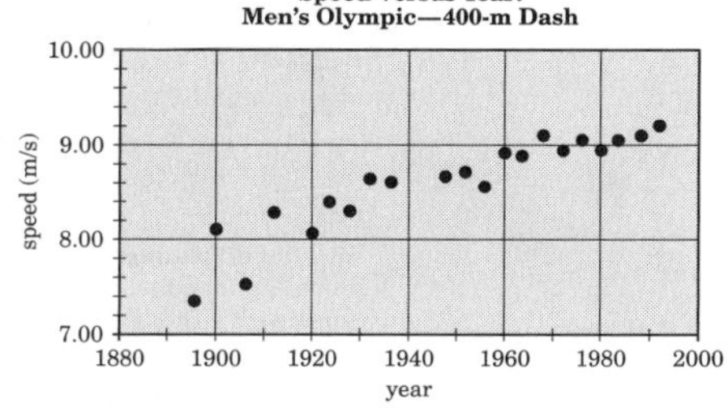

GOALS

In this activity you will:

- Sketch a best fit curve or trend line on a graph on which points have been plotted.
- Extrapolate graphs to predict trends beyond available data.
- Create graphs of data presented in tabular form.
- Calculate the average speed of a runner given distance and time.
- Calculate the time required for a runner to travel a specified distance given the runner's average speed.
- Interpret information presented in graphical form.

Activity Overview

Students in this activity use graphical analysis of authentic track records to explore the trend of average speed in the 400-m dash over the past 100 years. Students also analyze the speed vs. distance graph to determine what happens to average speed as the distance of the race increases.

Student Objectives

Students will:

- Sketch a best fit curve or trend line on a graph on which points have been plotted.
- Extrapolate graphs to predict trends beyond available data.
- Create graphs of data presented in tabular form.
- Calculate the average speed of a runner given distance and time.
- Calculate the time required for a runner to travel a specified distance given the runner's average speed.
- Interpret information presented in graphical form.

ANSWERS FOR THE TEACHER ONLY

What Do You Think?

Track records in particular events show clear trends over time.
Within reason, extrapolation allows prediction of future performances in track events.

Chapter 3

a) Copy the graph into your log. Sketch either a straight line or a smooth, curved line through the data points to show what you think is the shape of the graph. Do this for the points plotted from 1896 to 1996.

b) Comment in your log about what you see as the trend of average speed in the 400-m dash over the past 100 years.

c) Make the best guess you can to sketch how the graph would continue to the year 2020. This process of going beyond the data is called extrapolation. Try it.

d) According to your extrapolation, what will be the speed of the winning runner in the men's Olympic 400-m dash in the year 2020?

2. Look at the graph "Speed Versus Distance: Men and Women, Penn Relays." Notice that the average speed of runners is shown on the vertical axis and the distance of the race is shown on the horizontal axis. Be sure you understand that there are two sets of plotted points, one for men and one for women. Also, be sure you understand that the plotted points show how average speed varies with distance of the race for both men and women.

Speed Versus Distance:
Men and Women, Penn Relays Records

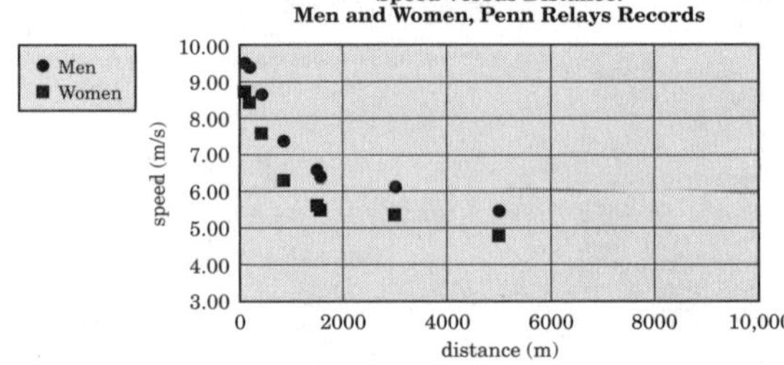

a) Copy the graph into your log. Sketch the shapes of the graphs for men and women by connecting the plotted points with either a straight line or a curved line, according to your choice.

ANSWERS

For You To Do

1. a) There is no "right" trend line or curve to sketch through the points; some estimates of the trends shown by the data would be better than others, depending on the methods used to establish "best fits" to the data.

   b) The average speed has increased over the past 100 years.

   c) Students will use their trend line or curve to extrapolate data.

   d) Computer-generated extrapolations (as shown in this Teacher's Edition) indicate a winning speed of about 9.6 m/s for the Olympic 400-m dash in the year 2020. Accept any reasonable answers from the students.

2. a) There is no "right" trend line or curve to sketch through the points; some estimates of the trends shown by the data would be better than others, depending on the methods used to establish "best fits" to the data.

Activity 2 Analysis of Trends

b) Comment on the trends of average speed for both men and women as the distance of races gets longer and longer.

c) Extrapolate from the graphs to predict what the record speed at the Penn Relays would be for 10,000-m races for men and women.

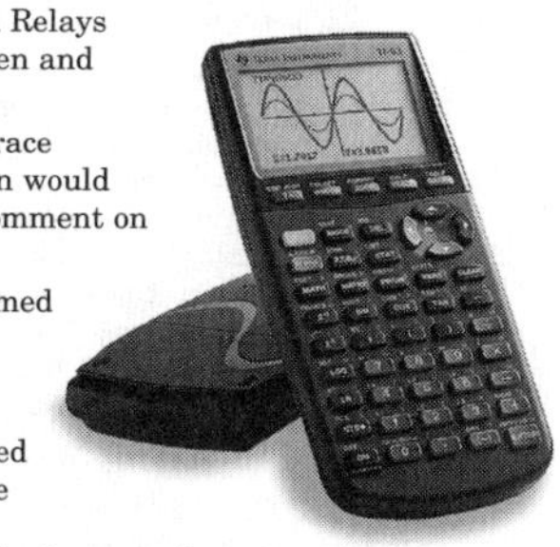

d) Try to use extrapolation to find a race distance for which men and women would run at the same average speed. Comment on your attempt.

3. The kind of data analysis you performed in this activity can be made much easier with a computer or graphing calculator. Basic distance and time data from track records can be entered into a computer spreadsheet software program or a graphing calculator. The computer or calculator can then be instructed to calculate speed from distance versus time data and to display graphs. The graphs can be analyzed to show trend lines, extrapolations, and other information. Your teacher or someone else familiar with computers and calculators may be able to help you use such devices as tools for data analysis.

Reflecting on the Activity and the Challenge

Your plan to help your school's team at the Penn Relays will only be as good as your knowledge of current performances in track and field. Also, knowledge of trends in how performances are changing with time in various events is also important.

Research shows that women are improving their track performances about twice as fast as men. This is because more and more women are participating in running as a sport than ever before. Certainly, you will want to consider the possibility of either a male or female athlete bringing home a medal from the Penn Relays for your school. Data can help you decide which athletes at your school will have the best chance.

Perhaps it's time to get some data about the performance of athletes at your school so that you can begin forming a strategy to help your team. Your teacher will help you with this.

ANSWERS

For You To Do *(continued)*

2. b) As the distance of the races increase, the average speed decreases.
   c) Students will use their trend line or curve to extrapolate data.
   d) Computer-generated extrapolations (as shown in this Teacher's Edition) indicate winning speeds of about 5.0 m/s for men and 4.3 m/s for women if a 10,000-m race for each gender were held at the Penn Relays. Accept any reasonable answers from the students.
3. Students use a computer spreadsheet or a graphing calculator.

ANSWERS

Physics To Go

1. From data on the school's track team, students will calculate the average speed for each event and superimpose the data on the graphs of speed vs. distance for the Penn Relays used in the activity. Students will critically compare school data to Penn Relays data.
2. Using a speed of 9.6 m/s obtained from computer-assisted extrapolation and the equation Time = Distance/Speed, Time = 400 m/9.6 m/s = 41 s. Student results may vary from this.
3. Using a speed for men of 5.0 m/s and a speed for women of 4.3 m/s, both obtained from computer-assisted extrapolation, and the equation Time = Distance/Speed, the times are:
   - for men, 10,000 m/5.0 m/s = 2000 s.
   - for women, 10,000 m/4.3 m/s = 2300 s.

   Student results may vary from the above.
4. Distance = Speed × Time = 10.14 m/s × 9.86 s = 100 m.
5. a) Use the equation Speed = Distance/Time

   1964: 400 m/52.00 s = 7.7 m/s
   1968: 400 m/52.00 s = 7.7 m/s
   1972: 400 m/51.08 s = 7.8 m/s
   1976: 400 m/49.29 s = 8.1 m/s
   1980: 400 m/48.88 s = 8.2 m/s
   1984: 400 m/48.83 s = 8.2 m/s
   1988: 400 m/48.65 s = 8.2 m/s
   1992: 400 m/48.83 s = 8.2 m/s

   b) Students provide graph.

   c) Until 1980 the average speed has been increasing. During the past few years the average speed has showed no significant change.

   d) Students extrapolate speed from their graph.

The Track and Field Championship

Physics To Go

1. Get distance and time data for the best performances of your school's boys' and girls' track teams for races of various distances.
   a) Calculate the average speed for each event.
   b) Plot the average speeds as data points on the Penn Relays graph used in **For You To Do.** Be sure to keep male and female data separate.
   c) Connect the points to show the shapes of the graphs for males and females.
   d) For which race events does the graph for males at your school come closest to touching the graph for men at the Penn Relays? What does this mean? Do the same analysis for females at your school and the Penn Relays.
2. In **For You To Do** you extrapolated from a graph to predict the speed of the winner of the men's 400-m dash in the Olympics in the year 2020. Predict the winning time for that race. (Hint: See **Physics Talk** in **Activity 1.**)
3. In **For You To Do** you extrapolated from two graphs to predict the average speeds of men and women who would win if a 10,000-m run were held at the Penn Relays. Predict the time of the winning man and woman.
4. A runner had an average speed of 10.14 m/s for 9.86 s. Calculate the distance the runner traveled.
5. The table gives the years and the winning times for the women's 400-m dash in the Olympics.
   a) Calculate the winning speeds.
   b) Plot a speed vs. year graph for the data.
   c) What is the trend of the average speeds over the past years?
   d) From your graph extrapolate what the winning speed will be in 2020.

Year	Time (seconds)
1964	52.00
1968	52.00
1972	51.08
1976	49.29
1980	48.88
1984	48.83
1988	48.65
1992	48.83

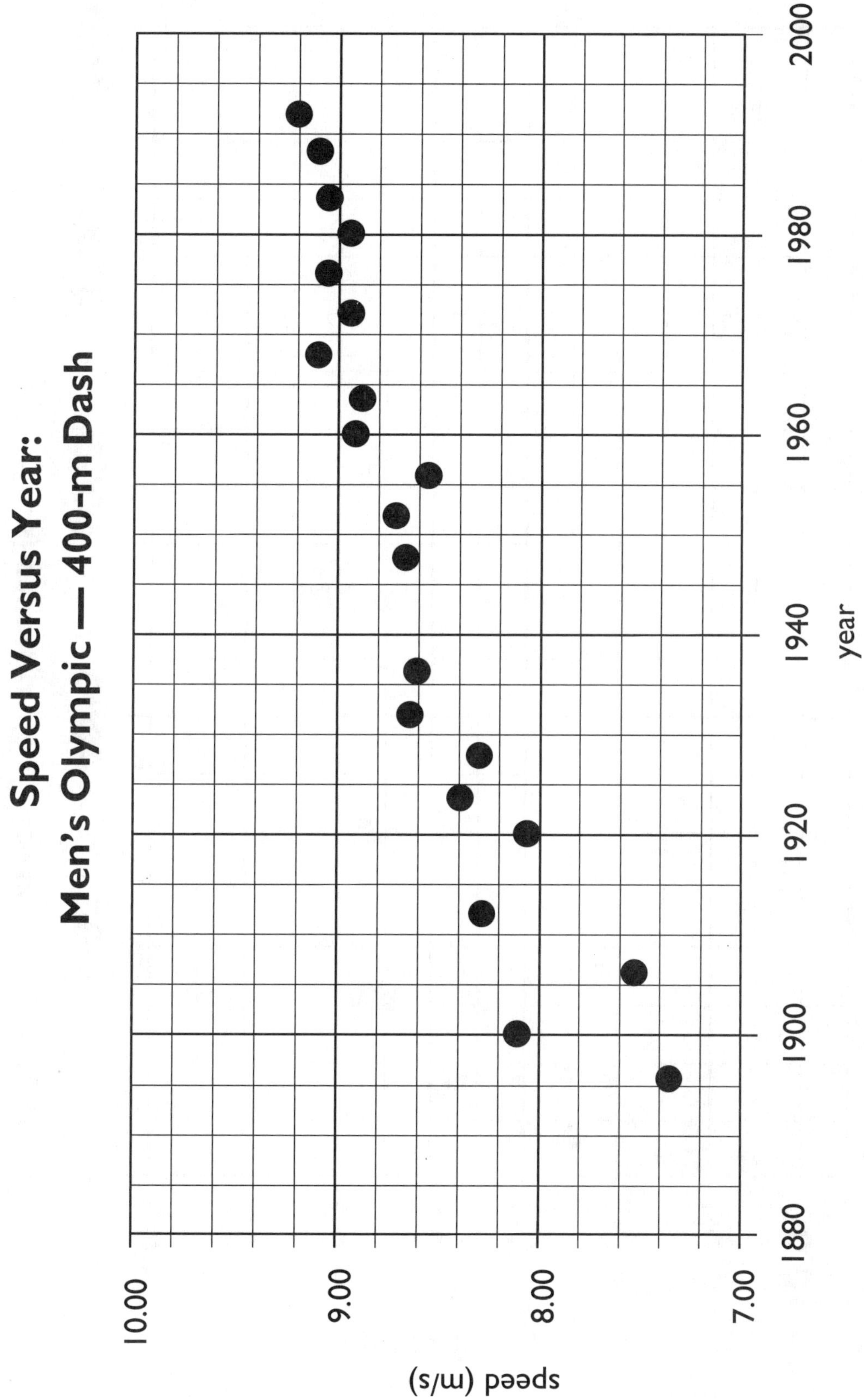

For use with *The Track and Field Championship*, Chapter 3, Activity 2: Analysis of Trends

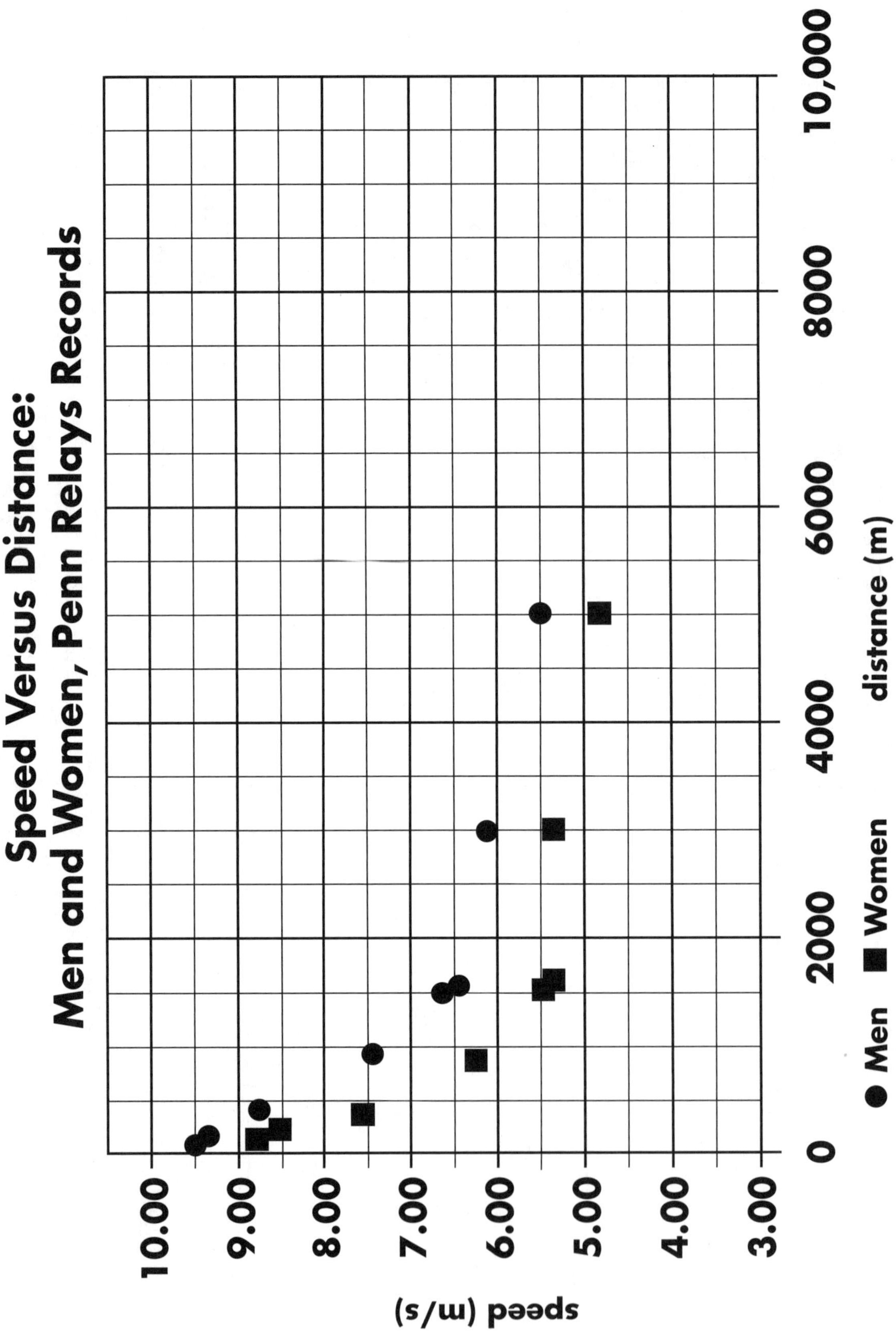

For use with *The Track and Field Championship*, Chapter 3, Activity 2: Analysis of Trends

Activity 2 A

Spreadsheet Games—Analysis of Long-Range Trends

FOR YOU TO DO

In this activity you will need a computer (Mac or MS DOS) and the spreadsheet program MS Excel.

1. Scientists often take a set of data and try to explain what happened in the past or predict what will happen in the future. A convenient method for doing this is to use a computer spreadsheet. Open the file track.xls and click the tab at the bottom of the screen that says "400-m dash." The winning time for each Olympic year is listed as well as the average speed of the runner during the race. To the right of the screen is a graph of Speed versus Year. As you can see from the graph, the speed of the runners has been generally increasing over the years.

   a) Can the speed of future races be predicted from the graph?

NOTE: When using the spreadsheet, to enter a number click in the box, type the number (do not include units or commas) and press <Enter>. If you get the message, "Locked cells cannot be changed," it means that you did not click in the proper box. Try again.

2. The first thing that has to be done is to try to determine how speed and time are related mathematically. In this exercise let's guess that the relationship is a simple straight line. With the aid of a computer, you can create the best straight line to represent this data.

   Click the mouse on the screen button named "Trend line." When you do this, the computer calculates the best-fit line. You will notice that the line is extended into the future and back into the past. This is called extrapolation.

   a) If this trend continues, what will be the speed of the winning runner in the year 2020?

3. Using the equation below, calculate the winning time in the year 2020?
   $$\text{distance (400 m)} = \frac{\text{time}}{\text{speed}}$$

   a) Record your calculations and answer in your log.

   b) Do you think this trend will continue? Explain your answer.

   c) Do you still think that the speed versus year graph is straight, or should it be curved?

   d) If you think the graph should be curved, sketch in your log what your guess is for the shape of the graph.

4. Open the file track.xls and click the tab at the bottom of the screen that says "Penn Relays." The winning time for each event is listed for both men and women. In addition, the average speed for each race has been calculated. To the right of the screen is a graph of Speed versus Distance for the races. It is easy to see from the graph that the speed of the runners generally decreased with increasing race distance. Record your predictions for the following questions:

   a) Can this data be used to predict unknown information? For instance, what will be the winning speed for a much longer race?

For use with *The Track and Field Championship*, Chapter 3, Activity 2: Analysis of Trends

b) If the distance is long enough, will men and women run at the same speed?

5. The first thing that has to be done to answer such questions is to try to determine how speed and race distance are related mathematically. In the 400-m dash activity, the data looked like a straight line. However, this data does not look at all straight. One of the advantages of spreadsheets is that they have the capability to try to calculate a "best fit" and extend the line past the data.

Click the mouse on the screen button named "Trend line." When you do this, the computer calculates the best-fit line. You will notice that the line is also extended to a race distance of 25,000 m; this is called extrapolation.

a) If this trend continues, is there a distance at which men and women will run at the same speed? Explain.

b) Based on the graph, what would be the record speeds (men and women) for a 10,000-m race? Explain.

c) Using the equation below calculate the time for this race.

$$\text{distance (400 m)} = \frac{\text{time}}{\text{speed}}$$

d) The equation calculates time in seconds. Convert this to minutes and seconds.

6. Compare the records at the Penn Relays to the world records.

a) Add the data points for the world records to the Penn Relays graph.

7. Click the mouse on the screen button named "World Records."

a) Based on the graph, do the Penn Relay runners get closer to the world records, further, or stay about the same?

b) Based on the graph, what should be the world record for the 25,000-m run? Determine speed from the graph and then calculate time using the above equation.

c) Convert the time to hours, minutes, and seconds.

Stretching Exercises

Use the spreadsheet to try to compare the best runner in the history of your school with the Olympic runners of the past.

- Look up your school record for the 400-m dash.
- To enter the value on the spreadsheet for the 400-m dash, click in the box pointed to by the arrow.
- Type the school record.
- Click the mouse on the screen button named "Plot Your Record." You should now see a dot on the graph. The position of this dot was determined by calculating the average speed for the runner and finding where that speed appeared along the trend line.
- Using the plotted point for your school record, determine what year your high school runner would have been competitive in the Olympic 400-m dash.

For use with *The Track and Field Championship*, Chapter 3, Activity 2: Analysis of Trends

Performance-Based Assessment Rubric: Reading and Interpreting Graphs

Criteria	Excelling	Good	Basic
Makes comparisons of changes in speed over time.	Working independently, is capable of identifying trends and describe how relay speeds are changing over time.	Requires some assistance to identify trends and describe how relay speeds are changing over time.	Can read graphing coordinates but is unable to make generalizations about how speeds in relays have changed over time.
Makes comparisons of two independent data sets.	Working independently, is able to compare race speeds for men and women at any given year.	Is able to group coordinates for two sets of data, but requires some assistance to compare race speeds for men and women at any given year.	Is unable to group coordinates for independent data sets and is unable to compare race speeds for men and women at any given year.
Compares rates of change for two independent data sets.	Working independently, is capable of comparing the rate at which race speeds are changing for men and women from graphical data.	Is able to explain rate of change, but requires some assistance in reading graphical data to compare the rate at which race speeds are changing for men and women.	Is unable to explain rate of change or make comparisons about how race speeds are changing for men and women from graphical data.
Uses extrapolation to make predictions from a single data set.	Working indepen-dently, is capable of making predictions by extrapolating graphical information to predict relay speeds for 10,000 m.	Requires some assistance to make predictions by extrapolating graphical information to predict relay speeds for 10,000 m.	Can read graphing coordinates but is unable to extrapolate information to predict relay speeds for 10,000 m.
Uses extrapolation to make predictions from two data sets.	Working independently, is capable of making predictions by extrapolating graphical information from two sets of data to find a race distance for which men and women would have the same average race speed.	Requires some assistance to make predictions by extrapolating graphical information from two sets of data to find a race distance for which men and women would have the same average race speed.	Is unable to make predictions by extrapolating graphical information from two sets of data to find a race distance for which men and women would have the same average race speed.

For use with *The Track and Field Championship*, Chapter 3, Activity 2: Analysis of Trends

ACTIVITY 3
Who Wins the Race?

Background Information

This activity continues to apply the concept of average speed, but the particular cases of motion chosen for analysis intentionally involve acceleration. Two major points about average speed should be made through the experiences provided by the activity:

In any race, the runner having the highest average speed wins the race; all runners travel the same distance, but the winner finishes in the least amount of time.

A particular runner's average speed for an entire race may or may not be representative of the runner's speed at various instants during the race.

It is important to maintain a distinction between an instant in time and a time interval. Be clear in your own mind, and encourage your students also to be clear, that the calculation of average speed involves a time interval, the elapsed time between two instants in time.

When this activity has students measure the average speed during a very small time interval which involves only a small part of an object's total "trip," the average speed begins to approximate the instantaneous speed which happened sometime during the small interval. For your information (not suggested to be shared with students at this time), an object's average speed during an infinitesimally small time interval is the object's instantaneous speed:

Instantaneous speed = limiting value, as Δt approaches zero, of the ratio $\Delta d/\Delta t$

where Δd is the distance travelled during the time interval Δt

For your information as the teacher (not recommended to be brought up to students at this time), the final speed of each car at the end of each 1-m run in **For You To Do, Step 4** should depend on the overall change in the height from the beginning to the end of each track.

Speed is proportional to $\sqrt{\Delta h}$,where Δh is the change in height.

Active-ating the Physics InfoMall

In addition to the references from the previous activities on speed and graphing, we need information on acceleration for this activity. Again, we can look at the list of hits for the search using "student difficult*" or "student understand*". On that list is "Investigation of student understanding of the concept of acceleration in one dimension," *American Journal of Physics*, vol. 49, issue 3.

This activity uses a ticker-tape timer extensively. You can search the InfoMall using "ticker" and "tape" and "timer" and get a short list of hits, many of which are not very useful. But notice that the book *Teaching Physics: A guide for the Non-specialist* (from the Book Basement) appears a number of times. This book describes the ticker-tape timer, but also points out some problems associated with the timer. Another hit on this list is "Resource Kit for the New Physics Teacher", found in the Pamphlet Parlor. If you look at the table of contents for this pamphlet, you will find that there are sections that you may find useful for other activities as well. One of the parts of this is a worksheet on the "Analysis of a falling body tape." Contained in this is a nice method for converting the tape to a graph of position versus time. Also in this search, you can find that there is an alternative to ticker-tapes.

Physics Lab Experiments and Computer Aids suggests that strobe photographs can be used to replace ticker-tapes. Depending on what equipment you have, you may or may not want an alternative. A little more searching (you may want to broaden your search by using only the words "ticker" and "timer") finds the article "Can pupils learn through their own movement? A study of the use of a motion sensor interface," in *Physics Education*, vol. 26, issue 6. This mentions that the motion sensor is compared favorably with the ticker-tape timers by students.

Planning for the Activity

Time Requirements

• One class period.

Materials Needed

For each group:

- calculator, basic
- dynamics carts
- incline track–1meter long
- roll of adding machine tape
- stick, meter, 100 cm, hardwood
- tape, masking, 3/4" x 60 yds.
- ticker tape timer, AC timer

Advance Preparation and Setup

The apparatus intended for this activity is Dynamic Cart tracks.

If you do not have this particular cart, there is high probability that your students, colleagues, or friends would have a collection which you may be able to borrow, or even inherit.

Teaching Notes

It would be worth class discussion of the questions presented in **What Do You Think?** to detect if any students do not strongly agree that the runner having the highest average speed is the winner of a race. If some students do not recognize this, it is clear they have not internalized the meaning of average speed and need clarification. Hopefully this activity will help them accomplish that.

The blackline cartoon shown on the following page of the Teacher's Edition may be reproduced to help emphasize the concept of average speed. Encourage students who enjoy drawing cartoons to make up their own cartoons to convey the misconceptions of average speed. Also, suggest that the students examine closely the humorous illustrations presented in the *Active Physical Science* books. Have the students identify which physics concept is being illustrated. We hope that you and your students will find many of the drawings helpful and informative, as well as amusing.

This activity should convey to students the value of split times for analyzing the performances of runners in races. Split times for runners would be a good basis for discussion among members of the class, the track coach, and members of the track team. Perhaps some students could volunteer to serve as timers for gathering split time data for the school's athletes in practice sessions or at track meets — volunteers for such purposes often are needed and welcomed by coaches.

Unless you are sure that all of the ticker-tape timers operate at the same frequency, you will want each group of students to carry a single timer (and steady power supply) with them from station-to-station to assure that the timer's frequency does not enter as a variable in the activity.

Each group first will make runs of the car, pulling a timer tape, on a horizontal surface. A sufficient number of timers and surfaces to accommodate all groups simultaneously would be ideal (the floor could be used as a surface for this part of the activity).

Groups should rotate among sloped track stations. To prevent need to wait at stations, duplicate sets of tracks could be used.

For the runs down the tracks, you should be aware that need may exist to use a guide for the timer strip at the bottom end of the first sections of Tracks 2 and 4. A smooth, cylindrical object such as a pen could be held in horizontal orientation, crossing above the track with room for the car to pass beneath it. The timer strip then would "feed" under the smooth cylinder without lifting as the car continues on the final section of the track.

Refer to *Active*-ating the Physics Infomall for further suggestions on the use of the ticker-tape timer. An alternate **Activity 3**, using a photogate, is presented in the pages following this activity in the Teacher's Edition.

Chapter 3

Cartoon by Larry Stone http://home.earthlink.net/~larrystone

"Seventy-five miles per *hour?* Impossible — I only left home five minutes ago."

For use with *The Track and Field Championship,* Chapter 3, Activity 3: Who Wins the Race?

Activity 3 Who Wins the Race?

GOALS

In this activity you will:

- Measure short time intervals in arbitrary units.
- Measure distances to the nearest millimeter.
- Use a record of an object's position vs. time to calculate the object's average speed during designated position and time intervals.
- Measure and describe changes in an object's speed.
- Relate changes in the speed of an object traveling on a complex sloped track to the shape of the track.

What Do You Think?

The fastest human can't go as fast as a car traveling at 25 miles per hour.

Who wins the race?

- **The runner with the highest finishing speed?**
- **The runner with the highest average speed?**
- **The runner with the greatest top speed?**

Take a few minutes to write answers to these questions in your *Active Physics* log. Discuss your answers with your small group to see if you agree or disagree with others. Be prepared to discuss your group's ideas with the entire class.

Activity Overview

In this activity students produce and use a distance versus time record of a cart moving on a variety of sloped tracks to explore changing speed.

Student Objectives

Students will:

- Measure short time intervals in arbitrary units.
- Measure distances to the nearest millimeter.
- Use a record of an object's position vs. time to calculate the object's average speed during designated position and time intervals.
- Measure and describe changes in an object's speed.
- Relate changes in the speed of an object traveling on a complex sloped track to the shape of the track.

ANSWERS FOR THE TEACHER ONLY

What Do You Think?

The runner with the highest average speed wins the race.

Chapter 3

ANSWERS

For You To Do

1. Student activity.

2. a)-c) As a result of the horizontal run of the car, students should recognize that equal spacing of the dots on the tape indicates that the car is moving with constant speed. If the students have pulled the cars at a relatively constant speed, the spacing between the dots will be the same all along the tape.

   d) If the spacing stays the same, the car is moving at a constant speed. If the spacing varies, the car must be speeding up or slowing down.

   e) Student answers will vary.

For You To Do

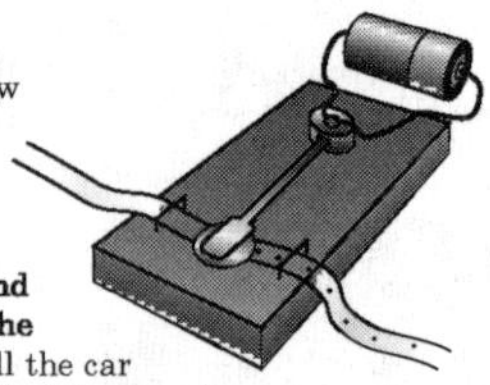

1. Your teacher will show you how to use a ticker-tape timer to record a toy car's **position** and the time it takes to move. Thread a piece of **paper tape** about 1 m long in the **timer, and** attach one end of the **tape to the** car. Turn on the timer, and pull the car at a nearly constant speed so **that the tape** is dragged completely through the timer.

Use only the tape provided. Do not substitute other paper.

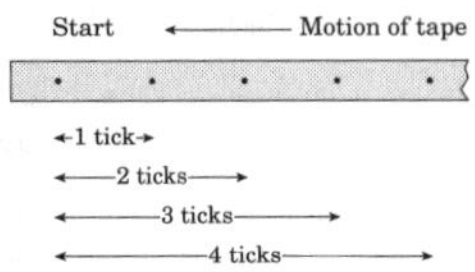

2. Examine the pattern of dots that the timer makes on the tape. Assume that the timer makes dots that are separated by equal amounts of time. The amount of time from one dot to the next will be called a "tick." Obviously, a tick in this case is some small fraction of a second. You will use the tick as a unit of time in this activity and not worry about converting it to seconds.
   a) Do you agree that the distance from one dot to the next on the tape is the distance that the car travelled during one tick of time? Check with your teacher if you have difficulty with this. When you understand this concept, record it in your log.
   b) Do you agree that if you measured the distance from one dot to the next in a unit such as centimeters, you would know the car's speed during that part of the motion in "centimeters per tick"? (Remember, average speed equals distance divided by, or per unit of, time.) When you understand this concept, record it in your log.
   c) Is the spacing between the dots about the same all along the tape, or does the spacing vary?
   d) What would it mean if the spacing stays about the same? If the spacing varies?
   e) Find the part of the tape that shows the beginning of the car's motion where the dots are far enough apart to be seen clearly, and mark one dot as the first dot for analysis. Also locate the last dot made before the tape left the timer at the end of the motion. Measure the distance from the first clear dot to the last dot in centimeters (to the nearest $\frac{1}{10}$ cm, or 1 mm). Also, count the number of tick time intervals (the total number of spaces between dots) from the first to the last dot. Record these measurements in your log.

f) Use the data from **Part (e)** to calculate the average speed of the car in "centimeters per tick." Record your work in your log.

3. Make another "run" with the car, using another paper tape, but this time try to pull the car to make it go faster and faster along its path.

a) Compare the pattern of dots on the tape for this run to when you tried to pull the car at constant speed. In your log, explain differences between the two records of motion in terms of position, speed, and time.

b) Choose a section near the middle of the "faster and faster" tape that is 5-tick intervals long (count 5 spaces between dots, not 5 dots) and mark the beginning and end of the 5-tick time interval. Measure the distance between the first and last dots of the interval. Calculate the average speed during the interval in "centimeters per tick." Record your work.

c) How would the average speed compare if you were to measure it for a similar interval earlier in the run? Later? How do you know? Write your answer in your log.

4. Now let's race! Your "runner" is going to race on each of four inclines as shown. You are going to use the techniques learned above to analyze each race.

Track 1: **This is a simple incline 1.00 m long, supported by a meter stick along its length to keep the track from sagging. One end is raised so that h = 0.10 m.**

Track 2: **This time the incline is 0.40 m long, with a 0.60-m level section for the remaining part. As for track 1, h = 0.10 m.**

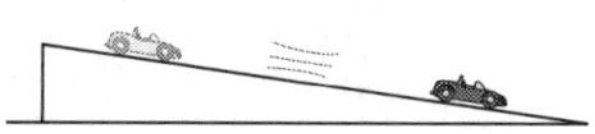

Track 3: **The setup is the same as for track 1, except that h = 0.15 m.**

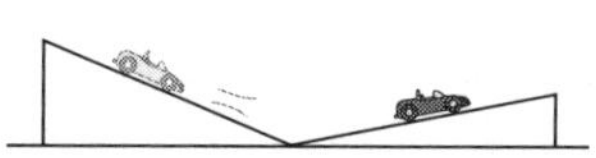

Track 4: **Set the first height at 0.20 m. At about 0.50 m along the track, slope the track up so that the second height is 0.10 m. Use a manila folder to smooth the transition between the inclines.**

ANSWERS

For You To Do

(continued)

f) Students will use the relationship Speed = distance/time to calculate the average speed along the horizontal track. The unit for speed will be cm/tick.

3. a) Students should recognize that increasing or decreasing spacing between the dots indicates, respectively, that the car is "speeding up" or "slowing down." Some students may volunteer the terms "acceleration" and "deceleration" to the latter cases, and that is fine. You may wish to point out that the formal terms will be introduced in **Activity** 5. Students should relate the ideas that when the dots are further apart, the car has travelled a greater distance, since the time between the dots is the same, the speed must have increased.

b)-c) The average speed will vary along the track. It would be slower at the beginning and faster toward the end.

Chapter 3

ANSWERS

For You To Do

(continued)

4. a) Accept any reasonable student answer. Have the students record their prediction in their logs before they begin the race.

5. a) For the final speed students will find the speed during the last few ticks at the end of the tape. For top speed, students will calculate the speed in the length of the tape where the ticks are closest together. On Track 1 and 3 the top speed and the final will be the same. Students will find the overall average speed by dividing the total distance travelled along the track by the total amount of time taken (the total number of ticks).

   b) Check that students have correctly answered **Part c)** in their logs before they proceed to complete this chart.

   c) Neglecting friction (an unrealistic ideal in this case):

   Tracks 1 and 2 should produce equal final speeds. Friction along the final horizontal section of Track 2 probably will cause a lower final speed for that track.

   Tracks 3 and 4 should produce equal final speeds, higher than the final speeds for Tracks 1 and 2.

   The highest top speed should be reached on Track 4 at the bottom of the first half of the Track, followed by Track 3 at the end of the run, where the top speed should equal the final speed. Tracks 1 and 2 should produce equal top speed, equal to the final speeds.

   The highest average speed should be found for Track 4, followed by, in decreasing order of average speed, Tracks 3, 2 and 1.

a) Predict which track will produce the winning result if your car is released and allowed to run 1 m along each track, starting from the top. Record your prediction in your log.

5. Run your toy car on each track. In each case:

   Allow the force of gravity to do the pulling by simply releasing the car at the start of the run.

   Have the car pull at least 1 m of tape through the timer. This may require adding a "leader" to the 1-m tape length to allow for the distance between the car and the timer at the start of the race.

   At the beginning of each run, the timer should be started and then the car should be released.

   After each run, mark the track number on the tape, mark the first clear dot made at the beginning of the run, and place a mark on the tape 1 m (100 cm) beyond the first clear dot.

   a) For each race, explain in your log how you will analyze the tapes to measure the final speed, the top speed, and the overall average speed in centimeters per tick.

   b) Make the necessary measurements, and do the calculations to fill in the speed values in a table similar to the one below.

Track Number	Final Speed	Top Speed	Average Speed
1			

   c) Which track produced the winning run in the big race? How can you tell? Explain your answer in your log.
   If a photogate timer is available, this is a preferable way in which to measure the velocities. If not, try to minimize the friction of the timer from the experiment.

Reflecting on the Activity and the Challenge

Now you can see that it is the details of what happens during a race that determines who wins. The distance of a race and the time taken to run it do not reveal what a champion does along the way to win races consistently.

Speed within most races varies. The sprinters who get up to top speed quickly and maintain their speed throughout a race often win. Those who start quickly and "fade" at the end of the race often lose.

Helping athletes at your school analyze their performances in terms of speed during parts of a race will be needed if they are to compete with the best runners.

Physics To Go

1. Describe a procedure that you could use to convert one "tick" of the timer used in this activity into seconds of time. How could you find out how many "ticks" equal one second?
2. What would the spacing of dots look like for a ticker-tape timer record of an object that is slowing down in its motion?
3. From what you observed and measured during this activity, describe how the speed of a toy car behaves as it travels:
   a) on a straight ramp that slopes downward
   b) on a level surface when the car already has some speed at the beginning
   c) on a straight ramp that slopes upward
4. Aisha and Bert are running at constant speeds, Aisha at 9.0 m/s and Bert at 8.5 m/s. They both cross a "starting line" at the same time. The "finish line" is 100 m away.
   a) How long does it take Aisha to finish the race?
   b) How long does it take Bert to finish the race?
   c) Where is Bert when Aisha crosses the finish line?
   d) By how many meters does Aisha finish ahead of Bert?
5. The Penn Relays women's high school record for the 1500-m run is 4 min, 24.0 s. The women's high school record for the mile (1609 m) run at the Penn Relays is 4 min, 49.2 s. In which race did the record holder have the greatest average speed in meters/second?
6. Salina runs the 200-m race for the school's track team. She runs the first 100 m at 9.0 m/s. Then she hears her classmates cheer, "GO, Salina, GO!" and runs the final 100 m at 10.0 m/s.
   a) Calculate the time for Salina to run the first 100 m.
   b) Calculate the time for Salina to run the final 100 m.
   c) Calculate Salina's average speed for the entire race.

ANSWERS

Physics To Go

1. A wall clock or wristwatch could be used to measure seconds of time as a tape is pulled through a ticker-tape timer. The number of dot intervals created during, for example, 10 s could be used to calculate how many dots are made by the timer per second. This process is called calibration.

2. An object slowing down would produce a pattern of dots which would have the distance from one dot to the next decreasing along the tape.

3. a) The speed increases steadily.

   b) The speed is constant, perhaps slowing down gradually due to friction.

   c) The speed decreases steadily.

4. a) 100 m / 9.0 m/s = 11.1 s

   b) 100 m / 8.5 m/s = 11.8 s
   c) (Bert's speed) × (Aisha's time) = 8.5 m/s × 11.1 s = 94.3 m from start line.

   d) 100 m - 94.3 m = 5.7 m

5. 1500 m / 264.0 s = 5.681 m/s
   1609 m / 289.2 s = 5.564 m/s
   The 1500 m runner had the highest average speed.

6. a) 100 m / 9.0 m/s = 11.1 s

   b) 100 m / 10.0 m/s = 10.0 s

   c) 200 m / (11.1 s + 10.0 s) = 200 m / 21.1 s = 9.48 m/s

Chapter 3

Activity 3 A

Who Wins the Race? [Using a Photogate]

FOR YOU TO DO

1. The speed of a car at a particular point on its track can be measured using a photogate timer. A photogate timer consists of a small light source and a light detector. When the light beam is interrupted, the detector can signal a timing device to start. When the light beam reaches the detector again, it can turn the timer off.

   A piece of cardboard 5.0 cm long attached to the top of a car will be used to interrupt the light beam as the car goes through the photogate.

   Set up a piece of track and the photogate on a horizontal surface, and try out the operation of the timing system. Be sure only the cardboard interrupts the photogate beam. Give the car a gentle push to give it a small speed.

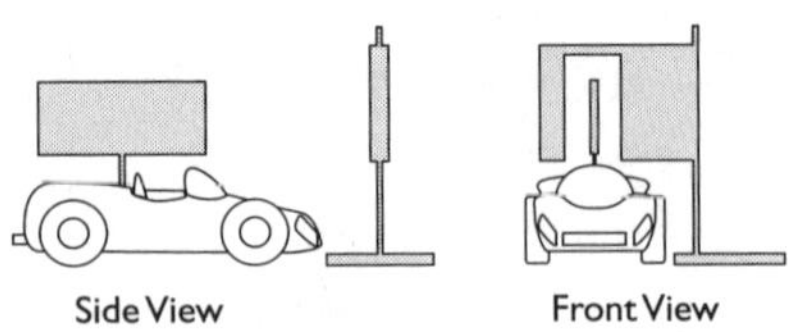

   a) Record the time for the 5.0 cm card to pass through the photogate system.

   b) Divide the length of the piece of cardboard (0.050 m) by the time interval reading from the photogate to calculate a value for the speed of the car as it passed through the photogate. Record this value.

2. Reset the system. Give the car a higher speed.

   a) Record the time and distance values, and calculate the speed.

   b) Does the calculation also show that the speed was higher this time? If not, review your procedures and find out what went wrong.

3. The speeds of cars at various points along four different 1-m tracks will be measured. Set up four tracks as shown at left.

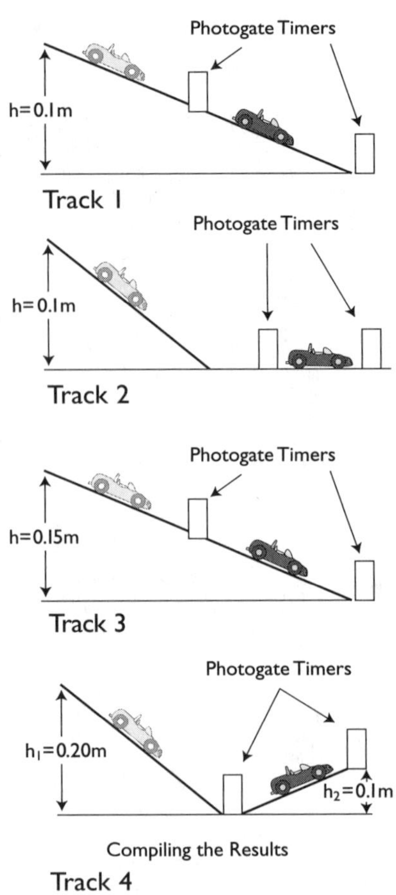

   a) Predict which track will produce the winning result if your car is released and allowed to run 1 m along each track, starting from the top. Will it be the track that produces the highest final speed? Will it be the track that produces the highest top speed? Record your prediction in your log.

4. Run the race.

   a) Record your results in the tables on the next page.

   b) Based on the results of your race, write a statement informing the track team about who wins the race and why. Include any hints you have for improving their performances for running.

Note: Use a manilla folder to smooth the transition.

For use with *The Track and Field Championship,* Chapter 3, Activity 3: Who Wins the Race?

One-Meter Sprints

TRACK	MID GATE TIME (SECONDS)	END GATE TIME (SECONDS)	DISTANCE (METERS)
1			0.050
2			0.050
3			0.050
4			0.050

TRACK	MID GATE SPEED (M/S)	END GATE SPEED (M/S)
1		
2		
3		
4		

Compiling the Results and Making Predictions

TRACK	TOP SPEED (M/S)	FINAL SPEED (M/S)	PLACE PREDICTIONS
1			
2			
3			
4			

TRACK	BEST TOP?	BEST FINAL?	BEST AVERAGE?	WINNER
1				
2				
3				
4				

Chapter 3

For use with *The Track and Field Championship,* Chapter 3, Activity 3: Who Wins the Race?

ACTIVITY 4
Understanding the Sprint

Background Information

Instantaneous speed is introduced in this activity as the slope of a distance versus time graph at a particular instant; this is approached only semi-quantitatively.

Prior to introduction of instantaneous speed, a histogram is used to allow comparison of the average speed of a sprinter during each 10-m interval of a 100-m dash. Be careful how you interpret the histogram; it is tempting to think of the average speed during each 10-m interval as the speed at a particular instant, such as the beginning of the interval — this is not true. The data allows calculation of only the average speed during each 10-m interval, and the particular instant(s) at which the average speed value may have occurred as the instantaneous speed during the interval cannot be determined from the histogram.

To satisfy yourself that you understand that the slope of the distance versus time graph at a particular instant gives the instantaneous speed at that instant, you may wish to measure the slope of the graph at one or more points. To do so at a particular time, such as 3 s:

- Identify the point on the curve corresponding to $t = 3$ s.
- Use a straightedge to estimate the orientation of a line tangent to (having the same direction as) the curve at $t = 3$ s, and draw the tangent.
- Mark two points on the tangent line (recommended: points corresponding to $t = 2$ s, $t = 4$ s, which establishes the horizontal "run" of the slope as 4 s - 2 s = 2 s.
- From each of the two marked points on the tangent line, draw a horizontal line to intersect the distance axis. The difference in the two intersections establishes the vertical "rise" of the slope (approximately 30 m - 10 m = 20 m in this example).
- Calculate the instantaneous speed at $t = 3$ s as the "rise divided by the run," approximately 20 m / 2 s = 10 m/s.

You may wish to use the same method to measure the instantaneous speed at another, earlier time when the slope obviously is different.

Taking the slope of a curve is another task that can be accomplished much easier, and more accurately, using a computer or graphing calculator.

Active-ating the Physics InfoMall

Do a search with keyword "sprint*". You will get a great list of references, including "Effect of wind and altitude on record performance in foot races, pole vault, and long jump," *American Journal of Physics,*vol. 53, issue 8, and "How Olympic records depend on location," *American Journal of Physics,* vol. 54, issue 6. Also in this list of hits is Chapter 2 Describing Motion in *The Fascination of Physics* from the Textbook Trove. This has an informative discussion of speed, velocity, and sprinting, including acceleration. A glance through this chapter finds more about strobe photographs. (This technique may be deserving of its own search on the InfoMall.) If you look at the questions at the end of this chapter in the textbook, you will find that question C9 compares distance runners with sprinters. If you click on the "answer" link, you will get the answer to the question, which can be found in the *Instructor's Guide for The Fascination of Physics* in the Study Guide Store. This search also finds some problems from the Problem Place that you may wish to check out and use for your class. You are urged to do this search, and others, yourself to find how much information is available on the InfoMall.

Planning for the Activity

Time Requirements

- One class period.

Materials Needed

For each lab group:

- calculator, basic
- graph paper pad of 50 sheets

Advance Preparation and Setup

No significant advance preparation is required for this activity. If you need to construct accelerometers for the next activity, you may wish to get started now. (See the **Materials Needed** for **Activity 5**.)

Teaching Notes

You may wish to have students work in their groups, but each student should produce and analyze the histogram and graph as directed in **For You To Do.**

You may wish to provide the students with a copy of the graph from student page 157 to complete.

Encourage students to get involved with the school's track coach and members of the track team. Suggest specific proposals that students could make which would be of mutual benefit to students for the **Chapter Challenge** and to the team for improved performances.

Activity Overview

In this activity students create and then compare a distance versus time graph and a bar graph of speed versus time, to analyze in detail the speed of a runner during a race.

Student Objectives

Students will:

- Calculate the average speed of a runner given distance and time.
- Produce a histogram showing the average speeds of a runner during segments of a race; analyze changes in the runner's speed.
- Produce a graph of distance versus time from split time data for a runner.
- Estimate the slope of a distance versus time graph at specified times.
- Recognize that the slope of a distance versus time graph at a particular time represents the speed at that time.

ANSWERS FOR THE TEACHER ONLY

What Do You Think?

In his world record 100-m dash, Carl Lewis reached his top speed about halfway into the race in terms of both time and distance. World class sprinters run fastest during the last part of a race; the first part of the race involves acceleration from rest and, therefore, lower average speed.

The Track and Field Championship

Activity 4 Understanding the Sprint

GOALS

In this activity you will:

- Calculate the average speed of a runner given distance and time.
- Produce a histogram showing the average speeds of a runner during segments of a race; analyze changes in the runner's speed.
- Produce a graph of distance versus time from split time data for a runner.
- Estimate the slope of a distance versus time graph at specified times.
- Recognize that the slope of a distance versus time graph at a particular time represents the speed at that time.

What Do You Think?

It was not believed to be humanly possible to run a mile in less than four minutes until Roger Bannister of England did it in 1954.

- **How much time does it take to get "up to speed"?**

Take a few minutes to write an answer to this question in your *Active Physics* log. Discuss your answer with your small group to see if you agree or disagree. Be prepared to discuss your group's ideas with the class.

For You To Do

1. Carl Lewis established a world record for the 100-m dash at the World Track and Field Championships held in Tokyo, Japan, in 1991. The times at which he reached various distances in the race (his "split times," or "splits") are shown in the table below.

Distance (meters)	0.0	10.0	20.0	30.0	40.0	50.0	60.0	70.0	80.0	90.0	100.0
Time (seconds)	0.00	1.88	2.96	3.88	4.77	5.61	6.45	7.29	8.13	9.00	9.86

Active Physics 156

ANSWERS

For You To Do

1. a)-b) (see chart)

Distance (m)	Time Interval (s)	Average Speed (m/s)
0.0 - 10.0	1.88	5.3
10.0 - 20.0	1.08	9.3
20.0 - 30.0	0.92	10.9
30.0 - 40.0	0.89	11.2
40.0 - 50.0	0.84	11.9
50.0 - 60.0	0.84	11.9
60.0 - 70.0	0.84	11.9
70.0 - 80.0	0.84	11.9
80.0 - 90.0	0.87	11.5
90.0 - 100.0	0.86	11.6

a) In your log, copy and complete the table to the right. Use subtraction to calculate the time taken by Carl Lewis to run each 10 m of distance during the race.

b) Calculate Lewis's average speed during each 10 m of the race. The values of the time interval and the average speed have been entered in the table for the first 10 m of the dash.

Distance (meters)	Time Interval (seconds)	Average Speed (meters/second)
0.0–10.0	1.88	5.32
10.0–20.0		
80.0–90.0		
90.0–100.0		

2. Use the data you created for the above table to make a bar graph to give you a visual display of Carl Lewis's average speed during each 10 m of his world-record 100-m dash. Use a piece of graph paper set up as shown to the right.

a) Tape or copy the bar graph in your log.

Carl Lewis's World Record 100-m Dash Average Speed, 10-m Intervals

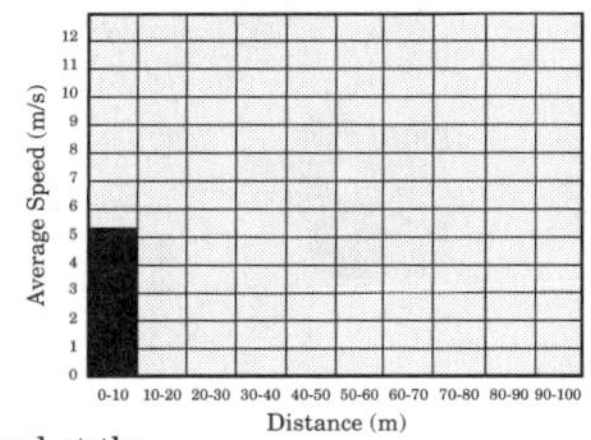

3. Analyze the bar graph to answer these questions:

a) At what position in the dash did Lewis reach top speed? How close can you state that position to the nearest meter? to the nearest 10 m? Explain your answer.

b) How well did Carl Lewis keep his top speed once he reached it? Did he seem to be getting tired at the end of the race? Give evidence for your answers.

c) Can you tell how fast Lewis was going at an exact position in the race, such as at 15.0 m or 20.0 m? Why or why not?

d) It took 9.86 s for Lewis to run the entire 100 m. Calculate his overall average speed. Draw a horizontal line across the bar graph at an appropriate height to represent the average speed for the entire race. Compare the height of each bar on the graph with the height of the line. Explain what the comparisons mean.

4. Use the "splits" given at the start of this activity to plot a graph of Carl Lewis's position versus time.

a) On a piece of graph paper, make a vertical distance scale from 0 to 100 m and a horizontal time scale from 0 to 10 s. Plot each position at the appropriate time and connect the points to show what you think is the shape of the graph. Tape or copy the graph in your log.

4. a)

Carl Lewis's World Record for the 100-m
Distance vs. Time

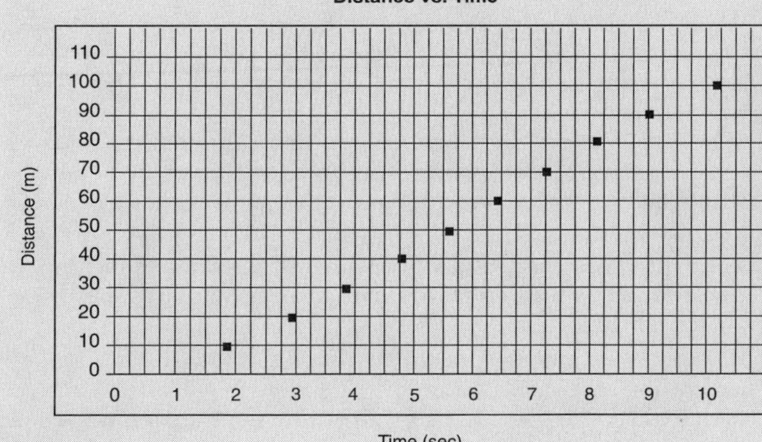

ANSWERS

For You To Do

(continued)

2. a)

Carl Lewis's World Record 100-m Dash Average Speed, 10-m Intervals

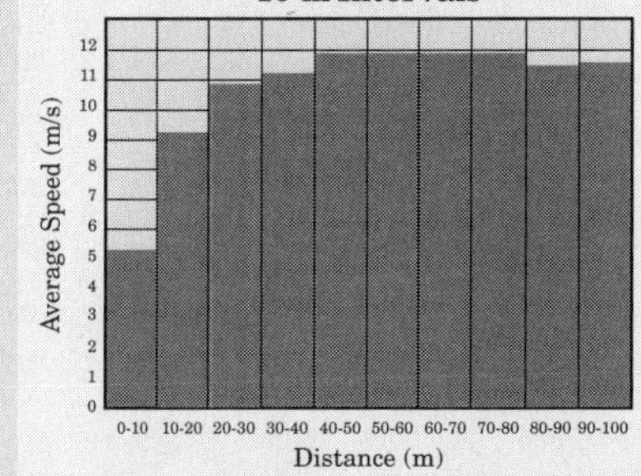

3. a) Lewis reached his top average speed per interval, 11.9 m/s, at the interval 40-50 m. The refinement of the histogram allows "resolution" only to the 10-m interval.

b) Lewis's speed "faded" only slightly during the final two 10-m intervals.

c) No. The histogram is an assembly of average speeds during 10-m intervals; it does not give information about the speed at exact positions along the 100 path.

d) Lewis's overall average speed for the dash was 100-m / 9.86 or s = 10.1 m/s. His average speed during the first interval was only about half of the overall average speed. At the second interval he had almost reached the overall average speed, and thereafter he exceeded the overall average speed.

Chapter 3

5. Compare the distance versus time graph with the bar graph of speed versus distance.

a) When the distance versus time graph is curving early in the run, do the bars on the bar graph change in height or are they fairly steady in height? What does this comparison mean? When the graph is climbing in a straight line, what is happening to the heights of the bars on the bar graph? What does this comparison mean?

b) Someone said, "The slope, or steepness, of a distance versus time graph at any instant is the speed at that instant." Do you believe this statement? Why or why not?

c) Describe the slope, or steepness, of the distance versus time graph each second (1.00 s, 2.00 s, 3.00 s, and so on) during Lewis's record run.

Reflecting on the Activity and the Challenge

In this activity you saw two different ways to analyze in detail the speed of a runner during a race. If you had split-time information for runners in sprint events for your school's track team, you could help the runners find out, for example, if they are "letting up" at the end of a race or how rapidly they are reaching top speed.

With more knowledge about details of their performances, your school's runners may find that they can improve parts of their races.

Physics Talk

$$\text{Speed} = \frac{\text{Distance traveled}}{\text{Time elapsed}}$$

On a distance versus time graph, the speed is equal to the slope of the graph.

ANSWERS

For You To Do *(continued)*

5. a) When the graph is curving early in the run (increasing in slope), the bars of the histogram at corresponding positions and times are increasing in height; both representations indicate that Lewis was accelerating.
When the graph is climbing on about a straight line, the bars of the histogram at corresponding positions and times are of fairly steady height; both representations indicate that Lewis had a fairly constant speed.

b) Formally (not expected of students here), the slope of the distance versus time graph at a particular instant gives the "instantaneous speed."

c) With some latitude for individual differences in visual impression, a sequence would be: 1 s : low, 2 s : higher, 3 s : even higher, 4 to 10 s : highest (about the same at all times 4 through 10 seconds).

Physics To Go

1. If you were to design a track for a toy car to run as in **Activity 3** to simulate Carl Lewis's 100-m dash, what shape would you design for the track?
2. For long distances, humans can run at a constant speed of about 6 m/s, pigs at about 4 m/s, and horses at 20 m/s. Sketch a distance versus time graph with three lines showing a person, a pig, and a horse running for 100 s.
3. Sketch a distance versus time graph for a person who is not moving at all.
4. In a 2 × 100-m relay race, Joan ran the first 100 m at a speed of 5 m/s and then Rami ran the next 100 m at a speed of 10 m/s. What was the average speed for the entire relay race? (Hint: It is not 7.5 m/s.)
5. Do you think it is possible for a runner to keep increasing speed for an entire race as a strategy to win?
6. Examine what excellent runners do in long-distance races. On the right is a chart of Eamonn Coghlan's split time and total time every 200 m in a mile (1609-m) race.
   a) Use the total times listed to calculate Coghlan's average speed for distances of 200, 400, 800, and 1000 m distances during his run.
   b) Compare Coghlan's average speeds for distances of 200, 400, 800, and 1000 m with the average speeds of Penn Relays record holders at the same distances. (Penn Relays record average speeds for various distances are listed in **Activity 1**.) What patterns do you see in the comparison? At what distances was Coghlan's speed getting closest to world-record speed?
   c) Use Coghlan's split times to identify the distance interval (0–200 m, 200–400 m, 400–600 m, and so on) when he had the greatest average speed. Also identify the interval of lowest average speed.
   d) You found out in this activity that Carl Lewis slowed down slightly near the end of his record 100-m dash. Did Eamonn Coghlan do the same thing near the end of his mile run? Use data in your answer.
   e) How could you use data about Coghlan's performance to give advice to members of your school's track team who enter long-distance events?

Distance (m)	Split Time (s)	Total Time (s)
0	0.00	0.00
200	29.23	29.23
400	29.87	59.10
600	30.09	89.19
800	30.25	119.44
1000	29.88	149.32
1200	29.90	179.22
1400	29.38	208.60
1600	29.55	238.15

Answers

Physics To Go

1. The track would have a downward slope for about the first half of the distance and a flat section for about the last half - the last section could have a slight upward slope near the very end of the track.

2.

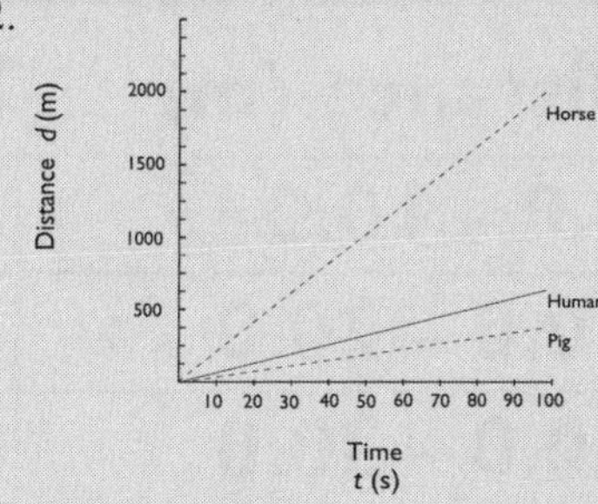

3. For zero speed, the distance versus time graph would correspond to the time axis.

4. Joan: 100 m / (5 m/s) = 20 s
   Rami: 100 m / (10 m/s) = 10 s
   Average speed = 200 m / 30 s
   = 6.7 m/s
   Be alert to a misconception: taking the "average" of average speeds, in this case 15 m/s, is not valid because the lower speed in this case was maintained for more time than the higher speed. Some students may find this tempting, but it is not a valid method.

5. Maintaining a significant acceleration for a significant amount of time is hardly possible.

6. a) 200 m : 6.84 m/s
   400 m : 6.76 m/s
   800 m : 6.70 m/s
   1000 m : 6.70 m/s
   b) All of Coghlan's average speeds are less than world record speeds for corresponding distances; however, a clear pattern exists which is that Coghlan's speeds get closer to world record average speeds as the distance increases (his speed up to 1000 m is 6.7 m/s; the world record average speed for that distance is 7.55 m/s).
   c) Coghlan had the greatest average speed when his split time was lowest, during the interval 0-200 m; he had the least speed when his split time was greatest, during the interval 600-800 m.
   d) No, Coghlan did not slow down near the end of the race; indeed, he accelerated, as shown by the decreased splits for the last two 200-m intervals.
   e) One piece of useful advice which could be derived from Coghlan's race would be to save some energy for a "kick" of acceleration near the end of the race.

Carl Lewis's World Record 100-m Dash Average Speed, 10-m Intervals

Distance (m)	Time Interval (s)	Average Speed (m/s)
0.0 - 10.0	1.88	5.3
10.0 - 20.0		
20.0 - 30.0		
30.0 - 40.0		
40.0 - 50.0		
50.0 - 60.0		
60.0 - 70.0		
70.0 - 80.0		
80.0 - 90.0		
90.0 - 100.0		

For use with *The Track and Field Championship*, Chapter 3, Activity 4: Understanding the Sprint

NOTES

ACTIVITY 5
Acceleration

Background Information

$a = \Delta v / \Delta t$ where Δv is the change in speed accomplished in the time interval Δt.

The standard unit of acceleration is m/s^2. If this unit is mysterious to you, see discussion of the unit in the **For You To Read** section of the student text for this activity.

It is possible that you will encounter acceleration expressed in nonstandard units such as "feet per second squared" (the acceleration due to gravity in the British system of units is 32 ft/s^2; some students may have heard about this number and may bring it up in discussion). Another nonstandard unit of acceleration used in this country for automobiles, with which some students may be familiar, is "miles per hour per second."

Regardless of the unit used to express acceleration, the meaning is the change in speed per unit of time.

Strictly speaking, the above definition applies to "average acceleration." The acceleration of runners in track events changes with time. For example, a runner in a 100-m dash typically begins a race with high acceleration, and then the acceleration decreases, falling to zero when the runner reaches and maintains a fairly constant "top speed." If the overall change in speed, Δv, during the early part of the race is the difference between zero speed (at the start) and the top speed, the average acceleration is Δv divided by the amount of time, Δt, taken to reach top speed. The runner's "instantaneous acceleration" (the acceleration at a particular instant) would at some times be less than the average acceleration and at other times more than the average acceleration. Just as average speed may or may not be representative of the detailed behavior of speed during a journey, average acceleration may or may not be representative of detailed variations in speed.

Do not assume that all cases of acceleration involve constant acceleration (a steady, constant rate of change of speed). Constant acceleration, also referred to as uniform acceleration, happens only under special conditions, including, for example, free fall and circular motion at constant speed.

Notice that the symbol "v" is used for speed in the definition of acceleration. This is because the term "velocity" is used in place of the term "speed" in more advanced treatments of motion where a directional property is included to treat velocity as a vector quantity. To avoid confusion for students who may study physics at a more advanced level in the future, it would be best for you to use the word "speed" and the symbol "v" and to avoid using the word "velocity."

Active-ating the Physics InfoMall

You may wish to search the InfoMall with keywords "acceleration" and "misconception*". This generates a slightly different list than the search done earlier. One interesting hit is "Physics that textbook writers usually get wrong," in *The Physics Teacher*, vol. 30, issue 7. A passage from this reference says that "Deceleration means a decrease in speed, so if the velocity is positive, a negative acceleration is a decrease in velocity, and therefore a deceleration. A ball thrown upward ($v > 0$) has an acceleration downward ($a < 0$) and therefore decelerates. However, the same ball, as it falls ($v < 0$), is subject to the same negative acceleration (a is downward), and therefore gains speed; it is not decelerated." This is often a cause of confusion with students. (See the **Physics To Go Questions**, especially 4 and 5.)

You may also find "Common sense concepts about motion," *American Journal of Physics*, vol. 53, issue 11, to be interesting. Perform a search using the keyword "accelerometer" for some information regarding these devices. Note that Chapter 6 of *Teaching High School Physics* (in the Book Basement) has a nice graphic of a simple cork accelerometer you can easily build yourself. Don't forget that we found some good references for acceleration earlier, including "Investigation of student understanding of the concept of acceleration in one dimension," *American Journal of Physics*, vol. 49, issue 3.

Planning for the Activity

Time Requirements

- One class period.

Materials Needed

For each group:

- dynamics carts
- accelerometer
- accelerometer, cork
- accelerometer-to-cart
- mass hanger
- steel pulley on mount
- weight set, slotted
- string, ball of, 225 ft.

Advance Preparation and Setup

If you do not have cork accelerometers, you may wish to construct one per group using the directions given at the end of this activity.

Other kinds of accelerometers may be substituted, but doing so will require that you alter the directions to students in **For You To Do** to comply with the behavior of the kind of accelerometer used.

Teaching Notes

Expect that some students will have difficulty grasping the meaning of acceleration. It is a "rate of a rate"— a compound abstraction that can be expected to present intellectual difficulty for young minds (and some old minds). Be patient, and hope that the variety of experiences which students will have involving acceleration will allow each to come to understand it.

Similar to the definition of acceleration, some students can be expected to have difficulty understanding the unit of acceleration. It is strongly suggested that you check their understanding of the mathematical steps involved in arriving at meters per second squared as the "short" version of the unit.

Activity Overview

In this activity students are introduced to acceleration using a simple accelerometer.

Student Objectives

Students will:

- Understand the definition of acceleration.
- Understand meters per second per second as the unit of acceleration.
- Use an accelerometer to detect acceleration.
- Use an accelerometer to make semi-quantitative comparisons of accelerations.
- Distinguish between acceleration and deceleration.

ANSWERS FOR THE TEACHER ONLY

What Do You Think?

Many races are won during the period of acceleration at the beginning of the race.

The runner that is out of the blocks first, that is, the one who has the quickest response time after the starting pistol is fired, has an advantage. All other things being equal, that runner will win the race. This is particularily important in short distance races.

Activity 5 Acceleration

GOALS

In this activity you will:

- Understand the definition of acceleration.
- Understand meters per second per second as the unit of acceleration.
- Use an accelerometer to detect acceleration.
- Use an accelerometer to make semi-quantitative comparisons of accelerations.
- Distinguish between acceleration and deceleration.

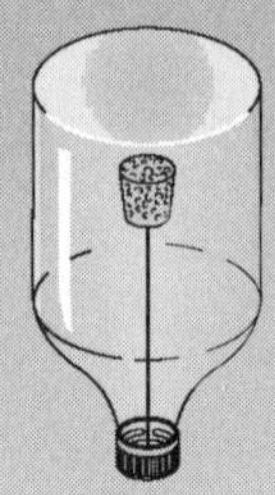

A cork attached to a string floats in a liquid-filled bottle.

Use only plastic bottles to construct accelerometers.

What Do You Think?

Accelerating out of the starting blocks is important if a runner is going to win a race.

- **If all runners in a dash have equal top speeds and none "fades" at the end of the race, what determines who wins?**
- **How is response time a factor in determining who wins the race?**

Write your answers to these questions in your *Active Physics* log. Be prepared to discuss your ideas with your small group and other members of your class.

For You To Do

1. In this activity, you will use an "accelerometer," a device for measuring acceleration. There are many kinds of accelerometers. The diagram below on the left shows a "cork accelerometer."

Explore how the accelerometer works by holding it in your hands in front of you so that you can look down to see the top of the cork in the bottle and liquid. Hold the accelerometer upright so that the cork is centered.

2. Take the accelerometer for a walk. Observe any movement of the cork, especially as you start from a resting position and speed up (accelerate). Walk at a fairly constant speed, and then slow down to a stop (decelerate). Try it a few times—starting, walking, and stopping at normal rates.
   a) Record your observations of the cork's movement.
3. Repeat the above walk and observe what happens if you start faster, if you walk faster at a constant speed, and if you stop faster.
   a) Record your observations in your log.

ANSWERS

For You To Do

1. Student activity.
2. a) While exploring the accelerometer, students should discover that the cork "leans" in the direction of acceleration, and, the greater the acceleration, the greater the amount of lean of the cork.
3. a) The greater the acceleration, the greater the amount of lean of the cork.

4. Repeat the walk in **Step 2,** but walk backward.
   a) Record your observations in your log.
5. Use your observations to answer the following questions in your log.

   Describe the *amount* and the *direction* the cork leans in each of the following situations:
   a) standing at rest
   b) low acceleration while walking forward
   c) high acceleration while walking forward
   d) low constant speed while walking forward
   e) high constant speed while walking forward
   f) high deceleration (slowing down) while walking forward
   g) low deceleration while walking forward
6. Someone said, "Deceleration while walking forward is the same as acceleration while walking backward."
   a) Do you agree or disagree? Use your observations of the accelerometer for your answer.
7. Set up a system to take the accelerometer for a ride on a cart that is being pulled by a falling weight, as shown in the diagram to the right. Hang a weight on the string and allow it to pull the cart as you observe the accelerometer. Record your answers to the questions below in your log.

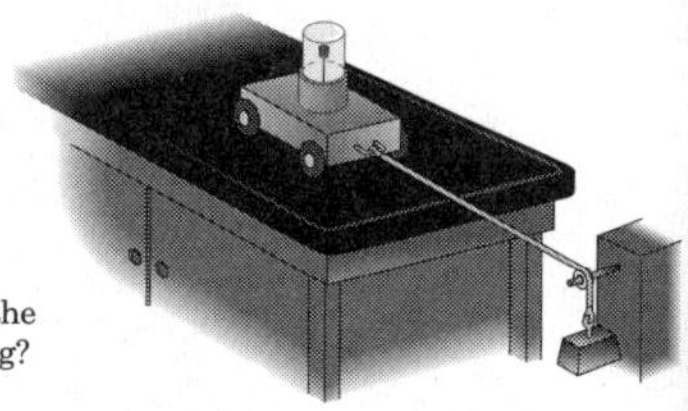

   a) Does the cart appear to accelerate? Does the accelerometer tell you that it's accelerating? How can you tell?
   b) Does the accelerometer show that the acceleration is constant or changing? How can you tell?

Keep area under the falling mass clear.

8. Repeat **Step 7** using a larger weight to pull the cart. Answer the following questions in your log:
   a) How does the acceleration for the large weight compare with the acceleration for the small weight? How can you tell?
   b) Which produced a more steady, constant acceleration: using the falling weights or walking with the accelerometer? What evidence do you have for your answer?

ANSWERS

For You To Do

(continued)

4. a) The cork leans in the direction of the acceleration. In this case, rearward.
5. a) Standing at rest; Zero lean.
   b) Low acceleration while walking forward; Low lean, forward direction.
   c) High acceleration while walking forward; High lean, forward direction.
   d) Low constant speed while walking forward; Zero lean.
   e) High constant speed while walking forward; Zero lean.
   f) High deceleration (slowing down) while walking forward; High lean, rearward direction.
   g) Low deceleration while walking forward; Low lean, rearward direction.
6. a) Yes. In both cases the cork leaned rearward.
7. a) The accelerometer indicates that the cart is accelerating because there is a lean.
   b) The accelerometer shows a lean of a fixed amount during each run.
8. a) The amount of lean should be greater for the greater acceleration provided by the larger weight.
   b) Students probably discovered earlier that it is very difficult to walk with constant acceleration; gravity does a much better job of providing constant acceleration.

PHYSICS TALK

Acceleration

The relationship between **acceleration,** speed, and time can be written as:

$$\text{Acceleration} = \frac{\text{Change in speed}}{\text{Time interval}}$$

Using symbols, the same relationship can be written as:

$$a = \frac{\Delta v}{\Delta t}$$

where
a = acceleration,
Δv = change in speed, and
Δt = the time interval, or change in time, for the change in speed to happen.

Sample Problem

A sprinter at the start of a race increases speed from 0 m/s to 5.0 m/s as the clock runs from 0 s to 2.0 s. Find the sprinter's acceleration.

Strategy: You can use the equation for acceleration.

$$a = \frac{\Delta v}{\Delta t}$$

You can the write this equation to show the change in speed from the initial to final speed and the time interval.

$$a = \frac{v_{final} - v_{initial}}{t_{final} - t_{initial}}$$

Givens:

$v_{initial} = 0$ m/s
$v_{final} = 5.0$ m/s
$t_{initial} = 0$ s
$t_{final} = 2.0$ s

Physics Words

acceleration: the change in velocity per unit time.

vector: a quantity that has both magnitude and direction.

Solution:

$$a = \frac{v_{final} - v_{initial}}{t_{final} - t_{initial}}$$

$$= \frac{5.0 \text{ m/s} - 0 \text{ m/s}}{2.0 \text{ s} - 0 \text{ s}}$$

$$= \frac{5.0 \text{ m/s}}{2.0 \text{ s}}$$

$$= 2.5 \text{ (m/s)/s or } 2.5 \text{ m/s}^2$$

Mathematically, meters per second per second is equal to meters per second squared. Therefore, when you use "m" and "s" as abbreviations for meters and seconds, you can shorten the unit of acceleration to m/s^2. The following ways of stating the unit of acceleration are the same:

(m/s)/s $\quad$ m/s^2

When speaking about acceleration, you can describe the unit as "meters per second squared" or "meters per second every second."

FOR YOU TO READ

Scalars and Vectors

Acceleration is defined as a change in velocity with respect to time. In everyday language, you probably use the words velocity and speed interchangeably. In physics, the two words have precise and distinct meanings.

Speed is a measure of how fast something is moving. A runner's speed may be 8 m/s or 9 m/s. Velocity is a measure of how fast and in what direction something is moving. A runner's velocity may be 8 m/s north or 9 m/s east.

Acceleration is a change in velocity with respect to time. Velocity can change when the speed changes or when the direction changes (or if both change). The accelerometer in the activity indicated acceleration when the speed of the cart changed. The accelerometer will also indicate acceleration when the cart changes direction. You can hold the accelerometer in your hand. As you rotate, you will notice that the cork leans toward the center of the circle indicating that there is an acceleration.

Quantities in physics that have magnitude (e.g., 8 m/s) and direction (e.g., east) are called **vectors.** Velocity is an example of a vector →

Chapter 3

quantity. Acceleration and **displacement** are also vectors. Displacement is a vector because a person can run 50 m north or 50 m toward the goal.

Some quantities don't have direction. You may be 17 years old. The temperature may be 66°. There may be 29 students in your physics class. None of these values have direction and are referred to as **scalars.** Distance and speed are also scalars. A distance of 1200 m or 3 cm does not have any direction. A speed of 22 m/s also has no direction and is a scalar. (This is why physicists mean different things by velocity and speed.)

The wear and tear on your shoes or the tires of your car can be a measure of the distance you have traveled. The odometer of your car measures the distance your car has traveled, in all directions. The speedometer on your car measures the speed of your car, while traveling in any direction.

Imagine you drive 30 mi. to the east at 30 mph and then drive 30 mi. west on your way back home at 30 mph. Your total distance traveled is 60 mi. and your average speed is 30 mph for two hours. Your total displacement will be zero and your average velocity will be zero. Your total displacement is zero because you ended up in the same position as you began. Your average velocity can be zero because you traveled at 30 mph east for the first hour and then you traveled 30 mph west (or –30 mph) for the second hour.

A runner in the 1600-m race covers a distance of 1600 m. Her displacement is 0 m if she ends up in the same place that she began. (This is the case if the race is run on an oval track.)

If she runs the 1600 m in 300 s, then you can calculate her average speed.

$$\text{Average speed} = \frac{\text{total distance}}{\text{total time}} = \frac{600\text{ m}}{300\text{ s}} = 2\text{ m/s}$$

You can also calculate her average velocity, which is in sharp contrast to her speed.

$$\text{Average velocity} = \frac{\text{total displacement}}{\text{total time}} = \frac{0\text{ m}}{300\text{ s}} = 0\text{ m/s}$$

Activity 5 Acceleration

The introduction of vector and scalar quantities may seem to make things more complicated than they need to be. As you analyze more complex situations, like throwing a javelin or high jumping, the use of vectors will make these descriptions easier. Mathematics is only introduced in physics when it simplifies analysis of a situation and makes things clearer.

Scalar Quantities	Vector Quantities
Distance	Displacement
Speed	Velocity
Time	Acceleration
Mass	Force (mass x acceleration)
	Momentum (mass x velocity)
Energy	
Work	

Reflecting on the Activity and the Challenge

You now know a lot more about acceleration. Since a race always begins with a speed of zero and ends with runners in motion, acceleration is part of every race. Depending on the distance of a race, acceleration may go up, go down, or disappear several times during the race. One runner may have more acceleration than another runner at the start of a race, or one runner's acceleration may change during a race. Do the athletes on your school's track team know about acceleration? If not, maybe they should, and perhaps you can help them.

Physics To Go

1. Is there anything in nature that has constant acceleration?
2. If Carl Lewis were to carry a cork accelerometer during the start of a sprint, describe what the accelerometer would do.
3. Did Carl Lewis accelerate for the entire 100 m of his world-record dash? Explain his pattern of accelerations.
4. If you are running, getting tired, and slowing down, are you accelerating? Explain your answer.
5. When you throw a ball straight up in the air, what is the direction of its acceleration while it is going up? While it is coming down?
6. What additional tips could you give your school's track team as a result of this activity?

Physics Words

scalar: a quantity that has magnitude, but no direction.

displacement: the difference in position between a final position and an initial position; it depends only on the endpoints, not on the path; displacement is a vector; it has magnitude and direction.

ANSWERS

Physics To Go

1. Objects in a state of free fall have constant acceleration due to gravity.
2. In principal, the cork would lean forward, but, in reality, it would waver wildly if carried by Lewis.
3. Carl Lewis accelerated during approximately the first half of his record dash. He also had minor deceleration and acceleration near the end of the dash.
4. Slowing down, also called deceleration, can be thought of as negative acceleration; an amount of speed is being subtracted per unit of time.
5. At all times during which a ball is in the air, the direction of its acceleration is downward.
6. Advice to sprinters: (a) accelerate to top speed as soon as possible, (b) maintain top speed to the end of the race.

ANSWERS

Physics To Go

(continued)

7. a) force → vector (it has direction)
   b) displacement → vector (it has direction)
   c) distance → scalar (no direction)
   d) acceleration → vector (it has direction)
8. 4 blocks west
9. 4 m east
10. 30 m north
11. a) False, with zero acceleration, the object CANNOT be speeding up.
    b) True, an object with a constant speed has zero acceleration.
    c) False, the object may be at rest, but it does not have to be at rest. It can be traveling at a constant speed.
    d) False, slowing down is a change in velocity and therefore there must be an acceleration.
12. $a = \Delta v/\Delta t = (9.26\ m/s)/(1.96\ s) = 4.72\ m/s^2$
13.

Time (s)	Acceleration due to gravity (m/s^2)	Final velocity (m/s)
1.0	9.8	9.8
2.0	9.8	19.6 (20)
3.0	9.8	29.4 (29)
4.0	9.8	39.2 (39)
5.0	9.8	49

14. $a = \Delta v/\Delta t = (30.0\ m/s)/(0.6\ s) = 50\ m/s^2$
15. $a = \Delta v/\Delta t = (11.5\ m/s - 11.9\ m/s)/(0.87\ s) = -0.46\ m/s^2$
16. a) $v = at = (9.8\ m/s^2)(0.5\ s) = 4.9\ m/s$
    b) Since he began at 0 m/s, his average will be $(4.9\ m/s + 0\ m/s)/2 = 2.45\ m/s$
    c) $d = vt = (2.45\ m/s)(0.5\ s) = 1.225\ m$

The Track and Field Championship

7. Which of the following terms represent scalar quantities and which are vector quantities? Explain your choice for each.
   a) force
   b) displacement
   c) distance
   d) acceleration
8. A student walks 3 blocks south, 4 blocks west, and 3 blocks north. What is the displacement of the student?
9. A person travels 6 m north, 4 m east, and 6 m south. What is the total displacement?
10. If a woman runs 100 m north and then 70 m south, what is her total displacement?
11. Which of the following statements about the movement of an object with zero acceleration are true and which are false? For each statement, explain why you indicated it to be true or false.
    a) The object may be speeding up.
    b) The object may be in motion.
    c) The object must be at rest.
    d) The object may be slowing down.
12. Carl Lewis began at rest and reached a speed of 9.26 m/s after 1.96 s of the race. Calculate his acceleration, assuming that it was constant during this time.
13. Objects in free fall that are not affected by air resistance have acceleration due to gravity of 9.8 m/s^2. Calculate the final velocity that a falling object will have after:
    a) 1.0 s b) 2.0 s c) 3.0 s d) 4.0 s e) 5.0 s
14. The javelin reaches a speed of 30.0 m/s within 0.6 s. What is the acceleration of the javelin?
15. During the 100-m run, Carl Lewis decreased his speed late in the race. His speed decreased from 11.9 m/s to 11.5 m/s in 0.87 s. What was his acceleration during that time interval?
16. A pole-vaulter manages to get over a 5.0-m bar. The acceleration due to gravity is 9.8 m/s^2.
    a) When he falls from that height, how fast will he be going after 0.5 s?
    b) What will be his average speed after 0.5 s?
    c) How far will he have fallen after 0.5 s?

Active Physics 166

Constructing a Cork Accelerometer

A cork accelerometer can be constructed from a jar having a tight-fitting lid, a cork on a string, and liquid. Some provision such as glue must be made for attaching the end of the string to the inside of the jar lid.

A plastic jar and lid is essential. Do not use a glass jar due to near certainty of breakage. A pint-size jar works well, and you should select one which has a relatively flat bottom (not depressed in the molding process) so that the cork can be seen through the bottom of the jar when the system is inverted (lid side down) for use.

A thicker-than-water liquid such as vegetable oil is preferred to slow down the response of the cork to acceleration.

Any cork which will float in the liquid will work; brightly colored fishing bobs work very well and have provision for attaching the string.

Modern multipurpose glue of the "Goop" variety works well to attach the end of the string to the inside of the jar lid.

The lid can be glued permanently to the jar to deter the student who can't resist the temptation of opening the jar.

It adds convenience to glue the jar lid to a piece of wood slightly larger than the lid so that the accelerometer can be clamped to a cart to prevent "flying accelerometers." Alternatively, the accelerometers may be placed in a heavy base such as a coffee can filled with sand, to prevent the accelerometer from flying off.

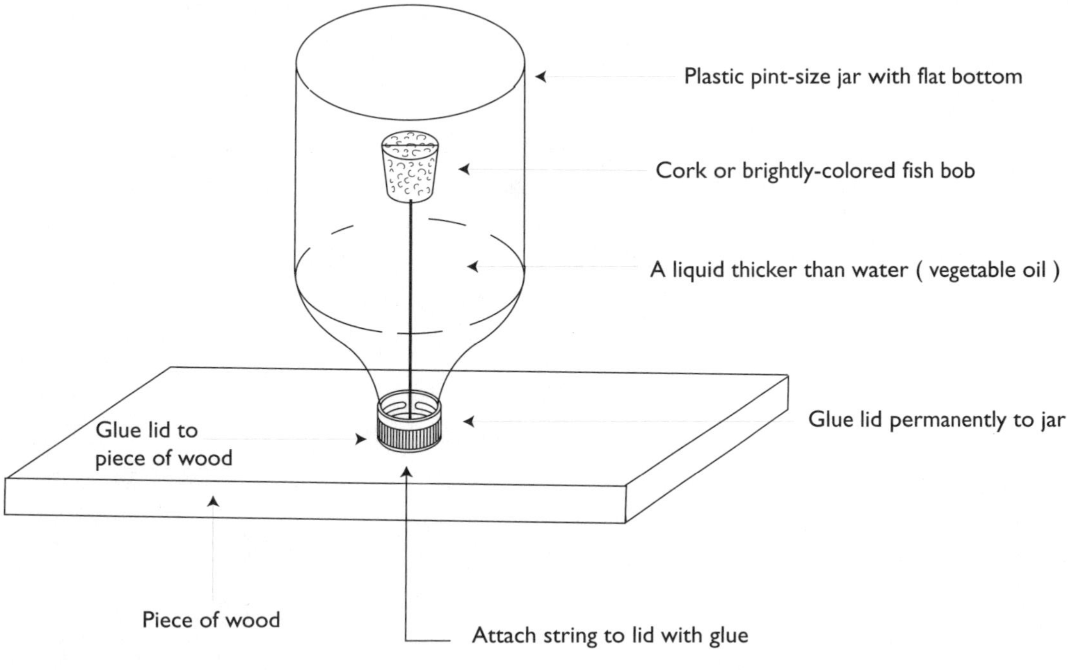

For use with *The Track and Field Championship,* Chapter 3, Activity 5: Acceleration

Activity 5 A

Investigating Different Types of Accelerometers

FOR YOU TO DO

1. A device that detects acceleration is an accelerometer. It can respond to changes in speed. One kind of accelerometer is made by enclosing a liquid in a container. A tubular bubble level fits that description, so it can be used as an accelerometer. When the bubble level is at rest on a horizontal surface, the bubble is in the middle. Try it.

2. Place the bubble level on a free-wheeling cart that is at rest.

3. Stick the level to the cart with clear tape or plastic clay. If necessary, adjust the level so that the bubble is in the middle when the cart is at rest.

4. Now pull the cart along the table so that the bubble stays in the middle after you get the cart into motion.

   a) Describe the kind of motion that keeps the bubble in the middle. Does it make any difference whether the motion is slow or fast?

   b) Which of the diagrams best describes the position of the bubble in the following situations:

   - As the cart is just put into motion to the right?
   - As the cart is brought to an abrupt halt when it is moving to the right?
   - As the cart is brought to an abrupt halt when it is moving to the left?

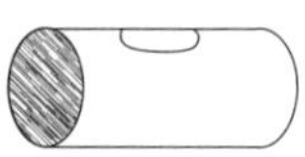

Bubble level at rest on a horizontal surface with the bubble in the center.

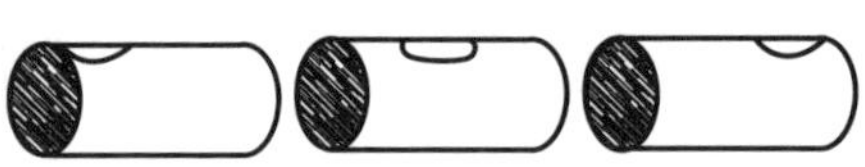

1 2 3

For use with *The Track and Field Championship,* Chapter 3, Activity 5: Acceleration

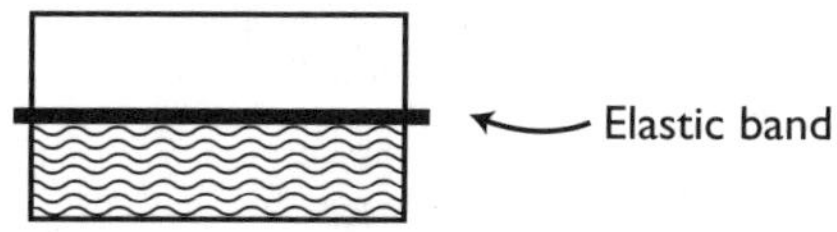

5. Another kind of accelerometer is made from two sheets of clear plastic with a thin space between them filled with colored water. An elastic band can be used as a reference to mark the position of the water surface when the accelerometer is a rest on a level surface.

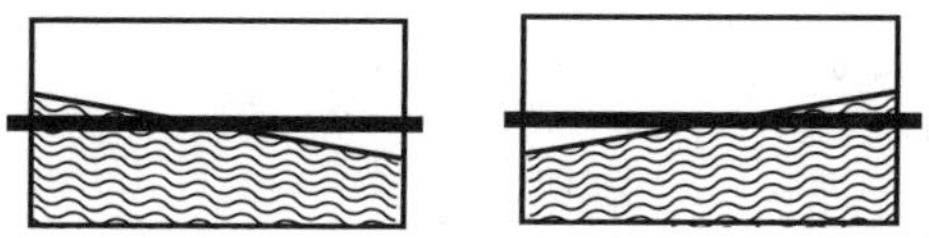

6. When this kind of accelerometer indicates acceleration, the water forms half an arrowhead. Try it on a cart.

   a) Does this "arrow" point in the direction of the acceleration or in the opposite direction?

   b) Compare the behavior of this accelerometer with the behavior of the bubble level accelerometer.

   c) Which bubble level accelerometer diagram (1, 2 or 3) is equivalent to each of these accelerometers (4, 5 or 6)?

7.An extremely simple pocket accelerometer can be constructed from a piece of string and a steel washer (one among many possibilities).

   a) Match the behavior of this kind of accelerometer (7, 8, 9) with the other two kinds of accelerometers (1, 2, 3) and (4, 5, 6). Which string and washer is equivalent to which bubble level diagram?

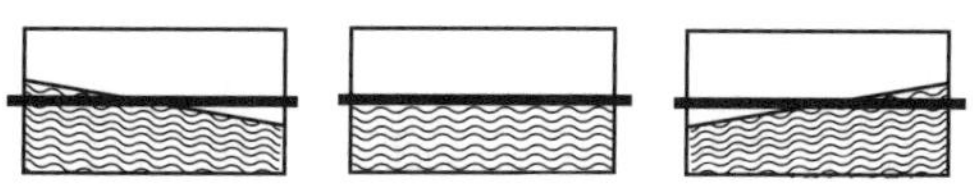

4 5 6

7 8 9

Chapter 3

Activity 5 B

Constant Acceleration

FOR YOU TO DO

1. Now that you can detect acceleration, you can try to achieve constant acceleration to see what affect this kind of motion might have on a race. Before you personally try for constant acceleration, it may be useful to observe some controlled accelerations. Set up a system so that a cart can be pulled along by a falling weight.

2. Put an accelerometer on the cart. Also set up a "Smart Pulley" for one of the string guides as shown.

   a) Does the falling weight accelerate the cart at a constant rate?

   b) Does changing the amount of weight used to pull the cart affect the acceleration?

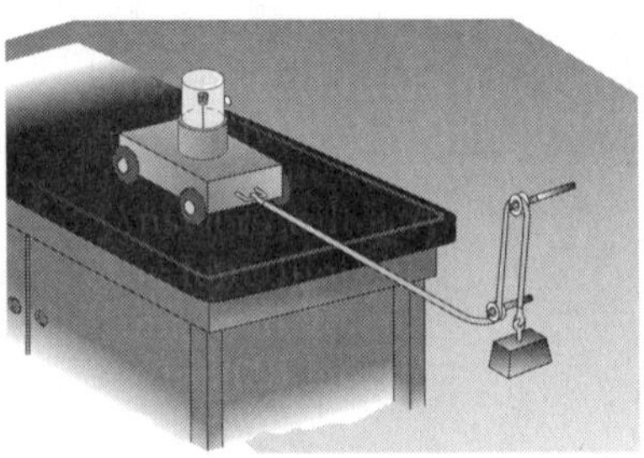

Smart Pulley

3. Nature runs the race with a constant acceleration. A falling object pulls a cart so that the speed of the cart increases at a constant rate. Although you may find it extremely difficult to run at constant acceleration, the falling mass does it well.

   Select a weight to pull the cart so that it accelerates at a comfortable pace. Now try to keep pace with the cart as it goes along.

   a) Can you keep up with the cart when the maximum available weight is used to pull it?

4. Without using a pace cart, try moving with constant acceleration over a prescribed course of several meters. To see how well you do at holding a constant acceleration, attach a string around your waist and guide it over a "Smart Pulley" system.

5. Run your race, and then look at the computer display of the acceleration.

   a) What should you look for on the graph of acceleration vs. time to show that you achieved constant acceleration?

   b) Is it possible to run a race at constant acceleration? Why or why not?

   c) While you are slowing down, are you accelerating?

For use with *The Track and Field Championship*, Chapter 3, Activity 5: Acceleration

ACTIVITY 6
Measurement

Background Information

Eratosthenes (276-195 B.C.) was a geographer and mathematician. He is best known for his relatively accurate determination of Earth's polar circumference as 250,000 stadia, or 46,290 km. (Current measure of polar circumference is 39,941 km.) The measurement was made by comparing the noontime shadow cast by the Sun in a well in Alexandria and the distance between Alexandria and Syene, where the noontime Sun cast no shadow. Pacing was a common method of distance measure, and ancient surveyors were trained to walk in equal paces.

Measurement arose from trading in early civilizations. Lengths were based on parts of the body, palms or hands, digits, and feet. The yard measure may have been first used in the building of Stonehenge, beginning as early as 2800 B.C.

The measurement of a quantity is a comparison to a standard. A meaningful measurement contains both a numerical value and the unit. In the 1790s, the French Academy of Sciences established the first standard unit of length, the meter. Prior to that, different people used different units of length, which varied from place to place. The standard meter was one ten-millionth of the distance from Earth's equator to either pole, and was represented by a platinum rod. About 100 years later, the meter was defined more precisely as the distance between two engraved marks on a particular bar of platinum-iridium alloy. One of these bars was designated the international standard and was kept at the International Bureau of Weights and Measures, near Paris, and the others were housed in scientific laboratories all over the world. By 1900, the meter was redefined as 1/650,763.73 wavelengths of an orange light emitted by the gas krypton-86. In 1983 the meter was redefined again as the length of path traveled by light in a vacuum during 1/299,792,458th of a second.

The most important system of units today is the Systeme International, or SI. The standard of length is the meter (m); mass, the kilogram (kg); and time, the second (s).

Random errors arise because every measurement is uncertain. Any measuring instrument has limited accuracy, and users are unable to read an instrument beyond some fraction of the smallest unit shown.

Good measurements include some estimate of the fraction of the smallest unit, and should state the estimated uncertainty. The uncertainty of a measurement is the range within which the actual value is likely to fall compared to the measured value. The convention for stating uncertainty is to write ± and the smallest unit after the measurement itself. For example, a length measured as 5.4 cm is appropriately written 5.4 cm ± 0.1 cm. Often, the approximation is not written, but is understood.

The percent error is found by dividing the difference between the measured value and the accepted value by the accepted value, and then multiplying the quotient by 100%.

In addition to their use in measurement, numbers that are defined or counted are used in physics. There is no uncertainty in these numbers. The number of sides of a triangle, for example, is a defined number. The number of pennies in a jar is a counted number.

Active-ating the Physics InfoMall

The InfoMall is helpful for this activity; maybe too helpful. Often, simple searches resulted in too many hits. Sometimes when the InfoMall search engine is used to find references to common topics, the number of references is too large for the program to handle. This activity generates many such situations. For example, search under the word "measurement" and you are sure to overload the software. For this reason, you will want to limit your search to suit your particular needs.

For example, if you do a compound search just in the Articles and Abstracts Attic with "measurement" and "systematic error" you will get many hits, but not so many that you will be overloaded. A quick glance through the search list shows three interesting articles: *American Journal of Physics*, vol. 19, "The Fundamental Problems of Experimental Physics," *Physics Today*, vol. 24, issue 9,"How accurately can temperature be measured?," and *The Physics Teacher*, vol. 28, "Systematic errors in an air track experiment."

Depending on your needs, a wider search can be conducted on the other stores for "random error" for example. Note that another way to expand a search is to use a "wild card" character; attach an asterisk to the end of a word, and the InfoMall will search for all words beginning with your word. For example, search for "error*" and you will get results for "errors" as well as "error".

Planning for the Activity

Time Requirements

- 1 class period

Materials Needed

For each group:

- stick, meter, 100 cm, hardwood
- graph paper pad of 50 sheets

Advanced Preparation and Setup

Find an appropriate area for the pacing exercise. A long corridor in school or a paved walkway on the school grounds would be good choices. Notify whomever necessary of the activity, for students may get enthused and make more noise than usual in the area.

Measure the actual length of the distance being paced so that you can judge the accuracy of the students' results.

Teaching Notes

The **What Do You Think?** questions are designed to stimulate thinking about the activity. Have students share their answers and accept all answers without correction.

Students will probably be more familiar with British units of length, the yard, foot, and inch. An inch is defined as precisely 2.54 cm. You may want to post conversion factors for those interested; however, exclusive use of metric units without conversion is recommended.

Students may make systematic errors in measuring. Watch for that type of error, and correct it, as you work with students in this activity. They will also make random errors, the size of which is based on the instruments they use.

Making estimates and using mental math are good scientific habits of mind. Model these skills in your teaching, and encourage students to check the reasonableness of their calculations before committing to them. Rounding to the nearest largest unit and performing math on the rounded units is a useful strategy. For example, to estimate the volume of a cube 1.8 cm on each side, round 1.8 cm to 2 cm, and multiply 2 cm x 2 cm x 2 cm, for 6 cm^3. The actual value will be a bit less than the estimate. You'll find that students rely on their calculators, often without realizing that estimation is a good way to compute.

Significant digits are the digits in a measurement that are known with certainty plus one digit that is uncertain. The last digit is the uncertain one. It is arrived at by estimating, and is partly a result of the quality and specificity of the measuring tool. The number of significant digits allowed in the result of a calculation depends on the number of significant digits in the data used to calculate the result. In addition and subtraction, the result should contain the last digit position common to all the data. In multiplication and division, the answer should have no more significant digits than found in the number with the fewest significant digits.

If possible, have students work in pairs for the **For You To Do** activity; one can actually pace the distance while the other serves as a recorder.

Make sure that the area used is free of debris and hazardous surfaces. Also make sure that your use of the area will not impede emergency exits or areas that must be free for access during an emergency.

Students will rely on their calculators for answers to computations. Emphasize that each of them is responsible for checking the reasonableness of an answer given by a calculator and making sure that they report the number to the correct number of significant digits. For example, 4.5 x 2.3 is 10.35; however, the correct answer in terms of significant digits is 10. with a decimal point, to show the significance of the zero in the ones place.

Activity 6 Measurement

GOALS

In this activity you will:

- Calibrate the length of a stride.
- Measure a length by pacing and with a meter stick.
- Identify sources of error in measurement.
- Evaluate estimates of measurements as reasonable or unreasonable.
- Measure various objects and calculate the error in each measurement.

What Do You Think?

In about 200 B.C. Eratosthenes calculated the circumference of the Earth. He used shadows cast by the Sun in two cities and a measurement of the distance between the two cities. The distance between the cities was found by pacing.

- **Two people measure the length of the same object. One reports a length of 3 m. The other reports a length of 10 m. Has one of them goofed? Why do you think so?**
- **What if the measurements were 3 m and 3.1 m?**

Record your ideas about these questions in your *Active Physics* log. Be prepared to discuss your responses with your small group and the class.

For You To Do

1. Select a cleared distance along the floor of the cafeteria, gym, hall, corridor, or classroom, or a paved area away from traffic out of doors.
2. Measure the length of your stride using a meter stick. Finding the length of your stride is an example of calibration—making a scale for a measuring instrument.

a) Record your measurement in your log.

ANSWERS

For You To Do

1. Student activity.
2. a) The length of a stride will vary considerably, as the humorous illustration in the text suggests.

Activity Overview

In this activity, students calibrate a measurement unit, a stride, and use it to pace off a selected length. They also explore sources of error in measurement, and use their common sense and mental math to determine if an estimate is reasonable.

Student Objectives

Students will:

- Calibrate the length of a stride.
- Measure a length by pacing and with a meter stick.
- Identify sources of error in measurement.
- Evaluate estimates of measurements as reasonable or unreasonable.
- Measure various objects and calculate the error in each measurement.

ANSWERS FOR THE TEACHER ONLY

What Do You Think?

The difference in the measurements is very large. One person has made a mistake.

Both measurements may be correct. It would depend on the quality and specificity of the instrument used.

ANSWERS

For You To Do
(continued)

3. a) The number of strides will depend on the length of the area being used.

4. a) Answers will vary depending on lengths of strides and distance to be measured. Students may use a calculator, but encourage them to estimate their answers to check their calculations.

5. a) Measurements will vary. Use your own measurements (advance preparation) to check the students' results.

   b) Students should appreciate that differences will exist. Accept any reasonable source of error. Students may suggest that stride lengths will vary among students, students may have miscounted, students' strides were not of consistent length.

   c) Accept all reasonable answers. Students may suggest repeating measurements and finding the average, or practicing walking with even strides.

6. a) Expect measurements to vary. Again, use your measurements to evaluate the accuracy of the students' measurements.

   b) Students' ideas will vary, but they should realize that some difference in measurements is inevitable.

7. a)-b) Students' ideas will vary, but they should realize that some difference in measurements is inevitable.

3. Count off the number of strides it takes you to cover the selected distance.
   a) Record this in your log.

4. Use the number of strides you took and the length of your stride to compute the distance.
   a) Record your calculations.

5. List the results of the measurements made by the entire class on the board.
   a) Do all the measurements agree?
   b) Why do you think there are differences among the measurements made by different students? List as many different sources of error as you can.
   c) Suggest a way of improving your measurements.

6. Measure the distance with a meter stick.
   a) Record your measurement in your log. Make a list of all the class measurements on the board.
   b) Can you develop a system that will produce measurements all of which agree exactly or will there always be some difference in measurements? Justify your answers.

7. Physicists identify two kinds of errors in measurement. Errors that can be corrected by calculation are called systematic errors. For example, if you made a length measurement starting at the 1 cm mark on a ruler, you could correct your measurement by subtracting 1 cm from the final reading on the ruler.

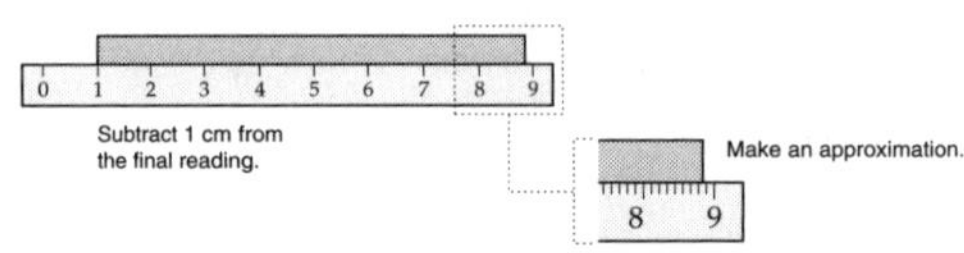

Errors that come from the act of measuring are called random errors. No measurement is perfect. When you measure something you make an approximation. Random errors exist in any measurement. Scientists provide an estimate of the size of the random errors in their data.
   a) Classify the sources of error you have listed as systematic or random.
   b) Estimate the size of each error.

Activity 6 Measurement

8. Sometimes a precise measurement is not needed. A good estimate will do. What is a good estimate? Use your common sense and prior knowledge to judge if an estimate is a "good," or reasonable, one. Determine if each is reasonable. Explain your answers.

> Example:
> - A single-serving drink container holds 5 kg of liquid.
>
> Use common sense and mental math to see if this is a good estimate. One kilogram of water takes up about 1 L of volume. Five kilograms of water would take up 5 L. Most drinks are like water in their density. A 5-L container is much bigger than a single serving.

a) A college football player has a mass of 100 kg.
b) A high school basketball player is 4-m tall.
c) Your teacher works 1440 min every day.
d) A poodle has a mass of 60 kg.
e) Your classroom has a volume of 150 m^3.
f) The distance across the school grounds is 1 km.

FOR YOU TO READ

Measurement and Track Records

Measurements are crucial in all sporting events. As you learned in this activity all measurements have some error associated with them. These measurement errors can bring into question whether a world record was actually broken.

Every four years, the summer Olympics take place in a different city. Each city builds a track to be 400 m. How accurately built is the 400-meter track? Could it be 1 cm longer in one city and 1 cm shorter in another city? If so, then a runner in the first Olympics may be running an extra distance of 2 cm on every lap. If the race is a 1600-m race, then one runner may run a total of 8 cm further. It takes at least 0.01 s to run 8 cm. If a runner "beats" the world record by only 0.01 s, it may have been because she was running on a shorter track, not because she was faster.

ANSWERS

For You To Do *(continued)*

8. a) Reasonable.

b) Unreasonable.

c) Unreasonable. (It just feels like you do.)

d) Unreasonable.

e) Answers will vary with size of classroom, but it is probably a reasonable measurement.

f) Probably unreasonable, but depends on the size of the school grounds.

ANSWERS

Physics To Go

1.a) Larger objects are best measured with the meter stick, while smaller objects can be best measured with the centimeter ruler.

b) The errors are likely to be random errors, but some systematic errors may occur. Discuss the measurements with the students.

2. Answers will vary. The errors will probably be random.

3. Answers will vary. For example, the length of this book is 25 cm long, and 26 cm long would be estimated lengths about which you could agree. The width of a room is 50 cm and 5 m are lengths about which you could not agree.

4. To the accuracy of 1 cm.

5. The measurement probably contains a random error to the degree of a percent of one barrel to a few barrels.

6. Answers will vary.

7.a) No; a 2-liter bottle of soft drink probably contains enough liquid to provide 8 normal servings.

b) Yes, this is reasonable. The distance from Boston to New York City is about 220 miles. A mid-size car can hold at least 10 gallons of gasoline and, in highway driving, should travel more than 22 miles per gallon.

Reflecting on the Activity and the Challenge

A measurement is never exact. When you make a measurement, you estimate. All measurements have random errors. You can try to minimize these errors but you cannot eliminate them. One decision you must make is how accurate a measurement you really need.

In this **Chapter Challenge**, you will be writing a track and training manual. You may want the runners to know that the inside lane is shorter than the outside lane. You may also want to inform them that some tracks are very slightly different than other tracks. It's easy to tell who was faster when two runners compete side by side. It's more difficult to tell who was faster when one runner competed on a different track in a different city.

Physics To Go

1. Get a meter stick and centimeter ruler. Find the length of five different-sized objects, such as a door (height), a table top, a large book, a pencil, and a postage stamp.
   a) Which measuring tool is best for measuring each object?
   b) Calculate the error in each measurement. What kind of errors are these?
2. Pace off the size of a room. Estimate your accuracy. Then check your accuracy with a meter stick or tape measure.
3. Give an estimated value about which you and someone else would agree. Then give an estimated value about which you and someone else would not agree.
4. An Olympic swimming pool is 50 m long. Do you think the pool is built to an accuracy of 1 m (49 m to 51 m) or 1 cm (49.99 m to 50.01 m)?
5. An oil tanker is said to hold 5 million barrels of oil. Do you think this measurement is accurate? How accurate?
6. Choose 5 food products. How accurate are the measurements on their labels?
7. Are these estimates reasonable? Explain your answers.
   a) A 2-L bottle of soft drink is enough to serve 12 people at a meeting.
   b) A mid-sized car with a full tank of gas can travel from Boston to New York City without having to refuel.

Assessment: Measuring

Skills Demonstrated	Yes	No	Comments
Measures stride length to one decimal place accuracy and records in log			
Counts the number of strides to cover selected distance and records in log			
Uses number of strides and stride length to compute the distance			
Measures distance using a meter stick correct to one decimal place			
Indicates possible sources of error			

Performance-Based Assessment

Teacher value for measured distance: ___________________

$$\% \text{ Error} = \frac{\text{Teacher value - Student value}}{\text{Teacher value}} \times 100$$

20 % or less = 3
21 % - 40% = 2
39% - 70% = 1
greater than 70% = 0

For use with *The Track and Field Championship*, Chapter 3, Activity 6: Measurement

ACTIVITY 7
Increasing Top Speed

Background Information

An alternate way of calculating a runner's speed is introduced in this activity.

Speed = Stride frequency × Stride length

Dimensional analysis presented in the **Physics Talk** section of the student text for this activity shows that the above equation is equivalent to:

Speed = Distance/Time

A caution is offered regarding the unit "strides per second" to express stride frequency and the unit "meters per stride" to express stride length in this activity. The caution is based on a subtle, sometimes misunderstood aspect of the definition of frequency.

The formal unit of frequency is the "hertz" (abbreviated Hz) which is defined as:

1 hertz = s^{-1} = $1/s$

A common misconception about the hertz as a unit of measurement arises when a frequency of, for example, 60 Hz is expressed as 60 "cycles per second," or 60 "vibrations per second," or, in the case of this activity, some number of "strides per second." While it is true that terms such as "cycles," "vibrations," "strides" or other descriptive nouns may enhance communication, it is essential to recognize that, to formally comply with the definition of frequency, a runner making, for example, 2 strides per second should have his frequency expressed as "2/s." Similarly, if the same runner makes strides which are of average length 3 m, the proper expression of stride length would be, simply, "3 m." Clearly, dimensional analysis using proper, formal units shows that stride frequency multiplied by stride length yields speed:

2/s × 3 m = 6 m/s

Carrying descriptors such as "strides" or "cycles" in expressions of frequency will do no harm here, but will if, in the future, the student uses such descriptors when performing dimensional analysis of complex equations involving frequency in areas such as electricity or quantum mechanics. For example, suppose a student wants to calculate the energy of a photon of light corresponding to a frequency of 4.0×10^{14} Hz using the below equation and, through bad habit, substitutes 4×10^{14} cycles/second as the frequency:

Energy of photon = Planck's constant × frequency

= (6.6×10^{-34} joule-second) × (4.0×10^{14} cycles/second)

= 2.6×10^{-19} joule-cycles

The student has a problem, because the answer should be in joules, not joule-cycles; this problem would not exist if the student had expressed the frequency properly, substituting 4×10^{14}/second for 4.0×10^{14} Hz as the frequency.

This may seem "nit picky," but the author actually has observed students struggle while trying to get "cycles" to cancel during dimensional analysis of calculations involving frequency. It seems a disservice to plant misconceptions which are avoidable; therefore, it is suggested that you call students' attention to "per second" and "meters" as the proper units for, respectively, stride frequency and stride length, in this activity.

Active-ating the Physics InfoMall

Does the information regarding the cheetah sound interesting? Search the InfoMall with the word "cheetah" and you get two hits. Both are interesting, but you should look at the second one, *Many Magnitudes: A Collection of Useful and Useless Numbers*, in the Utility Closet. This has a LONG list of interesting numbers that students may enjoy reading.

The relationship that velocity is the product of frequency and wavelength occurs often in physics. You can find this in most textbooks on the InfoMall. However, "frequency" is a term that occurs so often that you will need to limit your search (either to one or two stores, or by adding more keywords to your search) to avoid getting "too many hits."

Planning for the Activity

Time Requirements

- One class period.

Materials Needed

For the group:

- stopwatches
- tape measure, windup
- VCR & monitor

- *Sports* Content Video
- calculator, basic
- stick, meter, 100 cm, Hardwood
- Tape, Masking, 3/4" x 60 yds.
- Marker, Felt Tip

Advance Preparation and Setup

An area will be needed to serve as a walking "track." The area needs to be a minimum of 12 m in length, which may be larger than your classroom. If needed, arrange to use a larger room, hallway, or outdoor area. Noise can be expected. Tape (or some other material) must be able to be placed on the walking surface to mark intervals on two walking "lanes," each 12 m in length. One lane must be marked at 0.50-m intervals and the other at 0.75-m intervals. The entire class should be able to observe as volunteers walk along the lanes with controlled stride lengths.

A VCR player and monitor should be reserved to be used for this activity.

Teaching Notes

Data collection from the video, calculations based on the data and discussion of **Physics Talk** to verify the equation Speed = Stride Frequency × Stride Length is suggested as a whole-class activity. This will provide that all students will complete **Steps 1-5** of **For You To Do** at the same time and be ready to proceed as a total class to the next part of the activity which involves the walking with controlled stride lengths.

It would provide the most efficient use of time to have a few volunteers walk the lanes with controlled stride lengths while other volunteers measure time with a stopwatch. The data collected for the volunteers could be used as data for all class members. However, all students should try walking both lanes to have direct experience with controlling one's stride.

After walking the lanes, students may complete answers to the questions while working in groups, alone, or as homework.

Having just been introduced to the relationship between speed, stride frequency, and stride length, most students probably haven't thought about the relationship of stride length and stride frequency to acceleration. Perhaps you should point out that when a runner accelerates, either stride frequency, stride length, or both, must be changing. Encourage students to think of ways that they could actually observe, measure and do calculations involving these variables for track team members, and use the results to improve the performance of the school's athletes in races.

NOTES

Activity 7 Increasing Top Speed

GOALS

In this activity you will:

- Calculate the average speed of a runner given distance and time.
- Measure the frequency of strides of a running person.
- Measure the length of strides of a running person and calculate speed.
- Recognize that either the equation (Speed = Distance / Time) or the equation (Speed = Stride Frequency × Stride Length) may be used to calculate a runner's speed with equivalent results.
- Infer ways in which stride length and stride frequency can be adjusted by a runner to increase speed.

What Do You Think?

A cheetah can reach a top speed of 60 miles per hour (about 30 m/s).

- **What can a runner do to increase top speed?**

Record your ideas about this question in your *Active Physics* log. Be prepared to discuss your responses with your small group and the class.

For You To Do

1. Watch the video of a runner.
2. Use information from the video to answer the following questions. Record your data and show your calculations in your log.

a) Use the total distance traveled and the total time to calculate the runner's speed in yards per second.

$$\text{Speed (yards/second)} = \frac{\text{Distanc e (yards)}}{\text{Time (seconds)}}$$

(Yards/second are used since the running is being done on a football field.)

ANSWERS

For You To Do

1. Students watch video.

2. a) Distance = 50.0 yards
Time = 14.3 seconds
Speed = 50.0 yd/14.3 s
= 3.50 yd/s

Activity Overview

In this activity students run their own race, exploring the relationship between speed, stride length, and frequency.

Student Objectives

Students will:

- Calculate the average speed of a runner given distance and time.
- Measure the frequency of strides of a running person.
- Measure the length of strides of a running person.
- Recognize that either the equation (Speed = Distance / Time) or the equation (Speed = Stride Frequency × Stride Length) may be used to calculate a runner's speed with equivalent results.
- Infer ways in which stride length and stride frequency can be adjusted by a runner to increase speed.

ANSWERS FOR THE TEACHER ONLY

What Do You Think?

Students should understand that to win a race, they must have a faster average speed than their opponents. In a sprint, a runner may win against an opponent who has a faster top speed by exploding out of the starting blocks with tremendous acceleration. In a distance race, a runner may win against an opponent with a faster top speed by having greater stamina, running more efficiently, and being able to keep a reasonable speed for a longer time.

Even though a runner can compensate for not having the fastest top speed, the runner with the greatest top speed will clearly have an advantage. Students probably will not have given previous thought to stride length and stride frequency as factors which affect a runner's speed. They may look ahead to the title and early parts of **For You To Do** and give responses based on their first impressions of how these variables affect speed.

ANSWERS

For You To Do
(continued)

b) Number of strides = 37 strides
Time = 14.3 s
Stride Frequency = 37 strides / 14.3 s = 2.6 strides/s

c) Average stride length: 25 yd / 18 strides = 1.4 yd/stride

d) Speed = Stride Frequency × Stride Length
= 2.6 stride/s × 1.4 yd/stride
= 3.6 yd/s

3\. a)-b) The speeds determined by the two methods, 3.5 yd/s and 3.6 yd/s, differ by less than 3%; within error of measurement, they appear to be equal speeds.

4\. Student activity.

5\. a)-d) The answers involving walking with controlled stride length will depend on the frequency with which the individuals walk. Observers will need to count the number of strides (determined by number of marked intervals; 16 strides for the 0.75 m stride length), and a volunteer must time the duration of each 12-m walk.

b) Count the number of strides taken by the runner during the entire run. Use the number of strides and the total time to calculate the runner's stride frequency in strides per second.

$$\text{Stride frequency (strides/second)} = \frac{\text{Number of strides}}{\text{Time (seconds)}}$$

c) Calculate the average length of one stride for the runner. To do so, measure the length of several single strides and then calculate the average length per stride. The unit for your answer will be yards/stride.

d) Calculate the runner's speed using the following "new" equation for speed. Your answer will be in yards/second.
Speed = Stride Frequency × Stride Length.

3\. Compare the results of using the two equations for calculating speed.

a) Do the equations agree on the runner's speed? How good is the agreement?

b) How could you explain any difference in results?

4\. You will test the "new" equation at a "track." Your teacher will show you where to set up the track. Place marks at intervals of 0.75-m along a 12-m track.

5\. Starting from the "zero" mark, walk so that you step on each mark. Count the number of strides to complete the walk, and use a stopwatch to measure the total time.

a) Record your data in your log.

b) Calculate your speed using the distance/time equation.

c) Calculate your stride frequency.

$$\text{Stride frequency (strides/second)} = \frac{\text{Number of strides}}{\text{Time (seconds)}}$$

d) Calculate your speed using the stride frequency and stride length.
Speed = Stride Frequency × Stride Length.

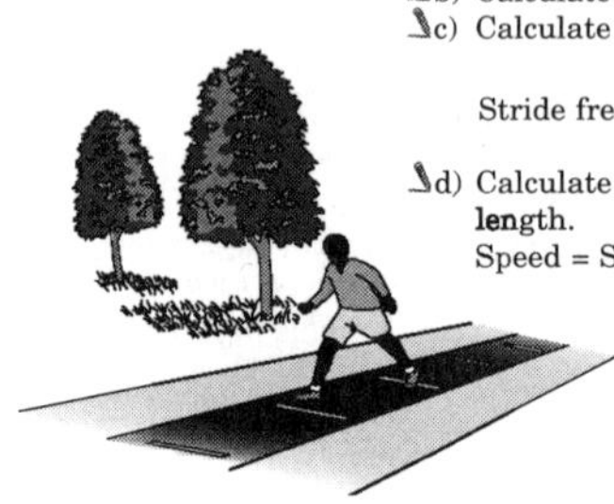

6. What happens to your speed if you change your stride length?
   a) Record what you think will happen to the speed if you decrease the length of the stride.
7. Mark the track at 0.50-m intervals and walk again, stepping on each mark. Again measure the time and count the number of strides.
   a) Record the data in your log.
   b) Calculate your speed using both equations.
   c) How well do your speeds calculated by both methods on this track compare?
8. Compare your performances on the two different tracks.
   a) Was your speed on the track which had 0.75-m stride lengths about the same as, or different from, your speed on the track which had 0.50-m intervals? Why?
   b) If you must step on each mark, what would you need to do to make your speed on the second track equal to your speed on the first track? Express your answer in numerical terms.

PHYSICS TALK

Calculating Speed Using Frequency and Length

If it is really true that Speed = Frequency × Stride Length, then it should also be true that the relationship produces an answer that has a unit of speed, such as meters/second. If frequency is measured in strides/second and if length is measured in meters/stride, then

$$\frac{\text{strides}}{\text{second}} \times \frac{\text{meters}}{\text{stride}} = \frac{\text{meters}}{\text{second}}$$

The equation produces an answer that has a unit of speed.

ANSWERS

For You To Do

(continued)

6. a) Have students enter their predictions before beginning the activity.

7. a)-c) The answers involving walking with controlled stride length will depend on the frequency with which the individuals walk. Observers will need to count the number of strides (determined by number of marked intervals; 24 strides for the 0.50 m stride length), and a volunteer must time the duration of each 12-m walk.

8. a)-b) Since a particular walker probably will not adjust frequency to offset the change in stride length from one track to the other, the speeds for an individual on the two tracks should not be expected to be equal. To make the speeds equal, the frequency on the track having the 0.50 m stride length would need to be 0.75/0.50 = 1.5 times higher than the frequency on the track having the 0.75 m stride length.

ANSWERS

Physics To Go

1. a) 1.8 strides/s × 2.0 m/stride = 3.6 m/s

   b) 200 m / 3.6 m/s = 56 m/s

2. (6.0 m/s) / 1.5 m/stride = 4.0 strides/s

3. 4.0 strides/s × 1.6 m/stride = 6.4 m/s

4. 2.0 strides/s × 0.65 m/stride = 1.3 m/s

5. The graphs generally show that speed decreases with age and that stride length also decreases with age, which is consistent with the equation:

   Speed = Stride Frequency × Stride Length.

   Trends of the effects of age on Stride Frequency could be explored by selecting pairs of data points for corresponding age and gender on the two graphs and then dividing the speed by the stride length to determine the frequency.

6. Hurdle events in track require that hurdlers establish, and sometimes adjust between hurdles, stride lengths so that a they arrive "on step" at a desired "launching point" in front of each hurdle. Missing the launching position may cause the hurdler to trip on the hurdle, which usually results in losing the race.

Reflecting on the Activity and the Challenge

Distance runners learn from their coaches how to make conscious changes in stride length and frequency to improve their performances. To increase speed, the "trick" is to increase one—either frequency or stride length—without decreasing the other. Good runners know how to increase either, or even both, when needed in a race. Think of experiments that you could do with members of your school's track team to help them to learn to use frequency and stride length to win races.

Physics To Go

1. A runner's stride length is 2.0 m and her frequency is 1.8 strides/s.
   a) Calculate her average speed.
   b) What would be her time for a 200-m race?
2. A runner maintains a constant speed of 6.0 m/s. If his stride length is 1.5 m, what is his stride frequency?
3. If the runner in **Question 2** increases his stride length to 1.6 m without changing his stride frequency, what will be his new speed?
4. If a marching band has a frequency of 2.0 strides/s and if each stride length is 0.65 m, what is the band's marching speed?
5. The graphs on the next page are reproduced from an article that reported the characteristics of runners who are 30 or more years of age.

   Write a statement that explains the information on the two graphs. Include your own inferences about what happens to stride frequency with increasing age.
6. The running events called "hurdles" present special problems involving stride length and frequency. In both 100- and 400-m races, runners must jump over hurdles placed at regular intervals along the track. Discuss special techniques that hurdlers must develop to make sure that they are ready to jump when they reach each hurdle.

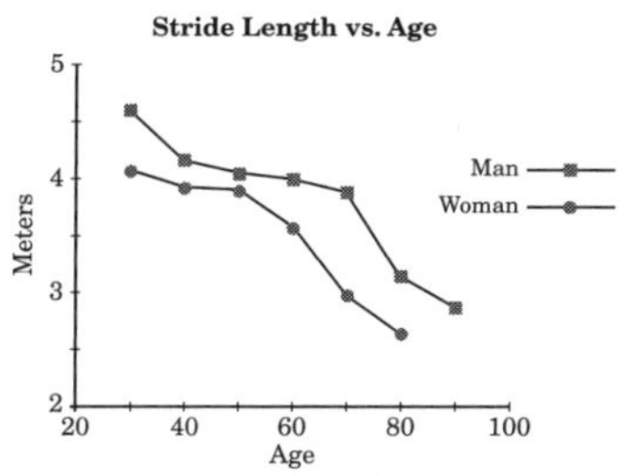

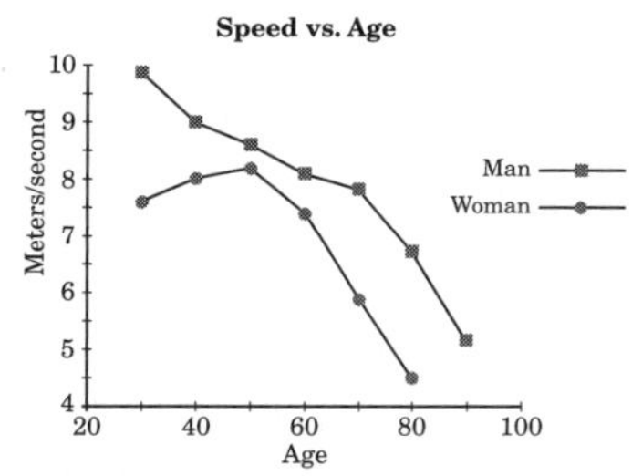

"Sprint Stride Kinematics."
Nancy Hamilton in *Log of Applied Biomechanics*, Volume 9, 1993, pp. 15–26

Stretching Exercises

Common sense suggests that physical characteristics, such as the length of a person's legs, may affect stride length. Do long-legged people have greater stride length? Can long-legged people maintain the same stride frequency when running as people with shorter legs? If so, do long-legged people have higher running speeds? Ask some classmates or friends in your neighborhood to serve as volunteers for an experiment designed to identify and test the effects of leg length, or other body characteristics, on stride length and frequency during walking and/or running.

ANSWERS

Stretching Exercises

Student activity.

ANSWERS

Alternative Solutions to Physics To Go

Alternate forms of solutions to **Physics To Go Questions 1, 2, 3** and 4 which comply with correct formal expression of frequency and length are shown in brackets. See **Background Information** for the Teacher regarding a caution about including descriptive terms such as "strides" in calculations involving frequency.

1. a) 1.8 strides/s × 2.0 m/stride = 3.6 m/s
[1.8/s × 2.0 m = 3.6 m/s]

   b) 200 m / 3.6 m/s = 56 s

2. (6.0 m/s) / 1.5 m/stride = 4.0 strides/s
[(6.0 m/s) / 1.5 m = 4.0/s]

3. 4.0 strides/s × 1.6 m/stride = 6.4 m/s
[4.0/s × 1.6 m = 6.4 m/s]

4. 2.0 strides/s × 0.65 m/stride = 1.3 m/s
[2.0/s × 0.65 m = 1.3 m/s]

Chapter 3

ACTIVITY 8
Projectile Motion

Background Information

New phenomena introduced in this activity include:

- free fall
- projectile motion

It is suggested that you read the **Physics Talk** section of the student text for this activity to see the meaning of terms used in the below **Background Information.**

Free Fall: All objects in a state of free fall, regardless of mass, shape, or other characteristics, experience a uniform (constant) acceleration of approximately 10 m/s/s in the downward direction. (Notice that the condition made in **Physics Talk** in the student text would not include objects which experience a significant force due to air resistance as being in a state of free fall.)

Projectile Motion: Inertia (an object in motion remains in motion unless a force acts to cause a change in motion) causes a projectile to retain any horizontal motion which it has at the instant it is launched, and it retains that motion in the form of constant speed.

Gravity causes a projectile to exhibit vertical motion which matches free fall, whether or not the projectile has simultaneous horizontal motion.

This activity does not approach projectile motion quantitatively, but it is clearly demonstrates that a coin which simply falls strikes the floor in the same amount of time as a coin launched horizontally from the same height at the same instant. The only way this can happen is if both objects fall downward with identical motions. The fact that one coin simultaneously moves horizontally (at constant speed) as it falls affects where it lands, not when.

This activity further demonstrates that a ball thrown upward does not have its (accelerated) upward and downward motion affected by any horizontal motion which it may have at the time of launch. The horizontal motion (constant speed) is independent of the vertically accelerated motion. If a student sitting in a stationary chair throws a ball upward, it lands in her hand; if the chair is moving at constant speed throughout the vertical toss, inertia demands that the ball also lands in the student's hand.

Quantitative treatment of projectile motion, with additional background for the teacher, is presented in the next activity.

Active-ating the Physics InfoMall

The obvious keywords to use for a search related to this activity are "projectile motion." This search is particularly rewarding. Look at the information from the textbooks *Modern College Physics* and *Foundations of Modern Physical Science*. Both have great discussions of projectiles, and they also have graphics that illustrate the independence of vertical and horizontal motion (Figure 3.4 in *Foundations,* and Figure 8A in *Modern College Physics*).

Our earlier search also produced the article "Aristotle is not dead: Student understanding of trajectory motion," in the *American Journal of Physics*, vol. 51, issue 4. Other nice references are *A Guide to Introductory Physics Teaching: Motion in Two Dimensions* in the Book Basement, and "Thoughts on projectile motion," in the *American Journal of Physics*, vol. 53, issue 2.

In **For You To Do, Step 1**: Equipment for a similar experiment can be found in the Cenco catalog (in the Catalog Corner). This item is on the list generated from the "projectile motion" search. It is also item number 76525 in the catalog, in case you wish to browse for yourself.

For **Step 4**, you should look at Figures 3.1 and 3.3 in Chapter 3 of *Foundations of Modern Physical Science* (in the Textbook Trove) for this activity.

Planning for the Activity

Time Requirements

- One class period.

Materials Needed

For each group:

- ball, tennis
- chair with wheels
- projectile launcher
- stick, meter, 100 cm, hardwood
- washer, metal, 3/4"

Advance Preparation and Setup

SAFETY PRECAUTION: A chair on wheels is used for a demonstration in this activity. You will want to locate a chair which is as stable and sturdy as possible to prevent a "crash." An office chair having a 4 or 5 wheel base and arm rests is suggested.

Plan to control the situation, preferably with pushing the chair as a student rides in the chair and throws a ball upward and then catches it when it falls. Do not let students play with the chair; it could be very dangerous.

Teaching Notes

Help students appreciate that both a shot put and a human body can be a projectile. Both will follow a path through the air called a trajectory.

Set an amount of time for students to complete the work with coins in the first part of **For You To Do**, allowing sufficient time for the demonstration during the last part of the class period.

If students have poor aim using a flick of a finger to send the moving coin in a direction so that it barely "ticks" the stationary coin, have them try using a meter stick as a cue stick to launch the coin.

Be sure to have each student commit to a prediction of where the ball will land relative to the student in the moving chair. Expect the misconception that some students will predict that the student will "move out from under the ball as the ball flies straight up and down." If this prediction is not offered (some may believe it, but won't volunteer it), plant it as a suggested possibility. The demonstration will present a discrepancy which may cause such students to discard their misconception.

Have the students practice throwing the ball vertically and catching it again with the chair at rest first. Have the student throw the ball to several different heights. Lay out a straight line course of about 3 to 5 m. When the chair is put into motion and gets up to speed, have the occupant throw the ball upward gently and catch it again. Post observers along the side of the course so that the horizontal distance (range) from release to catch can be marked and measured. Repeat the measurements to get a sense of the variation that can result from the differences from trial to trial.

In **Reflecting on the Activity and the Challenge** ask the class to develop a list of track and field events which involve projectile motion (long jump, high jump, hurdles, shot put, discus, javelin, hammer throw). Do fewer athletes pursue these events than others? Might these events offer higher probability of winning if less competition exists? Might some of these events have greater dependence on knowledge and technique and less dependence on physical ability than other events? Events involving projectile motion may have high potential to be improved through physics-based help.

You may wish to point out that Carl Lewis is one of the world's greatest long jumpers. The Jesse Owens 1936 Olympics performance in the sprints and the long jump also makes a very good story.

Activity Overview

Students explore the independence of vertical and horizontal components of a projectile's motion in this activity.

Student Objectives

Students will:

- Understand and correctly apply the term "free fall."
- Understand and correctly apply the term "projectile."
- Understand and correctly apply the term "trajectory."
- Understand and correctly apply, within physics context, the term "range."
- Observe that all projectiles launched horizontally from the same height strike a level surface in equal times, regardless of launch speed (including zero launch speed).
- Understand that the vertical and horizontal components of a projectile's motion operate independent of one another.
- Infer factors which affect the range of a projectile.
- Infer the shape of a projectile's trajectory.
- Infer ways in which stride length and stride frequency can be adjusted by a runner to increase speed.

ANSWERS FOR THE TEACHER ONLY

What Do You Think?

The initial velocity (speed and direction) determines the range of a projectile; a difference in the elevations of the launch and landing points also would affect the range.

A 100 mph pitch thrown horizontally by a major league player will hit the ground in the same amount of time as a 10 mph pitch thrown horizontally from the same height by a child. It is a near certainty that many students would never believe this type of fact. The work with coins should present a discrepancy for them which, hopefully, will cause them to confront their misconceptions about projectiles.

The Track and Field Championship

Activity 8 Projectile Motion

GOALS

In this activity you will:

- Understand and correctly apply the term "free fall."
- Understand and correctly apply the term "projectile."
- Understand and correctly apply the term "trajectory."
- Understand and correctly apply, within physics context, the term "range."
- Observe that all projectiles launched horizontally from the same height strike a level surface in equal times, regardless of launch speed (including zero launch speed).
- Understand that the vertical and horizontal components of a projectile's motion operate independent of one another.
- Infer factors which affect the range of a projectile.
- Infer the shape of a projectile's trajectory.
- Infer ways in which stride length and stride frequency can be adjusted by a runner to increase speed.

What Do You Think?

Some track and field events involve launching things into the air such as a shot put, a javelin, or even one's body in the case of the long jump.

- **What determines how far an object thrown into the air travels before landing?**

Record your ideas about this question in your *Active Physics* log. Be prepared to discuss your response with your small group and the class.

For You To Do

1. Place one coin at the edge of a table with about half of the coin hanging over the edge. Place another coin flat on the table and use a finger to shoot this coin across the tabletop to strike the first coin. Aim "off center" so that the coin at the edge of the table drops straight down and the projected coin leaves the edge of the table with some horizontal speed. Repeat the event as many times as needed to record your answers to the following question in your log:

176

Active Physics

a) Which coin hits the floor first? (Hearing is the key to observation here, although you may also wish to rely on sight.)

Make sure the path is clear before launching the coins.

2. Vary the speed of the projected coin.

a) How does its speed affect the amount of time for either coin to fall to the floor?

b) How does its speed affect how far across the floor the projected coin lands?

3. Use a box, stack of books, or a different table or countertop to vary the height.

a) Which coin hits the floor first?

b) How does increasing the height affect how far the projected coin travels horizontally as it falls?

4. Your teacher will supervise an activity in which one student sits on a chair that is moving at constant speed. While the chair is moving, the student will throw a ball straight up into the air and try to catch it when it comes down. The class will stand in a line along the track to observe the event, prepared to mark the horizontal distance (range) the ball travels from release to catch.

ANSWERS

For You To Do

1. a) Both coins will hit the ground at the same time.

2. a) The initial speed has no effect on the amount of time for either coin to fall to the floor.

   b) The coin with the greater initial velocity will have a greater range, it will travel further across the floor.

3. a) The coin with the initial higher elevation will hit the floor second, the lower elevated coin will hit first.

   b) Increasing the height has no effect on the distance that the coin travels.

ANSWERS

For You To Do
(continued)

4. a) It is important that students record their predictions before the activity is completed. This permits them to confront their misconceptions, if there are any. If students are not encouraged to commit to a prediction, they will often predict the "right answer" after the fact. Although students may learn to respond with the correct answer, they may still not have confronted their misconceptions.
   b) The ball should land in the student's hands as if the chair were not moving because, due to inertia, the ball retains its horizontal speed as, independently, it flies up and down with accelerated vertical motion due to gravity.
   c) Two factors affect the range of the ball: the vertical speed at which the ball is launched; the horizontal speed of the chair (and ball) when the ball is launched.
   d) The shape of the trajectory will be a parabola (some students may guess this as the shape — encourage them by asking what they know about parabolas, and why that may be the shape of the trajectory).

The Track and Field Championship

a) In your log write your prediction of what you think will happen.
b) Write in your log what you observed about the ball's trajectory (shape of the ball's path) and the ball's approximate range (horizontal distance) for trials in which you varied the speed of the chair and the launching speed of the ball.
c) According to your observations, what factors affect the range of the ball?
d) According to your observations, what is the shape of the trajectory?

Physics Words
projectile: an object traveling through the air.
trajectory: the path followed by an object that is launched into the air.

PHYSICS TALK
Projectiles and Trajectories

Physicists often work with objects that have been launched into the air in a state of "free fall." In a free fall, the main force acting on the object while it is in the air is the downward pull of the Earth's gravity.

An object launched into the air is called a **projectile**. Examples of projectiles are a javelin, a shot put, or a broad jumper. The path that the projectile follows when launched into the air is called the **trajectory**.

FOR YOU TO READ
Vector Components

By observing two falling coins and by tossing a ball in a moving chair, you gained evidence of two very important aspects of how thrown objects move in space. Since the shot put, the javelin, the hammer, and even the high jumper are objects thrown in the air, these two observations are crucial to helping the track team improve its performance.

The horizontally thrown coin and the dropped coin hit the ground at the same time, when there is no air resistance. Under careful observations, you find that this is always true – the horizontal motion of the coin does not affect its downward motion. If you were to take a picture of the coin every tenth of a second, you would observe the two coins as shown on the following page:

178

Active Physics

Activity 8 Projectile Motion

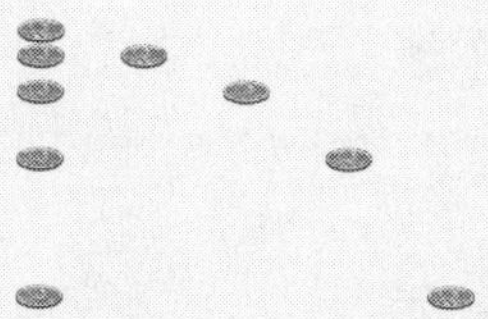

Both coins fall the same amount in each tenth of a second. The vertical motions are identical. The projected coin kept moving to the right, but its vertical motion was identical to the dropped coin.

Similarly, the projected coin has a constant speed to the right, when there is no air resistance. The vertical motion does not affect this constant horizontal speed.

At any point in its motion, the projected coin is moving down and to the right. You can draw its velocity at any time. After a short time, it has a small vertical speed and its constant horizontal speed. You can add these as vectors as you did with forces.

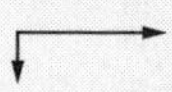

To add these two velocity vectors, you "complete the rectangle" and draw the diagonal. This diagonal is the resultant velocity vector.

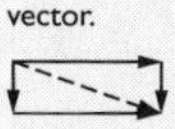

An alternative way to add these two vectors is to put them "tip to tail." By sliding one vector over (maintaining its length and direction), the resultant is then drawn from the tail of the first vector to the tip of the second vector.

As you notice, the resultant vector is identical in length and direction, independent of whether you added the two vectors by the "rectangle" or "tip-to-tail" method.

If you look at the velocity some time later, you notice that the coin is moving faster in the vertical direction but continues horizontally at the same speed. You can add the vectors to determine what happens to the resultant vector.

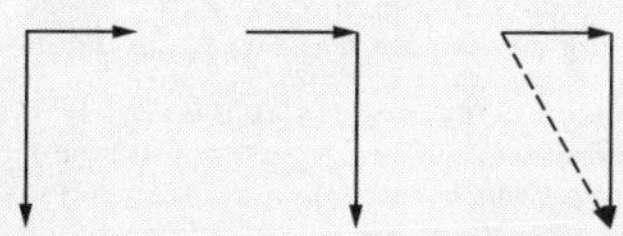

The resultant or total velocity has gotten larger and its direction has changed. The total velocity is in a more vertical direction.

If you took the velocity at any one point in the path, you could also use that resultant velocity vector to find the horizontal and vertical components. You draw the resultant velocity vector to the correct size and in the correct direction. Next, you draw horizontal and vertical axes from the tail of the vector. Then you draw a line from the tip of the vector to the horizontal axis. By doing this, you can see the horizontal and vertical vectors that can add together to produce this resultant.

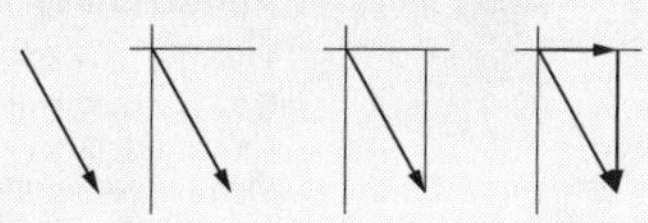

→

Chapter 3

If you were to take numerous velocity vectors representing the path of the object, you would notice two things. First, that the horizontal velocity components would always be equal. The second is that the vertical component increases as time goes on.

Sample Problem

a) A javelin is thrown at 20.0 m/s at an angle of 30° with respect to the horizontal. What is its velocity in the *x*-direction?

b) If the javelin were thrown at 40°, what is its velocity in the *x*-direction?

c) How far does each javelin travel after 3.0 s?

Strategy: You can solve the first two parts by drawing vector diagrams to scale and finding the *x*-components. In **Part (c)** you can find how far each javelin traveled by using the relationship:

Distance = velocity x time

$d = vt$

Solution:

a) The first vector must be 20 units long at an angle of 30°. (The scale is 1 unit = 1 m/s.)

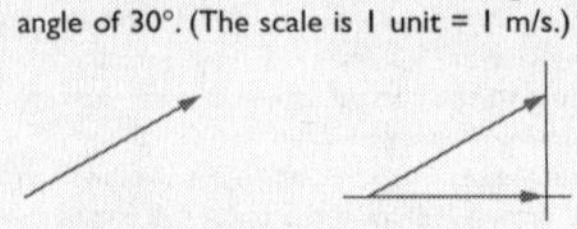

Measuring the *x*-component and using the scale, you find the *x*-component is 17.3 m/s.

b) The second vector is also 20 units long at an angle of 40°.

Measuring the *x*-component and using the scale, you find the *x*-component is 15.3 m/s.

c) The first trajectory has a horizontal velocity component of 17.3 m/s for 3.0 s. Its distance is:

$$d = vt$$
$$= (17.3 \text{ m/s})(3.0 \text{ s})$$
$$= 52 \text{ m}$$

The second trajectory has a horizontal velocity component of 15.3 m/s for 3.0 s. Its distance is:

$$d = vt$$
$$= (15.3 \text{ m/s})(3.0 \text{ s})$$
$$= 46 \text{ m}$$

Reflecting on the Activity and the Challenge

The first part of this activity (two falling coins) demonstrated that the time required for a coin to fall is independent of the horizontal speed. If two long jumpers rise to the same height, then they will remain in the air for identical times.

Activity 8 Projectile Motion

The second part of this activity (the rolling chair) showed that the faster the chair is moving, the farther the ball will travel horizontally. If a long jumper is able to increase horizontal speed, then the jumper will travel farther.

To maximize the distance an object travels, you should try to maximize the horizontal speed and maximize the height it can rise. How can you use these conclusions from the activity to improve the performance of your broad jumper?

Physics To Go

1. Draw a sketch of your two coins leaving the table. Show where each coin is at the end of each tenth of a second. Remember to emphasize that they both hit the ground at the same time.
2. Repeat the sketch of the two coins leaving the table, but this time have one of the coins moving at a very high speed.

3. It is said that a bullet shot horizontally and a bullet dropped will both hit the ground at the same time. Draw sketches of this (the bullet is like a very, very fast-moving coin).
4. a) Survey your friends and family members to find out which they think will hit the ground first, a bullet that is dropped, or a fast-moving bullet.

   b) Explain why you think people may believe that the two coins hit the ground at the same time, but that they have a more difficult time believing the same fact about bullets.
5. Use evidence from your observations of the two coins in this activity to prove that a 100 mile/hour pitch thrown horizontally by a major league player will hit the ground in the same amount of time as a 10 mile/hour pitch thrown horizontally from the same height by a child.
6. Use evidence from your observations of the ball and chair in this activity to show the truth of the statement, "A projectile's horizontal motion has no effect on its vertical motion, and vice versa."

ANSWERS

Physics To Go

1. Students provide sketch.
2. Students provide sketch.
3. Students provide sketch.
4. a)-b) Answers will vary. Encourage students to confront their own misconceptions when answering this question.
5. Both objects strike the ground at the same instant.
6. The ball would not have returned to the hands of the student riding in the chair if the horizontal and vertical motions were not independent of one another.

Physics To Go
(continued)

7. The long jumper should run as fast as possible to have great horizontal speed at launch, and then, upon launch, should jump straight upward with the greatest possible speed.

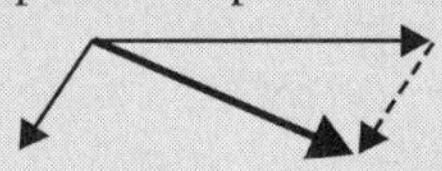

8. The times are equal. The dropped arrow and the shot arrow hit the ground at the same time. The vertical motion is independent of the horizontal motion.

9. Application of the Pythagorean Theorem yields:

   $(3.0 \text{ km/h})^2 + (2.0 \text{ km/h})^2 = v^2$

   $v = 3.6 \text{ km/h}$

   Using the tangent button on the calculator or a vector diagram, the angle is 34°.

7. The diagram to the left shows two forces acting on a point at the same time. Draw the vector that represents their resultant.

8. Above a flat horizontal plane, an arrow, *A*, is shot horizontally from a bow at a speed of 50 m/s. A second arrow, *B*, is dropped from the same height and at the same instant as *A* is fired. Neglecting air friction, how does the time *A* takes to strike the plane compare to the time *B* takes to strike the plane?

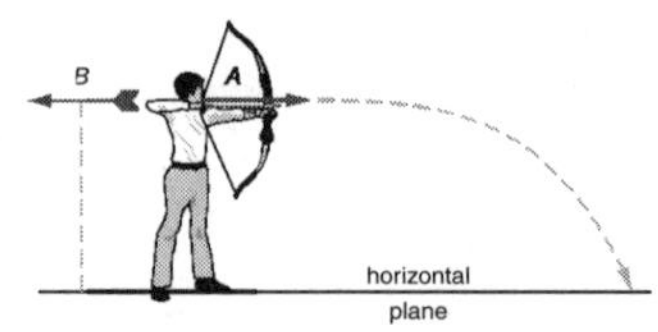

9. A swimmer jumps into a river and swims directly for the opposite shore at 2.0 km/h. The current in the river is 3.0 km/h. What is the swimmer's velocity relative to the shore?

10. A javelin is thrown at 15 m/s at an angle of 35° to the horizontal.
    a) What is its velocity in the x-direction?
    b) How far has the javelin traveled in 2.0 s?

11. A shot put is released at 12 m/s at an angle of 40° to the horizontal.
    a) What is its velocity in the x-direction?
    b) How far has the shot put traveled in 0.5 s?

12. Write a note to your school's track coach describing how the information you learned in this activity could help the team's long-jump athletes.

Physics To Go
(continued)

10.

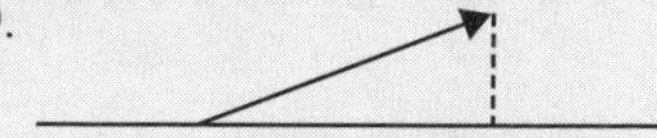

a) To find the x-direction component of the velocity, make a scale diagram and measure the x-component.

You can also use the cosine function:

$$vx = (15 \text{ m/s})(\cos 35°) = 12.3 \text{ m/s}$$

b) If it travels horizontally at 12.3 m/s for 2.0 seconds, the distance traveled is:

$$d = vt = (12.3 \text{ m/s})(2.0 \text{ s}) = 25 \text{ m}$$

11.

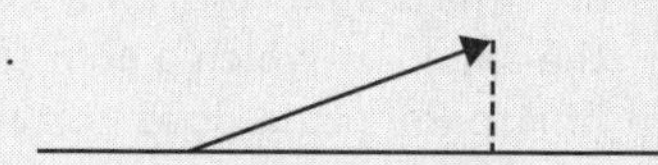

a) To find the x-direction component of the velocity, make a scale diagram and measure the x-component.

You can also use the cosine function:

$$vx = (12 \text{ m/s})(\cos 40°) = 9.2 \text{ m/s}$$

b) If it travels horizontally at 9.2 m/s for 0.5 s, the distance traveled is:

$$d = vt = (9.2 \text{ m/s})(0.5 \text{ s}) = 4.6 \text{ m}$$

12. Answers will vary. A good answer will include maximizing the horizontal speed and maximizing the time in the air. The long-jumper should run up as fast as possible and then jump vertically up without decreasing the horizontal speed. The time in the air is maximized and the large horizontal speed propels the long-jumper. High speed is the key – it is why the top sprinters are also the top long-jumpers.

Assessment Rubric: Two Falling Coins

Student understands that the time required for a coin to fall is independent of the horizontal speed.

Place a check mark (√) in the appropriate box.

Descriptor	Task accomplished	Task not accomplished
1. Student records experimental data demonstrating that the speed at which a coin travels along a horizontal plane does not affect the time it takes to fall.		
2. Student records experimental data demonstrating that the height at which a coin is dropped affects the time it takes to fall.		
3. Student is able to draw a sketch showing that the speed at which a coin travels along a horizontal plane does not affect the time it takes to fall.		
4. Student is able to draw a sketch showing that a bullet shot horizontally from a gun and a bullet dropped at the same time, from the same height, will hit the ground at the same time.		
5. Student successfully conducts a survey of friends and family to uncover misconceptions about falling bodies. "What will hit the ground first, a bullet that is shot from a gun or one that is dropped from the same height?" • Data is collected. • Data is compiled. • A conclusion is provided.		
6. Student proposes a hypothesis explaining why so many people hold the misconception.		
7. Student uses experimental evidence to explain why a pitch moving at 100 miles/hour will hit the ground at the same time as one thrown 10 miles/hour along the same trajectory.		

For use with *The Track and Field Championship*, Chapter 3, Activity 8: Projectile Motion

Assessment Rubric: The Rolling Chair

Student understands that the faster the chair moved, the farther the ball will travel horizontally.

Place a check mark (√) in the appropriate box.

Descriptor	Task accomplished	Task not accomplished
1. Student makes a prediction about how the velocity of the chair affects catching the ball.		
2. Student records experimental data in log book that shows the trajectory of the ball released from a moving chair.		
3. Student correctly identifies factors that affect the range of the tossed ball: • Height of the toss. • Velocity of the chair.		
4. Student observes and correctly records the path of the trajectory for the ball.		
5. Student uses experimental evidence to explain the statement: "A projectile's horizontal motion has no affect on its vertical motion."		
6. Student writes a note to the track coach explaining that if a long jumper is able to increase horizontal speed, then the jumper will travel farther.		

For use with *The Track and Field Championship*, Chapter 3, Activity 8: Projectile Motion

ACTIVITY 9
The Shot Put

Background Information

New material introduced in this activity includes:

- Measurement of the acceleration due to gravity.
- Calculation of an object's position and speed at any time after it enters a state of free fall.
- Quantitative treatment of the horizontal (constant speed) and vertical (free fall) components of the motion of a projectile.
- Modeling the trajectory of a projectile.
- Variables affecting the height and range of a projectile.

Acceleration Due to Gravity

All objects in a state of free fall, regardless of mass, shape, or other characteristics, experience a uniform (constant) acceleration, g, of approximately 10 m/s^2 in the downward direction. (Falling objects which experience significant air resistance are not considered to be in a state of free fall.)

For your information (not for students' at this time), Earth exerts an amount of downward gravitational pull on objects at or near its surface which is proportional to the mass of each object. The outcome of this is that all objects experience the same amount of force per unit of mass and, therefore, acceleration. For example, Earth's pull on a 2-kg rock is twice as much as the pull on a 1-kg rock; the result is that both accelerate at the same rate during free fall.

Active Physical Science "rounds off" g to 10 m/s^2, which is within about 2% of the average value, about 9.80 m/s^2, on our planet. The acceleration due to gravity depends on the radial distance from the center of Earth, and, therefore, the value of g varies with location by as much as 4 cm/s^2 due to terrain differences (mountains, valleys) and the fact that Earth is not a perfect sphere (the equatorial diameter is greater than the polar diameter). Some argue that local variations in g, if taken into account, would sometimes have a seemingly poor performance in a track and field event such as the shot put actually being better than a longer throw made at a location having a lower value of g. Physicists have determined that variations in g sometimes could make a difference in records in field events.

Distance, Speed, Acceleration, and Time in Free Fall

This activity follows arguments originally presented by Galileo regarding free fall. Galileo hypothesized that free fall involved constant, or uniform, acceleration.

From the definition of constant acceleration, $a = \Delta v/\Delta t$, he reasoned that the speed attained by an object falling from a "rest start" at any time during its fall would be $v = \Delta v = a(\Delta t)$. (The speed, v, at the end of a time of fall Δt would be equal to the change in speed, Δv, because the initial speed for a rest start is zero.)

Galileo further reasoned that, since the speed of a falling object increases uniformly (or at a constant rate) with time (his hypothesis), the average speed for a time of fall Δt would be half of the speed attained at the end of a time of fall Δt:

(Average speed at the end of time interval Δt):

$$v_{ave} = v/2 = a(\Delta t)/2$$

Finally, Galileo reasoned that the distance of fall and the end of a time of fall Δt would be, simply, the average speed multiplied by the time of fall:

(Fall distance at the end of time interval Δt):

$$d = [a(\Delta t)] / 2 \times (\Delta t) = (1/2)a(\Delta t)^2$$

The above reasoning was used as the strategy for developing the table in **Step 2** of **For You To Do** for this activity.

As an extension of the above for your understanding as the teacher, Galileo was limited in his ability to test his assumptions and reasoning. He did not have adequate instruments to measure acceleration and speed directly, so he devised a test of his thinking which involved what he could measure. From the last line in the above derivation, $d = (1/2)a(\Delta t)^2$, he solved for acceleration: $a = 2d / (\Delta t)^2$.

Rolling a sphere down a ramp to "dilute" the effect of gravity (he also assumed that a sphere rolling down a ramp also had constant acceleration, but less of it than an object in free fall) he used a crude timing device to measure the amounts of time for the sphere to roll, starting from rest, several measured distances along the ramp. Upon substituting pairs of distance and time measurements into the relation $a = 2d / (\Delta t)^2$, he obtained a fixed value for the right-hand side of the relation, proving that the acceleration indeed is constant.

The Trajectory of a Projectile

The steps used to develop the model of a trajectory in this activity provide a fine quantitative example of the independence of the vertical and horizontal components of the motion of a projectile. While it is not expected that students will predict the range, maximum height, time of flight, and other parameters of a projectile in terms of launch speed and direction, the following equations will prepare you to do so, if needed:

For a projectile launched horizontally at speed v from height h:

time of flight is $t = \sqrt{(2h/g)}$

range (horizontal distance) is: $R = vt = v\sqrt{(2h/g)}$

For the general case of a projectile launched from ground level at speed v at an angle θ above the horizontal, and traveling over flat ground: speed in the horizontal direction is $v_x = v \cos\theta$ (remains constant)

initial speed in the vertical direction is $v_{yo} = v \sin\theta$

speed in the vertical direction at time t is: $v_y = (v \sin\theta) - gt$

horizontal position at time t is $x = (v \cos\theta)t$

vertical position at time t is $y = (v \sin\theta)t - (1/2)gt^2$

time to reach maximum height is $t_{max} = (v \sin\theta)/g$

total time of flight is $t = 2t_{max}$ (see t_{max}, above)

range (horizontal distance) is $R = (v^2 \sin 2\theta)/g$

Note in the final of the above equations that $\sin 2\theta$ (and, therefore, also the range) has a maximum value of 1 when θ= 45 degrees (2θ = 90 degrees).

Active-ating the Physics InfoMall

Again, the title of the activity suggests a wonderful set of keywords: "shot put." There are articles on maximizing the put with calculus, without calculus, and with geometry. You may be interested in Figure 8F in the Projectiles section of *Modern College Physics* (in the Textbook Trove)

For You To Do, Step 10: Search the InfoMall for "trajectory model" to see another portable version of this. There is even a nice figure on this one! (This is Mb-17 in the *Demonstration Handbook for Physics,* found in the Demo & Lab Shop. It is in the Mechanics section, under Trajectory Model.)

Planning for the Activity

Time Requirements

- It is suggested that two class periods be allowed for this activity. Try to accomplish measurement of the acceleration due to gravity and completion of the table of free fall distances during the first class period. During the second class period, construct and manipulate the trajectory model.

Materials Needed

For the class:

- digital photogate
- picket fence
- trajectory model
- weight, small fishing

For each group:

- ball, tennis
- calculator, basic
- cards, unlined index
- chalk
- chalkboard
- ladder
- Post-it®-note
- stick, meter, 100 cm, hardwood
- string, ball of, 225 ft.
- ticker tape timer, AC timer

Teaching Notes

The method used to measure the acceleration due to gravity is left to your choice due to the variety of equipment available in schools. Whatever method you use, inform students that the value used in *Active Physical Science* will be the "rounded" value, 10 m/s^2.

It may be possible to begin the trajectory model during the end of the first class period; if so, go ahead and continue it during the second class period. Reserve as much time as possible for students to "match" the model trajectory by throwing a ball and to use the portable stick model to make inferences about how launch angle affects range.

Depending on the number of student groups you may have established earlier, assign each group to one or more rows of the table in **Step 2** for the production of clip and string assemblies as diagrammed in the student text; ten assemblies are needed, corresponding to two assemblies for each of the non-zero rows of the table. Duplicate assignments will do no harm and would have the advantage of checking one group's work against another.

As each group completes its assigned clip and string assembly invite the group to hang its assembly (or more than one, if assigned) from the pin(s) corresponding to the assigned time(s) of fall to the right hand side of the pin labelled zero. Also have the group place a mark (or Post-It) on the chalkboard (or other surface, as may apply) at the bottom end of the hanged assembly.

When all groups have placed their assemblies and have marked the chalkboard, remove (and save) the assemblies and then have a volunteer draw a smooth curve which connects the marks on the chalkboard.

Assemble the entire class and choose a volunteer to throw a ball horizontally to try to match the curve. The volunteer should stand on a ladder or stool and throw the ball horizontally (to the right hand side) from the zero time mark. Allow the volunteer to practice until the trajectory of the ball superimposes well against the drawn curve on the chalkboard.

Then ask, "When the curve is matched, what must be the horizontal launch speed? How many meters per second? How can we tell?" Call attention to the argument presented in the student text **Step 5 (a)** of **For You To Do** that the 0.40 m (40 cm) spacing of the pins on the tack strip was designed for a horizontal launch speed of 0.40 m/0.10 s = 4 m/s.

Next, use the clip and string assemblies again to construct a "mirror image" of the first trajectory to the left of the zero time mark to model. This will model an "arch" trajectory. Have a volunteer attempt to launch a ball from the bottom end of the mark corresponding to 0.50 s to the left of the zero mark to match the arch trajectory.

When the volunteer is able to match the trajectory with consistency, invite the class, to give the volunteer instructions to test the effects of launch angle and launch speed on the ball's trajectory.

Turn next to the "portable" 2.4-m Stick and String Model of a the same trajectory, prepared by you in advance. Explain that the model is identical to the one used above, but that an additional 0.10 s time interval with a fall distance of 1.8 m has been added to the model. Ask students to verify that the added fall distance is correct, and then follow the steps given in **For You To Do**, as a demonstration for the entire class. Be aware that the endpoint of the range of a projectile launched across level ground in this model is the point where the curve of the trajectory crosses the same vertical level as the "zero time"(launching) end of the stick; at inclinations of the stick exceeding about 45 degrees the longest (1.8 m) string no longer reaches down to the "launch level," so it will be necessary to extrapolate the curve of the trajectory.

Close the activity by discussing the content of **For You To Read.**

Knowledge of the physics of projectile motion presents a host of possibilities for helping the school's track team. Note the example given in the student text for **Reflecting on the Activity and the Challenge** which describes the improvement in an Olympic shot putter's performance based on advice from a physicist.

NOTES

Activity Overview

In this activity students create and compare mathematical and physical models of a projectile's motion to see if there is anything they can learn which they can use to improve performance in events where trajectories occur.

Student Objectives

Students will:

- Measure the acceleration due to gravity.
- Apply the equation Speed = $g\Delta t$ to calculate the speed attained by an object which has fallen freely from rest for a time interval Δt.
- Understand that the average speed at a specified time of an object which has fallen freely from rest is equal to half of the speed attained by the object at the specified time.
- Apply the equation (Distance = Average Speed X Δt) to calculate the distance travelled by an object which has fallen freely from rest for a time interval Δt.
- Use mathematical models of free fall and uniform speed to construct a physical model of the trajectory of a projectile.
- Use the motion of a real projectile to test a physical model of projectile motion.
- Use a physical model of projectile motion to infer the effects of launch speed and launch angle on the range of a projectile.

ANSWERS FOR THE TEACHER ONLY

What Do You Think?

Generally, for a fixed launch velocity (speed and angle), the range will be greater for a higher launch point (as when launching from a tower) and less for a lower launch point (as when launching toward a rising hill).

The optimum range across level ground is attained for a 45° launch angle.

The range of a projectile across level ground is governed by the equation: $R = (v^2 \sin 2\theta) / g$ where R is the range, v is the initial speed, θ is the launch angle measured from the horizontal, and g is the acceleration due to gravity. (It is not suggested that this equation be presented to students with the intent of mastery, if at all.)

The Track and Field Championship

Activity 9 The Shot Put

GOALS

In this activity you will:

- Measure the acceleration due to gravity.
- Apply the equation Speed = $g\Delta t$ to calculate the speed attained by an object which has fallen freely from rest for a time interval Δt.
- Understand that the average speed at a specified time of an object which has fallen freely from rest is equal to half of the speed attained by the object at the specified time.
- Apply the equation (Distance = Average Speed x Δt) to calculate the distance travelled by an object which has fallen freely from rest for a time interval Δt.
- Use mathematical models of free fall and uniform speed to construct a physical model of the trajectory of a projectile.
- Use the motion of a real projectile to test a physical model of projectile motion.
- Use a physical model of projectile motion to infer the effects of launch speed and launch angle on the range of a projectile.

What Do You Think?

A world record shot put of 23.12 m was set by Randy Barnes of the USA in 1990.

- **Will a higher or lower launch point of a projectile increase range?**
- **Will a particular launch angle increase range?**
- **Will a greater launch speed of a projectile increase range?**

Record your ideas about these questions in your *Active Physics* log. Be prepared to discuss your group's responses with your small group and the class.

For You To Do

1. Your teacher will provide you with a method of measuring the acceleration caused by the Earth's gravity for objects in a condition of free fall. One simple recommended method uses a picket fence and a photogate timer attached to a computer. The picket fence is dropped and the computer measures the time between black slats of the fence. The computer then displays the acceleration

184

Active Physics

due to gravity. A second method uses a ticker-tape timer and a mass. This second method requires more class time for the analysis of data.

a) In your log, describe the procedure, data, calculations, and the value of the acceleration of gravity obtained. The acceleration of gravity is used often, so it is given its own symbol, "g."

2. In your log, make a table similar to the following. Some data already have been calculated and entered in the table to help you get started.

Time of Fall (seconds)	Speed at End of Fall (m/s)	Average Speed of Fall (m/s)	Distance (m)
0.0	0	0.0	
0.1	1	0.5	
0.2	2		
0.3			
0.4			
0.5			

a) Calculate and record in the table the speed reached by a falling object at the end of each 0.10 s of its fall.

Example:
Assume $g = 10\ \text{m/s}^2$
Speed = acceleration × time
Speed at the end of 0.2 s = $10\ \text{m/s}^2 \times 0.2\ \text{s} = 2\ \text{m/s}$

b) Calculate and record the average of all of the speeds the object has had at the end of each 0.10 s of its fall. Since the object's speed has increased uniformly from zero to the speed calculated above at the end of each 0.10 s of the fall, the average speed will be one-half of the speed reached at the end of each 0.10 s of falling.

ANSWERS

For You To Do

1.a) Students provide procedure, data, and calculations for finding acceleration due to gravity.

2. a)-b) The completed table of calculated values of speed, average speed and distance at 0.10 s intervals for an object falling from rest is shown below. (See chart. Significant digits have not been used for clarity.)

Since "g" is limited to one significant figure, $10\ \text{m/s}^2$, the tabled values calculated using g also should be limited to one significant figure. Strict adherence would suggest that the fall distance 1.25 m should be rounded to 1 m, but it perhaps is not advisable to distract students with that detail at this time – use your judgment on whether or not to bring this up.

Time of Fall (seconds)	Speed at End of Fall (m/s)	Average Speed of Fall (m/s)	Distance (m)
0.0	0	0.0	0.00
0.1	1	0.5	0.05
0.2	2	1.0	0.20
0.3	3	1.5	0.45
0.4	4	2.0	0.80
0.5	5	2.5	1.25

Example:

Average speed = $\frac{\text{Zero + speed at end of time interval}}{2}$

Average speed during 0.2 s of fall = $\frac{(0 \text{ m/s} + 2 \text{ m/s})}{2}$

= 1 m/s

(Significant digits have not been used for clarity.)

c) Calculate and record the distance the object has fallen at the end of each 0.10 s of its fall. To do this, use the familiar equation Distance = Average speed × time.

Example:

The average speed during 0.2 s of falling is 1 m/s.

Distance = Average speed × time
= 1 m/s × 0.2 s
= 0.2 m

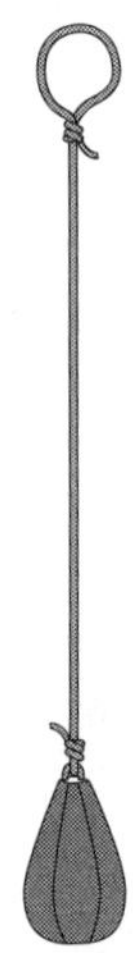

3. The table you have completed is a mathematical model of an object falling freely from rest. Now you will change the mathematical model into a physical model. Your teacher will assign your group a particular row in the data table for information about the falling object.

   Assemble a string and weight, as shown in the diagram, for the distance of fall assigned to your group.

4. Place a small label on the weight, showing your group's name and the time of fall assigned to your group.

5. Your teacher will place a horizontal row of pins labeled 0.0 s, 0.1 s, 0.2 s, etc. along the top edge of a chalkboard in your classroom. The times noted on the labels correspond to the instants for which you calculated distances of fall in the table. The horizontal distance from one pin to the next in the row is 40 cm. The 40-cm spacing of the pins is a model of the positions an object would have every 0.10 s if it traveled along the horizontal row of pins at constant speed.

ANSWERS

For You To Do *(continued)*

2.c) See data table on previous page.

3.-4. Student activities.

Activity 9 The Shot Put

a) Calculate the horizontal speed by dividing the distance traveled, 40 cm, during each time interval, by 0.1 s. (Dividing a number by 0.1 is equivalent to multiplying the number by 10). Show your calculation and the result in your log.

b) Hang your string and weight assembly from the pin corresponding to the time assigned to your group. Place a small mark on the chalkboard at the bottom end of the string and weight assembly.

6. A volunteer from the class should draw a smooth curve which connects the marks on the chalkboard. A volunteer should try to match the path, a trajectory, by throwing a tennis ball horizontally from the point corresponding to time = 0.0 s. To match the trajectory, the ball will need to be thrown horizontally at the speed calculated in **Step 5 (a)** above. This may require a few practice tries.

   a) Write your observations in your log.

7. Create a "mirror image" of the trajectory by moving the 0.1–0.5 s pins to positions 40 cm to the left of the 0.0 s pin. Hang the string and weight assemblies, mark the chalkboard and connect the points to create the second half of an "arch-shaped" model of a trajectory.

8. A volunteer should try to throw a ball to match this trajectory. Have another person prepared to catch the ball.

   a) What conditions seem to be necessary to match the trajectory? Write your observations in your log.

   b) When a volunteer is able to match the trajectory, the class should agree upon and give the volunteer instructions to test, one at a time, the effects of launch speed and launch angle on the range of the projectile. Write your observations in your log.

9. Your teacher will show you a "portable" version of the row of pins used in **Step 5** above. It is different in two ways: one additional time interval of 0.1 s has been added (the row is 40 cm longer), and a string to show the distance of fall, 1.8 m, at a time of 0.6 s has been added.

ANSWERS

For You To Do *(continued)*

5. a) 400 cm/s or 4 m/s.
   b) Students mark end of string assembly on chalkboard.

6. a) Students record their observations about a volunteer throwing the ball.

7. Student activity.

8. a) Students throw the ball upward with the angle and speed to mirror the trajectory in **Step 6**.

   b) Students record their observations for a number of different angles and speeds, as suggested by the students.

9. Teacher activity. You will display portable model of stick and string model of the Trajectory of a Projectile.

Chapter 3

10. Rest the end of the stick corresponding to 0.0 s on the tray at the bottom of the chalkboard while inclining the stick at an angle of 30°.

a) Is the path indicated by the bottom ends of the string and weights assemblies a "true" trajectory? Have a volunteer try to match it. Record your observations.

b) Repeat for angles of 45°, 60° and other angles of interest. Record your observations (it may be necessary to rest the lower end of the model on the floor to prevent the upper end from hitting the ceiling of the room).

c) What angle of inclination predicts the greatest range for the projectile? Record your observation.

d) Incline the stick to 90° (straight up)—do this outdoors if the ceiling is not high enough. What is being modeled in this case? Record your thoughts.

FOR YOU TO READ

Modeling Projectile Motion

The activities you have just completed demonstrate that a projectile has two motions that act at the same time and do not affect one another. One of the motions is constant speed along a straight line, corresponding to the amount of launch speed and its direction. The second motion is downward acceleration at 10 m/s^2 caused by Earth's gravity, which takes effect immediately upon launch. The trajectory of a projectile becomes simple to understand when these two simultaneous motions are kept in mind.

This activity also demonstrates the main thing that scientists do: create models to help you understand how things in nature work. In this activity you saw how two kinds of models, a mathematical model (the table of times, speeds, and distances during falling) and a physical model (the evenly spaced strings of calculated lengths) correspond to reality when a ball is thrown. For a scientific model to be accepted, the model must match reality in nature. By that requirement, the models used in this activity were good ones.

Technology as a Tool

Trajectories of projectiles can be modeled using a computer or graphing calculator. Your teacher, or someone else familiar with such devices, may be able to help you find computer software or enter equations into a calculator. These tools will allow you to manipulate variables such as launch angle, launch speed, launch height, and range to enhance your ability to explore and understand projectile motion.

ANSWERS

For You To Do *(continued)*

10. a) Yes.

b) All are "true trajectories."

c) 45°.

d) A ball thrown straight up.

Activity 9 The Shot Put

In this activity, you analyzed the motion of a shot put by looking at the horizontal motion and the vertical motion independently.

You were able to find the vertical distance traveled by first finding the average velocity and then multiplying that average velocity by the time. The vertical distance traveled can be found in one step by using the equation

$$y = \frac{1}{2}at^2$$

where a is the acceleration due to gravity (9.8 m/s^2 on Earth).

The value of 9.8 m/s^2 is often rounded up to be 10 m/s^2, as it was in the **For You To Do** section.

You now have a means to analyze two-dimensional motion mathematically. The analysis of two-dimensional motion begins with the recognition that the horizontal and vertical components are independent of one another, as you discovered in the activity. The horizontal velocity always remains the same. The vertical velocity always increases with time.

	Horizontal Component	Vertical Component
Position	Position $x = v_x t$ where v_x is the horizontal component of the velocity, t is the time, and x is the horizontal displacement.	$y = \frac{1}{2}at^2$ where a is the acceleration due to gravity (9.8 m/s^2 on Earth), t is the time, and y is the vertical displacement.
Velocity	The horizontal velocity is a constant. There is no net force in the horizontal direction. With no force, there is no acceleration.	$v_y = at$ where a is the acceleration due to gravity (9.8 m/s^2 on Earth), t is the time and v_y is the vertical velocity.
Acceleration	No acceleration in the x-direction.	Acceleration due to gravity in the y-direction.

During a broad jump the athlete runs and then travels in a parabola (a bowl-shaped curve). The faster she runs, the faster is her horizontal velocity. She must jump in the air to get height so she can stay in the air longer. She does this without slowing down the horizontal velocity. It's easier to analyze her motion by looking at the second half of her trajectory.

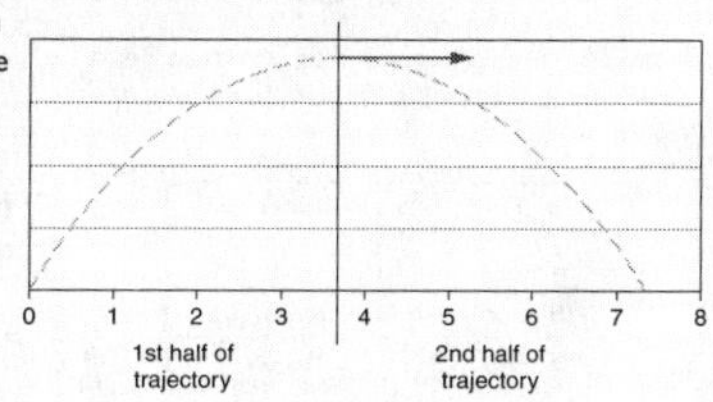

→

Chapter 3

By focusing on the second half of the trajectory, the jumper appears to jump horizontally from a certain height. Suppose the height is 1.8 m. Look at the following sample problem.

Sample Problem

A broad jumper jumps horizontally from a height of 1.8 m with a horizontal velocity of 6.0 m/s. Where will she land?

Strategy:

Step 1: Use the vertical motion information to determine the time in the air.

If she fell straight down from 1.8 m or jumped horizontally from 1.8 m, her vertical motion would be identical. Using the equation $y = \frac{1}{2}at^2$, you can find the time she is in the air.

Solution:

If she were in the air 1.0 s, the vertical distance would be

$$y = \frac{1}{2}at^2$$

$$y = \frac{1}{2}(9.8 \text{ m/s}^2)(1.0 \text{ s})(1.0 \text{ s})$$

$$= 4.9 \text{ m}$$

Since she falls from only 1.8 m, she must be in the air for less than 1 s. You can try some other times:

Time (s)	Distance (m)
0.3	0.441 or 0.4
0.4	0.784 or 0.8
0.5	1.225 or 1.2
0.6	1.764 or 1.8
0.7	2.401 or 2.4

To fall from 1.8 m, it would take a bit less than 0.6 s.

If you use algebra, you can rearrange the equation to solve for time.

$$y = \frac{1}{2}at^2$$

$$t = \sqrt{\frac{2y}{a}}$$

Solving for time:

$$t = \sqrt{\frac{2d}{a}}$$

$$= \sqrt{\frac{2(1.6 \text{ m})}{9.8 \text{ m/s}^2}}$$

$$= 0.57 \text{ s}$$

Strategy:

Step 2: If she has a horizontal velocity of 6.0 m/s and she is in the air for 0.57 s, where will she land? Her horizontal motion can be found by recognizing that distance equals velocity times time.

Solution:

$$x = v_x t$$

$$= (6.0 \text{ m/s})(0.57 \text{ s})$$

$$= 3.42 \text{ m or } 3.4 \text{ m}$$

The jumper moves 3.4 meters on the way down. Her total jump is twice this value since she moves horizontally on the way up as well.

$$x \text{ total} = 6.8 \text{ m}$$

You can solve lots of problems like this. A set of problems was solved where the jumper left the ground with the same total velocity but changed the angle. What was found was that the longest jump occurred when the athlete left the ground at an angle of 45°.

Let's see if this makes sense. If the athlete jumps straight up, she maximizes her time in the air but has no horizontal velocity. She will be in the air a long time, but won't go anywhere horizontally. If the athlete jumps straight out at a very small angle, she has a large horizontal component, but is not in the air very long. If she leaves the ground at 45°, she is in the air for quite some time and still has a large horizontal velocity. This angle of 45° gives the maximum range.

Reflecting on the Activity and the Challenge

The information learned about projectile motion in this activity applies not only to the shot put but to any track and field event that involves throwing things into the air (including the self-launching of a human body as in the hurdles, long jump, or high jump). It has been reported that one Olympian who competed in the shot put increased his range in that event by nearly 4 meters, based on suggestions made by a physicist. You are now a physicist specializing in projectile motion. How will you help your school's track team?

Physics To Go

1. If the launching and landing heights for a projectile are equal, what angle produces the greatest range? Why?
2. Compared to a launch angle of 45°, what happens to the amount of time a projectile is in the air if the launch angle is:
   a) Greater than 45°?
   b) Less than 45°?

Answers

Physics To Go

1. 45°.

2. a) More time.

   b) Less time.

3. For a constant launch speed, what angle produces the same range as a launch angle of:
   a) 30°?
   b) 15°?

4. If you launch a projectile from a high building, the angle for greatest range is less than 45°. Explain why this is true.

5. If a shot putter releases the projectile 2 m above level ground, what angle produces the greatest range? 45°? More than 45°? Less than 45°? Why? How could you find the exact angle?

6. Analysis of performances of long jumpers has shown that the typical launch angle is about 18°, far less than the angle needed to produce maximum range. Why do you think this occurs?

7. You are familiar with Carl Lewis as a medal-winning sprinter. But he is also an Olympic gold medalist in the long jump. Why do you think he's successful in both events?

8. The diagram shows a ball thrown toward the east and upward at an angle of 30° to the horizontal. Point X represents the ball's highest point.
   a) What is the direction of the ball's acceleration at point X? (Ignore friction.)
   b) What is the direction of the ball's velocity at point X?

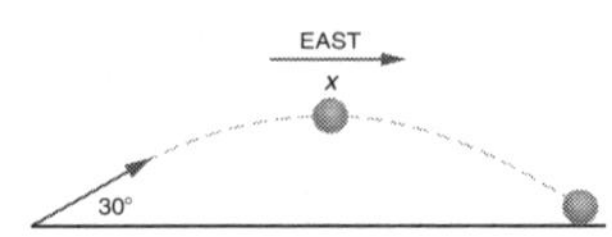

9. The diagram at the top of the next page shows a baseball being hit with a bat. Angle θ represents the angle between the horizontal and the ball's initial direction of motion. Which value of θ would result in the ball traveling the longest horizontal distance?

Physics To Go

(continued)

3. a) The complement of the angle: 60°.

   b) 75°.

4. Mathematical proof of the statement is beyond the scope of this physics course, but it is true that the maximum range of a projectile launched from a high building — or a shot putter's hand above ground level — is produced by using a launch angle of less than 45°. The 2.4-m Stick and String Trajectory Model used in **For You To Do** can be used to demonstrate this phenomenon empirically. To do the demonstration, place the "launch" end of the stick on the tray at the bottom of a chalkboard with the stick elevated at 45°. Slide the inclined stick until the end of the 1.8-meter string hangs beyond the end of the tray and mark the position of the bottom end of the 1.8 m string on the wall; it should reach a level between 10 and 11 cm below the level of the chalk tray. Hold a meter stick in horizontal orientation against the wall at the level of the mark. Then depress the angle of the stick slowly and observethat the curve of the trajectory (as imaginary "filled in" between the bottom ends of the last two strings) crosses the meter stick at a greater range. It is a subtle, yet observable demonstration which answers the question.

5. Less than 45°, depending on the height of launch and the speed. As noted in **Problem 4** above, calculation of the exact angle is complex, but possible (in some college level physics textbooks this is referred to as the "extended projectile problem"). The optimum launch angle for most shot putters probably would be between 40° and 45°.

6. The horizontal running speed of a long jumper is much greater than the initial vertical speed that the jumper can attain; therefore, the angle is far less than 45 °.

7. Carl Lewis can run fast, which is half of the requirement for a good long jumper. Apparently, he also jumps very well vertically.

8. a) The acceleration at point X is the acceleration due to gravity. Its direction is vertically down.

   b) The velocity at point X is horizontal. At the highest point, the ball is neither moving up nor moving down.

9. The angle for the longest horizontal distance is 45° when the object leaves the ground and returns to the ground. Since the baseball is 1 m above the ground, the optimum angle will be a very small amount less than 45° since the ball already has some vertical displacement.

Activity 9 The Shot Put

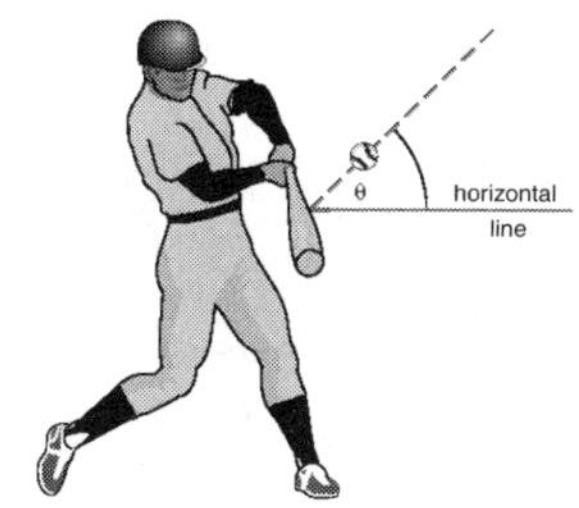

10. Four balls, each with mass m and initial velocity v, are thrown at different angles by a ball player. Neglecting air friction, which angular direction produces the greatest projectile height?

11. An object is thrown horizontally off a cliff with an initial velocity of 5.0 m/s. The object strikes the ground 3.0 s later.
    a) What is the vertical speed of the object as it reaches the ground?
    b) What is the horizontal speed of the object 1.0 s after it is released?
    c) How far from the base of the cliff will the object strike the ground?

12. The diagram below shows a ball projected horizontally with an initial velocity of 20.0 m/s east, off a cliff 100 m high.
    a) During the flight of the ball, what is the direction of its acceleration?

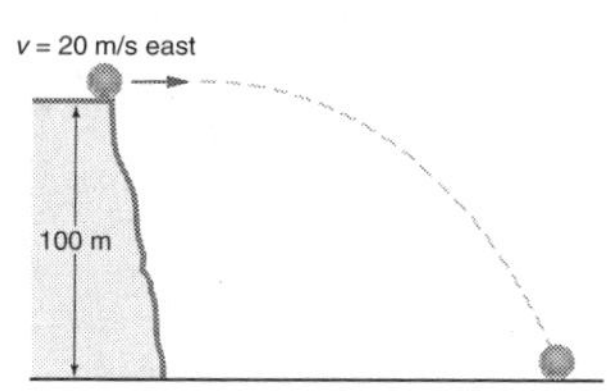

    b) How many seconds does the ball take to reach the ground?

193

Physics To Go

(continued)

10. The ball thrown the closest to 45° will travel the farthest. The ones thrown at 35° and 55° will travel identical distances. The 55° has less horizontal velocity but is in the air longer than the 35° ball.

11. a) The vertical motion of the horizontally thrown ball will be identical to the vertical motion of a dropped ball.

    $v_y = at = (9.8 \text{ m/s}^2)(3.0 \text{ s})$
    $= 29.4 \text{ m/s}$

    b) 5.0 m/s. The horizontal speed does not change because there is no force in the horizontal direction.

    c) $x = v_x t = (5.0 \text{ m/s})(3 \text{ s})$
    $=15 \text{ m}$

12. a) The acceleration is the acceleration due to gravity. Its direction is vertically down.

    b) The ball will take the same amount of time to reach the ground as an object dropped from 100 m.

    $d = 1/2\, at^2$

    $t = 4.5 \text{ s}$

Chapter 3

Measuring the Acceleration Due to Gravity

You must select a method of measuring the acceleration due to gravity based on the equipment available at your school. Several possibilities exist:

- Sonic ranger connected to a computer or graphing calculator with software for analysis of motion data.
- Picket fence and timer.
- Stroboscopic Photography.
- Calibrated ticker-tape timer.

Manufacturers manuals and physics laboratory manuals abound which give detailed directions for measuring "g." The trick is to match the method to the equipment available at your school. It is possible that the mathematics teachers at your school have graphing calculators (and, possibly, motion sensors for the calculators) and that you haven't heard about their existence. If that is the case, by all means approach the teachers about using the calculators; if necessary, your budget may allow purchasing motion sensing attachments for the calculators.

For use with *The Track and Field Championship*, Chapter 3, Activity 9: The Shot Put

NOTES

Four-meter Tack Strip for Trajectory Model

If you have a chalkboard a minimum of 4 m wide which has a cork strip across the top for tacking material to the strip, you are in business and need to do nothing more than prepare labels and pins to be placed at 40 cm intervals along the strip (see label specifications in equipment list). If you do not have such a setup, you need to create something which will serve the same purpose.

If you have a chalkboard without a tacking strip you could erect a wood strip across the top edge of your chalkboard (the strip can be in sections; it need not be one continuous strip). If the chalkboard is less than 4 m, it will probably serve if the expanse of wall on which it is mounted is 4 m wide.

If you do not have a chalkboard, or if yours is just too small, use a 4-m length of freezer paper taped to a wall, oriented about as a chalkboard would be on the wall and place a wood strip across the top edge of the paper (or tape labels and clip and string assemblies to the wall without using a wooden strip).

Whatever you use, it should appear as shown below:

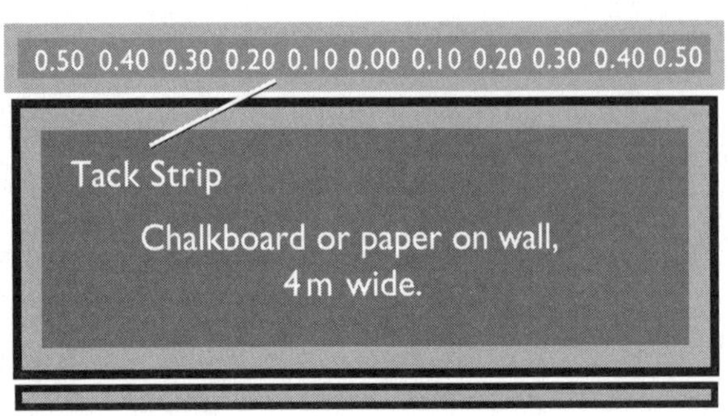

When the clip and string assemblies are in place for the first half of the model of the trajectory, the appearance of the model should be (lengths of clip & string assemblies not shown to exact scale) as shown:

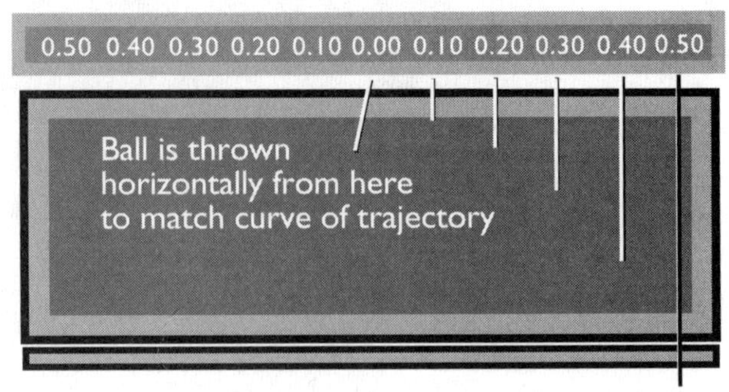

The appearance of the model for the second half of the trajectory should be as shown:

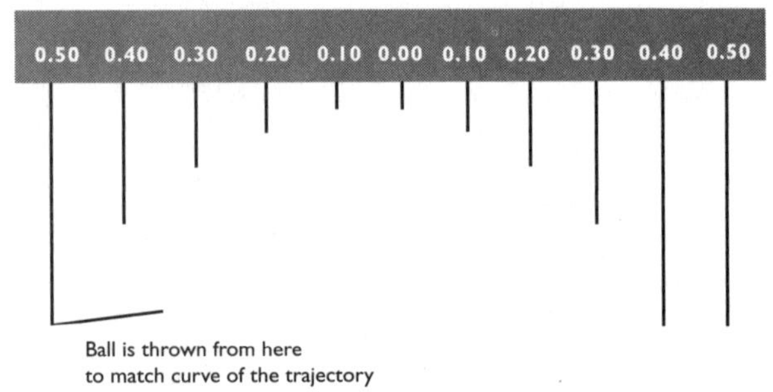

The portable trajectory model will have the same appearance as the right-hand half of the above model, except the strip on the wall is replaced by a longer (2.4 m instead of 2.0 m) stick, and a label for a time of 0.60 seconds is added with a string of length 1.8 meters at the 1.60 second mark to represent the distance of fall at that time.

For use with *The Track and Field Championship*, Chapter 3, Activity 9: The Shot Put

2.4-meter Stick and String Model of the Trajectory of a Projectile Launched at a Speed of 4.0 m/s

• An 8-foot (2.44 m) length of 2"x2" lumber serves as the stick • Washers may be used to weight string ends
• Place time labels at 0.4 m (40cm) intervals along stick

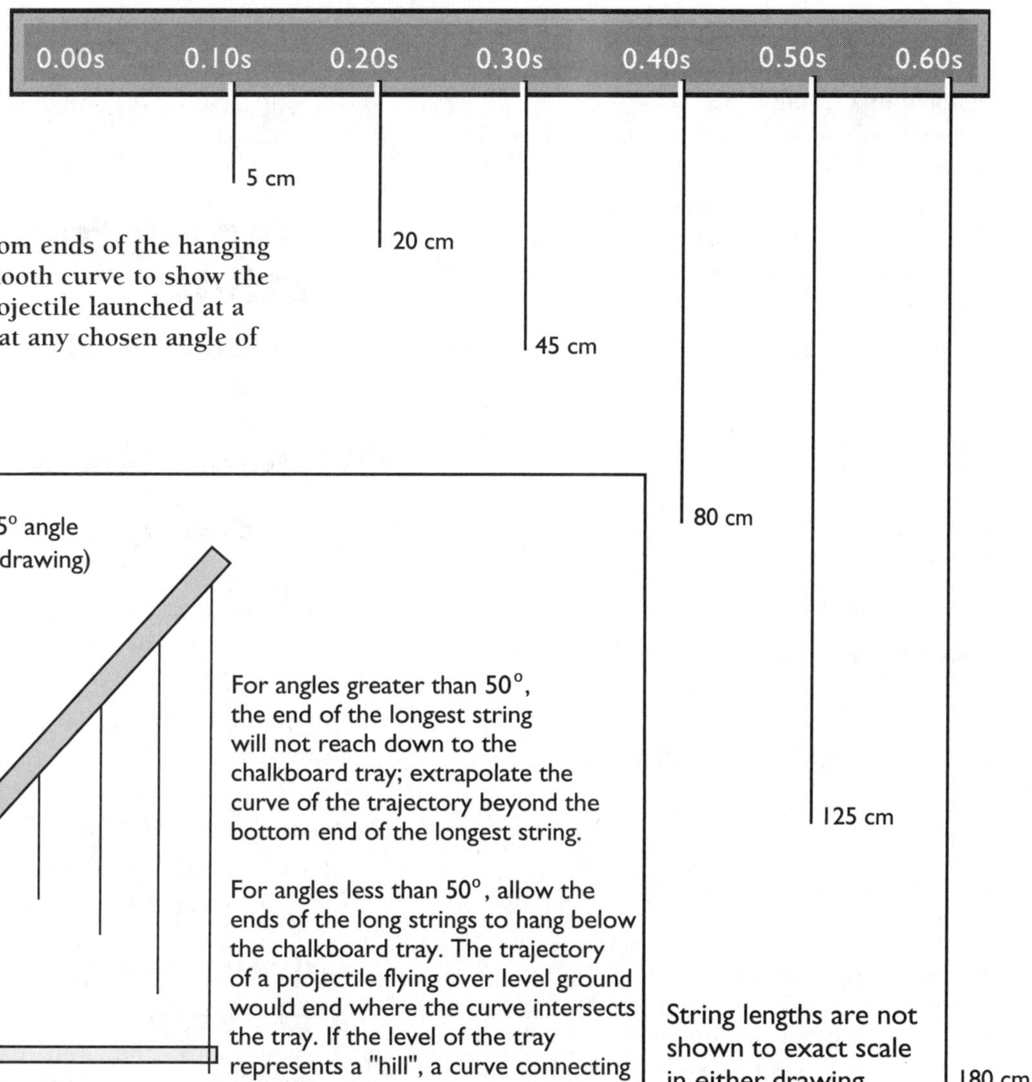

Connect the bottom ends of the hanging strings with a smooth curve to show the trajectory of a projectile launched at a speed of 4.0 m/s at any chosen angle of inclination.

Model inclined at 45° angle
(1/2 scale of above drawing)

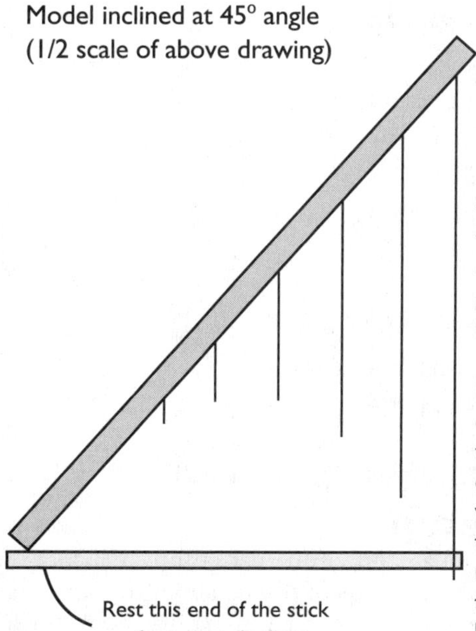

Rest this end of the stick on the tray at the bottom edge of a chalkboard and incline the stick at the desired angle.

For angles greater than 50°, the end of the longest string will not reach down to the chalkboard tray; extrapolate the curve of the trajectory beyond the bottom end of the longest string.

For angles less than 50°, allow the ends of the long strings to hang below the chalkboard tray. The trajectory of a projectile flying over level ground would end where the curve intersects the tray. If the level of the tray represents a "hill", a curve connecting the ends of the strings which hang below tray level shows how the trajectory would continue into a "valley."

String lengths are not shown to exact scale in either drawing.

Chapter 3

For use with *The Track and Field Championship*, Chapter 3, Activity 9: The Shot Put

ACTIVITY 10
Energy in the Pole Vault

Background Information

Concepts involving energy are introduced in this activity, including:

- Energy in three forms: kinetic energy, gravitational potential energy, and the energy stored in a spring.
- Transformations among and conservation of the above listed forms of energy.

The treatment of the joule as the unit of energy in this activity is "soft" because groundwork for defining the joule in a mechanical context has not yet been established.

At this stage without benefit of a definition of the joule, dimensional analysis can be used to satisfy yourself (and, only if it seems necessary, your students) that all of the equations given in the activity for forms of energy at least have the same unit:

Kinetic Energy = $1/2\ mv^2$

Unit: $(\text{kg})(\text{m/s})^2 = (\text{kg})(\text{m}^2)/(\text{s}^2)$

Gravitational Potential Energy = mgh

Unit: $(\text{kg})\ [\text{m}/(\text{s})^2]\ \text{m} = (\text{kg})(\text{m}^2)/(\text{s}^2)$

Potential Energy Stored in a Spring = $(1/2)kx^2$

(the spring constant, k, has a unit N/m which is equivalent to kg/s^2)

Unit: $(\text{kg/s}^2)\ \text{m}^2 = (\text{kg})(\text{m}^2)\ /(\text{s}^2)$

As shown above, all of the forms of energy defined in the activity share the same unit, $(\text{kg})(\text{m}^2)/(\text{s}^2)$. Since, for your information, 1 joule is the amount of work done when 1 newton of force is active through one meter of distance (1 joule = 1 newton × 1 meter) and since one newton is the force which will cause 1 kg to accelerate at 1 m/s^2, one joule also can be expressed as $(1\ \text{kg m/s}^2 \times 1\text{m}) = (\text{kg})(\text{m}^2)/(\text{s}^2)$. Therefore, the equations are legitimate because the unit $(\text{kg})(\text{m}^2)/(\text{s}^2)$ is equivalent to the standard energy unit, the joule.

The above information will prepare you to address questions about units, if they arise.

Active-ating the Physics InfoMall

In this final **Chapter 3** activity, the title once again suggests the search keywords: "pole vault." And again you will not be disappointed. Try it and see.

In **For You To Read**, we are introduced to Potential and Kinetic Energy. For more information, please consult any of the textbooks in the Textbook Trove. Also, you may wish to conduct searches of the InfoMall. Warning: you will want to limit your search to only a few stores or by using more keywords. If you search for "potential energy" you will get more than 1600 hits in the Textbook Trove alone!

Planning for the Activity

Time Requirements

- One class period.

Materials Needed

For each group:

- ball, steel, 1"
- cards, unlined index
- clamp, c-clamp
- flexible ruler
- marker, felt tip
- rubberbands
- ruler
- starting ramp 20 degrees
- tape, masking, 3/4" x 60 yds.
- washer, metal, 3/4"

Advance Preparation and Setup

You will need a supply (one per group) of 1-foot plastic rulers, or strips of plastic about the size of a ruler (the acetate strips used in many schools for demonstrating static electricity when the strips are rubbed with cloth would be ideal — if you do not have such strips, you may wish to see if other teachers have them). The rulers should be "floppy," having a thin, rectangular cross section which allows them to be bent with relative ease; safety caution: plastic rulers which are quite rigid may break and cause injury; do not use them for this activity.

You will also need marbles or ball bearings of mass such that, when rolled at reasonable speeds, they will produce reasonable, measurable amounts of bend in a clamped, flexible ruler upon striking the ruler near the tip.

You will need starting ramps which can be used to provide the range of speeds for the marbles; if you have no ramps, the groove between pages of an open book will serve as a ramp.

It is suggested that, in advance, you determine how the rulers will be clamped in the two orientations required for this activity (see diagrams in student text). The particular way of clamping the rulers will depend on the nature of the furniture in your classroom.

Teaching Notes

The activity is in two parts. The first part (**Steps 1 to 3 of For You To Do**) simulates the vaulter transferring kinetic energy to the pole, causing the pole to bend; the second part simulates transfer of the potential energy stored in the bent pole to the vaulter as gravitational potential energy.

Before beginning the activity, conduct a brief discussion of restoring forces, explaining that it is believed that interatomic forces in solid objects (such as rulers) behave as miniature springs which connect atoms. When an object is deformed by application of an external force, as when bent, some of the springs are compressed and others are stretched. When the external force is removed, the springs relax to normal length, providing a "restoring force" which returns the object to its original shape. In some cases the external force exceeds the "elastic limit" of the material, causing the springs to break, and permanently deforming the object or breaking it into parts.

Comments on the transformation of kinetic energy into energy stored in a bent ruler (**Steps 1 to 3 of For You To Do**):

It may not be clear to students that this part of the activity is meant to simulate the part of the pole vault event in which the vaulter "plants" the pole in the pit and "hangs on," stopping his running and causing the pole to bend as he and the pole rise due to rotation of the pole, the end of the pole in the pit serving as the axis of rotation. The main transformation of energy active in this part of the pole vault is the transfer of the kinetic energy of the vaulter into potential energy stored in the bent pole. This is what is being simulated.

Keep in mind that the procedures ask students to find only a "semi-quantitative" relationship between the speed of the rolling ball and the amount of deflection of the tip of the ruler (e.g., "The faster the ball, the more the ruler bends.")

Some students may choose to be more quantitative when establishing the speed of the ball and choose, for example, ramps heights (or lengths along a straight inclined ramp) in a ratio such as : 1 : 2 : 3. Such thinking should be encouraged. However, the same students may hold the common misconception that ramp heights in a simple ratio such as 1 : 2 : 3 will produce speeds in the same simple ratio. If you detect that misconception, you should move to correct it. The information below would be helpful for that purpose:

Speed of ball leaving ramp:

$$v = \sqrt{2gh}$$

Therefore, the speed of the ball as it leaves the ramp is proportional to $\sqrt{h}$, and balls launched from heights in the ratio 1 : 2 : 3 would have speeds in the (same order) ratio of the square roots of 1, 2, and 3, approximately 1.0 : 1.4 : 1.7. The ramp heights which would produce speeds in the ratio 1 : 2 : 3 would be heights in the ratio of the squared values of 1, 2, and 3, the sequence 1 : 4 : 9.

Going on to consider the kinetic energy of the ball as it leaves the ramp, recall that the kinetic energy of the ball is proportional to the squared value of the ball's speed. Therefore, ramp heights in the ratio 1 : 2 : 3, producing speeds in the ratio 1 : 1.4 : 1.7 (as explained above), would result in kinetic energies in the ratio of the squared values of 1.0, 1.4 and 1.7, the ratio 1 : 2 : 3.

Therefore, selecting ramp heights (or distances along a straight ramp) in a 1 : 2 : 3 ratio results in kinetic energies, not speeds, in the ratio 1 : 2 : 3.

Finally, the kinetic energy of the ball is transformed into energy stored in the bent ruler (only for an instant, until the ruler "snaps back"):

$$1/2mv^2 = 1/2\ kx^2$$

Solving for x, the deflection of the ruler:

$$x = v\ \sqrt{m/k}$$

Since m is constant for a chosen ball and since k is a constant for a chosen ruler, the deflection of the ruler, x, is proportional to the speed of the ball, v.

Chapter 3

Therefore, in the likely scenario where students choose ramp heights in the ratio 1 : 2 : 3 (resulting in ball speeds in the ratio 1.0 : 1.4 : 1.7) the defection of the tip of the ruler should be in the same ratio as the speeds, 1 : 1.4 : 1.7.

Maintain awareness as you interact with students during this activity (which is recommended) that some students will desire and perhaps even try to have the data "come out" to match their preconceptions (often misconceptions); keep them honest!

The actual data for this part of the activity will depend on the mass of the ball used, the speeds imparted to the ball, and the "stiffness"(k value) of the ruler used. Data for a sample case is not given due to the small likelihood that the case would resemble the conditions which you will have.

Comments on the transformation of the energy stored in a bent ruler into gravitational potential energy (**Step 4** of **For You To Do**):

Be sure that students understand that this part of the activity is intended to simulate that part of the pole vault in which the energy stored in the bent pole is transferred to increase the gravitational potential energy of the vaulter, "catapulting" the vaulter's body upward, hopefully high enough to clear the bar.

The relationship between the displacements of the end of the clamped ruler and the height to which a coin resting on the ruler should rise is:

$1/2\ k\ x^2 = mgh$

Solving for h, $h = k\ x^2 / 2\ mg$

Therefore, for a chosen ruler (constant k), chosen coin (constant m), and since g is a constant, the height to which the coin should fly should be proportional to the squared value of x. This suggests that displacements of the ruler (x values) of 1, 2 and 3 cm should result in heights of the coin which have relative values fitting a ratio of the squares of x,

1 : 4 : 9.

Depending on the stiffness of rulers available to you, you could adjust the x values specified in the student text to another set of values which would give reasonable amounts of ruler bend and coin height; for example you could specify x values of 0.5, 1.0 and 1.5 cm or you could use 2, 4, 6 cm. In any case, choose x's which are in the ratio 1 : 2 : 3.

Again, sample data is not given because it would vary greatly with conditions.

Do not assume that students can absorb the energy equations presented in **Physics Talk** without your help. Help students link data gathered during **For You To Do** to the equations. The joule as the unit of energy is somewhat casually introduced in the sample calculation presented in **For You To Read** because rigorous definition of the joule is reserved until force and work are formally defined in another chapter. The joule is not needed for any of the **Physics To Go** problems and should not be emphasized at this time.

NOTES

Activity Overview

In this activity students examine the conservation of energy as it occurs in a pole vault.

Student Objectives

Students will:

- Understand and apply the equation Kinetic Energy = $1/2\ mv^2$.
- Understand and apply the equation Gravitational Potential Energy = mgh.
- Recognize that restoring forces are active when objects are deformed.
- Understand and apply the equation Potential Energy Stored in a Spring = $1/2\ kx^2$.
- Understand and measure transformations among kinetic and potential forms of energy.
- Conduct simulations of transformations of energy involved in the pole vault.

ANSWERS FOR THE TEACHER ONLY

What Do You Think?

Forms of energy involved in pole vault: kinetic energy, potential energy stored in a spring, gravitational potential energy (work, done by the vaulter using arm and upper body muscles is involved, but not mentioned in the activity).

Human limitations on the kinetic energy and the amount of work which the vaulter is able to provide as input limit the height.

The Track and Field Championship

Activity 10 Energy in the Pole Vault

GOALS

In this activity you will:

- Understand and apply the equation Kinetic Energy = $1/2\ mv^2$.
- Understand and apply the equation Gravitational Potential Energy = mgh.
- Recognize that restoring forces are active when objects are deformed.
- Understand and apply the equation Potential Energy Stored in a Spring = $1/2 kx^2$.
- Understand and measure transformations among kinetic and potential forms of energy.
- Conduct simulations of transformations of energy involved in the pole vault.

What Do You Think?

You would need a fence more than 20 feet high to keep the world champion pole vaulter (6.14-m record) out of your yard.

- **If good vaulters have a 25-foot pole, why can't they vault 25 feet?**
- **What factors do you think limit the height vaulters have been able to attain?**

Take a few minutes to write answers to these questions in your *Active Physics* log. Compare your answers with the answers given by others in your group. Be prepared to represent your group's ideas to the entire class.

For You To Do

1. Carefully clamp a ruler in a vertical position so that the clamp is near the bottom end and the top end extends a few centimeters above the edge of a tabletop. Tape a pen or pencil to the surface of the ruler near the top end of the ruler so that the writing end of the pen extends to one side of the top end of the ruler. If the top end of ruler moves as it is bent, the pen will move with it.

194

Active Physics

ANSWERS

For You To Do

1. Student activity.

2. Set up a ramp as shown in the diagram. Three different starting points on a ramp will be used to roll a ball across the tabletop at three different speeds. Each time the ball rolls, it will strike the ruler near the top end, causing the ruler to bend. A marking surface held in contact with the tip of the pen or pencil will be used to measure the deflection.

3. Roll the ball at low, medium and high speeds to bend the ruler. In each case, measure the amount of deflection of the end of the ruler as indicated by the length of the pen mark.
   a) If the rolling ball represents the running vaulter and the ruler represents the pole in the model, how does the amount of bend in the pole depend on the vaulter's running speed? Record your data and response in your log.

4. Carefully clamp a ruler flat-side down to a tabletop so that two-thirds of the ruler's length extends over the edge of the table.

5. Place a penny on the top surface of the ruler at the outside end.

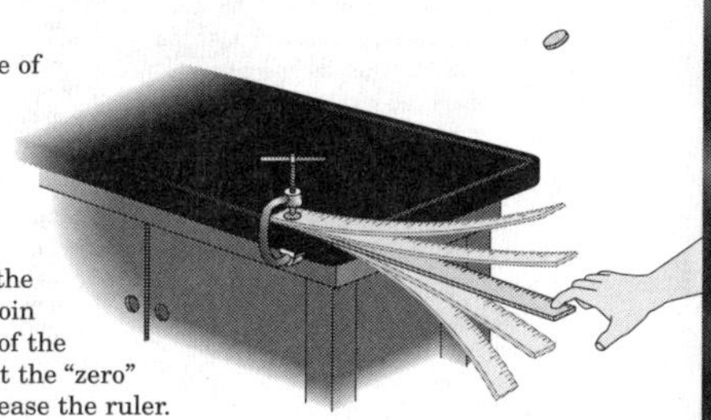

6. Use a second ruler to measure a 1 cm downward deflection of the outside end of the clamped ruler; that is, bend the ruler downward. Prepare to measure the maximum height to which the coin flies upward using the position of the coin when the ruler is relaxed at the "zero" vertical position of the coin. Release the ruler.
   a) Record in your log the height to which the coin travels.
   b) Repeat **Step 6** above for ruler deflections of 2 cm and 3 cm, and in each case, record the maximum height of the "vaulted" coin.
   c) How does the height to which the coin travels seem to be related to the amount of deflection of the ruler? (Remember, the coin is a projectile, in this case.)

Do not deflect the ruler excessively.

ANSWERS

For You To Do *(continued)*

2. Student activity.

3.a) The greater the speed of the vaulter, the greater the bend of the pole.

6.a)-c) See **Teaching Notes** re: possible quantitative analysis. The greater the amount of deflection of the ruler, the greater the height of the coin.

Chapter 3

FOR YOU TO READ

Conservation of Energy in the Pole Vault

The pole vault is a wonderful example of the Law of Conservation of Energy. The forms of energy are changed, or transformed, from one to another during a vault, but, in principle, the total amount of energy in the system of the vaulter and the pole remains constant. Food energy provides muscular energy for the vaulter to run, gaining an amount of **kinetic energy.** Some of the vaulter's kinetic energy is used to catapult the vaulter with an initial speed upward and the remaining kinetic energy is converted into work done on the pole to cause it to store an amount of **spring potential energy.** As the bent pole straightens, its potential energy is delivered to the vaulter to increase the vaulter's gravitational potential energy.

In making measurements of the ruler's deflection and the height of the coin, you continued your study of the conservation of energy – the most important principle of science. Richard Feynman, an American physics giant of the 20th century, provides a story that may help you to understand energy conservation. In his story, a child plays with 28 blocks. Every day the child's mother counts the blocks and always finds the total to be 28. On one occasion, she only finds 27 blocks, but then realizes that one block is hidden in a box. By finding the mass of the box and its contents, she can determine that 1 block must be inside, if she knew the mass of the empty box. On another day, she finds only 25 blocks, but can see that the water in a pail is higher than expected. By measuring the height difference, and knowing something about the original height of the water and the volume of a block, she determines that 3 blocks are below the surface of the water. On a third day, she finds 30 blocks! She then remembers that a friend came over to visit and decides that 2 of the blocks belong to the friend. Feynman equates "counting the blocks" with measuring the total energy. There were 28 blocks and there will always be 28 blocks. If there are 28 units of energy, then there will always be 28 units of energy.

In the activity, the energy came from deflecting the ruler. The muscles in your arm applied a force over a distance and deflected the ruler and the ruler gained spring potential energy. As you let go of the ruler it tossed the coin into the air. The ruler lost some of its energy and the moving coin gained kinetic energy. The coin rose up in the air, slowing down along the way. The kinetic energy was transformed as the coin rose, and there was a gain in gravitational potential energy. As the coin descended, the gravitational potential energy became kinetic energy once again.

If you were to measure the energy, you might find that the work done by your muscles in deflecting the ruler may have been equal to, say three joules. If that were so, then the three joules of work would have created three joules of spring potential energy that would have become three joules of kinetic energy that would then have become three joules of gravitational potential energy. The energy comes from external work and this energy then remains constant.

You can calculate the work done by your muscles by multiplying the force applied and the distance over which the force was applied.

$$W = F \cdot d$$

The dot between the F and the d is to signify multiplication of the force and displacement when they are (at least partly) in the same direction.

There are also equations that can help you to measure the energies.

Spring potential energy:

$$PE_{spring} = \frac{1}{2}kx^2$$

where k is the spring constant, and x is the amount of bending.

Gravitational potential energy:

$$PE_{grav} = mgh$$

where m is the mass of the object, g is the acceleration due to gravity, h is the height through which the object is lifted.

Kinetic energy:

$$KE = \frac{1}{2}mv^2$$

where m is the mass of the moving object and v is the speed of the object.

All of the energies are expressed in joules. Energy is a scalar. It has no direction.

A second example of energy transformations that you may be familiar with is in the sport of archery. In archery, you pull on the bowstring and the arrow flies through the air. If you shoot the arrow straight up, you can look at energy conservation in the following way: Your arm does work on the bow by pulling on the string and stretching the bow. The bow now has spring potential energy. The string is let go. The bow loses its spring potential energy and the arrow

gains kinetic energy. If the arrow goes up, the kinetic energy of the arrow now becomes gravitational potential energy. As the arrow descends, the gravitational potential energy becomes kinetic energy. If the arrow sticks up in the ground and comes to rest, all of the kinetic energy transfers to the ground and becomes heat energy. These ideal transformations of energies occur when there are no external forces, like air resistance, depleting energy from the original amount of energy.

The force to stretch a string is a Hooke's Law force.

$$F = kx$$

This is Hooke's Law. It says that the larger the stretch of a spring, the larger the force required. When you pull a bowstring, you need almost no force to pull it a tiny bit. As you stretch it, the force you must apply gets greater. The average force will be halfway between the zero force (to start the stretch) →

Chapter 3

and the final force for the last stretch. The final force is kx. The initial force is 0. The average force is:

$$\bar{F} = \frac{kx + 0}{2} = \frac{1}{2}kx$$

The total stretch of the spring is x. The work done is:

$$W = F \cdot d = \left(\frac{1}{2}kx\right)x = \frac{1}{2}kx^2$$

which is the expression for the spring's potential energy.

You can also calculate the work done to lift an object of mass m up through a distance h. Here the applied force must be equal to the weight of the object mg.

$$W = F \cdot d = mgh$$

which is the expression for the gravitational potential energy.

Also, you can calculate the work done in accelerating an object from an initial velocity v_i to a final velocity v_f with a constant force.

$$\begin{aligned} W &= F \cdot d \\ &= mad \\ &= \frac{1}{2}mv_f^2 - \frac{1}{2}mv_i^2 \end{aligned}$$

which is the expression for the change in kinetic energy.

The work equations and the energy equations are identical. This is where the statement "energy is the ability to do work" originates.

Sample Problem I

Your teacher gives you a pop-up toy. When you push down on it, it sticks to the desk for a moment and then pops into the air.

a) If the toy has a mass of 100.0 g and leaps 1.20 m off the table, how much potential energy does it have at its point of maximum height? (Use $g = 9.80\ \text{m/s}^2$.)

Strategy: The toy has a type of energy that depends on its position in the Earth's gravitational field. You can use this "potential" energy to do some work or to change to some other form. So you need to use the formula for gravitational potential energy.

Givens:

$m = 100.0\ \text{g} = 0.100\ \text{kg}$

$h = 1.20\ \text{m}$

Solution:

$$\begin{aligned} PE_{grav} &= mgh \\ &= (0.100\ \text{kg})(9.80\ \text{m/s}^2)(1.20\ \text{m}) \\ &= 1.18\ \text{J} \end{aligned}$$

b) When the toy jumps off the desk, with what speed does it leave?

Strategy: At the point where it jumps off the desk, the toy has its maximum amount of kinetic energy. This is what becomes the potential energy at the peak. Because energy is conserved, these two values will be equal—kinetic energy at the bottom equals the potential energy at the peak.

Givens:

$PE_{grav} = 1.18\ \text{J}$

Solution:

$$\begin{aligned} PE_{grav} &= KE \\ KE &= \frac{1}{2}mv^2 \end{aligned}$$

You can use algebra to rearrange the equation to solve for v.

$$v = \sqrt{\frac{KE}{\frac{1}{2}m}}$$

$$= \sqrt{\frac{1.18\ \text{J}}{\frac{1}{2}(0.100\ \text{kg})}}$$

$$= 4.90\ \text{m/s}$$

c) If you push the toy down 2.00 cm to make it stick to the desk, what is the spring constant of the spring in the toy?

Strategy: Where did the kinetic energy to make the toy leap off the desk come from? It came from doing work on the spring and storing a different type of potential energy: elastic potential energy. How much elastic potential energy did the toy store? 1.18 J! So again you can use conservation of energy to solve this problem!

Givens:

$x = 2.00\ \text{cm} = 0.0200\ \text{m}$

Solution:

$$PE_{spring} = \frac{1}{2}kx^2$$

You can use algebra again to rearrange the equation to solve for k.

$$k = \frac{PE_{spring}}{\frac{1}{2}x^2}$$

$$= \frac{1.18\ \text{J}}{\frac{1}{2}(0.0200\ \text{m})^2}$$

$$= 5900\ \text{N/m}$$

This sounds like a really large value for such a little toy, but remember that this is newtons per meter and the spring in the toy is only compressed a few centimeters.

d) What force was needed to compress the spring the 2.00 cm?

Strategy: Now that you know the compression and the spring constant, it is possible to find the amount of force required to press down on the spring.

Givens:

$k = 5900\ \text{N/m}$
$x = 0.0200\ \text{m}$

Solution:

$$F = kx$$

$$= (5900\ \text{N/m})(0.0200\ \text{m})$$

$$= 118\ \text{m or } 120\ \text{N}$$

Sample Problem 2

At what height, above the surface of Earth, could a tennis ball ($m = 57$ g) be dropped to give it the same kinetic energy it has when traveling at 45 m/s? (Neglect air resistance.)

Strategy: As an object is dropped from a height, its gravitational potential energy decreases and its kinetic energy increases. Assume that you are looking for the vertical position, which will yield a speed of 45 m/s the instant *before* the ball touches the ground. The problem can be solved in one step using the conservation of energy ($PE_{grav} = KE$).

Givens:

$m = 57\ \text{g} = 0.057\ \text{kg}$
$v = 45\ \text{m/s}$
$g = 9.8\ \text{m/s}^2$

→

Chapter 3

Solution:

$$PE_{grav} = KE$$

$$mgh = \frac{1}{2}mv^2$$

You can use algebra to rearrange the equation to solve for h. Notice that in solving the problem in one step, you do not have to take into account the mass of the ball.

$$h = \frac{v^2}{2g}$$

$$= \frac{(45 \text{ m/s})^2}{2 \times 9.8 \text{ m/s}^2}$$

$$= \frac{2025 \text{ m}^2\text{s}^2}{19.6 \text{ m/s}^2}$$

$$= 103.3 \text{ m or } 100 \text{ m}$$

When the ball is traveling at 45 m/s, it has a kinetic energy of 58 J. (You can calculate this:)

$$\frac{1}{2}mv^2 = \frac{1}{2}(0.57 \text{ kg})(45 \text{ m/s}^2)$$

If the tennis ball were positioned at a location 100 m above Earth, the gravitational potential energy of the ball would also equal 58 J.

Reflecting on the Activity and the Challenge

In this activity you were told that throughout the event of pole vaulting, energy changes from one form to another, but the total amount of energy in the system at all instants remains the same. (A small amount of energy may be lost from the system of the vaulter and pole by making a dent in the end of the pit, which stops the pole or by generating heat in the pole as it bends.)

This will be important for the athletes at your school to know. They will need to understand what they must do to gain the greatest height in a pole vault. By examining the equations for kinetic and potential energy, they will also be able to appreciate why there is a limit to the height that somebody can vault. The faster someone runs, the more kinetic energy they have. This kinetic energy makes the pole bend. The more kinetic energy there is, the more bend in the pole is expected. Potential energy stored in the bent pole will be transformed to increase the vaulter's gravitational potential energy. The more bend in the pole, the higher the vaulter goes. One key to success in the pole vault is to have the most kinetic energy. Another key to success is to bend the pole as much as possible. You may be able to use this knowledge to help a pole vaulter in your school improve performance.

Physics Words

kinetic energy: the energy an object possesses because of its motion.

spring potential energy: the internal energy of a spring due to its compression or stretch.

Physics To Go

1. Describe the energy transformations in the shot put.
2. Describe the energy transformations in the high jump.
3. Assume that a vaulter is able to carry a vaulting pole while running as fast as Carl Lewis in his world record 100-m dash. Also assume that all of the vaulter's kinetic energy is transformed into gravitational potential energy. What vaulting height could that person attain? (Hint: Use the equation $\frac{1}{2}mv^2 = mgh$.)
4. Why doesn't the length of the pole alone determine the limit of vaulting height?
5. Some poles lose a significant amount of energy to heat as they flex. Use the Law of Conservation of Energy to explain how this would affect performance.
6. The women's pole vault world record as of spring 1997 was 4.55 m, set by Emma George. What do you estimate was Emma's speed prior to planting the pole? Use conservation of energy for your prediction.
7. Sergei Bubka held the world record for the pole vault as of spring 1997 at 6.14 m. How did Sergei's speed compare with Emma George's speed?
8. A 2.0-kg rock is dropped off a 100-m high cliff.
   a) Without using kinematics, calculate the speed the rock would be going when it got to the bottom of the cliff.
   b) Can you do this if you do not know the mass of the rock?
9. A bow is strung with a bowstring that has a spring constant of 1500 N/m.
   a) If you pull it back 25 cm, how much work are you doing on the string?
   b) If the string is pushing against an arrow that has a mass of 0.10 kg, how fast would the arrow be going when it left the bow?

ANSWERS

Physics To Go

1. The shot putter imparts a launching speed to the projectile, and therefore kinetic energy, by giving the shot two speeds which, since they are in the same direction, add together. One of the motions is provided by the spinning motion of the shot putter before release, and the other is provided by the thrusting action of the shot putter's arm. In both cases, the athlete does work, a form of energy (force times distance), which is transformed into the kinetic energy which the ball has upon release. The horizontal component of the shot's speed is maintained while the projectile is in the air; the vertical component of the speed can be used to calculate the part of the kinetic energy which will be transformed into gravitational potential energy, allowing prediction of the height to which the shot will rise at the peak of its flight.
2. The high jumper uses the leg muscles to jump upward with an initial kinetic energy which is transformed into potential energy, causing the jumper to rise to a height which hopefully is enough to clear the bar.
3. Lewis's average speed in the dash: 10.1 m/s.

   Solving for h in the equation given in the problem: $h = v^2/2\,g$
4. The vaulter's kinetic energy, determined by running speed, plus the amount of work which the vaulter does using the arms to lift the body determine the maximum height to which the vaulter can rise, regardless of the length of the vaulting pole.
5. The amount of heat generated during the bending of the pole would "rob" part of the potential energy can be stored in the pole by causing it to bend less than it would if no heat were generated and by causing the pole to straighten with less "straightening" speed than if no heat were generated; overall, heating of the pole would reduce the vaulter's height.
6. $v = \sqrt{2gh} = \sqrt{2 \times 10\,\text{m/s}^2 \times 4.6\,\text{m}} = \sqrt{92\,\text{m}^2/\text{s}^2} = 9.6\,\text{m/s}$
7. Notice that Emma's height was rounded to two significant figures in the above substitution.

   $v = \sqrt{2gh} = \sqrt{2 \times 10\,\text{m/s}^2 \times 6.1\,\text{m}} = \sqrt{120\,\text{m}^2/\text{s}^2} = 11\,\text{m/s}$

   Notice that while Sergei's speed was only about 15% greater than Emma's speed, his vault height, 6.1 m compared to Emma's 4.6 m, was about 32% higher, this due to the squared effect of speed on kinetic energy; squaring the ratio of speeds verifies that the ratio of heights should be 1.32 : 1.00, as shown below

   $[(6.1\,\text{m/s}) / (4.6\,\text{m/s})]^2 = (1.15)^2 = 1.32$. Since g is known to only two significant figures (10 m/s^2), round Lewis's speed to 10 m/s when substituting in the above equation: $h = (10\,\text{m/s})^2 / (2 \times 10\,\text{m/s}^2) = (100\,\text{m}^2/\text{s}^2) / (20\,\text{m/s}^2) = 5.0\,\text{m}$

(Answers to 8 – 11 on next page)

Chapter 3

ANSWERS

Physics To Go

(continued)

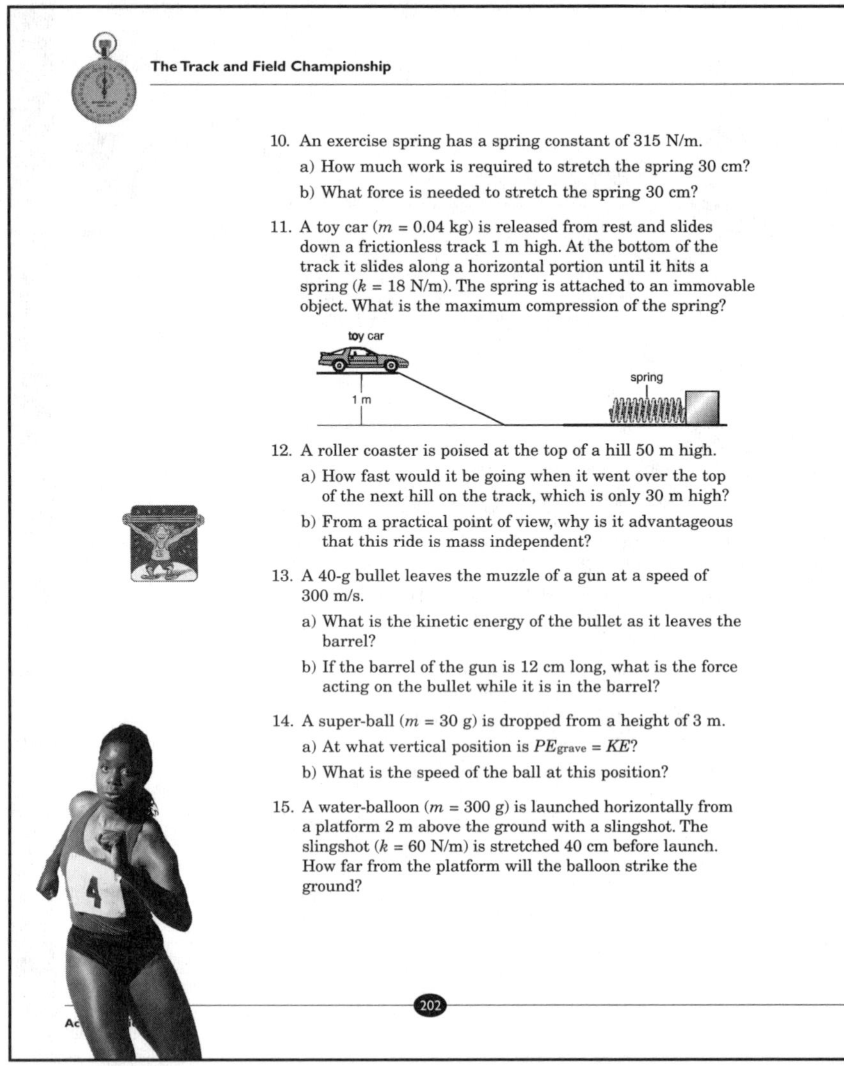
The Track and Field Championship

10. An exercise spring has a spring constant of 315 N/m.
    a) How much work is required to stretch the spring 30 cm?
    b) What force is needed to stretch the spring 30 cm?
11. A toy car (m = 0.04 kg) is released from rest and slides down a frictionless track 1 m high. At the bottom of the track it slides along a horizontal portion until it hits a spring (k = 18 N/m). The spring is attached to an immovable object. What is the maximum compression of the spring?

12. A roller coaster is poised at the top of a hill 50 m high.
    a) How fast would it be going when it went over the top of the next hill on the track, which is only 30 m high?
    b) From a practical point of view, why is it advantageous that this ride is mass independent?
13. A 40-g bullet leaves the muzzle of a gun at a speed of 300 m/s.
    a) What is the kinetic energy of the bullet as it leaves the barrel?
    b) If the barrel of the gun is 12 cm long, what is the force acting on the bullet while it is in the barrel?
14. A super-ball (m = 30 g) is dropped from a height of 3 m.
    a) At what vertical position is $PE_{grave} = KE$?
    b) What is the speed of the ball at this position?
15. A water-balloon (m = 300 g) is launched horizontally from a platform 2 m above the ground with a slingshot. The slingshot (k = 60 N/m) is stretched 40 cm before launch. How far from the platform will the balloon strike the ground?

202

8. Use conservation of energy. The gravitational potential energy (GPE) at the top of the cliff is equal to the kinetic energy (KE) at the bottom.

$$\text{GPE} + \text{KE} = \text{GPE} + \text{KE}$$

$$mgh + 0 = 0 + \tfrac{1}{2}mv^2$$

$$v = \sqrt{2gh}$$

$$= \sqrt{2(9.8 \text{ m/s}^2)(100 \text{ m})}$$

$$= 44 \text{ m/s}$$

   b) Yes, the speed is independent of the mass. All objects fall at the same rate. In the energy equations, you can see that the mass "drops" out.

9. a) $W = 1/2\ kx^2 = 1/2\ (1500\ \text{N/m})((0.25\ \text{m})^2 = 47\ \text{J}$

   b) The work becomes spring potential energy (SPE). This SPE becomes kinetic energy (KE) of the arrow.

$$KE = 1/2\ mv^2$$

$$47 \text{ J} = 1/2\ (0.1 \text{ kg})v^2$$

$$v = 31 \text{ m/s}$$

10. a) $W = 1/2\ kx^2 = 1/2\ (315\ \text{N/m})(0.30\ \text{m})^2 = 5.7\ \text{J}$

    b) $F = kx = (315\ \text{N/m})(0.30\ \text{m}) = 95\ \text{N}$

11. The gravitational potential energy (GPE) of the car will become kinetic energy (KE) of the car. This KE will then do work on the spring and compress it. The compression of the spring will store this energy as spring potential energy (SPE).
    You can therefore compare the GPE to the SPE and not concern yourselves with the KE.

$$mgh = \frac{1}{2}kx^2$$

$$x = \sqrt{2gh/k}$$

$$= \sqrt{\frac{2(9.8 \text{ m/s})(100 \text{ cm})}{18 \text{ N/cm}}}$$

$$= 10 \text{ cm}$$

(Answers to 12 - 15 on next page)

PHYSICS AT WORK

Erv Hunt

Coach to the Champions

In 1998, Erv Hunt began his 25th season as the head coach of the University of California at Berkeley's track and field team. He has brought more than just a reputation as one of the foremost track and field coaches in the United States to his job. He has brought to his team the experience of having served as the US men's head track and field coach at the 1996 Olympic Games in Atlanta.

"We have a much better understanding of the science of running now, the physics of it, than we use to," states Erv, "and that is one of the reasons runners have improved over the years." Erv believes that it's important when training to understand pace work and intervals during the different stages of a race. He begins each training season with his team in a slow and relaxed manner. "I want the athletes to know what it feels like to run correctly without tensing up," he says, "then they can slowly pick up the pace. I work on breaking down their running techniques, creating strategies, at different intervals and seeing what works with a sense of control. Too often runners begin to lose control mentally and physically when they're running fast. I want them to have the same sense of control at fast paces that they have during the slower paces."

To be around and train some of the great athletes during his career has been an incredible experience for Erv. "For someone to reach that level of excellence makes them somewhat a different breed of person," and Erv has had the privilege to work with some of the best.

12. a) Since energy is conserved, you can compare the GPE and the KE at all locations.

$$GPE + KE = GPE + KE$$
$$mgh + 0 = GPE + 1/2\ mv^2$$
$$1/2\ mv^2 = mg(50\text{ m} - 30\text{ m})$$
$$v = \sqrt{2g\Delta h}$$
$$v = \sqrt{2(9.8\text{ m/s}^2)(30\text{ m})}$$
$$v = 24\text{ m/s}$$

b) The mass of the roller coaster varies because the number of passengers may change.

13. a) $KE = 1/2\ mv^2$
$KE = 1/2\ (0.04\text{ kg})(300\text{ m/s})^2$
$KE = 1800\text{ J}$

b) The KE of the bullet is equal to the work done on the bullet by the gun.
$W = Fx$
$F = W/x = (1800\text{ J})/0.12\text{ m})$
$= 15{,}000\text{ N}$

14. a) The GPE and the KE will be equal when the ball has 1/2 the original GPE. This will be when it is at a position 1/2 the original height.

b) The KE at this position can be found by the change in GPE.

$$1/2\ mv^2 = mg(3.0\text{ m} - 1.5\text{ m})$$
$$v = \sqrt{2g\Delta h}$$
$$v = 2(9.8\text{ m/s}^2)(1.5\text{ m})$$
$$v = 5.4\text{ m/s}$$

15. To find where the balloon lands, you must know the horizontal speed of the balloon and the time that it will be in the air: $x = vxt$
The speed can be found from energy considerations. The SPE becomes KE of the balloon.The time can be found by recognizing that the balloon will be in the air the same time that a balloon dropped from 2 m will be in the air.

Speed: $1/2\ kx^2 = 1/2\ mv^2$

$v = \sqrt{kx^2/m}$
$v = \sqrt{(60\text{ N/m})(0.4\text{ m})^2/(0.300\text{ kg})}$
$v = 5.7\text{ m/s}$

Time: $d = 1/2\ at^2$
$t = 0.64\text{ s}$

Horizontal distance:
$x = v_x t$
$x = (5.7\text{ m/s})(0.64\text{ s})$
$x = 3.6\text{ m}$

Chapter 3 Assessment

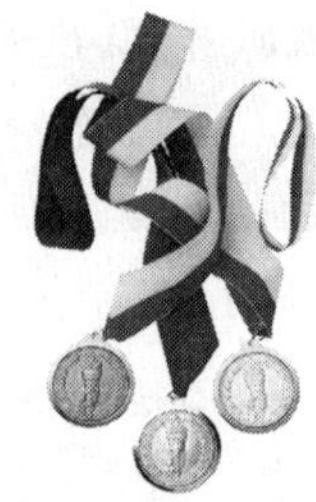

In this chapter you studied physics as it relates to track and field events. Now you have the knowledge to write a physics manual about track and field training for your high school team to help improve its performance. The manual should:

- **help students compare themselves with the competition**
- **include a description of physics principles as they relate to track events**
- **provide specific techniques to improve performance**

Review and remind yourself of the grading criteria that was agreed on by the class at the beginning of the chapter. You may have decided that some or all of the following qualities should be graded in your track and field manual:

- **physics principles**
- **inclusion of charts**
- **past records**
- **relevant equations**
- **definitions**
- **specific techniques**

Any advice you give should be understandable to athletes who have not studied physics. You can describe any activities you have done to explain how you know that the technique works, but you should not tell so much about each activity that the reader becomes bored.

Physics You Learned

Distance, Time, Speed

$$\text{Speed} = \frac{\text{Distance traveled}}{\text{Time taken}}$$

Speed vs. Time graphs

Histograms

Average speed, Instantaneous speed

Distance vs. Time graphs

Acceleration

Accelerometers

Speed = Frequency × Stride Length

Trajectory motion

Horizontal and Vertical motion

Optimum angle for trajectories

Potential energy

Kinetic energy

Olympic Pole Vault: Actual Heights and Heights Predicted From Average Speed in Men's 100-meter Dash, 1896 to 1992

Year	Time	Speed for 100 m race	Predicted Height (m)	Predicted Height (ft)	Actual Height (ft)	Actual Height (m)
1896	12	8.33	3.54	11.69	10.83	3.28
1900	11	9.09	4.22	13.91	10.83	3.28
1904	12	9.09	4.22	13.91	11.48	3.48
1908	10.8	9.26	4.37	14.43	12.17	3.69
1912	10.8	9.26	4.37	14.43	12.96	3.93
1920	10.8	9.26	4.37	14.43	13.42	4.07
1924	10.6	9.43	4.54	14.98	12.96	3.93
1928	10.8	9.26	4.37	14.43	13.77	4.17
1932	10.3	9.71	4.81	15.87	14.15	4.29
1936	10.3	9.71	4.81	15.87	14.27	4.32
1948	10.3	9.71	4.81	15.87	14.10	4.27
1952	10.4	9.62	4.72	15.57	14.92	4.52
1956	10.5	9.52	4.63	15.27	14.96	4.53
1960	10.2	9.80	4.90	16.18	15.42	4.67
1964	10	10.00	5.10	16.84	16.73	5.07
1968	9.95	10.05	5.15	17.01	17.71	5.37
1972	10.14	9.86	4.96	16.38	18.04	5.47
1976	10.06	9.94	5.04	16.64	18.04	5.47
1980	10.25	9.76	4.86	16.03	18.96	5.74
1984	9.99	10.01	5.11	16.87	18.85	5.71
1988	9.92	10.08	5.18	17.71	19.77	5.99
1992	9.96	10.04	5.14	16.97	19.02	5.76

For use with *The Track and Field Championship*, Chapter 3, Activity 10: Energy in the Pole Vault

Alternative Chapter Assessment

1. A sprinter runs the 100-m race in 9.80 s. Calculate her average speed for the race.

2. At the 50-m mark in the above race, do you think that the elapsed time was less than, equal to, or more than 4.90 s? Explain your answer.

3. Who wins the race?

   a) the runner with the highest finishing speed

   b) the runner with the best average speed

   c) the runner with the greatest top speed

4. The following table gives the split times for a 100-m race run by Leroy Burrell in the 1991 World Championships:

Distance(m)	Time(s)
10	1.83
20	2.89
30	3.79
40	4.68
50	5.55
60	6.41
70	7.28
80	8.12
90	9.01
100	9.88

   a) Fill in the table below by calculating the time it takes Leroy Burrell to run each interval and his average speed for each interval.

Interval (m)	Time(s)	Speed(m/s)
0-10	________	________
10-20	________	________
20-30	________	________
30-40	________	________
40-50	________	________

50-60	________	________
60-70	________	________
70-80	________	________
80-90	________	________
90-100	________	________

b) At what distance did Leroy stop accelerating?

c) What is Leroy's average speed for the entire race?

5. A runner is holding a cork accelerometer. Draw the position of the accelerometer for each phase of the race.

   a) The runner is accelerating to the right.

   b) The runner is running at constant speed.

6. A runner has a stride length of 1.0 m. During the race, he takes 26 strides in 13 s. What is his average speed?

7. A person standing on a 50.0-m high cliff throws a ball out horizontally with an initial speed of 10 m/s. At the same instant, she drops her mitt. Which hits the ground first? Why?

8. What launching angle produces the greatest range if the launch and landing heights are the same?

9. At what angle with respect to the horizontal should the shot put be launched if the starting height is 2.0 m, to have the greatest range?

10. The following chart shows the times for three runners to run each of the five 20.0 m intervals in a 100-m race.

Runner	0-20.0 m	20.0-40.0 m	40.0-60.0 m	60.0-80.0 m	80.0-100 m
A	2.62 s	2.60 s	2.53 s	2.52 s	2.49 s
B	2.60 s	2.58 s	2.53 s	2.45 s	2.52 s
C	2.62 s	2.61 s	2.53 s	2.48 s	2.50 s

Answer the following questions based on this information:

a) Which runner had the fastest speed in any one time interval? How fast was that runner moving in that interval?

b) Which runner was running the fastest at the end of the race? How fast was that runner moving?

c) Which runner wins the race? Explain your answer using the concept of speed in your explanation.

11. Determine the maximum possible height that a pole vaulter could clear when a speed of 6.0 m/s is reached prior to planting the pole. How will the possible height change if the maximum speed is 12 m/s? Assume all the kinetic energy is converted to potential enery and that the vaulter does no work while rising.

12. Explain the effect on a runner's speed if attempting to double the stride length also cuts the frequency in half.

13. An accelerometer indicates that an acceleration to the left has occurred. Discuss and explain two possible motions that would result in this observation.

14. a) A ball is thrown into the air. Describe the speed and acceleration at the following places:

i) on its way up

ii) at the peak of its path

iii) while falling back down

b) How and where would the speed and acceleration change if a human were jumping into the air and coming back down?

Alternative Chapter Assessment Answers

1. $v = d/t = 100\ \text{m}/9.80\ \text{s} = 10.2\ \text{m/s}$

2. Since it takes 50 m or more to reach maximum speed when starting from rest, the elasped time will be more than the time at the end of the race.

3. b)

4. a)

Leroy Burrell		1991	
Distance(m)	Time(s)	Interval time	Speed(m/s)
10	1.83	1.83	5.46
20	2.89	1.06	9.43
30	-3.79	0.9	11.11
40	4.68	0.89	11.24
50	5.55	0.87	11.49
60	6.41	0.86	11.63
70	7.28	0.87	11.49
80	8.12	0.84	11.90
90	9.01	0.89	11.24
100	9.88	0.87	11.49

b) 60-m mark

c) v= 10.1 m/s

5. Students draw accelerometer with cork

a) leaning to the left

b) straight up and down

6. v = 2.0 m/s

7. They hit at the same time because the horizontal motion is independent of the vertical motion.

8. 45°.

Chapter 3

9. 41°.

10.a) runner B speed = 8.16 m/s during 60.0-80.0 interval

b) runner A speed = 8.03 m/s during last interval

c) runner B

The total time for B is 12.68 s. Average speed for B = 100 m/12.68 s = 7.89 m/s.

The total time for C is 12.74 s. Average Speed = 7.85 m/s.

The total time for A is 12.76 s. Average Speed = 7.84 m/s.

The runner with the greatest average speed will win the race.

11. Using $1/2\ mv^2 = mgh$ and solving for h the speed of 6.0 m/s will result in h = 1.8 m, if v is doubled h will quadruple giving a new height of 7.2 m.

12. Doubling the frequency alone would double the speed, cutting the frequency in half would cut the speed in half, since both are changed, the overall speed will remain exactly the same as what it was.

13. An acceleration to the left could occur under the following conditions:

object is moving to the left and speeding up

object is moving to the right and slowing down

14. a) Going up — speed is up and decreasing in magnitude, acceleration is constant and directed down.

At peak of path — speed is zero and acceleration is still constant and down.

Going down — speed is down and increasing in magnitude, acceleration is still constant and directed down.

b) A human once in the air behaves exactly like the ball in terms of velocity and acceleration.

Chapter 4

THRILLS AND CHILLS

Chapter 4–Thrills and Chills
National Science Education Standards

Chapter Challenge

A roller coaster company requires help from a student population to modify a designed roller coaster. The students must first decide for whom the new roller coaster will be designed. The roller coaster design will include hills, curves, and vertical loops specifically adapted to the chosen population of riders. Students will be required to analyze their roller coaster in terms of both thrills and safety. This will require an analysis of energy considerations as well as forces and accelerations.

Chapter Summary

The roller coaster is an excellent application of conservation of mechanical energy where students can study work, gravitational potential energy, kinetic energy, and energy lost to heat. The thrills of roller coasters emerge from the accelerations. In creating a roller coaster, students will also be learning about forces, force diagrams, Newton's Second Law and circular motion. The experience of creating a roller coaster will engage students in the following content from the National Science Education Standards.

Content Standards

Unifying Concepts

- System, order and organization
- Evidence, models and explanations
- Constancy, change and measurement

Science as Inquiry

- Abilities necessary to do science inquiry
- Understanding about scientific inquiry

Science and Technology

- Abilities of technological design
- Understanding about science and technology

History and Nature of Science

- Science as a human endeavor
- Nature of scientific knowledge

Science in Personal and Social Perspectives

- Personal and community health
- Natural and human-induced hazards
- Science and technology in local, national, and global challenges

Physical Science

- Motion and forces
- Conservation of energy
- Interactions of energy and matter

Key Physics Concepts and Skills

Activity Summaries	Physics Principles
Activity 1: The Big Thrill Students create sketches of a roller coaster from two perspectives. They then experience the kinesthetic feel of a roller coaster while in a chair. To better describe roller coasters, they learn about velocity and acceleration including how to measure velocity with a stopwatch, ruler and photogates. Finally, they learn to calculate acceleration.	• Displacement • Velocity • Acceleration
Activity 2: What Goes Up and What Comes Down Through an inquiry activity, students investigate the relationship between the angle of an incline and the final speed of a steel ball descending that incline. From this activity and the follow-up analysis, students arrive at the principle of the conservation of mechanical energy.	• Gravitational potential energy • Kinetic energy • Conservation of energy
Activity 3: More Energy Students explore a pop-up toy by investigating the compression of a spring and the height that the toy rises. This extends the concept of the conservation of energy to include the potential energy of a spring. On the roller coaster, the energy is not supplied by a spring but by an electrical motor that lifts the coaster to its highest point.	• Spring potential energy • Conservation of energy
Activity 4: Your "At Rest" Weight Students first learn how to distinguish mass from weight. They then explore the behavior of springs and Hooke's Law with a graphical analysis of the effect of force on the stretch of a spring. From an understanding of Hooke's Law, students can explain how a spring scale measures weight.	• Mass • Weight • Hooke's Law • Springs as scales to measure weight
Activity 5: Weight on a Roller Coaster Students analyze how weight changes in an ascending and descending elevator. The application of Newton's First and Second Laws of Motion can explain why the apparent change takes place as the elevator accelerates. The use of force vectors help to make sense of the elevator as a "cheap" roller coaster ride.	• Force vectors • Weight changes during acceleration • Newton's First Law of Motion • Newton's Second Law of Motion
Activity 6: On the Curves Students investigate circular motion. The analysis of centripetal forces and accelerations are applied to roller coaster design as students calculate the required centripetal forces for horizontal and vertical curves on the coaster. Students first become aware of some of the safety features required in roller coasters.	• Circular motion • Centripetal acceleration • Centripetal forces • Normal forces
Activity 7: Getting Work Done To get the roller coaster started, work must be done to provide the coaster with gravitational potential energy. Students investigate the relationship between work, force, and displacement as they drag carts up different inclines. The work-energy theorem is then stated as a way to tie together this activity with the concept of the conservation of mechanical energy in earlier activities. Students also calculate the power required for a roller coaster as it ascends a hill.	• Conservation of energy • Work • Power
Activity 8: Vectors and Scalars Students view a roller coaster as both an energy ride and as a force ride. The thrills of the roller coaster come from the accelerations. Energy as a scalar quantity cannot adequately describe the accelerations and students are introduced to vector addition to better appreciate how forces contribute to roller coaster fun.	• Scalars • Vectors • Vector addition • Energy as a scalar • Force as a vector
Activity 9: Safety is Required but Thrills are Desired Students investigate what may happen to passengers and the roller coaster if accelerations are too large. This leads to explorations of safety aspects of the roller coaster design including forces on the wheels, on the tracks and the need for strong materials. Students also discuss psychological means of making the roller coaster appear less safe and more fun.	• Circular motion • Centripetal acceleration • "g" forces

GETTING STARTED WITH EQUIPMENT NEEDED TO CONDUCT THE ACTIVITIES.

Items needed—not supplied in Material Kits

Preparing the equipment needed for each activity in this chapter is an important procedure. There are some items, however, needed for the chapter that are not supplied in the It's About Time material kitpackage. Many of these items may already be in your school and would be an unnecessary expense to duplicate. Please read carefully the list of items to the right which are not found in the supplied kits and locate them before beginning activities.

Items needed—not supplied by It's About Time:

- **Blindfold**
- **Chair with Wheels**
- **Some Nickels**
- **Piece of Paper**
- **Pennies**

Chapter 4

Equipment List For Chapter 4

PART	ITEM	QTY	ACTIVITY	TO SERVE
SS-7722	Ball Of String, 225 ft.	1	2, 6	Group
BS-1608	Batteries, AA	2	6	Group
BA-0004	Batteries, AAA	2	1, 2, 3	Group
HO-0001	Cast Iron Right Angle Holder	1	1, 2, 4, 7	Group
CS-3246	C-Clamp for Photogate	2	1, 2, 3	Group
DC-0601	Dynamics Cart	1	7	Group
GS-5016	Goggles, Safety	1	6	Group
GS-7706	Graph Paper Pad of 50 Sheets	1	2, 4	Group
SH-7212	Large Ringstand	1	1, 2, 4, 7	Group
MH-2621	Mass Hanger	1	4, 5	Group
PM-0220	Plumb Bob W/String	1	2	Group
PT-0001	"Pop-up" Toy	1	3	Group
NS-1210	Post-Its®	1	8	Group
PM-0521	Protractor	1	2, 7	Group
RS-6140	Rubber Stopper w/Hole	1	6	Group
WH-9012	Slotted Weight Set	1	4, 5	Group
SS-2303	Spring Scale, 0-10 Newton Range	1	5, 7	Group
HO-0001	Springs for Hooke's Law-Set Of 5	1	4	Group
BA-0003	Steel Ball	1	1, 2	Group
ST-0002	Steel Rod, 1/2" x 9" (To Act As Crossarm)	1	1, 2, 4, 7	Group
SM-1676	Stick, Meter, 100 cm, Hardwood	1	1, 2, 3, 4, 7	Group
SS-7778	Stopwatches	1	1	Group
TS-2662	Tape, Masking, 3/4" x 60 yds.	2	3, 4, 8	Group
CA-0100	Toy Car, Battery Operated	1	6	Group
TR-0001	Track For Roller Coaster Demonstration	1	1, 2, 7	Group
TR-0002	Track Support	1	1, 2, 7	Group
BE-0001	Velocimeter	1	1, 2, 3	Group
ITEMS NEEDED – NOT SUPPLIED BY IT'S ABOUT TIME				
	Blindfold	1	1	
	Chair with Wheels	1	1	
	Some Nickels	4	3	
	Piece of Paper	1	4	
	Pennies	100	8	

Organizer for Materials Available in Teacher's Edition

Activity in Student Text	Additional Material	Alternative / Optional Activities
ACTIVITY 1: The Big Thrill, p. 208	Terminator Express, p. 414	
ACTIVITY 2: What Goes Up and What Comes Down, p. 218	Data Tables, p. 431 Gravitational Potential Energy and Kinetic Energy Tables, p. 432	
ACTIVITY 3: More Energy, p. 231		
ACTIVITY 4: Your "At Rest" Weight, p. 239	Stretch of a Spring Graph, p. 459	
ACTIVITY 5: Weight on a Roller Coaster, p. 253	Summary of Changes to Bathroom-Scale Readings on an Elevator, p. 476	
ACTIVITY 6: On the Curves, p. 266	Diagram of Roller Coaster Car in Different Positions in a Loop, p. 500 Forces on a Roller Coaster Car on a Level Track, at the Bottom of a Loop, and at the Top of a Loop, p. 501	
ACTIVITY 7: Getting Work Done, p. 286		
ACTIVITY 8: Vectors and Scalars, p. 297	Concepts For Roller Coaster Chapter, p. 529 Sample Concept Map 1, p. 530 Sample Concept Map 2, p. 531	
ACTIVITY 9: Safety is Required but Thrills are Desired, p. 309		

Meets the standard of excellence. **5**	AN EXCELLENT ROLLER COASTER DESIGN STUDY CONTAINS ALL OF THE FOLLOWING REQUIREMENTS:
	● New design of roller coaster
	  • Model and sketch both depict the same roller coaster
	    - model clearly depicts the entire ride and includes features like forces, velocities, accelerations, and thrill positions
	    - sketch shows both a top view and a side view
	  • Creative design
	  • Design is suited to the population of your choice
	  • Name of roller coaster is "catchy"
	  • The roller coaster has all the required elements
	    - 2 hills
	    - one horizontal curve
	  • Appropriate changes of heights, radius of hill bottoms, hill tops, and/or curves
	  • New elements added and/or the order of the elements of the ride changed
	  • Technology—rejected designs and why they were rejected
	  • Explanation of why design is an improvement
	  • Additional thrills of your ride noted
	    - sounds and scenery to enhance the thrills of a roller coaster ride
	● Energy considerations are taken into account
	  • The GPE and KE are given at 5 locations
	    - KE and velocity are shown to be equal at equal elevations
	    - GPE = mgh
	    - KE = $1/2mv^2$
	  • The work required by the motor to lift the roller coaster to the top of the first hill is calculated.
	    - $W = F \cdot d$
	    - $P = W/t$
	  • Mention is made about work done by air resistance and friction and its impact on roller coaster motion even though your analysis will be for a frictionless ride
	  • Conservation of energy is described as it relates to roller coasters
	  • Conservation of energy expands to other systems (springs, heat, light)
	    - SPE = $1/2kx^2$
	  • Energy as a scalar
	● Force considerations are taken into account
	  • At four crucial locations forces are depicted and described
	    - normal force
	    - gravitational force (weight)
	      changes in weight
	    - How a spring scale works
	    - Hooke's law $F = kx$
	  • At four locations, the centripetal force responsible for the circular motion is identified
	    - $F = mv^2/R$
	  • Force as a vector
	    - forces add as vectors
	● The roller coaster is safe
	  • Speeds are provided at five locations
	  • No acceleration is greater than 4 g's
	  • All vertical curves are possible (the roller coaster does not leave the track)
	  • Other safety features are explored and reported
	● Describe when energy or force considerations are more appropriate for analyzing aspects of roller coaster rides.

Approaches the standard of excellence.
 	
4
	A GOOD ROLLER COASTER DESIGN STUDY CONTAINS ALL OF THE FOLLOWING REQUIREMENTS:
	• New design of roller coaster
	  • Model and sketch both depict the same roller coaster
	    - model clearly depicts the entire ride and includes features like forces, velocities, accelerations, and thrill positions
	    - sketch shows both a top view and a side view
	  • Creative design
	  • Design is suited to the population of your choice
	  • Name of roller coaster is "catchy"
	  • The roller coaster has all the required elements
	    - 2 hills
	    - one horizontal curve
	  • Appropriate changes of heights, radius of hill bottoms, hill tops, and/or curves
	  • New elements added and/or changing the order of the elements of the ride
	  • Technology—rejected designs and why they were rejected
	  • Explanation of why your design is an improvement
	  • Additional thrills of your ride noted
	    sounds and scenery to enhance the thrills of a roller coaster ride
	  • Energy considerations are taken into account
	  • The GPE and KE are given at 3 locations
	    KE and velocity are shown to be equal at equal elevations
	    - $GPE = mgh$
	    - $KE = 1/2mv^2$
	  • The work required by the motor to lift the roller coaster to the top of the first hill is calculated.
	    - $W = F \cdot d$
	    - $P = W/t$
	  • Mention is made about work done by air resistance and friction and its impact on roller coaster motion even though your analysis will be for a frictionless ride
	  • Conservation of energy is described as it relates to roller coasters
	  • Conservation of energy expands to other systems (springs, heat, light)
	    - $SPE = 1/2kx^2$
	  • Energy as a scalar
	• Force considerations are taken into account
	  • At two crucial locations forces are depicted and described
	    - normal force
	    - gravitational force (weight)
	      changes in weight
	    - How a spring scale works
	    - Hooke's Law $F = kx$
	  • At two locations, the centripetal force responsible for the circular motion is identified
	    - $F = mv^2/R$
	  • Force as a vector
	    - forces add as vectors
	• The roller coaster is safe
	  • Speeds are provided at two locations
	  • No acceleration is greater than 4 g's
	  • All vertical curves are possible (the roller coaster does not leave the track)
	  • Other safety features are explored and reported
	• Describe when energy or force considerations are more appropriate for analyzing aspects of roller coaster rides.

Meets an acceptable standard. **3**	A SATISFACTORY ROLLER COASTER DESIGN STUDY CONTAINS ALL OF THE FOLLOWING REQUIREMENTS: ● New design of roller coaster • Model and sketch both depict the same roller coaster - model clearly depicts the entire ride and includes features like forces, velocities, accelerations, and thrill positions - sketch shows both a top view and a side view • Creative design • Design is suited to the population of your choice • Name of roller coaster is "catchy" • The roller coaster has all the required elements - 2 hills - one horizontal curve • Appropriate changes of heights, radius of hill bottoms, hill tops, and/or curves • New elements added and/or changing the order of the elements of the ride • Technology—rejected designs and why they were rejected • Explanation why your design is an improvement • Additional thrills of your ride noted sounds and scenery to enhance the thrills of a roller coaster ride ● Energy considerations are taken into account • The GPE and KE are given at 3 locations - KE and velocity are shown to be equal at equal elevations - GPE = mgh - KE = $1/2mv^2$ • The work required by the motor to lift the roller coaster to the top of the first hill is calculated. - $W = F \cdot d$ - $P = W/t$ • Mention is made about work done by air resistance and friction and its impact on roller coaster motion even though your analysis will be for a frictionless ride • Conservation of energy is described as it relates to roller coasters • Conservation of energy expands to other systems (springs, heat, light) - SPE = $1/2kx^2$ • Energy as a scalar ● Forces considerations are taken into account • At two crucial locations forces are depicted and described - normal force - gravitational force (weight) changes in weight - How a spring scale works - Hooke's Law $F = kx$ • At two locations, the centripetal force responsible for the circular motion is identified - $F = mv^2/R$ • Force as a vector - forces add as vectors ● The roller coaster is safe • Speeds are provided at two locations • No acceleration is greater than 4 g's • All vertical curves are possible (the roller coaster does not leave the track) • Other safety features are explored and reported • Describe when energy or force considerations are more appropriate for analyzing aspects of roller coaster rides.

Below acceptable standard and requires remedial help.
 	
2
	ROLLER COASTER DESIGN STUDY CONTAINS THE FOLLOWING REQUIREMENTS:
	• New design of roller coaster
	  • Model and sketch both depict the same roller coaster
	    - model clearly depicts the entire ride and includes features like forces, velocities, accelerations, and thrill positions
	    - sketch shows both a top view and a side view
	  • Creative design
	  • Design is suited to the population of your choice
	  • Name of roller coaster is "catchy"
	  • The roller coaster has all the required elements
	    - 2 hills
	    - one horizontal curve
	  • Appropriate changes of heights, radius of hill bottoms, hill tops, and/or curves
	  • New elements added and/or changing the order of the elements of the ride
	  • Technology—rejected designs and why they were rejected
	  • Explain why your design is an improvement
	  • Additional thrills of your ride noted
	    - sounds and scenery to enhance the thrills of a roller coaster ride
	• Energy considerations are taken into account
	  • The GPE and KE are given at only one location
	    - KE and velocity are shown to be equal at equal elevations
	  • The work required by the motor to lift the roller coaster to the top of the first hill is mentioned
	  • Conservation of energy is described as it relates to roller coasters
	• Force considerations are taken into account
	  • At two crucial locations forces are depicted and described
	    - normal force
	    - gravitational force (weight)
	• The roller coaster is safe
	  • No acceleration is greater than 4 g's

Basic level that requires remedial help or demonstrates a lack of effort.
 	
1
	A ROLLER COASTER DESIGN STUDY CONTAINS THE FOLLOWING REQUIREMENTS:
	• New design of roller coaster
	  • Model and sketch both depict the same roller coaster
	    - model clearly depicts the entire ride and includes features like forces, velocities, accelerations, and thrill positions
	  • Design is suited to the population of your choice
	  • The roller coaster has some of the required elements
	    - 2 hills
	    - one horizontal curve
	  • Explanation why your design is an improvement
	• Energy considerations are taken into account
	  • The GPE and KE are given at only one location
	  • The work required by the motor to lift the roller coaster to the top of the first hill is mentioned
	  • Conservation of energy is described as it relates to roller coasters
	• Force considerations are taken into account
	  • At one location forces are depicted and described
	    - normal force
	    - gravitational force (weight)
	• The roller coaster is safe
	  • No acceleration is greater than 4 g's

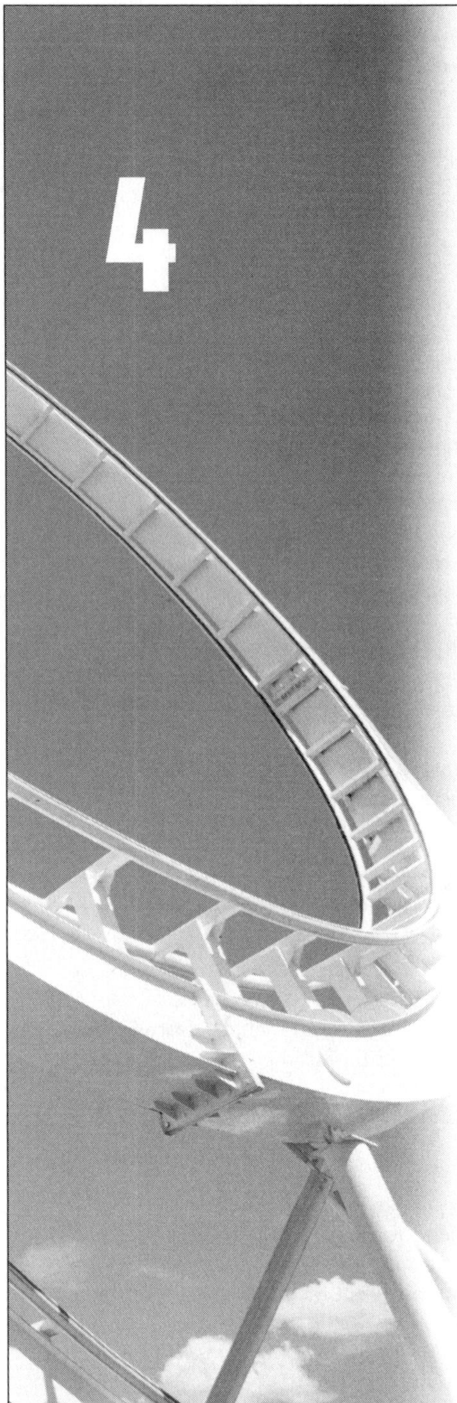

4

Thrills and Chills

Scenario

You are excited and scared as you sit back into the seat. You pull the safety restraints into place. The next thing you know, you are beginning a slow but steady ascent into the sky. Then, with a sudden jolt, you reach the top. This is where the thrill or nightmare begins. You hurtle down the track at ever-increasing speeds. You are flung against one side of your seat as you scream around a curve. You shriek as you hang upside down, fortunately, firmly secured to your seat. All the time, your stomach has no idea where it is or you are. Finally, you come to rest where you began. What a ride! Want to go again?

Roller coasters have been enjoyed for many years. However, the roller coaster that may appeal to you, may not appeal to your parents or other friends and relatives.

206

Chapter and Challenge Overview

Thrills and Chills introduces students to the design of a roller coaster ride. Whether you are a roller coaster fanatic or someone who anticipates a roller coaster ride as much as a trip to the dentist, there is a thrill in watching the riders shriek as the coaster whips around turns and riders flip upside down.

The roller coaster ride is designed to produce the most thrills with the least danger. In the **Chapter Challenge**, students are asked to improve upon a given roller coaster design. In keeping with the *Active Pysical Science* philosophy of ensuring that all students help define the content and tasks in their education, we ask students to think about for whom they will design the ride. Some students will immediately want to build the roller coasters for daredevils. Others, however, may choose to design a roller coaster for children. Others may choose to provide a ride for physically challenged individuals. Another group may work on the roller coaster for senior citizens.

The roller coaster has been used quite extensively to illustrate the conservation of mechanical energy. The thrill of the roller coaster has little to do with energy conservation and everything to do with forces and accelerations. In this chapter, students view the roller coaster as both an energy-transformation device and as a force and acceleration ride. When they are asked to create new roller coasters, the demands of safety provide a constraint on their imagination and help them to learn more physics than they would otherwise.

Designing a roller coaster begins with being able to draw a schematic of the coaster. Since the coaster is a three-dimensional ride, the students learn how to create pairs of drawings to show both the hills and the horizontal turns. They will then learn about gravitational potential energy and kinetic energy through inquiry exercises. After a synthesis of the concepts of work, energy, and energy conservation, students then turn their attention to forces and vectors. Finally, students look specifically at the required safety aspects of all roller coaster rides.

Sections of the chapter on forces and acceleration will provide a useful review of these concepts if they have been introduced previously. Having the students learn to look at the roller coaster from both an energy perspective and a force perspective is similar to the way in which a physicist would approach the design. There are times when it is preferable to apply knowledge about forces to understanding a problem. There are other times when energy conservation is a more straightforward approach.

Challenge

A roller coaster, called the Terminator Express, has been designed for an amusement park. Your challenge is to take the roller coaster design and modify it for a select group of riders. For instance, you may decide that you will modify the roller coaster so that young children can experience the thrill of a roller coaster in a safe and non-threatening way. You may prefer to design the roller coaster for adults that are a bit squeamish about the big hills and sharp turns. They want to experience the thrill of a roller coaster ride, but are ready to pass on the death-defying action. On the other hand, you may choose to design a roller coaster for daredevils that are ready to handle any thrill you can provide. You may also wish to design a roller coaster for people who are physically challenged or visually impaired. However, in each situation, you must ensure that the roller coaster is safe and nobody will be in danger.

Criteria

You will make the design based on whatever group you choose. Your design will have to meet certain criteria. All designs must have certain components including:

- **at least two hills**
- **one horizontal curve**

In addition, you will have to provide evidence that the ride is safe. Safety data will include the height, speed, and acceleration of the roller coaster at five designated locations. Finally, you will have to calculate the work required to get the roller coaster rolling.

You will present your design as both a model and a written poster or report.

Make a list of criteria for each of the required elements of the roller coaster. What should a roller coaster design that deserves an "A" look like? Are there other elements that you think should be included in a roller coaster grade? How much should each element be worth? Discuss these questions with your small group and the class.

Active Physical Science

ACTIVITY 1
The Big Thrill

Background information

It is suggested that the **Physics Talk** section of the student text for **Activity 1** be read before proceeding in this section.

Velocity is defined as the change of displacement with respect to time. This definition can be written as an equation: $v=\Delta d/\Delta t$. If the units for displacement are meters and the units for time are seconds, then the units for velocity are m/s (meters per second). Velocity is a vector quantity. It has both a magnitude and a direction.

Although we recognize speed and velocity as we move about in cars, trains, planes or walking, there are misconceptions that we are well aware of and we must listen intently to our students to catch these misconceptions and help them confront them. For example, students may confuse position with velocity. When one car passes another, students may think that at the instant one car passes another, their speeds are identical.

Acceleration is defined as a change in velocity with respect to time. This definition can be written as an equation: $a=\Delta v/\Delta t$. If the units for velocity are m/s and the units for time are seconds, then the units for acceleration are (m/s)/s (meters per second every second). It is often abbreviated as the equivalent m/s^2 (meters per second squared) but student understanding will be increased if the units are stated as m/s every second. This will continue to reinforce the definition of acceleration.

Acceleration is a vector quantity. It has both a magnitude and a direction.

Acceleration as a change of velocity with respect to time, while velocity is, in itself, a change of displacement with respect to time leads to confusion. It's very difficult for students to grasp a change of a change. In calculus, we recognize a change of a change as the second derivative of the function.

Planning for the Activity

Time Requirements

- Allow about one 40-minute period to complete this activity.

Materials Needed

For each group:

- batteries, AAA
- blindfold
- c-clamp for photogate
- cast iron right angle holder
- chair with wheels
- large ringstand
- steel ball
- steel rod, 1/2" x 9" (to act as crossarm)
- stick, meter, 100 cm, hardwood
- stopwatches
- track for roller coaster demonstration
- track support
- velocimeter

Advance Preparation and Setup

Ensure that the photogate timers are working properly.

Ensure that the students know how to start, stop, and reset the stopwatches.

Teaching Notes

The swivel chair exercise is best done as a class demonstration. This will ensure safety. It is worthwhile to give a few students a "ride" in the chair. Their reactions will all be different and will help convince those who didn't get a "ride" that it is indeed the accelerations that provide the thrill.

Use an overhead of the Terminator Express for the side view and top view to assist in classroom discussion. A Blackline Master of the views is provided at the end of this activity.

Show the students a sample calculation of velocity and acceleration.

Sample problems on the board can highlight who understands the calculations. Students can earn 3 points when solving a physics problem:

1 point for equation

1 point for correct substitution of values with units

1 point for correct answer with units

Activity Overview

Students create sketches of a roller coaster from two perspectives. They then experience the kinesthetic feel of a roller coaster while sitting in a chair. To better describe roller coasters, students then learn about velocity and acceleration including how to measure velocity in the lab with a stopwatch and ruler as well as with a photogate. Finally, they learn how to calculate acceleration.

Student Objectives

Students will:

- Be able to draw and interpret a top view and a side view of a roller coaster ride.
- Conclude that thrills in roller coaster rides come from accelerations and changes in accelerations.
- Define acceleration as a change in velocity with respect to time and recognize the units of acceleration.
- Be able to measure and calculate velocity and acceleration.

Answers for the Teacher Only

What Do You Think?

The **What Do You Think?** question will give students an opportunity to tell about their own experiences on roller coaster rides. If some of the students have not been on a roller coaster, you may wish to have them view a movie that has a roller coaster ride as a film segment.

Thrills and Chills

Activity 1 The Big Thrill

GOALS

In this activity you will:

- Be able to draw and interpret a top view and a side view of a roller coaster ride.
- Conclude that thrills in roller coaster rides come from accelerations and changes in accelerations.
- Define acceleration as a change in velocity with respect to time and to recognize the units of acceleration.
- Be able to measure and calculate velocity and acceleration.

What Do You Think?

The tallest wooden roller coaster has a height of about 66 m (218 ft.). The tallest steel roller coaster is 128 m (420 ft.) high. This is as tall as a 40-story high rise!

- **Which part of the roller coaster produces the loudest screams? Why?**

Record your ideas about these questions in your *Active Physics* log. Be prepared to discuss your response with your small group and the class.

Active Physics 208

For You To Do

Part A: Sketch of the Roller Coaster

1. Sketch a roller coaster with a first hill of 15 m that quickly descends to 6 m and then turns to the left in a big circle (radius of 10 m) and then descends back to the ground.

2. Compare your roller coaster design to those drawn by other people on your design team.
   a) Which sketch do you like the best? Provide three reasons why you prefer that sketch.

3. Create two sketches for the same roller coaster. The first sketch should be a side view. The second sketch should be a view from the sky.
   a) What are the advantages of having two sketches?

4. Below is the roller coaster that has been designed by the professional team that is asking for your help. It is called the Terminator Express. There are two views of the roller coaster. The first view is a side view. The second view is a view from the sky.
   a) Sketch the side and top view in your *Active Physics* log.

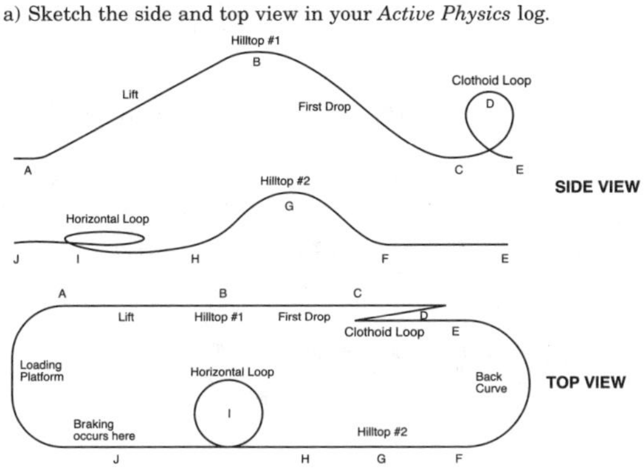

ANSWERS

For You To Do

Part A: Sketch of the Roller Coaster

1.–2. Some students will have difficulty drawing a three-dimensional diagram of a roller coaster ride. Let the students try their best at this. Their opportunity to invent ways in which to draw this and the struggle to do this adequately will set the stage for their appreciation and understanding of the two perspective drawings that will follow.

3. a) Once again, we are not interested in a "correct" answer as much as we are in providing an opportunity for students to give it their best attempt. Eliminating this step will make the interpretation of the two diagrams that are provided more difficult.

4. a) Having the students actually sketch the diagrams in the text is preferable to giving them a copy of the diagram at this time.

ANSWERS

For You To Do
(continued)

5. Ensure that lab partners reverse rolls so that each has a chance to move a finger around the roller coaster. Encourage the students to participate.

Part B: Roller Coaster Fun

1. Although observing a video of people in a roller coaster is a "safe" way of getting a sense of the roller coaster ride, the kinesthetic sense of being shifted about is more memorable. You should have a blindfolded student sit in a swivel chair with wheels. Have the student wear a bicycle helmet and hold onto the arms of the chair. Move the chair at a constant speed and then quickly change directions. Play with sudden stops, quick turns, large increases in speed. Have the other students record the rider's reactions. Maintain the safety of the rider at all times. Your students will find that the rider will smile, laugh, or make sounds during the accelerations.

2. a) Any change in direction or speed will probably produce a reaction, regardless of the direction of the push.

3. a) The velocity does not produce a large reaction by the rider. It is the changes in the velocity. These changes could have been increases or decreases in the speed or a change in direction of the speed or a combination of the two simultaneously.

5. The roller coaster car begins from the loading platform and then rises along the lift. It arrives at the top of hilltop #1 and then makes its first drop. It then goes into a vertical loop. This clothoid loop (it has a big radius at the bottom and a small radius at the top) allows the riders to be safely upside down. The car then goes along the back curve, rises over hilltop #2 and then swings into a horizontal loop. The brakes are applied and the roller coaster comes to a stop.

   Have one team member read this description as you move your finger along the roller coaster track

   Repeat the procedure with the top view.

Part B: Roller Coaster Fun

The rider must be wearing a bicycle helmet.

1. You will now blindfold someone in your group in order to observe the thrilling parts of a roller coaster ride. The blindfolded person will sit in a chair with wheels.
   a) Write down the safety concerns when one of your team members is blindfolded and you will be pushing him or her. What could go wrong? How can you prevent this?

2. Push on the chair of the blindfolded team member. While the person is moving give the blindfolded team member another push. Continue the pushing with pushes from different directions.
   a) Notice when the blindfolded team member smiles or laughs or exhibits some emotion. Is it when the push comes from the back or the front or the side?

3. A person's velocity is a measure of the speed and direction of the person. The person's velocity may have been 1.2 m/s north or 1.5 m/s toward the door or 1.0 m/s toward the window. In each case there is a magnitude (1.2 m/s, 1.5 m/s, 1.0 m/s) and a direction (north, toward the door, toward the window).
   a) Was the velocity responsible for the "rider's" reactions? Did the blindfolded person react more when they were going faster as they were gently pushed or pulled or when they were moving in a certain direction?

4. The change in a person's velocity over time is referred to as acceleration. Suppose a person were moving at 1.5 m/s toward the door and someone pushed him or her and made the person move at 1.3 m/s toward the window. There is acceleration because there was a change in speed and a change in direction.

   Acceleration also occurs if the person changed velocity from 1.1 m/s north to 1.4 m/s north. Here the acceleration is due to a change in speed.

   There would also have been acceleration if the person changed velocity from 1.3 m/s east to 1.3 m/s south. Here the acceleration is due to a change in the direction, with no change in speed.

   a) Was acceleration responsible for the reactions of the blindfolded person? Did they react more when they accelerated?

5. Acceleration is a change in velocity in a specific time. The change from 1.1 m/s north to 1.5 m/s north may have taken 1 s.

   The change in velocity is 0.4 m/s in one second. There are a number of ways in which this can be stated:

   0.4 m/s(meters per second) in one second

   0.4 m/s every second

   0.4 m/s per second

   0.4 (m/s)/s

   0.4 m/s^2

Answers

For You To Do
(continued)

4. Students are provided with the definition of acceleration as a change in velocity with respect to time. Accelerations occur with a change in speed or a change in direction or a combination of the two.
   a) Students can now identify their response to **Step 3** in a more precise way. It is acceleration that causes the thrill of the roller coaster.
5. Irrespective of what form for the units of acceleration are used, physicists think of acceleration as that change in velocity with respect to time.

Chapter 4

ANSWERS

For You To Do
(continued)

Part C: Measuring Velocities and Calculating Accelerations

1. a) You would need something to measure distance and something to measure time. Distance measurement instruments could be rulers, meter sticks or tape measures. A shoe could also be measurement instrument. Time could be measured with a clock, a stopwatch, or your heartbeat. (Heartbeats are not particularly accurate. You can explain that this is because the exciting experiments in physics make your heart race faster.)

2. a) Students should make the measurement and use the format for calculations that include the equation, substitution into the equation with units, and calculation of the result with units. For instance, if the ball has moved 25 cm in 4.0 s, a student would write the calculation as follows:

$$v = \frac{\Delta d}{\Delta t}$$

$$v = \frac{25 \text{ cm}}{4.0 \text{ s}}$$

$$v = 6.3 \text{ cm/s}$$

3. a) Students should repeat the calculation. Insist the students write the equation, and show the calculation with the units.

4. The photogate is an accurate clock. It can measure the time. If the length of the object starting and stopping the clock is used as input, then the photogate timer can give a value in terms of velocity rather than time. The photogate measures time, but it can calculate the speed in the same way that the students did.

   a) You have to know the length of the object.

5. a) The speed of the large steel ball is 3 cm/s.

6. a) The photogate timer will calculate the speed. Students could also calculate the speed by writing the equation, substituting the measured values of distance and time into the equation, and then displaying their answer with units.

Thrills and Chills

Be prepared to stop the ball at the end of the track so it does not roll onto the floor. If the ball rolls onto the floor pick it up right away. If you cannot find the ball, tell your teacher right away.

Part C: Measuring Velocities and Calculating Accelerations

1. The value 1.5 m/s north is a velocity. The velocity 1.5 m/s tells you that the object can travel 1.5 m in 1 s.

   a) If an object were moving across the table, what instruments would you need for measurements to determine if the object were traveling at 1.5 m/s?

2. Place a track flat on the top of your table. You will set a steel ball moving along the track.

   a) Calculate and record the velocity of the object using a ruler and a stopwatch. The equation for calculating velocity is

   $$v = \frac{\Delta d}{\Delta t}$$

   where v is the velocity,

   d is the displacement, and

   t is the time.

   The symbol Δ (delta) signifies "change in."

   (Remember: a velocity must have a direction.)

3. Decrease the speed of the steel ball.

   a) Calculate and record its velocity again.

4. Your teacher will demonstrate the use of a photogate timer. The timer starts when an object breaks the beam. The timer stops when the beam is no longer broken. The elapsed time can be measured very accurately.

   a) To determine the velocity of the steel ball, what additional information would you (or the computer) have to know?

5. A large steel ball with a diameter of 6 cm passes through a photogate. The elapsed time recorded on the photogate timer is 2 s.

   a) Calculate the speed of the ball. (Since the speed is requested, you do not have to worry about the direction of motion. Speed is a scalar—it has no direction. Velocity is a vector—it has direction.)

6. Use the photogate timer to help you find the speed of the steel ball traveling down the track.

212

Active Physics

---

a) Record the speed of the steel ball traveling down the track.

7. Create a slope for a steel ball to travel down. Have a ball travel down the slope.

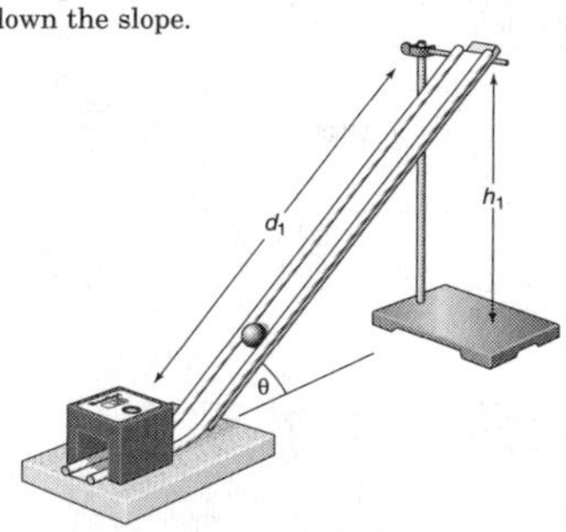

a) Measure and record the speeds at two different points.

b) Calculate the acceleration of the ball from the two speed measurements. (You will also have to measure the time between velocity measurements.) Acceleration is the change in velocity with respect to time. The equation to calculate acceleration is

$$a = \frac{\Delta v}{\Delta t}$$

where a is the acceleration,

v is the velocity, and

t is the time.

The symbol Δ (delta) signifies "change in," and can be calculated by subtraction.

Part D: Acceleration on the Roller Coaster—Pulling g's

1. On a roller coaster, you often feel heavier or lighter as you whip around curves. The accelerations take a toll on your body. This is often called "pulling g's." The Terminator Express has a number of places where you will be "pulling g's." Try to imagine a ride on the roller coaster shown.

ANSWERS

For You To Do
(continued)

7. a) The speed can be measured near the top of the ramp and near the bottom of the ramp. The increase in speed will be visible to the students.

b) The calculation of acceleration requires the students to know the speed at each location AND the time between the measurements (i.e., the time for the cart to move down the incline).

Part D: Acceleration on the Roller Coaster—Pulling g's

1. a) The students should be able to recognize where accelerations (changes in speed or changes of direction) are taking place in the depicted roller coaster. There are changes in speed from: A to B, B to C, C to D to E, G to H and during braking. There are changes in direction from: C to D to E, E to F, during I.

Chapter 4

ANSWERS

Physics Talk

The distinction between distance and displacement and between speed and acceleration becomes particularly important when we make calculations. Provide practice by having student groups come up with an example of a distance and a displacement, a speed and a velocity, as well as accelerations. Have one group listen to the example and respond with the correct terminology.

a) Copy the drawing below into your *Active Physics* log. Indicate where the passengers might feel light, and where they might feel heavy. Also indicate on the diagram where you think the coaster is speeding up and where it is slowing down.

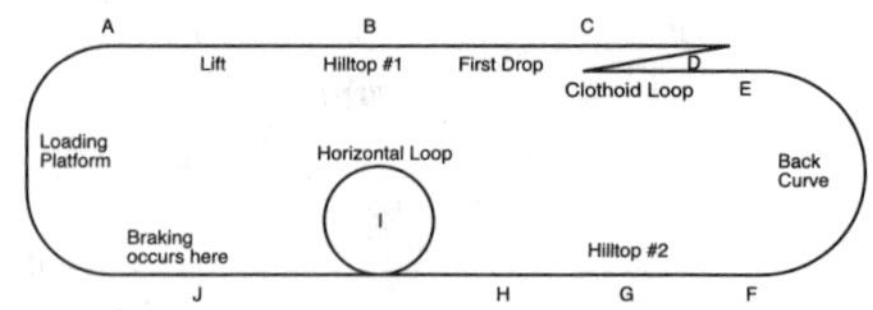

TOP VIEW

Physics Words

scalar: a quantity that has magnitude, but no direction.

vector: a quantity that has both magnitude and direction.

PHYSICS TALK

Measuring Velocity and Acceleration

In this activity you were introduced to some terms that you would need to understand in order to redesign the Terminator Express.

Distance is measured with a piece of string or a tape measure along a path. The unit used is usually a meter. For example, an object traveled 3 m. Distance is a **scalar** quantity. It has no direction.

Displacement is measured with a meter stick with direction included. It depends only on the endpoints, not on the path. An object traveled 3.5 m east. Displacement is a **vector.** It has magnitude (3.5 m) and direction (east).

In the diagram to the right, the curve represents the path of an object. The straight line represents the displacement.

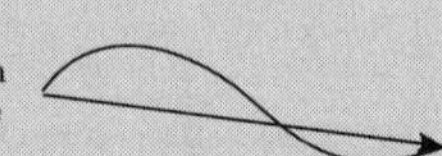

Time is measured with a stopwatch or other type of watch or clock.

Activity 1 The Big Thrill

Speed is the change in distance per unit time. The object's speed may be 4 m/s. This means that the object moves 4 m every second. Speed is a scalar. It has no direction.

Velocity is the change in displacement per unit time. The object's velocity may be 4 m/s south. Velocity is a vector. It has magnitude (4 m/s) and direction (south).

The equation to calculate velocity is

$$v = \frac{\Delta d}{\Delta t}$$

where v is the velocity,

d is the displacement, and

t is the time.

The symbol Δ (delta) signifies "change in."

For a person walking one lap around a city block, the distance is equal to the perimeter of the city block. The speed is equal to this distance divided by the time to complete the walk. The displacement equals 0 (since the person ended up where she started) and the velocity equals 0 as well.

Acceleration is the change in velocity per unit time. An object's acceleration may be 5 m/s every second. This means that the object changes its speed by 5 m/s every second. The speed will increase from 0 m/s to 5 m/s to 10 m/s to 15 m/s with each change requiring one second. 5 m/s every second is also written as 5 m/s^2 (five meters per second squared).

The equation to calculate acceleration is

$$a = \frac{\Delta v}{\Delta t}$$

where a is the acceleration,

v is the velocity, and

t is the time.

The symbol Δ (delta) signifies "change in."

Physics Words

speed: the change in distance per unit time; speed is a scalar; it has no direction.

velocity: speed in a given direction; displacement divided by the time interval; velocity is a vector quantity; it has magnitude and direction.

acceleration: the change in velocity per unit time.

Chapter 4

ANSWERS

Physics To Go

1. Students should notice that this roller coaster does not descend back to the ground. They can suggest ways in which to make this happen with additional design features.

2. The biggest thrill will probably be in the clothoid loop. This is where the riders are experiencing an acceleration due to a change in speed with respect to time and an acceleration due to a change in direction. They are also upside down for a few moments.

3. a) City A has the greater speed. City B has moved a much smaller distance in the 24 hours.

   b) The speed is approximately 1700 km/hr (close to 1000 mi/hr).

$$v = \frac{\Delta d}{\Delta t} = \frac{40{,}000 \text{ km}}{24 \text{ hrs}} = 1700 \text{ km/hr}$$

   c) Thrills are produced by changes in speed.

Thrills and Chills

Reflecting on the Activity and the Challenge

Velocity and acceleration are a big part of roller coaster fun. Traveling at a big speed is not enough to give a big thrill. The thrills come from accelerating around the curves and along the straight segments. This acceleration changes your speed and your direction as you ride along the path of the coaster. Additional thrills come from changes in acceleration. More rapid changes require greater accelerations. In designing your variation of the Terminator Express, you will want to ensure that the speeds and accelerations are right for your riders. You are required to have hills and turns, but the loop may be too much for your riders.

Physics To Go

1. Draw a top view and a side view of a roller coaster with the following characteristics: The roller coaster car begins from the loading platform and then rises along the lift. It arrives at the top of hilltop #1 and makes its first drop. It then climbs hill #2 which is half the height of hill #1. The car then goes along the back curve, rises over hilltop #3, and swings into a horizontal loop. The coaster then comes out of the loop onto a level plane. The brakes are applied and the roller coaster comes to a stop.

2. Identify where the biggest thrill will be in the Terminator Express roller coaster. Explain why this will be the big thrill.

3. Speed doesn't produce thrills. Living on the Earth, you already have a big speed.

   a) The Earth makes a complete revolution once every 24 h. City A is close to the equator and travels a large circumference in 24 h. City B is close to the Arctic Circle and travels a small circumference in 24 hours. Which city has the greater speed?
   b) The circumference of (path around) the Earth's equator is about 40,000 km. It requires one day or 24 h to complete one revolution. Calculate the speed you are traveling on Earth.
   c) Why don't you get a big thrill going at such a high speed?

216

Active Physics

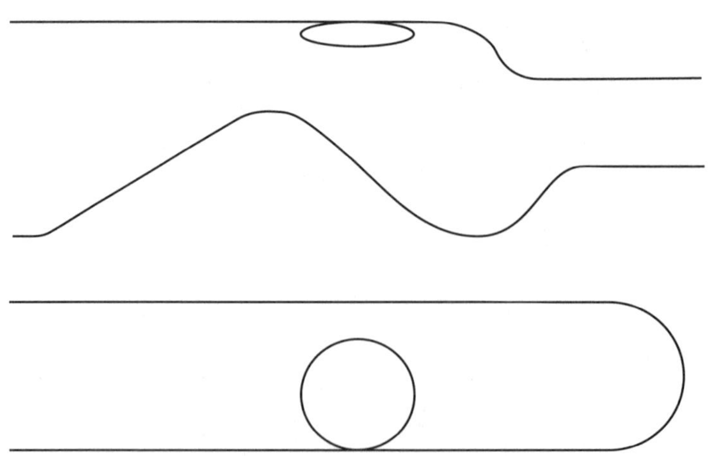

SIDE VIEW

TOP VIEW

4. A roller coaster rider changes from a speed of 4 m/s to 16 m/s in 3 s. Calculate the magnitude of the acceleration of the ride.

5. Identify the following situations as one of distance, displacement, speed, velocity, or acceleration. Indicate which is a vector and why.

   a) a car traveling at 50 km/h
   b) a student riding a bike at 4 m/s toward home
   c) a roller coaster ride whips around a left turn at 5 m/s
   d) a roller coaster car rises 12 m
   e) a train ride takes you 150 km

6. Suppose you were designing a roller coaster for young pre-school children.
   a) Describe two changes you would make to the Terminator Express roller coaster. Explain why you would make these changes.
   b) Draw the top and side view of the roller coaster with these additional changes.

7. A cart is 10 cm long. It travels through a photogate in 2 s. Calculate the cart's speed.

8. A second cart is 5 cm long. If it were traveling at the same speed as the cart in **Question 7**, what would the photogate timer record as the elapsed time?

Stretching Exercise

Investigate roller coasters on the Internet. Which are the most modern? What are some innovations in newer roller coasters? What features from historic coasters have been retained? Compare wooden and steel coasters.

ANSWERS

Physics To Go
(continued)

4. $a = \frac{\Delta v}{\Delta t} = \frac{16 \text{ m/s} - 4 \text{ m/s}}{3 \text{ s}} = \frac{12 \text{ m/s}}{3 \text{ s}} = \frac{4 \text{ m/s}}{\text{s}}$

5. a) speed—a scalar (no direction)

   b) velocity—a vector (speed with a direction)

   c) acceleration—a vector (a change in velocity with respect to time)

   d) displacement—a vector (it has direction)

   e) distance—a scalar

6. a) You may want to have smaller heights, gentler curves, and smaller changes in acceleration.

   b) Students' drawings will vary. However, their drawing should show the changes suggested.

7. $v = \frac{\Delta d}{\Delta t} = \frac{10 \text{ cm}}{2 \text{ s}} = 5 \text{ cm/s}$

8. Going at the same speed, half the distance will take half the time or 1 s.

   $v = \frac{\Delta d}{\Delta t}$

   $5 \frac{\text{cm}}{\text{s}} = 5 \frac{\text{cm}}{\Delta t}$

   $\Delta t = 1 \text{ s}$

Chapter 4

Terminator Express

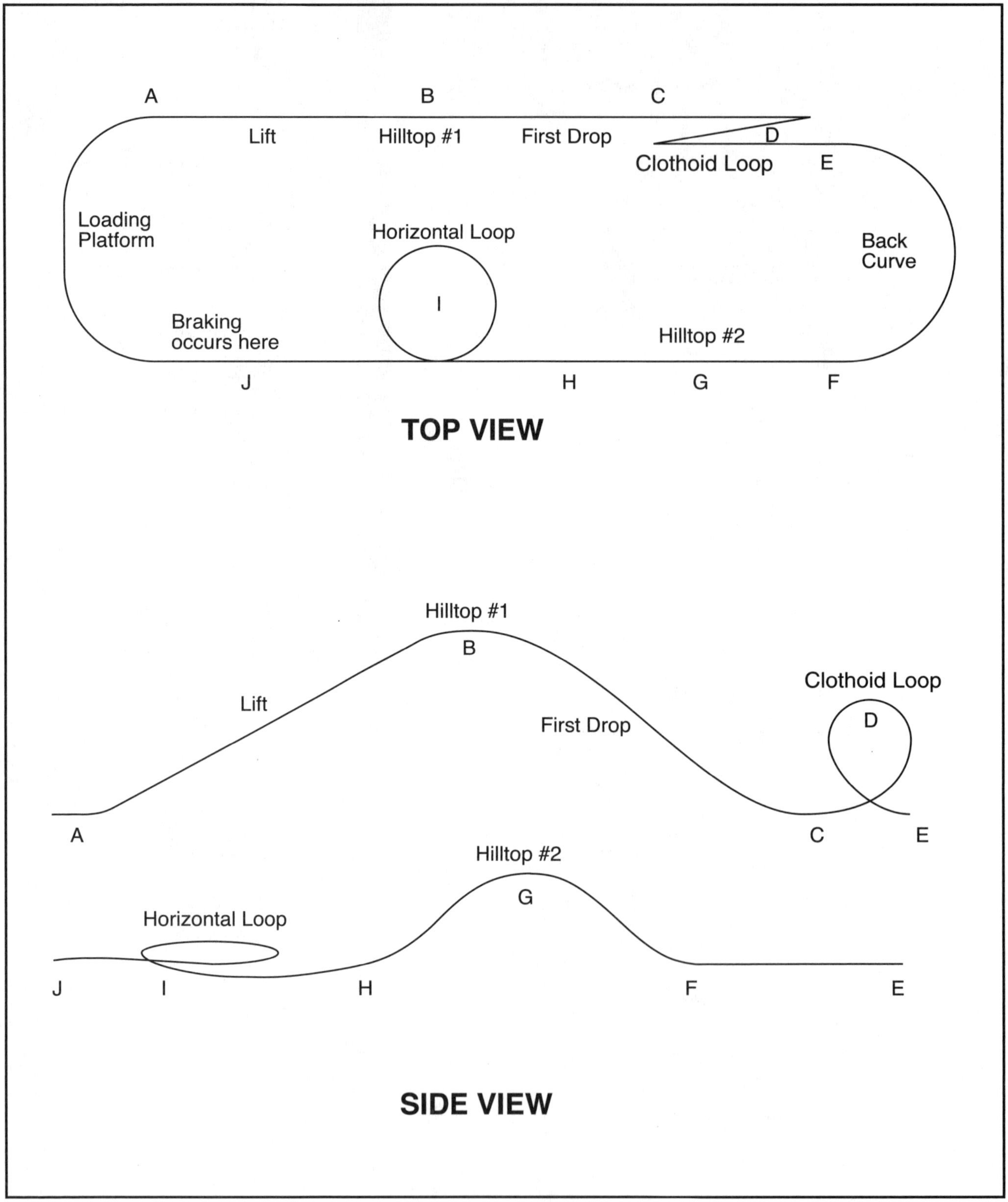

For use with *Thrills and Chills*, Chapter 4, Activity 1: The Big Thrill

NOTES

ACTIVITY 2
What Goes Up and What Comes Down

Background Information

It is suggested that you read the **For You To Read** and **Physics Talk** sections in the student text for **Activity 2** before proceeding in this section.

The conservation of energy is arguably the most important principle in science. The roller coaster ride is a wonderful application of the conservation of mechanical energy. If we ignore loss of energy due to friction producing sound and heat, we can state that the changes in the kinetic energy (and speed) of the roller coaster are due to changes in the gravitational potential energy of the roller coaster.

Energy is measured in joules (J). We can find the equivalent units for joule by looking at any equation for energy.

$GPE = mgh = (kg)(m/s^2)(m) = (kg)(m^2/s^2)$

$1\ J = 1\ kg \cdot m^2/s^2$

$KE = 1/2\ mv^2 = (kg)(m/s)^2 = (kg)(m^2/s^2)$.

$1\ J = 1\ kg \cdot m^2/s^2$

Energy is a state function. This means that we can compare the energy at any two points on the roller coaster without concern for what happened at other points on the roller coaster. The speed of the roller coaster at any one point depends only on the height of the roller coaster at that point and not on whether the roller coaster is moving up or down or is level at that point.

If we know the total energy at any point on the roller coaster, we know that every other point has an identical energy. If we know the height at that point, we can easily calculate the gravitational potential energy. From there, we can then calculate that the remainder of the energy is kinetic energy. Knowing the kinetic energy and the mass of the cart allows us to find the speed.

Equating the total energy at any point to the total energy at another point, we see that the mass in the equation "drops out."

At Point A	At Point B

$$GPE + KE = GPE + KE$$

$$mgh + 1/2\ mv^2 = mgh + 1/2\ mv^2$$

$$gh + 1/2\ v^2 = gh + 1/2\ v^2$$

The physical significance of the mass "dropping out" of the equation is that the speed of the roller coaster at any point is not dependent on the mass of the roller coaster.

The conservation of energy can also be applied to a ball falling or (as in the **Stretching Exercise**) a skateboarder in the vert. It can also be used to explain the changes in speed of a pendulum or a playground swing.

Planning for the Activity

Time Requirements

- Allow about two class periods to complete this activity.

Materials Needed

For each group:

- ball of string, 225 ft.
- batteries, AAA
- c-clamp for photogate
- cast iron right angle holder
- graph paper pad of 50 sheets
- large ringstand
- plumb bob with string
- protractor
- steel ball
- steel rod, 1/2" x 9" (to act as crossarm)
- stick, meter, 100 cm, hardwood
- track for roller coaster demonstration
- track support
- velocimeter

Advance Preparation and Setup

Be sure that all photogate timers work properly.

Teaching Notes

In this inquiry activity, students are given lots of freedom to set up the tracks and to determine the heights from which to release the steel ball. It may take a bit of time for them to get their first measurements, but they will find subsequent measurements to be straightforward.

As you circulate about the room, students may need some guidance in creating the two additional data charts as well as in completing the graphs.

The **For You To Read** section is crucial to understanding the activity in terms of conservation of energy. You should read this section aloud with the students and review the equations and the units on the board as you proceed. The chart and similar charts that you can create will give you a way to check for student understanding. (This chart is available as a Blackline Master at the end of this activity.) You should first ensure that all students know that the total energy (GPE + KE) remains the same for a specific ball going down a specific ramp.

All students should be able to calculate the GPE and KE. Some students may have difficulty in calculating the speed because of the introduction of the square-root function. You should speak to a math teacher to find out how the students have approached this in math class.

The calculation of GPE and KE and the resulting calculation of speed at different points of the roller coaster will be crucial in the student design of their own roller coaster and safety aspects of the ride. Be sure to tie the activity back to the **Chapter Challenge** and get students thinking about how they may use this information in their roller coaster design.

Activity Overview

Through an inquiry activity, students investigate the relationship between the angle of an incline and the final speed of a steel ball descending that incline. From this activity and the follow-up analysis, students arrive at the principle of the conservation of mechanical energy.

Student Objectives

Students will:

- Measure the speed of an object at the bottom of a ramp.
- Recognize that the speed at the bottom of a ramp is dependent on the initial height of release of the object and independent of the angle of incline of the ramp.
- Complete a graph of speed versus height of the ramp.
- Define and calculate gravitational potential energy and kinetic energy.
- State the conservation of energy.
- Relate the conservation of energy to a roller coaster ride.

ANSWERS FOR THE TEACHER ONLY

What Do You Think?

Most students will correctly recognize that the steeper roller coaster will produce the bigger thrill. Most students will incorrectly attribute this thrill to a faster speed. In fact, and as the students will discover in this activity, the speed at the bottom of the inclines is identical. The steeper slope allows the cart to reach the identical top speed in less time. It is the acceleration, the rate of change of speed, that produces the thrill.

Activity 2 What Goes Up and What Comes Down

GOALS

In this activity you will:

- Measure the speed of an object at the bottom of a ramp.
- Recognize that the speed at the bottom of a ramp is dependent on the initial height of release of the cart and independent of the angle of incline of the ramp.
- Complete a graph of speed versus height of the ramp.
- Define and calculate gravitational potential energy and kinetic energy.
- State the conservation of energy.
- Relate the conservation of energy to a roller coaster ride.

What Do You Think?

The steepest angle of descent on a wooden roller coaster is 70°. The steepest angle of descent on a steel roller coaster is 90°.

Two roller coaster slides are shown in the illustration below.

- **Which roller coaster will give the bigger thrill? Why?**

Record your ideas about these questions in your *Active Physics* log. Be prepared to discuss your response with your small group and the class.

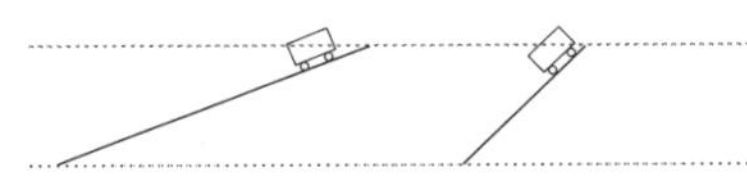

Activity 2 What Goes Up and What Comes Down

For You To Do

In this activity, you will investigate the speed at the bottom of a roller coaster slide. You will use a steel ball and a track to determine if a pattern exists between the placement of the ball and its speed at the bottom of the track. A pattern for speed will allow you to predict the speed for a new roller coaster.

1. The basic setup for this inquiry investigation is a track and a steel ball (or cart). You can measure distances to the nearest tenth of a centimeter with a ruler and speeds with a photogate timer.

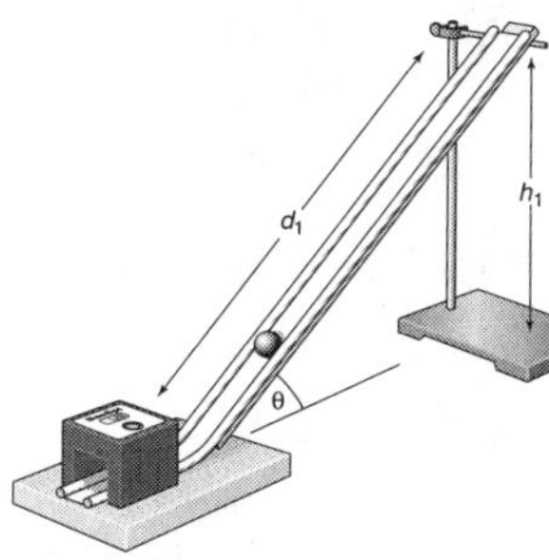

Be prepared to stop the ball at the end of the track so it does not roll onto the floor. If the ball rolls onto the floor pick it up right away. If you cannot find the ball, tell your teacher right away.

2. Your first step is to determine the speed of the steel ball at the bottom of the incline when the ball is placed at different points along the track. You should not vary the angle of the track. It will be useful to record the speed as a function of both the distance (*d*) and the height (*h*). (Although you can find the height if you know the distance, it's easier to make the extra measurement than to do the calculation using trigonometry.)

ANSWERS

For You To Do

1. Students should acquaint themselves with the apparatus generally and how they will measure the distances along the incline and the speed at the bottom of the incline.
2. The data table can be copied into students' logs and the students can record the height, distance, and speed. This table is provided as a Blackline Master at the end of this activity. Be sure to remind students to write down the angle of the track at the top of the chart as you move about the classroom. Some students will know how to use a protractor and can give the correct angle for the track. If you do not wish to have the students use protractors, they can record the angle as small, medium, or large.

ANSWERS

For You To Do
(continued)

2. a) Students complete the data table in their log.

3. a) This will require two additional data tables like the one in **Step 2**. In the first data table, the values for height will be identical as the table in **Step 2**. In the second data table, the values for distance will be identical to the table in **Step 2**.

4. a) Students should recognize that the same height produces the same speed at the bottom, irrespective of the angle of the track. They will also recognize that the same distance on the steeper track results in a greater speed.

5. a) As a test for their understanding, students should create a chart that has the same heights and predict the same speeds at the bottom.

   b) Students record their measurement in their data tables.

   c) Although the predicted and measured speeds may not be identical, point out to the students as you circulate that they are very close.

a) Complete the data table in your *Active Physics* log for at least four different positions.

Angle of track =

Height (m)	Distance (m)	Speed at bottom (m/s)

3. Change the angle of the track and repeat the investigation. First use the same heights and then use the same distances. This will require eight measurements of speed.
   a) Create two data charts, height and speed, and distance and speed. Complete the measurements.

4. Review the data from the first and second tracks.
   a) Is there a pattern between the heights and the speed or between the distances and the speed or both? Describe the pattern.

5. Change the angle of the track again.
   a) Construct a data chart where you can predict the speeds from the distances OR heights that you have chosen.
   b) Add a column to your chart so that it now has one column for predicted speeds and one for measured speeds. Complete the measurements for your chart.
   c) How good were your predictions?

6. Complete the same investigation using a curved track. Measuring the distance along a curved track will require some ingenuity. Measuring the height is similar to the straight track.

a) Construct a data table that includes both predicted speeds and measured speeds.

b) Write a summary statement comparing the speed at the bottom of a curved track and the speed at the bottom of a straight track.

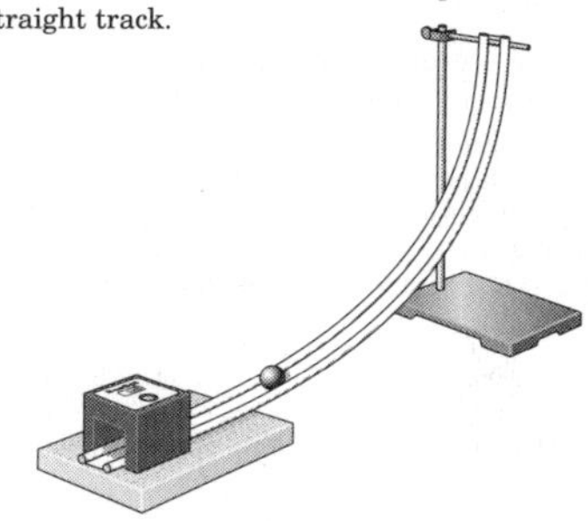

7. Complete a similar investigation using a pendulum. Measuring the distance the pendulum bob travels will require some ingenuity. Measuring the height is similar to the straight track.

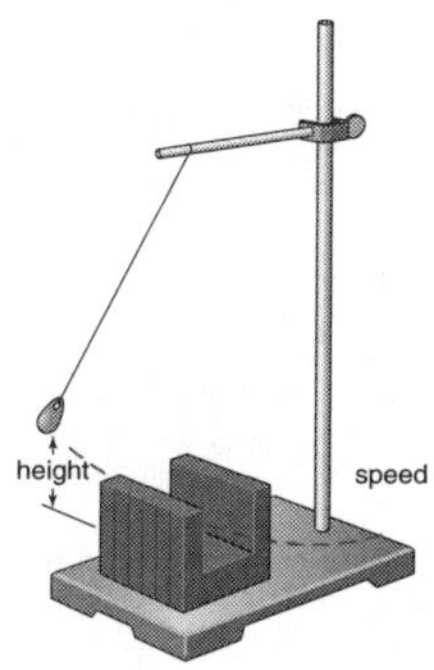

Answers

For You To Do
(continued)

6. As an additional part of the inquiry, students have to invent a way to measure the distance traveled along a curved track.

   a) Students construct a data table.

   b) Once again, students should find that the height of the ramp determines the final speed as it did for the straight tracks.

7. a) Students construct a data table.

   b) A pendulum's speed at the bottom is also determined by the height from which the pendulum is released.

Chapter 4

ANSWERS

For You To Do
(continued)

8. Student graphs should have a title, axes labeled with value and units, and the axes should be drawn with a ruler.

   a) The first graph is similar to many graphs students have created in the past.

   b) The second graph has v^2 on the y-axis. This has units of m^2/s^2.

9. You may have to help individual student teams or have a brief "math refresher" of the equation for a straight line $y = mx + b$. If the students have not studied straight-line graphs in math class, then you may wish to point out that the distinguishing features of a straight-line graph are the slope and the y-intercept.

   a) Students may need help with this calculation.

   b) The calculation of the slope should yield the same value of 19.6 m/s^2 for all student groups. As will be shown later, this is twice the acceleration due to gravity.

a) Construct a data table that includes both predicted speeds and measured speeds.

b) Write a summary statement comparing the speed at the bottom of the swing for a pendulum and the speed at the bottom of a straight track.

8. A valuable way in which to analyze data is with a graph. Take any data set (straight track, curved track or pendulum) and construct two graphs.

   a) The first graph should have height on the x-axis and speed at the bottom on the y-axis.

   b) The second graph should have height on the x-axis and the speed squared on the y-axis. This will require you to calculate v^2 for each speed.

9. Graphs with curves are difficult to interpret. It's hard to tell if the curve is part of a circle, ellipse, hyperbola, parabola, or none of these. Graphs with straight lines are much easier. The equation for all straight lines is

   $$y = mx + b$$

   where m is the slope of the graph and

   b is the y-intercept.

   In your straight-line graph, $(\text{speed})^2$ is the y-variable and height is the x-variable. You can now write an equation for the graph.

   $$y = mx + b$$

   Since the graph intersects the origin, the value of the y-intercept, b, is 0.

   The equation for the graph becomes

   $$y = mx + 0 \text{ or } y = mx.$$

   Substituting for the variables in your graph:

   $$(\text{speed})^2 = \text{slope (height) or}$$

   $$v^2 = mh$$

   a) Calculate the slope of your graph and record its value.

   b) Compare the value of your slope with those of other groups in the class.

FOR YOU TO READ

Gravitational Potential Energy and Kinetic Energy

By varying the slope of the incline and measuring speeds, you were able to find that the speed at the bottom of a track is determined not by the length of the incline, but by the initial height of the incline. Two carts sliding down inclines will have the same final speed if they both start from the same height.

The carts shown in the diagram will have identical speeds at the bottom of the inclines. The second one will get there sooner, but will arrive with the same speed as the first one. (You are assuming that friction is so small that you can ignore it.)

This connection between the height and the speed was valid for different inclines, for curved tracks, and for a pendulum.

The concept of energy can be used to describe this relationship. In your activity, the steel ball or cart at the top of the incline has **gravitational potential energy** (GPE). (Gravitational potential energy is the energy an object has as a result of its position in a gravitational field.) The cart at the bottom of the hill has **kinetic energy** (KE). (Kinetic energy is the energy an object possesses because of its motion.) GPE is dependent on the height of the cart above the ground. The KE is dependent on the speed of the cart. A larger change in GPE is associated with a larger KE.

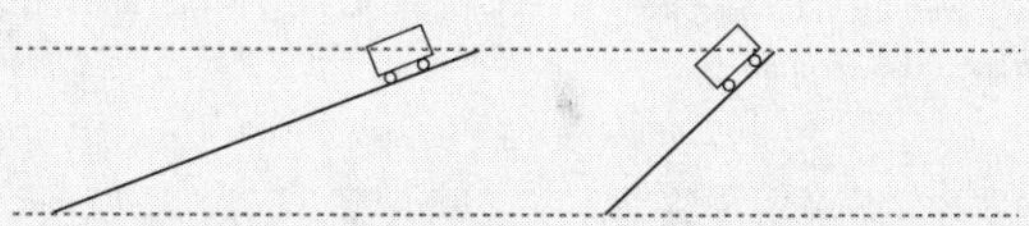

→

Physics Words

gravitational potential energy: the energy a body possesses as a result of its position in a gravitational field.

kinetic energy: the energy an object possesses because of its motion.

The equation for gravitational potential energy is

$$GPE = mgh$$

where *m* is the mass of the cart,

g is the acceleration due to gravity, and

h is the height above the ground.

You could just as easily have defined *h* as the height above the lab table. You will only concern yourself with the change in height. In this activity, the mass of the steel ball or cart and the acceleration due to gravity remained the same throughout the observations.

The equation for kinetic energy is

$$KE = \frac{1}{2}mv^2$$

where *m* is the mass of the cart, and

v is the velocity of the cart.

The unit for energy is a **joule** (symbol, J). Both GPE and KE are measured in joules. The chart below shows some calculations for a roller coaster car of mass 200 kg and an initial height of 20 m. Notice, that at the top of the roller coaster there are lots of joules of GPE and zero joules of KE. At the bottom of the incline, there are zero joules of GPE and lots of joules of KE. At the two other positions listed, there are some joules of GPE and some joules of KE.

Without knowing the velocity, it would seem that you could not calculate the KE. However, the sum of the GPE and KE must always be 40,000 J. This is because that was the total energy at the beginning. When the roller coaster was at a height of 20 m, there was no movement and 0 J of KE. All of the energy at this point was the 40,000 J of GPE. This 40,000 J becomes very important for this roller coaster. The sum of GPE and KE must always be 40,000 J at any point on the roller coaster. (There is, of course, in real life, some loss of energy to the

Mass of car = 200 kg and g = 10 m/s^2 (approximate value)			
Position of car (height) (m)	**GPE (J) = mgh**	**KE (J) = 1/2 mv^2**	**GPE + KE (J)**
Top (20 m)	40,000	0	40,000
Bottom (0 m)	0	40,000	40,000
Halfway down (10 m)	20,000	20,000	40,000
Three-quarters way down (5 m)	10,000	30,000	40,000

environment due to friction that must be taken into consideration. However, we will neglect that for now.)

At the bottom of the roller coaster (see line 2 on the chart), there are 0 J of GPE. To total 40,000 J, there must be 40,000 J of kinetic energy KE at the bottom.

Halfway down (see line 3 on the chart), the KE must equal 20,000 J, so that the sum of the GPE (20,000 J) and the KE (20,000 J) once again equals 40,000 J.

Three-quarters of the way down (see line 4 on the chart), the KE must equal 30,000 J. The sum of the GPE (10,000 J) and the KE (30,000 J) once again equals 40,000 J.

Given any height, you can determine the GPE and then determine the KE. In this roller coaster, the GPE and KE must equal 40,000 J.

In a higher roller coaster, the GPE and KE might equal 60,000 J. You can still calculate the GPE at any height and then find the corresponding KE.

A word of caution is necessary when discussing the conservation of energy. The sum of the GPE and KE only remains the same if there are no losses of energy due to friction, sound, or other outside sources and no additions of energy from motors.

The conservation of energy provides a way to find the kinetic energy if you know the change in height of the roller coaster. However, it also allows you to find the speed of the roller coaster. Using algebra, you can calculate the speed.

$$\text{Energy (at the bottom)} = \text{Energy (at the top)}$$

$$\text{KE (bottom)} + \text{GPE (bottom)} = \text{KE (top)} + \text{GPE (top)}$$

$$\text{KE (bottom)} + 0 = 0 + \text{GPE (top)}$$

$$\text{KE (bottom)} = \text{GPE (top)}$$

$$\frac{1}{2}mv^2 = mgh$$

$$v^2 = 2gh$$

where g is the acceleration due to gravity.

In the above equation, the *m* cancelled out. This means that the speed is independent of the mass of the car. (It doesn't matter whether the roller coaster car has 2 or 4 passengers.)

Since $g = 9.8\ \text{m/s}^2$ then $v^2 = 2gh$. You can see why your graph of v^2 versus h was a straight line. The slope of the line should equal: $2 \times 9.8\ \text{m/s}^2 = 19.6\ \text{m/s}^2$.

Chapter 4

Physics Words

joule: the SI unit for work and all other forms of energy; one joule (1 J) of work is done when a force of one newton moves an object one meter in the direction of the force.

PHYSICS TALK

Calculating Kinetic Energy from Gravitational Potential Energy

The gravitational potential energy is defined as
GPE = mgh

The kinetic energy is defined as KE = $\frac{1}{2}mv^2$

In a system like your roller coaster, the sum of the GPE and KE is constant.

$$GPE + KE = \text{constant}$$

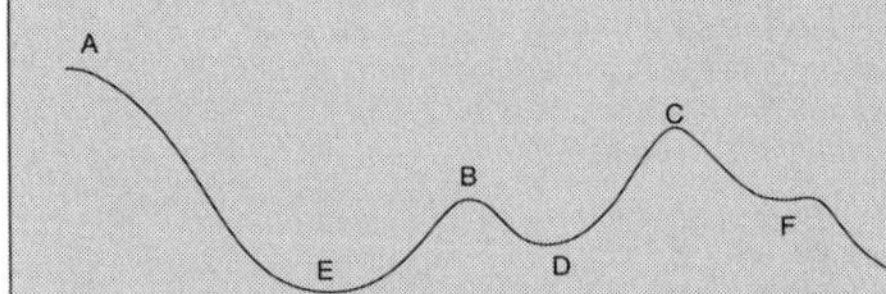

In the roller coaster above, if the GPE at point A is 30,000 J and the KE at point A is 0 J, then the total energy at point A and each and every other point on the roller coaster is 30,000 J. The total energy is also 30,000 J at points B, C, D, E, and F and every point in between. Since points B and F have the same height, the roller coaster car must also have the same GPE. That implies that the roller coaster car also has the same KE and is therefore going at the same speed at both points.

The height determines the GPE. The total energy at every point is the same. If you know the GPE, you can easily find the KE. The KE informs you about the speed.

Activity 2 What Goes Up and What Comes Down

Reflecting on the Activity and the Challenge

In designing a roller coaster, it is necessary to know how the speed of the roller coaster will vary. Knowing that the sum of the GPE and KE is constant is crucial in finding the speed at each point on the ride. How to calculate the GPE and KE will also be important when you want to ensure safety. You cannot let the roller coaster fall off the tracks, nor can you build a roller coaster that injures people. If your roller coaster begins on top of a tall hill, the height of that hill will determine the speed for each and every point along the roller coaster.

Physics To Go

1. For which track is the speed the greatest at the bottom? (Assume no friction.)

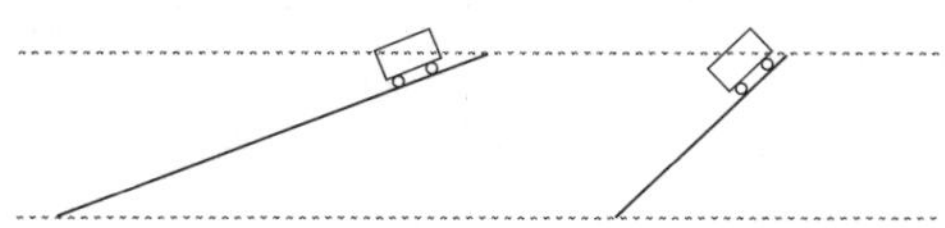

2. For which position is the speed the greatest at the bottom?

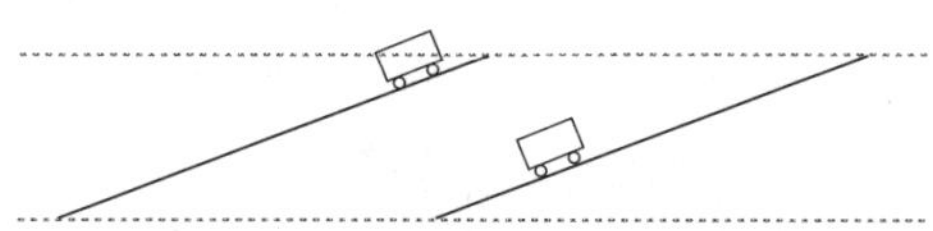

3. State the conservation of energy as it applies to roller coasters. Include in your statement GPE, KE, *mgh*, and $\frac{1}{2}mv^2$.

ANSWERS

Physics To Go

1. Both carts will have the same speed at the bottom since they both start from the same height.

2. The first cart will have the greater speed because it starts from a higher elevation.

3. As the roller coaster is lifted to its highest point, it gains gravitational potential energy (GPE = mgh). Once it starts its descent, the roller coaster loses gravitational potential energy and gains kinetic energy (KE = $1/2\ mv^2$). A gain in kinetic energy is a gain in speed. The total energy of the roller coaster at every point is identical. This conservation of energy equation can be stated mathematically as the sum of the kinetic and gravitational potential energy at one point on the roller coaster is equal to the sum of the kinetic and gravitational potential energy at any other point on the roller coaster.

ANSWERS

Physics To Go
(continued)

4. See table below.
5. See table below.
6. a) GPE = mgh = (0.2 kg)(10 m/s^2)(0.75 m) = 15 J
   b) KE at bottom is equal to GPE at top = 15 J
   c) The GPE and KE will be equal when the GPE has half its original value. This occurs at half its original height or 0.375 m (or 0.37 m) above the ground.
7. The roller coasters will have identical speeds at the bottom. However, the acceleration on the steeper slope will be greater, and therefore will produce the biggest thrill. Remind the students that this was the **What Do You Think?** question at the beginning of the activity. Have their responses changed as a result of the activity?
8. The speed of the roller coaster will not change. When calculating the speed from the KE and GPE equations, the mass "drops out" of the equation for speed.

4. Complete the chart below for a roller coaster.

Mass of car = 200 kg and g = 10 m/s² (approximate)			
Position of car (height) (m)	**GPE (J) = *mgh***	**KE (J) = $\frac{1}{2}mv^2$**	**GPE +KE (J)**
Top (30 m)	60,000	0	
Bottom (0 m)			
Halfway down (15 m)			
Three-quarters way down (7.5 m)			

5. Complete the chart below for a roller coaster.

Mass of car = 300 kg and g = 10 m/s² (approximate)			
Position of car (height) (m)	**GPE (J) = *mgh***	**KE (J) = $\frac{1}{2}mv^2$**	**GPE +KE (J)**
Top (25 m)			
Bottom (0 m)			
Halfway down (10 m)			
Three-quarters way down (5 m)			

6. A pendulum is lifted to a height of 0.75 m. The mass of the bob is 0.2 kg.
   a) Calculate the GPE at the top.
   b) Find the KE at the bottom.
   c) At what position will the GPE and the KE be equal?
7. What do you think now? Two roller coaster slides are shown below. Which will give the biggest thrill? Why?

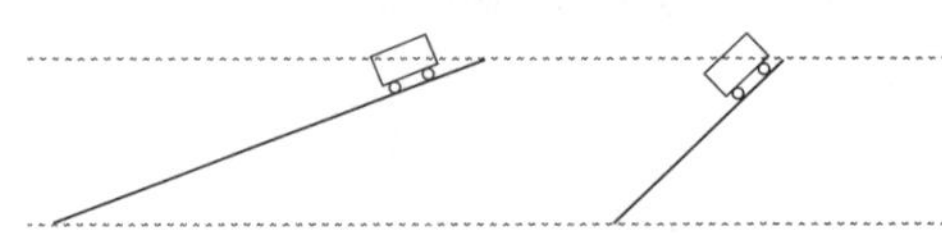

Table 4.

Position of car	GPE	KE	GPE + KE
Top	60,000	0	60,000
Bottom	0	60,000	60,000
Halfway down	30,000	30,000	60,000
Three-quarters way down	15,000	45,000	60,000

Table 5.

Position	GPE	KE	GPE + KE
Top	GPE = mgh = 75,000	0	75,000
Bottom	0	75,000	75,000
Halfway down (h = 10 m)	GPE = mgh = 30,000	45,000	75,000
Three-quarters way down (h = 10 m)	15,000	60,000	75,000

8. A roller coaster ride in the early morning only has 6 passengers. In the afternoon it has 26 passengers. Will the speed of the roller coaster change with more passengers aboard? Explain your answer.
9. Below is a side view of a roller coaster that starts from rest at position A.

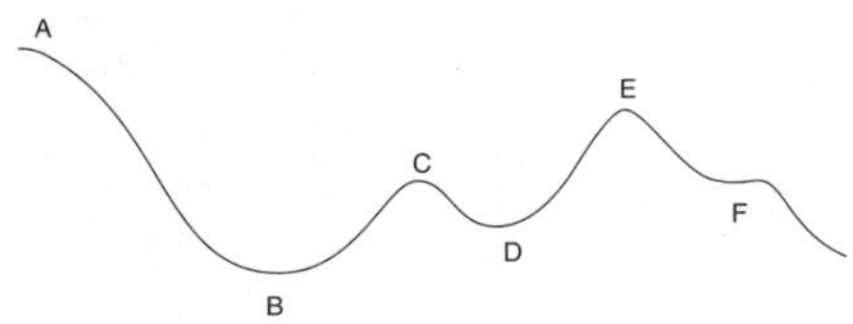

a) At which point is the roller coaster car traveling the fastest? Explain.
b) At which two points is the roller coaster car traveling at the same speed? Explain.
c) Is the roller coaster car traveling faster at E or D? Explain.

10. Below is a side view of a roller coaster that starts from rest at position A.

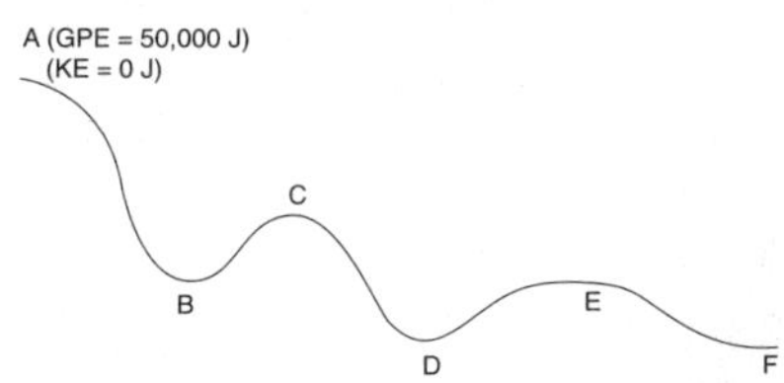

a) Determine plausible values for the GPE and KE at points B, C, D, E, and F.
b) At which two places is the roller coaster traveling at the same speed. Explain using GPE, KE, and speed in your explanation.

ANSWERS

Physics To Go

(continued)

9. a) The roller coaster is traveling fastest at B. That is where it has the least GPE and must have the largest KE.

   b) C and F. The roller coaster has the same speed because it is at the same height. The same height has the same GPE and therefore has the same KE.

   c) The roller coaster is faster at D than E because at D it has less GPE and therefore more KE.

10. a) See table below.

    b) The roller coaster is traveling at the same speed at points B and E. The GPE at these points is equal and therefore the KE is also equal. Since the KE is equal, the speeds must also be equal.

Table 10. a)

Position	GPE (J)	KE (J)	GPE + KE (J)
A	50,000	0	50,000
B	20,000	30,000	50,000
C	25,000	25,000	50,000
D	0	50,000	50,000
E	20,000	30,000	50,000
F	0	50,000	50,000

ANSWERS

Physics To Go
(continued)

11. Mass of car = 200 kg and $g = 10\ m/s^2$

See table below.

11. Complete this chart for a modified Terminator Express roller coaster. You may wish to try this on a spreadsheet.

Mass of car = 200 kg and g = 10 m/s² (approximate)					
Location	**Height (m)**	**Speed (m/s²)** $v = \sqrt{2gh}$	**GPE (J)** mgh	**KE (J)** $\frac{1}{2}mv^2$	**Total energy (J)** GPE + KE
Top					
Bottom of hill					
Top of 1st hill					
Top of loop					
Horizontal loop					

Stretching Exercise

As a skateboarder practices on the vert, there are constant changes in the gravitational potential energy GPE and the kinetic energy KE. Research the size of the vert and report back on how the conservation of energy plays an integral part in this sport. You may also wish to make measurements of skateboarders in the vert.

Table 11.

Location	Height (m)	Speed (m/s)	GPE (J)	KE (J)	GPE + KE (J)
Top	30	0	60,000	0	60,000
Bottom of hill	0	24	0	60,000	60,000
Top of 1st hill	20	14	40,000	20,000	60,000
Top of loop	25	10	50,000	10,000	60,000
Horizontal loop	10	20	20,000	40,000	60,000

What Goes Up and What Comes Down Data Tables

Angle of track =		
Height (m)	Distance (m)	Speed at bottom (m/s)

Angle of track =	
Height (m)	Speed at bottom (m/s)

Angle of track =	
Height (m)	Speed at bottom (m/s)

Angle of track =		
Height (m)	Predicted speed (m/s)	Speed at bottom (m/s)

Angle of track =		
Distance (m)	Predicted speed (m/s)	Speed at bottom (m/s)

Angle of track =			
Height (m)	Distance (m)	Predicted speed (m/s)	Speed at bottom (m/s)

For use with *Thrills and Chills*, Chapter 4, Activity 2: What Goes Up and What Comes Down

Gravitational Potential Energy and Kinetic Energy

Mass of car = 200 kg and g = 10 m/s^2 (approximate value)			
Position of car (height) (m)	**GPE (J) = *mgh***	**KE (J) = 1/2 *mv*^{2}**	**GPE + KE (J)**
Top (20 m)	40,000	0	40,000
Bottom (0 m)	0	40,000	40,000
Halfway down (10 m)	20,000	20,000	40,000
Three-quarters way down (5 m)	10,000	30,000	40,000

Mass of car = and g = 10 m/s^2 (approximate value)			
Position of car (height) (m)	**GPE (J) = *mgh***	**KE (J) = 1/2 *mv*^{2}**	**GPE + KE (J)**
Top (20 m)			
Bottom (0 m)			
Halfway down (10 m)			
Three-quarters way down (5 m)			

For use with *Thrills and Chills*, Chapter 4, Activity 2: What Goes Up and What Comes Down

NOTES

ACTIVITY 3
More Energy

Background Information

It is suggested that you read the **For You To Read** and **Physics Talk** sections in the student text for **Activity 3** before proceeding in this section.

In this activity, the principle of the conservation of energy is expanded to include the spring potential energy. When a spring is compressed, it has a spring potential energy. If the spring is released, the energy of the spring becomes kinetic energy of the object.

The expanded equation for conservation of energy would now be:

(SPE + KE + GPE) at point A =
(SPE + KE + GPE) at point B.

The spring potential energy can be calculated using the following equation:

SPE = $1/2kx^2$ where k is the spring constant and the x is the compression (or stretch) of the spring. The details of finding the spring constant k from Hooke's Law is not a part of this activity.

Once again, all energies are measured in joules.

In more complicated systems, some of the lost energy (e.g., heat, sound) would also be included in the energy equation. To use these energies into the equation, it would be necessary to derive a formula that would help us with each energy calculation.

This expansion of the principle of conservation of energy can also explain jumping on a trampoline where the spring potential energy would be a combination of the stretch of the springs supporting the trampoline as well as the stretch of the trampoline fabric itself. The equations can also be used to explain the conservation of energy in a pole vault. In the pole vault, the spring potential energy is the bend of the pole. All muscles in the body behave in some manner like a spring compressing.

Planning for the Activity

Time Requirements

- One class period is required to complete this activity.

Materials Needed

For each group:

- batteries, AAA
- c-clamp for photogate
- "Pop-up" toy
- some nickels
- stick, meter, 100 cm, hardwood
- tape, masking, 3/4" x 60 yds.
- velocimeter

Advance Preparation and Setup

Ensure that all photogate timers are working properly.

Check to make sure the pop up toys work well.

Teaching Notes

Students may have difficulty measuring the speed or height of the pop-up toy. With patience, they will invent methods and procedures to help them. They may want to remark about the consistency of their results in their student log.

The mathematical analysis of the lab requires them to recognize that the total energy is conserved. Knowing the total energy at one point allows them to know the kinetic energy at all points.

During the post-lab discussion, students can discuss why their results may not have been as good or consistent as other groups.

Although roller coasters are not brought to their height through the use of springs, the idea of additional energy being required would be the same. In a real roller coaster, the electrical energy of the motor brings the coaster to its highest point. Once at the top of the hill, the roller coaster is then released and the GPE and KE vary throughout the ride.

You may want to discuss "Rube Goldberg" devices in terms of the exchange of energy. Some of these devices have springs, levers, balls moving up and down, and they are a wonderful exchange of energies.

Activity 3 More Energy

GOALS

In this activity you will:

- Measure the kinetic energy of a pop-up toy.
- Calculate the spring potential energy from the conservation of energy and using an equation.
- Recognize the general nature of the conservation of energy with heat, sound, chemical, and other forms of energy.

What Do You Think?

The concept of a "lift hill" for a roller coaster was developed in 1885. This was the initial hill that began a roller coaster ride. A chain or a cable often pulled up the train to the top of this hill.

- **How does the roller coaster today, get up to its highest point?**
- **Does it cost more to lift the roller coaster if it is full of people?**

Record your ideas about these questions in your *Active Physics* log. Be prepared to discuss your response with your small group and the class.

Activity Overview

Students explore a pop-up toy in terms of investigating the compression of a spring and the height that the toy rises. This extends the concept of the conservation of energy to include the potential energy of a spring. On the roller coaster ride, the energy is not supplied by a spring but rather by a motor. The electrical motor lifts the roller coaster to its highest point.

Student Objectives

Students will:

- Measure the kinetic energy of a pop-up toy.
- Calculate the spring potential energy from the conservation of energy and using an equation.
- Recognize the general nature of the conservation of energy with heat, sound, chemical, and other forms of energy.

ANSWERS FOR THE TEACHER ONLY

What Do You Think?

Many students think that a motor is used for the entire roller coaster ride rather than at just the beginning of the ride to bring the roller coaster to its highest point. Although it may seem obvious that a roller coaster filled with people will be more difficult to lift, some students may think that the mass of the coaster does not matter since it did not make a difference to the roller coaster's speed as it descended.

Chapter 4

ANSWERS

For You To Do

1. a) The heights will vary.

   b) A range of heights would be a way of describing consistency.

2. a) Determining the kinetic energy requires a determination of the mass of the toy and the speed with which it leaves the table. (KE = 1/2 mv^2). The mass can be measured with a balance. The speed could be measured with a photogate timer.

   A second way in which to calculate the KE would be to equate the KE with the gravitational potential energy the toy has at the top of its jump. The GPE at the top would be identical to the KE at the bottom. Since GPE = mgh, we need a balance to measure the mass and a ruler to measure the height.

3. a) Students record their results using both methods.

   b) The two KE values should be approximately the same.

5. The heavier pop-up toy has more mass. It may have the same KE when it jumps, but with more mass it will have less velocity. With less velocity it will not go as high.

Eye protection must be worn during this activity. Have team members step back before the toy is released.

For You To Do

1. Everybody loves those little pop-up toys. You press the plunger, place it on a table, and "pop!" it flies into the air. In this inquiry investigation, you will determine the kinetic energy, KE, of the pop-up toy when it leaves the ground.

   Play with the toy to get a sense of how high it jumps.

   a) What is the approximate height of a jump?

   b) How consistent is the pop-up toy from one jump to the next?

2. Discuss among your group two distinct methods you can use to determine the KE when the pop-up toy leaves the ground. One method will require the photogate timer. The second method will require a meter stick to measure the height of the jump.

   a) Record your two methods in your *Active Physics* log. Since another team may want to understand what you have done, be quite careful to list all the steps. Indicate how all measurements are completed, and what is recorded or calculated.

3. Complete the investigation using both methods.

   a) Record your results. If during the experiment you changed your procedure, you should also record any changes here.

   b) Compare the KE determinations from the two methods.

Be sure the nickels are secure on the toy. Retape after every two to three trials.

4. Tape some nickels to the top of the pop-up toy in order to approximately double its mass. The mass of a nickel is approximately 5 g or 0.005 kg. (A dime is *not* 10 g; a quarter is *not* 25 g; and a penny is *not* 1 g.)

5. Repeat the investigation and find the KE of the pop-up toy as it leaves the ground.

   a) Why do you think that the heavier pop-up toy behaved differently? Use the terms GPE and KE in your explanation.

6. Answer the following questions in your *Active Physics* log:

a) What is the toy's KE and GPE when it sits on the table?
b) What happens to the toy's GPE and KE as it rises from the table?
c) If the total energy of the toy is conserved, where does this KE and GPE come from as it rises?
d) Where is the toy when its KE and when its GPE is greatest?

7. The pop-up toy had KE and GPE as it rose above the table. While it was sitting there, it also had spring potential energy, SPE. This SPE was converted to KE which then became GPE as the pop-up toy ascended. Using the concept of conservation of energy from the last activity, you notice that before popping up, there was all SPE. Just after popping up, there was all KE. When reaching the highest point, there was all GPE. The total energy at all other points was the same as the total SPE before popping, the total KE just after popping or the total GPE at its peak. You can show this in a chart. Notice that total energy is conserved. You now have spring potential energy SPE in addition to GPE and KE.

a) Complete the chart in your log with other reasonable values for SPE, KE, GPE and the sum in the respective columns.

Position above table (m)	SPE (J)	KE (J)	GPE (J)	SPE + KE + GPE (J)
At rest on table: height = 0 m	20	0	0	20
Just after popping: height = 0 m	0	20	0	20
At peak: height = 0.30 m	0	0	20	20
1/2 the way up: height = 0.15 m			10	
With the spring only partially opened: height = 0 m				
Some other position: height = ? m				

ANSWERS

For You To Do

(continued)

6. a) The KE as it sits on the table = 0 J. The GPE as it sits on the table = 0 J.

b) As it rises from the table, the KE and GPE both increase. After it leaves the table, the KE decreases as the GPE increases.

c) The energy comes from the compressed spring.

d) The KE is greatest when it just leaves the table. The GPE is greatest at the peak of the jump.

7. a) See table below.

Table 7. a)

Positon above table (m)	SPE (J)	KE (J)	GPE (J)	SPE + KE + GPE (J)
At rest on table: height = 0 m	20	0	0	20
Just after popping: height = 0 m	0	20	0	20
At peak: height = 0.3 m	0	0	20	20
1/2 the way up: height = 0.15 m	0	10	10	20
With the spring only partially opened: height = 0 m	10	10	0	20
Some other positon: height = ? m	Students' answers will vary.			20

FOR YOU TO READ

Conservation of Energy

Kaitlyn, Hannah, and Nicole share an apartment. Hannah keeps a bowl by the door filled with quarters that she can use for the washer and dryer at the laundromat. On Tuesday, Hannah counts her money and finds that she has 24 quarters or $6.00 in quarters. This is just the right amount for her laundry on Saturday. On Wednesday morning, Nicole comes rushing up to the apartment because she needs some quarters for the parking meter. She takes three quarters from the bowl and replaces them with six dimes and three nickels. The total money in the bowl is still $6.00. On Wednesday afternoon, Kaitlyn needs to buy a fifty-cent newspaper from the machine that takes all coins but pennies. Kaitlyn takes two quarters from Hannah's bowl and replaces these coins with fifty pennies. The total in the bowl is still $6.00.

Wednesday night, Hannah comes home and notices that her bowl is filled with quarters, pennies, nickels, and dimes. She knows that it still adds up to the $6.00 that was there in the morning, but also knows that she cannot do her laundry without all the money in quarters. Her roommates agree to exchange all the coins with quarters the next day.

The money in the bowl could represent the energy in a system. The total amount of energy may have been 600 J. As the coins in the bowl change from quarters to pennies to dimes and nickels and back to quarters, the energy in the system can vary from kinetic energy to gravitational potential energy to spring potential energy in any combinations.

If Kaitlyn had taken the two quarters and not replaced them with pennies, then the total money would be less. The loss in money due to Kaitlyn would have resulted in that money being somewhere else. In some systems, energy is lost as well. A bouncing ball does not get to the same height in each successive bounce. Some of the energy of the ball becomes sound energy and heat energy. These can be measured and will

show that the energy left the system but did not disappear. In the pop-up toy and the roller coaster, the total energy can be GPE, KE, SPE, but the sum of the energies must always be the same.

As you followed the changes in Hannah's bowl of money, you knew that there were ways to measure the total amount of money. Fifty pennies is identical in value to two quarters. Scientists look for all the energies in a system. There is electrical energy, light energy, nuclear energy, sound energy, heat energy, chemical energy, and others. Each one is able to be calculated using measurements. All the energies are measured in joules. The total number of joules must always remain the same.

In a real roller coaster, the roller coaster has all its energy as GPE (gravitational potential energy) as it sits on the highest hill. Most of this energy becomes KE (kinetic energy) as the roller coaster is released. Some small amount of the energy is converted to heat and a smaller part to sound.

Where does the roller coaster get all of that GPE that drives the rest of the ride? Something has to pull the roller coaster up to the top of the hill. The energy to pull the roller coaster is usually electric. The electrical energy comes from a power plant (which burns oil, gas, coal, or uses nuclear energy or water's potential energy as it comes crashing down) or from a local generator that may use gasoline.

After the cars are pulled to the top of the hill, the roller coaster is a closed system. Energy no longer enters the system. The total energy of the roller coaster remains the same except for losses due to heat and sound. In introductory physics, you can usually ignore heat and sound in the first analysis.

Chapter 4

Physics Words

spring potential energy: the internal energy of a spring due to its compression or stretch.

PHYSICS TALK

Calculating Spring Potential Energy

In this activity you extended the conservation of energy principle to include the **spring potential energy**. It is possible to calculate the spring potential energy.

The equation for spring potential energy is

$$SPE = \frac{1}{2}kx^2$$

where k is the spring constant and

x is the amount of stretch or compression of the spring.

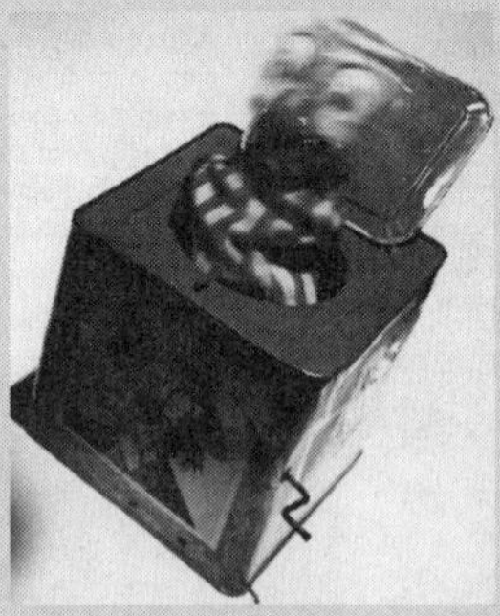

A spring that is difficult to compress or stretch will have a large spring constant k. That spring will "pack" more SPE for an identical compression than a spring that is easy to compress.

The total energy of a spring toy that can jump into the air is the sum of the SPE, the GPE, and the KE. Once the spring is compressed, you can consider it a closed system and the sum of these three energies, GPE, KE, and SPE must remain constant.

$$GPE + KE + SPE = \text{constant}$$

$$mgh + \frac{1}{2}mv^2 + \frac{1}{2}kx^2 = \text{constant.}$$

Reflecting on the Activity and the Challenge

There are other energies—heat, sound, chemical, etc. In your analysis of the roller coaster, you may decide to ignore heat and sound, but you had better mention this in your,report. In the actual construction, it will be important to take into account that a small amount of energy is being dissipated (lost).

The roller coaster uses electrical energy to get the cars to the top of the hill. This is similar to using the chemical energy of your body to compress the pop-up toy so that you can watch it jump.

Once the energy is in the spring of the pop-up toy, you can observe it as a closed system. The SPE becomes KE, which becomes GPE. In the same way, once the cars are on top of the hill, the GPE can become KE.

Physics To Go

1. Complete the chart with other reasonable values for SPE, KE, GPE, and the sum in the respective columns.

Position above table (m)	SPE (J)	KE (J)	GPE (J)	SPE + KE + GPE (J)
At rest on table: height = 0 m	25			
Just after popping: height = 0 m				
At peak: height = 0.60 m				
1/2 the way up: height = 0.30 m				
With the spring only partially opened: height = 0 m				
Some other position: height = ? m				

ANSWERS

Physics To Go

1. See table below.

Table 1.

Positon above table (m)	SPE (J)	KE (J)	GPE (J)	SPE + KE + GPE (J)
At rest on table: height = 0 m	25	0	0	25
Just after popping: height = 0 m	0	25	0	25
At peak: height = 0.60 m	0	0	25	25
1/2 the way up: height = 0.30 m	0	12.5	12.5	25
With the spring only partially opened: height = 0 m	10	15	0	25
Some other positon: height = ? m	Students' answers will vary.			25

Chapter 4

ANSWERS

Physics To Go
(continued)

2. If the mass were increased, the SPE would remain the same and therefore the total energy would remain the same. The GPE would remain the same because it is based on the height only. Therefore, the KE would remain the same as well. However, the spring toy with the same GPE would not reach the same height as before nor would it reach the same speed.

3. Electrical energy is sent to the muscles ordering them to contract. The chemical energy allows the muscles to contract. The contracted muscles push the ball into the air. It begins with a large kinetic energy. As it rises, the kinetic energy is transformed into gravitational potential energy. When it reaches its highest point, the ball has maximum gravitational energy and is momentarily at rest with zero kinetic energy. As it descends it picks up speed and kinetic energy as it loses gravitational potential energy.

2. How would the chart values in **Question 1** change if some extra mass were attached to the pop-up toy?

3. You throw a ball into the air and catch it on the way down. Beginning with the chemical energy in your muscles, describe the energy transformations of the ball.

4. Why can the second hill of the roller coaster not be larger than the first hill?

5. Why does the roller coaster not continue forever and go up and down the hills over and over again?

6. A roller coaster of mass 300 kg ascends to a height of 15 m. How much electrical energy was required to raise the cars to this height?

7. A roller coaster has a mass of 400 kg and a speed of 15 m/s.
   a) What is the KE of the roller coaster?
   b) What will be the GPE of this roller coaster at its highest point?
   c) How high can the roller coaster go with this much energy?

8. A ball is thrown upward from Earth's surface. While the ball is rising, is its gravitational potential energy increasing, decreasing, or remaining the same?

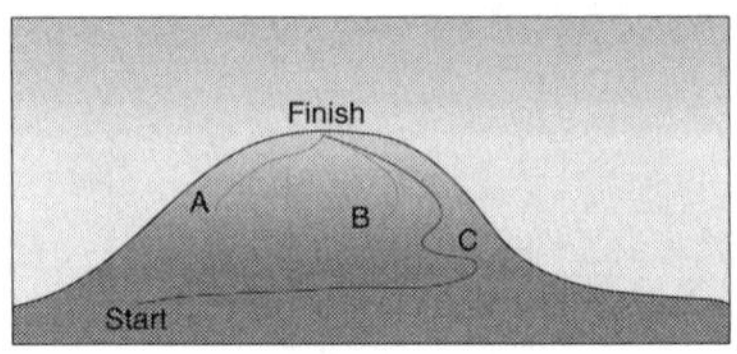

9. Three people of equal mass climb a mountain using paths A, B, and C shown in the diagram.

Along which path(s) does a person gain the greatest amount of gravitational potential energy from start to finish: A only, B only, C only, or is the gain the same along all paths?

4. For the roller coaster to reach the second hill, it would require additional gravitational potential energy. The second hill has more gravitational potential energy than the first. The total energy of the ride cannot increase.

5. In real life, the roller coaster is constantly losing energy due to friction with the track and with the air. This energy loss to the roller coaster is heard as sound or becomes heat energy.

6. The electrical energy required will be equal to the gravitational energy.

   GPE = mgh = (300 kg)(10 m/s^2)(15 m) = 45,000 J

7. a) KE= 1/2 mv^2 = 1/2 (400 kg) (15 m/s)^2 = 45,000 J

   b) GPE = 45,000 J

   c) GPE = mgh

   45,000 J = (400 kg)(10 m/s^2)(h)

   h = 11 m

8. As it rises, the GPE is increasing.

9. The gain is the same along all paths.

ACTIVITY 4
Your "At Rest" Weight

Background Information

It is suggested that you read the **For You To Read** and **Physics Talk** sections in the student text for **Activity 4** before proceeding in this section.

Students often have a great deal of difficulty in distinguishing mass and weight. That is probably because in our experience we have created a type of conversion. When we want to know how far we can throw something, we first "heft" it. We lift it up and down to determine its weight. Knowing the weight, we can then deduce its corresponding mass and know how hard it will be to push or throw. We also tend to confuse mass and weight because in common English usage, we use the same terms like ounces or pounds for both mass and weight. In physics, the distinction between mass and weight is crucial to our understanding. Mass is inertia–a resistance to motion. It is also the amount of "stuff." Weight is a force. It is the force that the Earth exerts on a given mass. Newton's Second Law tells us that $F = ma$. When the acceleration is equal to the acceleration due to gravity, the force is called the weight. $F = ma$ becomes $w = mg$, where g is the acceleration due to gravity (9.8 m/s^2). The unit for mass is the kilogram. The unit for weight is the newton. From Newton's Second Law, we can see that $1\ N = 1\ kg \cdot m/s^2$. It is sometimes useful to use alternate units for g of N/kg. Since an object resting on the table has mass and weight, it might clear up some misunderstandings if students are calculating the weight by multiplying the mass by the acceleration due to gravity in units of N/kg rather than the equivalent m/s^2.

Hooke's Law states the relationship between the force that a spring can exert on an object and the stretch of the spring. The larger the stretch (or compression) of the spring, the larger the exerted force. This is written mathematically as $F = -kx$. The negative sign indicates that when you stretch a spring to the right, the force is to the left. The force and the displacement (stretch) are in opposite directions.

Hooke's Law is so precise that we have used the spring for measuring weights of food, gold, and many other materials throughout the ages. We either hang objects from the scale and stretch the spring or place objects on the scale and compress the spring.

Hooke's Law is also used as a first approximation when we don't know the force of exertion by a material. If a piece of rubber was to be stretched, the first guess for the force would be Hooke's Law–the force is directly proportional to the stretch.

Springs can stretch beyond their ability to restore themselves to their original shape. As they stretch to this point, they no longer follow Hooke's Law. Graphing the data, one would notice that the graph of force versus displacement (stretch) would begin to diverge from the straight line of the earlier data.

Planning for the Activity

Time Requirements

One to two periods may be required to complete this activity. The length of time will vary depending on the students' comfort level with graphing.

Materials Needed

For each group:

- graph paper pad of 50 sheets
- mass hanger
- piece of paper
- slotted weight set
- springs for Hooke's Law-set of 5
- tape, masking, 3/4" x 60 yds.

Advance Preparation and Setup

Ensure that the given weights are able to stretch the spring, but not beyond its point of elasticity.

You may want to consider how the students can affix the ruler so that it does not shift as they add weights to the spring.

Teaching Notes

When calibrating weights of objects, there is a temptation to have each student calculate their own weight. This can be embarrassing for students as weight can be a very personal attribute. As an alternative, have students pick out a rap artist, a sports figure, a movie star, and have them guess the weight in pounds and convert that weight to kilograms and then to newtons.

The Hooke's Law activity works very well. The graph is a very predictable straight line. The best-fit line tends to go directly through almost all of the points. As you move about the room, be sure that the students are measuring the stretch accurately. You must also ensure that the students do not move the ruler during trials. There is a Hooke's Law apparatus that has the ruler next to the spring and the spring has a platform suspended from it on which to hang the weights.

Rather than giving each team two springs, teams can exchange springs. Students with masses of 100 g, 200 g, and 500 g may at first think that they can only take three data points. They should be encouraged to think about combining the masses and they will then realize that they can get data for 300 g, 600 g, 700 g, and 800 g, as well.

NOTES

Activity 4 Your "At Rest" Weight

Activity 4 Your "At Rest" Weight

GOALS

In this activity you will:

- Distinguish between mass and weight.
- Calculate weight in newtons.
- Measure the effect of weight on the stretch of a spring.
- Graph the relationship between weight and stretch of a spring.
- Use a spring to create a scale and explain how Newton's Second Law is used in the creation of the scale.
- Calculate spring forces using Hooke's Law.

What Do You Think?

A canary and an elephant have enormous differences in weight. The elephant may weigh more than 10,000 times as much as the canary.

- **Can you use the same scale to weigh a canary and an elephant?**
- **How does a bathroom scale work?**

Record your ideas about these questions in your *Active Physics* log. Be prepared to discuss your response with your small group and the class.

239 Active Physical Science

Activity Overview

Students first learn how to distinguish mass from weight. They then explore the behavior of springs and Hooke's Law with a graphical analysis of the effect of force on the stretch of a spring. From an understanding of Hooke's Law, students can explain how a spring scale measures weight.

Student Objectives

Students will:

- Distinguish between mass and weight.
- Calculate weight in newtons.
- Measure the effect of weight on the stretch of a spring.
- Graph the relationship between weight and stretch of a spring.
- Use a spring to create a scale and explain how Newton's Second Law is used in the creation of a scale.
- Calculate spring forces using Hooke's Law.

ANSWERS FOR THE TEACHER ONLY

What Do You Think?

It will be interesting to find out what students think about how a scale works. The same scale could not be used for an elephant and a canary. The canary scale must be able to differentiate between grams, while the elephant scale must be good to the nearest kilogram. Although you could build an elephant scale to the precision needed for a canary scale, it would be prohibitively expensive and serve no obvious purpose.

The bathroom scale is similar to the spring that the students used in the pop-up toy. The greater the force applied, the more the spring is compressed.

Chapter 4

ANSWERS

For You To Do

Part A: Mass and Weight

1. a) They would both hit at the same time. All objects fall at the same rate.

Thrills and Chills

For You To Do

Part A: Mass and Weight

1. There is a ride at the amusement park today in which all you do is drop straight down. If you were to record your motion, you would find that your speed increases by 9.8 m/s every second. This value of 9.8 m/s every second is the acceleration due to gravitation. All objects near the surface of the Earth fall at this same rate of change of velocity with respect to time.

   You have Galileo to thank for this insight. As the story goes, he dropped two objects from the leaning tower of Pisa and observed them hitting the ground at the same time. The story may not be true, but Galileo did perform many experiments with balls rolling down inclined planes. The "dropping experiment" has been repeated many, many times with very precise equipment and with the effects of air resistance minimized or eliminated.

   a) If you were at the leaning tower of Pisa and dropped a baseball and a bowling ball at the same time, which would hit the ground first? Explain your answer.

240

Active Physics

b) If you dropped a baseball and a piece of paper, which would hit the ground first? Explain your answer.

c) How would you modify the statement, "All objects fall at the same acceleration" to account for your observation of a baseball and a piece of paper?

2. Produce at a supermarket is often priced by weight. Apples may cost 79 cents per pound and watermelon may cost 22 cents per pound.

a) What is a "pound"?

b) How would you define weight?

3. In physics, **weight** is defined as the force of the Earth's gravity on an object. The weight is the mass of an object multiplied by the acceleration due to gravity. Large masses are heavy and small masses are light. In the metric system, the mass of an object is measured in kilograms. The acceleration due to gravity is measured as 9.8 m/s every second or 9.8 m/s^2. Using this, you can calculate the weight of a student with a mass of 50 kg.

weight = (mass) × (acceleration due to gravity)

$$w = ma_g$$
$$= (50 \text{ kg})(9.8 \text{ m/s}^2)$$
$$= 490 \text{ kg·m/s}^2$$
$$= 490 \text{ N (newtons)}$$

The unit for weight is a newton (N). Weight is a force and has the same units as force. Recall that force is numerically equal to mass times acceleration ($F = ma$). Weight is the force due to gravity where the acceleration is the acceleration due to gravity. This can be written using symbols as a_g or g. $F = ma_g$ or $F = mg$.

a) Compare the weights of a gymnast with a mass of 40 kg and a football player with a mass of 110 kg.

4. The newton is a good metric unit for weight. However, you may wish to compare this to a pound, with which you are more familiar. Each kilogram weighs 2.2 lb. A 220-lb. football player has a mass of 100 kg. The weight of 100 kg, according to the equation $w = mg$, is 980 N.

Physics Words

weight: the vertical, downward force exerted on a mass as a result of gravity.

ANSWERS

For You To Do
(continued)

1. b) The baseball would hit the ground first. Air resistance slows the paper down.

   c) All objects fall at the same acceleration unless air resistance is appreciable.

2. a) A pound is a unit of measure. It tells how heavy something is.

   b) Weight is how heavy something feels.

3. a) $w = mg = (40 \text{ kg})(9.8 \text{ m/s}^2) = 392 \text{ N}$ (or about 400 N)

   $w = mg = (110 \text{ kg})(9.8 \text{ m/s}^2) = 1078 \text{ N}$ (or about 1100 N)

ANSWERS

For You To Do
(continued)

4. a) (11 lbs) × (1 kg/2.2 lbs) × (9.8 m/s²) = 49 N

   b) (0.25 lbs) × (1 kg/2.2 lbs) × (9.8 m/s²) = 1.1 N

4. a) 1500 pounds is approximately 6000 1/4 lb-burgers.

   1500 lb. = 6000 N

Part B: The Properties of Springs

2. a) When you release it, it springs back to its original shape.

3. a) When you release it, it springs back to its original shape.

a) Find the weight in newtons of a bowling ball that weighs 11 lb.

b) Find the weight in newtons of a $\frac{1}{4}$ lb-burger (a meat patty that has a weight of $\frac{1}{4}$ lb).

5. The weight of the $\frac{1}{4}$ lb-burger is close to 1 N. In a country that uses metric measurements, a burger restaurant could call their $\frac{1}{4}$ lb-burger a "Newton Burger." You can use this as an approximate way to determine how much something weighs in newtons if you know the weight in pounds. A 50-lb. person has the equivalent weight of 200 $\frac{1}{4}$ lb-burgers. Therefore the 50-lb. person has an approximate weight of 200 N.

a) Find the approximate weight in newtons of a roller coaster car that weighs 1500 lb.

Part B: The Properties of Springs

1. You can use a set of masses to determine the properties of springs. Make a spring from a piece of paper. Take a piece of $8\frac{1}{2} \times 11$ paper and fold it like an accordion.

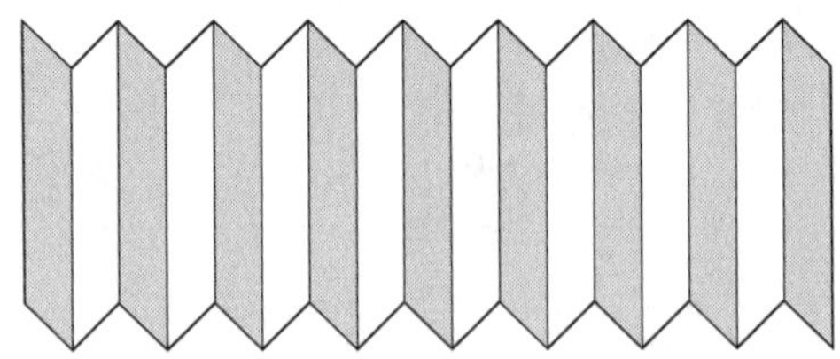

2. With a small force, stretch the paper spring slightly.

a) Record what happens when you release it.

3. With a small force, compress the paper spring slightly.

a) Record what happens when you release it.

4. Does the paper spring return to its original size and shape as the force of the stretch increases? Try it.

   a) Record your observations.

5. A metal spring has the same properties as the paper spring. The metal spring is better able to restore itself to its original shape than the paper spring. However, the metal spring can also be stretched past its load limit and you should be careful not to do this.

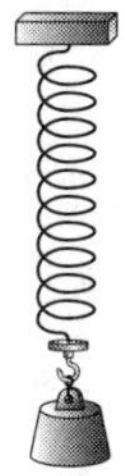

Be careful when placing and removing masses that the spring does not snap anyone. Have one person hold the bottom and top of the spring as another person adds or removes each mass.

6. The stretch of a metal spring can be measured precisely. Secure the spring vertically. You are now ready to measure its stretch with a given set of weights.

   You will have to convert from grams to kilograms to newtons. If a 100-g mass is used, this is equivalent to 0.1 kg. The mass of 0.1 kg has a weight of 0.98 N. This can be written as a single equation:

$$
\begin{aligned}
w &= mg \\
&= (100 \text{ g})\left(\frac{1 \text{ kg}}{1000 \text{ g}}\right)(9.8 \text{ m/s}^2) \\
&= 0.98 \text{ kg} \cdot \text{m/s}^2 \\
&= 0.98 \text{ N}
\end{aligned}
$$

   Notice that you converted grams to kilograms by multiplying by 1 kg/1000 g. In math class, you learned that you could always multiply a number by 1 and not change its value. For instance, $27 \times 1 = 27$. Since 1 kg is equal to 1000 g, the fraction 1 kg/1000 g has an equivalent numerator and denominator and the fraction equals 1. When you use this fraction in the equation, the 100 g gets converted to 0.1 kg. The gram units "drop out" and you are left with kilogram units. This dimensional analysis was done because you want the mass in kilograms.

ANSWERS

For You To Do
(continued)

4. a) If you stretch it too far, it won't return to its shape.

ANSWERS

For You To Do
(continued)

6. a) Students create a data table in their log. See table below.

7. a) Data will vary.

   b) The graph should be a straight line.

   c) Students predict stretch for a given weight. They should predict the stretch for a weight that they can test in the following step.

8. Students should have been able to predict the stretch with considerable accuracy.

9. a)–b) The graph will again be a straight line. The slope of the line will be different.

10. a)–b) The graph should be a straight line with a third slope.

11. a) Students complete the fourth column in their data table.

    b)–d) The weight divided by the stretched distance should give the same value for each pair of data of a given spring.

a) In your *Active Physics* log, create a data table with four columns. Label the first three columns mass, weight, and stretch of spring. The fourth column will be left blank for the time being.

7. Measure the stretch of the spring for different weights.

   a) Record your measurements in the data table.

   b) Plot a graph with the weights on the x-axis and the stretch of the spring on the y-axis.

   c) From your graph, predict what the stretch would be for a weight that you have not tried but between the weights that you have measured. This type of prediction from a graph is called interpolation.

8. Test your prediction by measuring the stretch of the spring for that weight.

   a) How good was your prediction?

9. Repeat the investigation for a second spring that looks different than the first. The spring may have larger or smaller coils or the coils may be closer together or further apart. You should have a new data table, a new graph, and a new interpolation.

   a) Describe how the springs differ in physical appearance.

   b) Describe how the springs differ in terms of the data chart and corresponding graph.

10. Invent a graph for a third spring.

    a) Sketch the invented spring's graph.

    b) Write a description of a spring that would have such data. The description should include the ease or difficulty of stretching the spring.

11. Return to your first data chart. Divide the weight by the stretched distance for each measurement.

    a) Record these values in the fourth column.

    b) What do you notice about these calculated values?

Table 6. a)

Mass	Weight	Stretch of spring	Weight ÷ stretch

c) Repeat for the calculations for the second data chart.

d) What value might your invented spring have for this column?

Part C: The Spring as a Weighing Machine

1. The spring stretches a different amount for each weight. Create a scale for weighing objects using one of the two springs that you have previously used. A scale has a spring and an arrow that points to the weight of the hanging object.

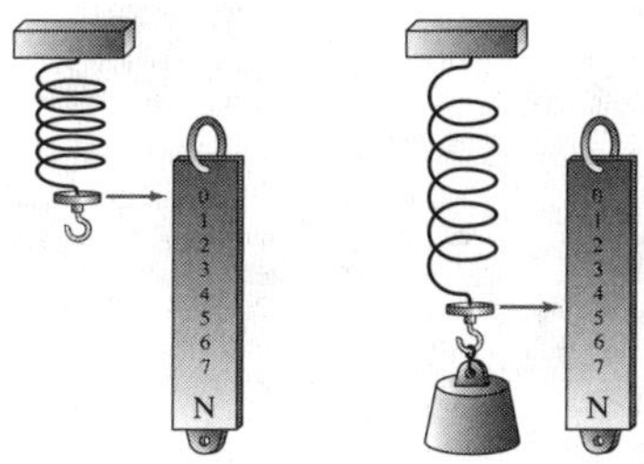

2. Choose three known masses. Measure their weight on your scale.

a) Record your values.

3. Choose two objects of unknown weight. Measure their weight on your scale.

a) Record the object and its weight.

ANSWERS

For You To Do
(continued)

Part C: The Spring as a Weighing Machine

2. a) Students' values will vary, depending on the known masses chosen.

3. a) Answers will vary. Encourage students to predict the weight before taking their measurements.

Chapter 4

FOR YOU TO READ

Hooke's Law Descibes the Restoring Force a Spring Exerts

Many springs have the property that the stretch of the spring is directly proportional to the force applied to it. This means that if you double the force, the stretch of the spring doubles. If you triple the force, the stretch of the spring triples. And if you make the force 2.7 times larger, the stretch of the spring is 2.7 times as large.

Springs that behave in this way are said to obey Hooke's Law. Sir Robert Hooke discovered this property of springs. The law explains very simply and precisely what restoring force a spring exerts if it is stretched. The more you stretch a spring, the larger the restoring force by the spring. You can describe this relationship in words or with a graph or with a mathematical equation. The equation for Hooke's Law is:

$$F = -kx$$

where F is the force,

x is the displacement, and

k is the spring constant.

The spring constant k is an indication of how easy or difficult it is to stretch or compress a spring.

The negative sign in the equation indicates that the pull by the spring is opposite to the direction it is stretched or compressed. Stretch a spring down and it tries to pull up. Stretch a spring to the right and it tries to pull to the left. Compress a spring and it tries to push.

You can calculate the value of the spring constant k by measuring the force exerted by the spring and the stretch of the spring.

Sample Problem 1

A 3.0-N weight is suspended from a spring. The spring stretches 2.0 cm. Calculate the spring constant.

Strategy: If a 3.0-N weight is suspended at rest from the spring, the spring must be applying a force of 3.0 N. If the spring were applying a force of less than 3.0 N, the weight would accelerate down. If the spring were applying a force of more than 3.0 N, the weight would accelerate up. When the force of gravity on the mass is 3.0 N down and the spring exerts a force of 3.0 N up, then the mass has no net force on it and it remains at rest once the friction brings it to rest. In this activity, you always took your measurements when the mass had stopped moving.

Givens:

$F = 3.0$ N

$x = 2.0$ cm

Solution:

$$F = kx$$

$$k = \frac{F}{x}$$

$$= \frac{3.0 \text{ N}}{2.0 \text{ cm}}$$

$$= 1.5 \text{ N/cm}$$

With a set of data points for the weight and the stretch, you would find that all values of k are the same or constant. And so k is the spring constant.

As you determined in the activity, you can also record the data on a graph. The graph can also be used to determine the spring constant k.

If the data are graphed so that the force is on the y-axis and the stretch is on the x-axis, then the spring constant will be the slope of the graph. You can see this if you compare the equations for Hooke's Law and the equation for a straight line (when the y-intercept is zero).

From here on, the negative sign in the equation will be omitted. In mathematics, you would say that you are calculating the absolute value of F using the absolute value of the stretch x. You must always keep in mind that the direction of the force is opposite to the direction of the stretch.

Hooke's Law: $F = kx$

Straight line: $y = mx + b$ (where $b = 0$)

Since the force is on the y-axis and the stretch x is on the x-axis, the slope m of the straight-line graph is k.

If the graph is constructed so that the force is on the x-axis and the stretch is on the y-axis, then the slope will be the reciprocal of the spring constant $1/k$. You can see this by comparing the equations for Hooke's Law and the equation for a straight line. In this case, we will write the equation for the straight line as $Y = MX + B$ so that the X-axis won't be confused with the stretch x, which is being placed on the Y-axis.

Hooke's Law: $F = kx$

$$x = \left(\frac{1}{k}\right)F$$

Straight line: $Y = MX$

Since the force is on the x-axis and the stretch x is on the y-axis, the slope m of the straight-line graph is $1/k$.

→

Chapter 4

ANSWERS

For You To Read

The graph of the stretch of a spring vs. force is available as a Blackline Master at the end of this activity. You may wish to make an overhead of the graph to show students when you discuss calculating slope.

Sample Problem 2

Weights are hung from a spring and the stretch is measured. The data collected is shown in the graph below. Calculate the spring constant from the graph.

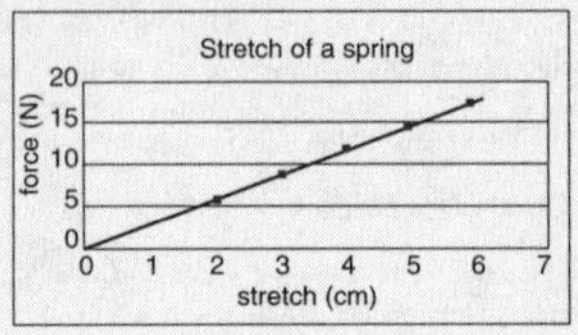

Strategy: Since the force is on the *y*-axis and the stretch is on the *x*-axis, you can compare the equations for a straight line and Hooke's Law.

Hooke's Law: $F = kx$

Straight line: $y = mx$

The slope of the graph will be equal to the spring constant *k*.

Givens:

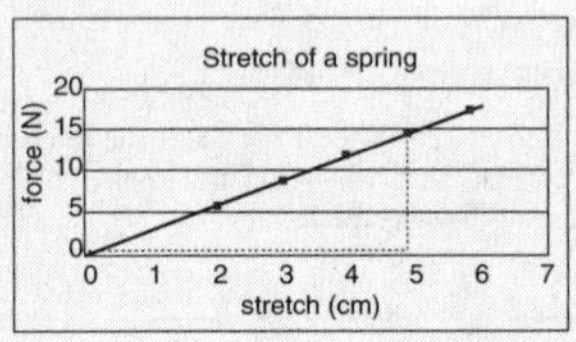

Solution:

$$\text{Slope} = \frac{\text{rise}}{\text{run}}$$

$$= \frac{15\text{N}}{5.0\text{ cm}}$$

$$= 3.0\text{ N/cm}$$

Stretch and Compress

You began the activity by both compressing and stretching a paper spring. You then made measurements on a stretched spring. Conducting an investigation with a compressed spring would produce similar results.

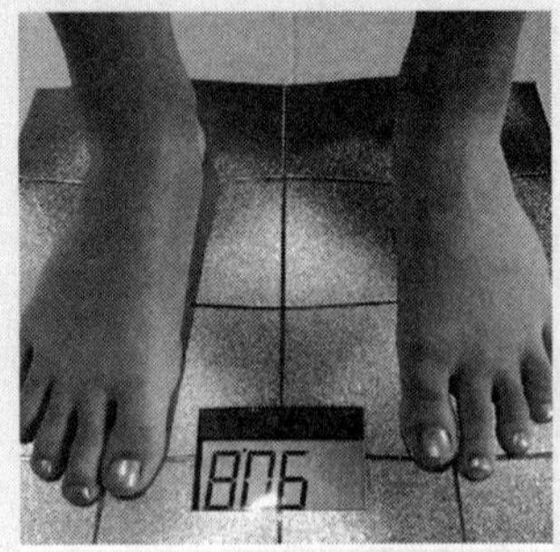

Many bathroom scales work by compressing the spring. Inside the bathroom scale is a spring. When you step on the scale, the spring compresses just enough to provide a force equal to your weight. The more weight, the more compression of the spring is required. The top of the spring is connected to a scale that has been calibrated.

Activity 4 Your "At Rest" Weight

As the spring gets compressed, the arrow points to a different number corresponding to the compression and force of the spring.

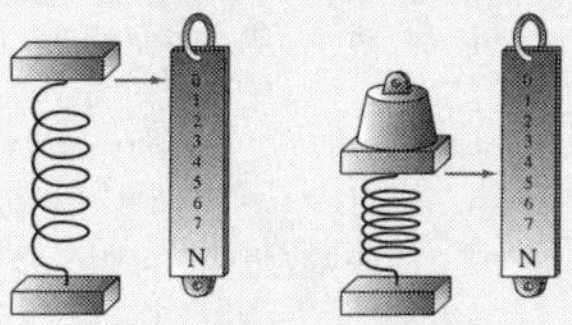

The scale does not read the weight of the object. The scale reads the compression of the spring. Of course, under normal circumstances the compression of the spring provides a force equal to your weight. You can then say that the scale reads your weight.

PHYSICS TALK

Hooke's Law

The restoring force of a spring is proportional to the stretch of a spring. The larger the stretch of a spring, the larger the force exerted by the spring.

The mathematical description of the force of a spring is called **Hooke's Law**. The equation is:

$$F = -kx$$

where F is the force,

x is the displacement, and

k is the spring constant.

The spring constant k is an indication of how easy or difficult it is to stretch or compress a spring. The negative sign in the equation indicates that the pull by the spring is opposite to the direction it is stretched or compressed. The spring has a tendency to return to its original shape.

→

Physics Words

Hooke's Law: the distance of stretch or compression of a spring is directly proportional to the force applied to it.

Thrills and Chills

A person steps on a scale. The scale moves a bit and then comes to rest. There are two forces acting on the person at rest. There is the force of gravity pulling down on the person. There is also the force of the spring pushing up on the person. These two forces are equal. The net force on the person is zero and the person remains at rest.

The sum of the forces = ma (Newton's Second Law)

The sum of the forces = 0 (The acceleration is zero)

Force due to gravity + force due to the spring = 0

$$mg \quad + \quad (-kx) \quad = 0$$

Reflecting on the Activity and the Challenge

Part of the fun of the roller coaster is the sensation you get as your weight appears to change at the peaks and valleys of the ride. In order to understand these apparent weight changes, you must understand how a scale works. All spring scales are based on Hooke's Law. The stretch (or compression of a spring) is directly proportional to the weight pulling or pushing on the spring. The more pull, the more is the stretch. Mathematically, you can write Hooke's Law as $F = kx$ where F is the weight (Force of Gravity) and x is the stretch. (The minus sign has been left out, but this requires you to remember that the force is in the opposite direction from the stretch.) The spring constant k is constant for a specific spring. The scale measures the compression of the spring. This force, when everything is at rest is equal to your weight. When the roller coaster starts moving, the scale will read many different values. You may want to include the weight changes in your design for your roller coaster. To do this, you may want to describe how a spring scale works.

Physics To Go

1. Objects near the surface of the Earth accelerate at a rate of 9.8 m/s every second. Based on this information, how fast will an object be going after it has fallen each of the following lengths of time?
   a) 2 s
   b) 5 s
   c) 10 s
2. Objects near the surface of the Moon accelerate at a rate of 1.6 m/s every second. Based on this information, how fast will an object be going after it has fallen each of the following lengths of time?
   a) 2 s
   b) 5 s
   c) 10 s
3. Calculate the weight of the following objects:
   a) a football player with a mass of 100 kg
   b) a toddler with a mass of 10 kg
   c) an adult with a mass of 60 kg
4. Use the approximation that a $\frac{1}{4}$-lb burger is one newton. Write down the approximate weights in newtons of the following objects:
   a) a 130-lb student
   b) a 1000-lb roller coaster cart
   c) a 50-lb child
5. Weights were hung from a spring and the stretch of the spring was measured. The data chart is given to the right.
   a) Graph the data with the weight on the x-axis and the stretch on the y-axis.
   b) Draw the best-fit line through the data points. (Do not connect the dots; draw the best-fit line.)
   Find the slope of the graph.
   What is the meaning of the slope?

Weight (n)	Stretch (cm)
0.0	0.0
0.3	2.0
0.7	4.6
1.2	8.0
2.0	13.0
2.4	16.2
3.1	21.0

ANSWERS

Physics To Go

1. a) $v = at = (9.8\ \text{m/s}^2)(2\ \text{s})$
   $= 19.6$ m/s (or about 20 m/s)

   b) $v = at = (9.8\ \text{m/s}^2)(5\ \text{s})$
   $= 49$ m/s (or about 50 m/s)

   c) $v = at = (9.8\ \text{m/s}^2)(10\ \text{s})$
   $= 98$ m/s (or about 100 m/s)

2. a) $v = at = (1.6\ \text{m/s}^2)(2\ \text{s})$
   $= 3.2$ m/s (3 m/s)

   b) $v = at = (1.6\ \text{m/s}^2)(5\ \text{s})$
   $= 8.0$ m/s (8 m/s)

   c) $v = at = (1.6\ \text{m/s}^2)(10\ \text{s})$
   $= 16$ m/s (20 m/s)

3. a) $w = mg = (100\ \text{kg})(9.8\ \text{m/s}^2)$
   $= 980$ N (1000 N)

   b) $w = mg = (10\ \text{kg})(9.8\ \text{m/s}^2)$
   $= 98$ N (100 N)

   c) $w = mg = (60\ \text{kg})(9.8\ \text{m/s}^2)$
   $= 590$ N (600 N)

4. a) 130 lbs = 520 quarter-pounds
   = 520 N

   b) 1000 lb = 4000 N

   c) 50 lb = 200 N

5. a)–b) Slope = $\Delta y/\Delta x = 20\ \text{cm}/3\ \text{N} = 6.7$ cm/N (7 cm/N)

   This is $1/k$.

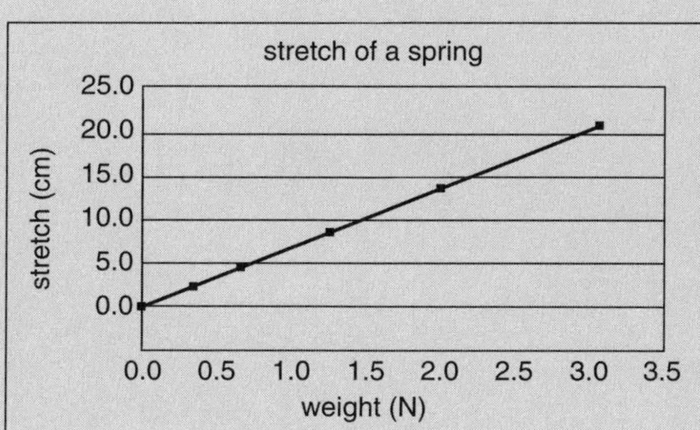

Chapter 4

ANSWERS

Physics To Go
(continued)

5. c) If the spring has a steeper slope, it will be easier to stretch.

6. As the weight pulling on the spring increases, the stretch of the spring increases in proportion. (Double the weight, you double the stretch.)

7. $F = -kx$

   $k = F/x = 12\text{ N}/3.0\text{ cm}$
   $= 4.0\text{ N/cm}$

8. The 15.0 N/cm spring requires 15 N to stretch it one centimeter while the 10.0 N/cm requires only 10 N to stretch it one centimeter.

9. The spring constant will be the change in *F* divided by the change in stretch.

   $k = F/x = 2.9\text{ N}/ 2.0\text{ cm}$
   $= 1.95\text{ N/cm (2.0 N)}$

10. A bathroom scale has a spring attached to the top and bottom. When you stand on the scale, the spring is compressed. The compression is proportional to your weight. The weights of the corresponding compessions are displayed as the numbers on the scale.

c) Invent a graph for a second spring. Sketch the invented spring's graph. Write a description of a spring that would have such data. The description should include the ease of difficulty of stretching the spring.

6. When Robert Hooke first described the relationship that has come to be known as Hooke's Law, he wrote "as the force, so the stretch." Explain in a full sentence or two what Hooke meant by this. (Hooke wrote this as a footnote in Latin with the letters all mixed up. This allowed him to keep his discovery a secret for a while.)

7. A weight of 12 N causes a spring to stretch 3.0 cm. What is the spring constant *k* of the spring?

8. Two springs have spring constants of 10.0 N/cm and 15.0 N/cm. Which spring is more difficult to stretch?

9. Calculate the spring constant *k* from the graph of a stretched spring below.

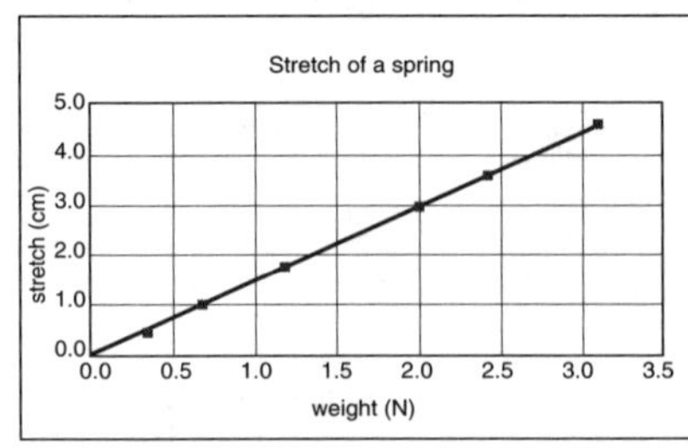

10. Describe how a bathroom scale works.

Stretching Exercise

Get permission to take apart a bathroom scale. Investigate the parts. Create sketches to explain how the scale works and the function of all of the parts. When your explanation is complete, put the scale back together.

Stretch of a Spring Graph

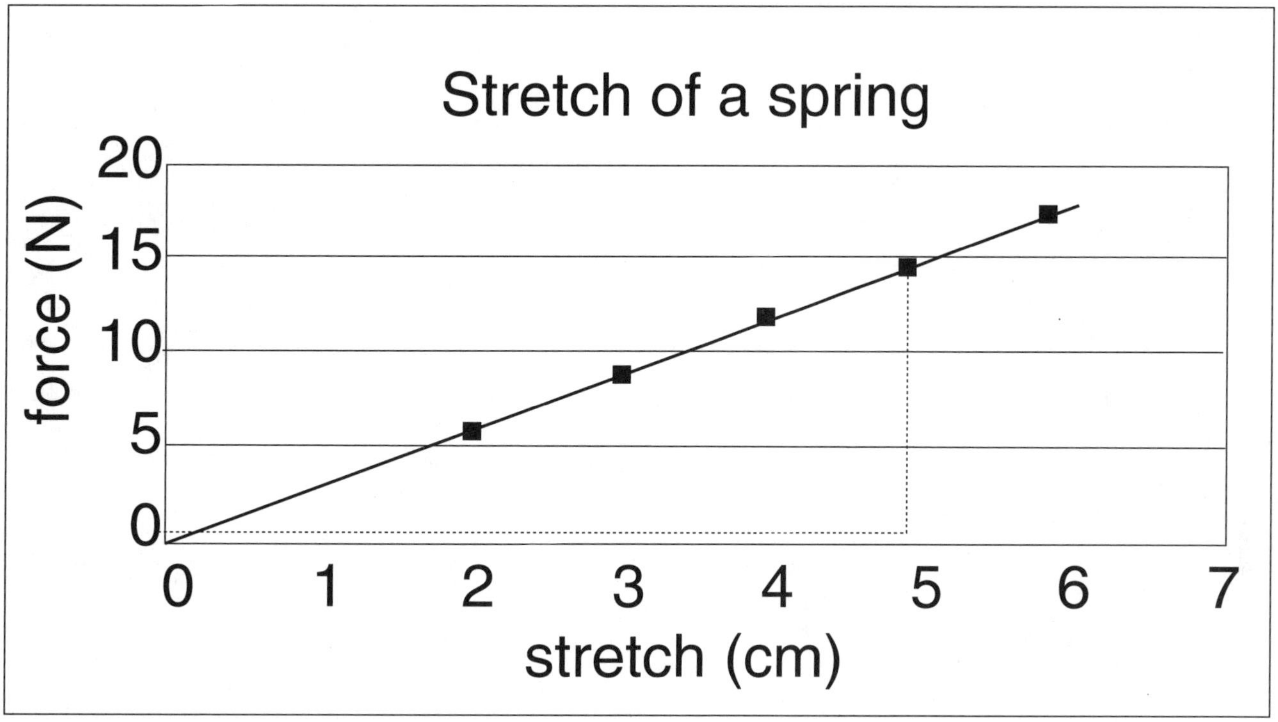

$$\text{Slope} = \frac{\text{rise}}{\text{run}}$$

$$= \frac{15\text{N}}{5.0\text{ cm}}$$

$$= 3.0\text{ N/cm}$$

For use with *Thrills and Chills*, Chapter 4, Activity 4: Your "At Rest" Weight

NOTES

ACTIVITY 5
Weight on a Roller Coaster

Background Information

It is suggested that you read the **For You To Read** and **Physics Talk** sections in the student text for **Activity 5** before proceeding in this section.

Newton's First Law informs us that objects that are rest remain at rest and objects in motion remain in motion unless acted upon by a force. This leads us to recognize that "at rest" and "constant velocity" are similar. Newton's First Law tells us that there is no way to distinguish "at rest" from "constant velocity." This certainly mirrors our experiences in trains, cars, and planes that travel at high speeds. Except for the occasional bump, sway, or air turbulence, sitting in a moving vehicle is similar to sitting in that same vehicle while it is at rest.

A scale measures the upward force that the spring is exerting on the mass. The force of gravitation (the weight) of the object pulls the object toward the Earth. The scale applies an equal force up so that you remain stationary. As you step onto the scale, the scale begins to descend. You are compressing the spring in the scale. As you compress the spring, the restoring force that the spring can exert gets larger (Hooke's Law). As you continue to descend the spring gets more compressed and the spring force gets larger. You find that you descend a bit too much and the spring now exerts a force larger than your weight and you begin to ascend. You finally come to rest when the spring force is equal to the weight. The motion would be easier to see if you stepped onto a trampoline. The trampoline would bend and you would descend. You would then move up and down until the trampoline came to rest. Without friction, you would continue to move up and down forever. On a scale, there are often magnetic brakes to allow the scale to come to rest quickly.

Newton's Second Law can also explain the relationship between the forces and your motion. Newton's Second Law states that $\Sigma F=ma$. If there is zero acceleration (you are at rest or moving at a constant speed), then the net force must be zero. In the case of the spring scale, the force up provided by the scale must be equal to the force down provided by gravitation (your weight).

For you to accelerate up or down while in an elevator requires a net force. The only two forces acting on you, while in the elevator, are gravitation pulling you down and the force of the scale pushing up. The gravitation force does not change. For you to accelerate up, the force of the scale pushing up must increase. For you to accelerate down, the force of the scale pushing up must decrease.

This is explored both in the activity and in the mathematical analysis in the Student Edition.

Planning for the Activity

Time Requirements

Allow one class period to complete the activity. Another class period will probably be required to work through the mathematical problems. Note that for many classes, the mathematical questions are optional.

Materials Needed

For each group:

- mass hanger
- slotted weight set
- spring scale, 0-10 newton range

Advance Preparation and Setup

Ensure that the mass provides a suitable reading with the spring scale.

Teaching Notes

It will take some practice for students to get the mass moving at a constant speed and to read the scale while it is moving at constant speed. As you circulate about the room, ensure that the students are taking the time to get a good reading. If you notice that students do not agree about the scale readings, help them to look only at the scale while it is moving at constant speed. You may want to suggest that a student block the scale from view (with a file folder) while the block begins to move or when the block begins to stop.

While analyzing the forces on the block or the person in the elevator, the students will want to think about the cable and the forces on the elevator itself. Remind them that the object of interest is the

Chapter 4

block or the person on the elevator. The elevator does not apply a force to them. The spring scale applies a force to them. It is true that the elevator applies a force to the spring scale and supports the spring scale, but that is not of interest to us as we analyze the forces on the block. The creation of the force diagrams of the block should help in this regard.

While reviewing the mathematics of the elevator (an optional topic for many classes), it is best to always begin with Newton's Second Law and then to solve for the force by the scale. There are other approaches to finding the scale reading, but beginning with Newton's Second Law is preferred. All students, irrespective of their algebra skills, should be able to identify whether the elevator is at rest, moving at constant velocity, or accelerating up or down and to correctly predict the relative reading on the spring scale (larger, equal, smaller than the weight.)

Activity 5 Weight on a Roller Coaster

GOALS

In this activity you will:

- Recognize that the weight of an object remains the same when the object is at rest or moving (up or down) at a constant speed.
- Explore the change in apparent weight as an object accelerates up or down.
- Analyze the forces on a mass at rest, moving with constant velocity, or accelerating by drawing the appropriate force vector diagrams.
- Mathematically predict the change in apparent weight as a mass accelerates up or down.

What Do You Think?

As the roller coaster moves down that first hill, up the second hill, and then over the top, you feel as if your weight is changing. In roller coaster terms this is called airtime. It is the feeling of floating when your body rises up out of the seat.

- **Does your weight change when you are riding on a roller coaster?**
- **If you were sitting on a bathroom scale, would the scale give different readings at different places on the roller coaster?**

Record your ideas about these questions in your *Active Physics* log. Be prepared to discuss your response with your small group and the class.

Activity Overview

Students analyze how weight changes in an ascending and descending elevator. The application of Newton's First and Second Laws of Motion can explain why the apparent change takes place as the elevator accelerates. The use of force vectors help to make sense of the elevator as a "cheap" roller coaster ride.

Student Objectives

Students will:

- Recognize that the weight of an object remains the same when the object is at rest or moving (up or down) at a constant speed.
- Explore the change in apparent weight as an object accelerates up or down.
- Analyze the forces on a mass at rest, moving with constant velocity or accelerating by drawing the appropriate force vector diagrams.
- Mathematically predict the change in apparent weight as a mass accelerates up or down.

ANSWERS FOR THE TEACHER ONLY

What Do You Think?

Encourage your students to describe how they "feel" on a roller coaster. Some students may also compare this feeling to a car traveling at a high speed over the crest of a hill. Once again, it is important in the **What Do You Think?** question to elicit prior understandings, not to reach closure on the question.

When riding on a roller coaster, your weight doesn't change, but your apparent weight changes.

If you were sitting on a scale, you would notice a different weight on the scale.

ANSWERS

For You To Do

Part A: Moving the Mass at a Constant Speed

1. a) Let's assume that the student is hanging a 1-kg mass. The weight will be 9.8 N.

For You To Do

In this activity, you will investigate the weight changes you feel when you are on a roller coaster. You will use the spring scale for your observations. However, you will explain what you observe with both the spring scale and the bathroom scale.

Part A: Moving the Mass at a Constant Speed

1. Hang a mass from the spring scale and note the force. When the mass is not moving the acceleration equals 0 and therefore the force of gravity pulling the mass down and the force exerted by the spring pulling the mass up must be equal. The force of the spring equals the force of gravity. You could also just say, "the spring is measuring the weight."

   a) Record the weight of the mass in your log.

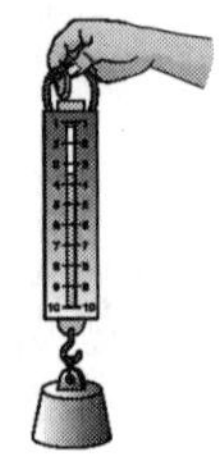

2. With your arm extended down, move the mass up until your arm is as high as you can reach. Once you start the mass moving, you want to keep lifting it at a constant speed. Your lab partners will try their best to read the spring scale during the time that the mass is moving at constant speed. Ignore the readings when you first start moving the mass and when you stop it. You will return to those observations later.

You may have to repeat this a few times so that you can lift the mass at a constant speed so that your lab partners can observe it.

a) Record your finding in your log.

3. The observations in **Step 2** may have been difficult for you to make accurately. The spring scale should have displayed the same reading when the scale moved at constant speed as it did when it was suspended at rest. The same result will occur in an elevator. If you are on a bathroom scale, the scale will compress and display your weight when the elevator is at rest. It will display the same weight when the elevator is moving at a constant speed between floors.

   Repeat the observation with the spring scale moving down at a constant speed. Is the weight once again the same? Please remember, you are only interested in the weight reading as the mass descends at a constant speed, not when you first get it to move.

   a) Record your findings.

4. **Newton's First Law** states that "An object at rest remains at rest and an object in motion remains in motion, unless acted upon by an unbalanced force." **Newton's Second Law** states that "An accelerating object must have a net force acting upon it: $F = ma$." When the mass is being lifted at constant speed, there is no acceleration since acceleration is defined as a change in speed with respect to time. If there is no acceleration, then there is no net force.

   a) Draw a box in your log and draw arrows to show the forces on the box when it is not accelerating.

5. In physics, the arrows you drew are called vectors. Check your drawing to see if the arrows (vectors) have these features:

Physics Words

Newton's First Law of Motion: an object at rest stays at rest and an object in motion stays in motion unless acted upon by an unbalanced, external force.

Newton's Second Law of Motion: if a body is acted upon by an external force, it will accelerate in the direction of the unbalanced force with an acceleration proportional to the force and inversely proportional to the mass.

ANSWERS

For You To Do
(continued)

2. a) When the mass is moving up at a constant speed, the weight is still 9.8 N.

3. a) When the mass is moving down at a constant speed, the weight is still 9.8 N.

4.

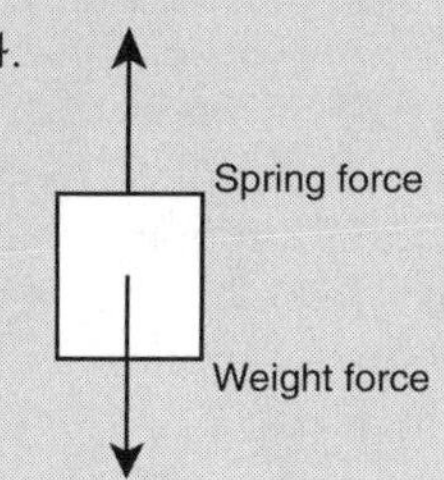

Chapter 4

ANSWERS

For You To Do
(continued)

5. a) The weight vector is down because the Earth is pulling the mass down.

   b) The spring vector is up because the spring is applying a force up.

   c) No, there are no other force vectors acting on the box. (A force may be applied to the spring to hold it in place, but this is not a force on the box—the spring applies the force to the box.)

   d) Yes, they are the same length because the forces are equal and opposite.

6. The diagram is identical to that in Step 4.

   Newton's Second Law states that when there is no acceleration ($a = 0$), then there is no net force ($\Sigma F = 0$). For the sum of the forces to be equal to zero, then the weight down must be equal to the spring force up.

a) Was the weight vector drawn down? Why did you draw it this way?

b) Was the force of the spring vector drawn up? Why did you draw it this way?

c) Were there any other force vectors? What were they representing?

d) Were the weight vector and the spring vector equal in length? The size of the vectors implies the size of the force. If the forces are equal, then the lengths of the vectors should be the same. If needed, change your force diagram so that the vectors represent the size of the forces.

6. Now consider the box if it were moving down.

a) Draw a second box with the force vectors when the box is moving down at a constant speed. Provide an explanation using Newton's First Law and Newton's Second Law (similar to **Step 4**) as a rationale for your diagram.

Part B: Accelerating the Mass

1. It's now time to return to the scale readings when you first start moving the mass. With your arm extended down, accelerate the mass up until your arm is as high as you can reach. Your lab partners will try their best to read the spring scale during the time that the mass starts to move. Once again, you may have to repeat this a few times so that you can lift as others observe.

a) Record your observation in your *Active Physics* log.

b) Use Newton's Second Law ($F = ma$) to make sense of the observation in your log.

c) Draw a box representing the mass and draw the force vectors acting on the box as it first begins to move.

2. Check your drawing to see if the force vectors have these features:

a) Was the weight vector drawn down? Why did you draw it this way?

b) Was the force of the spring vector drawn up? Why did you draw it this way?

Part B: Accelerating the Mass

1. a) When the mass is first lowered, the scale reads a lower reading. In the example above, instead of reading 9.8 N, the scale reads 6.0 N.

   b) Since the mass is accelerating down, then Newton's Second Law informs us that there must be a net force down. Since the weight of the object (the pull of the Earth on the object) is identical to what it was before, then the spring force must decrease. This decrease in the spring force gives a lower scale reading.

   c)

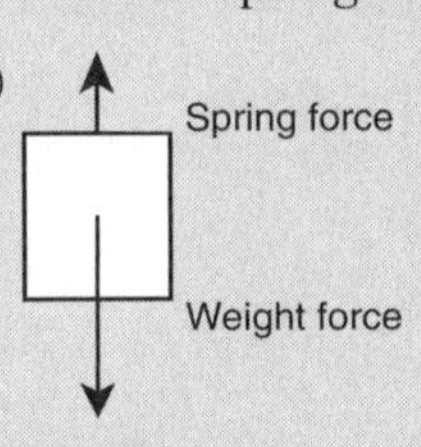

2. a) The weight vector is down because the Earth pulls the object down.

   b) The spring pulls the mass up.

c) Were there any other force vectors? What were they representing?

d) Were the weight vector and the spring vector equal in length? The size of the vectors implies the size of the force.

3. Since the box is accelerating up, the force of the spring must have been larger than the force of gravity. Newton's Second Law indicates that acceleration up requires a net force up. In your force-vector diagram, the vector representing the spring scale should be larger than the force of the gravity vector.

a) If required, modify your diagram.

4. The spring scale displays a value larger than the mass's weight on Earth. Suppose you were standing on a bathroom scale in an elevator. The elevator begins to move up.

a) How would the reading on the bathroom scale compare to your weight at rest?

5. Predict what would happen to the scale reading when the mass stops moving upward.

a) Record your prediction in your log.

b) Repeat the observations for the moments when the mass stops its upward motion. Describe your observation in your log.

6. Begin with the mass high in the air and suspended by the spring scale.

a) If you begin to lower the mass, in which direction is the acceleration?

b) Draw a force-vector diagram that has a net force in the direction of the acceleration.

c) Do you predict that the scale will read a value higher or lower than it does when the mass is at rest? Record your prediction.

d) Try it out. Record your observations in your log.

ANSWERS

For You To Do
(continued)

2. c) No, there are no other forces acting on the box.

d) No, the weight vector is larger because there is a net force down that provides the acceleration down.

3. a) Students may need to modify their diagrams.

4. a) The reading of the bathroom scale is the reading of the spring, not your weight. The bathroom scale would read a number less than your actual weight. If you typically weigh 700 N, the spring scale may read only 650 N (your apparent weight during the acceleration).

5. a) Make sure that students record their predictions before making their observations.

b) While you stop moving, the scale will read more than the original weight. This is because to stop, the mass will require an upward acceleration. An upward acceleration requires a net upward force. Thus, the spring scale force will be greater than the weight.

6. a) As you begin to lower the mass, the acceleration is down.

b)

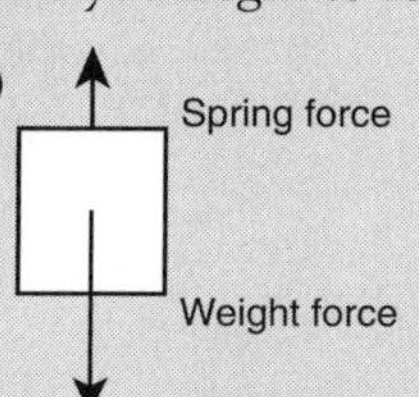

c) Students should predict that the scale will have a lower reading.

d) The scale will have a smaller reading, because the spring force is less than the weight.

ANSWERS

For You To Do
(continued)

7. See chart below.

   This chart is available as a Blackline Master at the end of this activity.

8. Similarities between roller coaster rides and elevators:
   - Both roller coasters and elevators involve motion.
   - Both roller coasters and elevators involve accelerations
   - Passengers on both roller coasters and elevators experience weight changes.

   Differences between roller coaster rides and elevators:
   - Elevators travel in one-dimension (up and down) while roller coasters travel in three-dimensions.
   - Elevators are programmed to have small accelerations while roller coasters are designed to have large accelerations.
   - Elevators have accelerations in the vertical direction while roller coasters have accelerations to the sides as well.
   - You remain vertical in elevators while you can be upside down in roller coasters.

7. As a summary of what changes occur to the spring scale reading (or the bathroom scale reading in an elevator), complete the following chart in your log. Some responses are provided.

	Acceleration (up, down, zero)	Scale reading (larger, smaller, equal to weight)
1. Elevator at rest on bottom floor	Zero	Equal
2. Elevator starts moving up	Up	Larger
3. Elevator moves up at constant speed		
4. Elevator comes to rest on top floor		
5. Elevator is at rest at top floor		
6. Elevator begins to move down	Down	Smaller
7. Elevator moves down at constant speed		
8. Elevator comes to rest on bottom floor		
9. Elevator at rest on bottom floor		

8. Riding in an elevator is similar to riding in a roller coaster. Although the physics is the same, the elevator ride does not have the excitement of a roller coaster.

   a) Compare elevator rides and roller coasters by providing three similarities and three differences.

7. Chart

	Acceleration (up, down, zero)	Scale reading (larger, smaller, equal to weight)
1. Elevator at rest on bottom floor	Zero	Equal
2. Elevator starts moving up	Up	Larger
3. Elevator moves up at constant speed	Zero	Equal
4. Elevator comes to rest on top floor	Down	Smaller
5. Elevator is at rest at top floor	Zero	Equal
6. Elevator begins to move down	Down	Smaller
7. Elevator moves down at constant speed	Zero	Equal
8. Elevator comes to rest on bottom floor	Up	Larger
9. Elevator at rest on bottom floor	Zero	Equal

Activity 5 Weight on a Roller Coaster

FOR YOU TO READ

Forces Acting during Acceleration

As the roller coaster moves you about, you feel in your stomach and from the car seat that things are happening. These, however, are more than just feelings. These changes can be measured. You can use physics to explain these changes.

If you were sitting on a scale on a level roller coaster, at rest or moving with a constant velocity, the scale reading would be equal to your weight. The force of the Earth pulling on you (your weight), shown as a blue vector in the diagram, would be equal to the force of the compressed spring within the bathroom scale, shown as a red vector. A force diagram shows this.

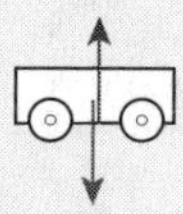

If the person on the roller coaster weighs 600 N, then the bathroom scale would have to be providing a force of 600 N. Any smaller and the person would accelerate down. Any larger and the person would accelerate up. When you first sit on the scale, the compression is too little and you do move down. The spring compresses and provides a larger force but you continue to move down. You go past the compression you need and the spring then pushes up. You go back up and down and up and down and continue this movement until the spring's force is exactly equal to your weight.

As the elevator or roller coaster starts moving up, there is acceleration up. (Remember that acceleration is a change in velocity with respect to time.) For you to accelerate up, there must be an unbalanced force pushing you up. Since you are in contact with the bathroom scale then it must be the bathroom scale that is pushing you up. (Yes, the elevator or roller coaster is pushing on the scale, but in physics you only have to worry about the forces on you for this situation.)

The scale reading will be greater than your weight. The force of the Earth pulling on you (your weight), shown as a blue vector, would be less than the force of the compressed spring within the bathroom scale, shown as a red vector. A force diagram shows this.

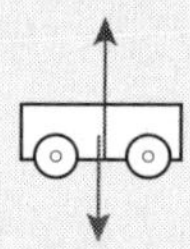

If the person on the roller coaster weighs 600 N, then the bathroom scale would have to be providing a force of greater than 600 N.

Another way of looking at the situation is to look first at the forces. According to the vector diagram, the force of the scale is larger than the force of gravity. The net force is therefore up and the object will accelerate up according to Newton's Second Law, $F = ma$.

Thrills and Chills

PHYSICS TALK

Calculating Acceleration

You can calculate the acceleration of an elevator if you know the weight of the object and measure the force of the spring scale during the acceleration. Assume that a person on a scale weighs 600.0 N and the force of the scale on the person is 700.0 N up.

$$\text{Net force} = 700.0\ \text{N} - 600.0\ \text{N}$$
$$= 100.0\ \text{N}$$

If the weight of the object is 600.0 N, you can calculate its mass:

$$\text{Weight} = ma_g \quad \text{where } a_g = 9.8\ \text{m/s}^2.$$
$$600.0\ \text{N} = m\,(9.8\ \text{m/s}^2)$$
$$m = 61\ \text{kg}$$

Knowing the mass and the net force, you can calculate the acceleration using Newton's Second Law:

$$F = ma$$
$$100.0\ \text{N} = (61\ \text{kg})\,a$$
$$a = 1.6\ \text{m/s}^2$$

Similarly, you can calculate the reading of the spring scale if you know the acceleration of the elevator.

Sample Problem

An elevator at the top floor begins to descend with an acceleration of 2.0 m/s^2. What will a bathroom scale read if a 50.0 kg person is standing on the scale during the acceleration?

Strategy: Since the elevator is accelerating down, the net force must be down. The vector forces would look like this:

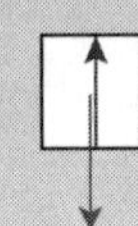

Activity 5 Weight on a Roller Coaster

Newton's Second Law states that

$\Sigma F = ma$

where ΣF means "the vector sum of the forces" or "the net force."

Since the weight is greater than the force of the spring and in the opposite direction, you can write this as:

Weight – force by spring = ma

Givens:

$m = 50.0$ kg

$a = 2.0$ m/s^2

Solution:

Weight $= ma_g$

$= (50.0 \text{ kg})(9.8 \text{ m/s}^2)$

$= 490$ N

Weight – force by spring = ma

$490 \text{ N} - F_s = (50.0 \text{ kg})(2.0 \text{ m/s}^2)$

$F_s = 390$ N

The scale would read 390 N instead of the person's weight, which is 490 N.

$F_s = 390$ N

If the elevator accelerates up, the scale reads a value higher than the weight of 490 N. There is really no limit to the upward acceleration though the person would become unconscious if the acceleration were greater than eight times the acceleration due to gravity. This has been experimentally determined by test pilots.

If the elevator accelerates down, the scale reads a value lower than the weight of 490 N.

There is a lower limit on the weight. If the elevator accelerates down at $9.8\,\text{m/s}^2$ (e.g., the cable broke and the elevator is in free fall), the scale would not push at all and its reading would be 0 N.

$$\Sigma F = ma$$

$$\text{Weight} - \text{force by spring} = ma$$

$$(50.0\text{ kg})(9.8\,\text{m/s}^2) - F_s = (50.0\text{ kg})(9.8\,\text{m/s}^2)$$

$$490\text{ N} - F_s = 490\text{ N}$$

$$F_s = 0\text{ N}$$

This is what is experienced when someone skydives out of a plane and has not yet pulled the cord on the parachute. It is also felt for a few moments in the amusement park ride where you are in free fall.

Reflecting on the Activity and the Challenge

A ride in an elevator is a lot like a ride in a roller coaster. The elevator moves up and down, but at too slow a pace and with too little acceleration to provide a great deal of excitement. When the elevator is at rest or moving up or down at constant velocity, your weight readings are identical. That's because at rest or moving at a constant velocity requires no net force.

The force of the scale up on you is equal to the force of your weight down. The bathroom scale denotes the value of the force up on you. When objects accelerate (change their velocities), there is a net force. When the elevator accelerates up, you also accelerate up. This is because the Earth pulls down on you less than the scale pushes you up. The scale reads a larger weight than before. You also feel as if you weigh more. When the elevator accelerates down, you also accelerate down. This is because the force of the scale up on you is less than the force of your weight down. The scale reads a smaller weight than before. You also feel as if you weigh less. If the elevator cable were to break, you would have only the force of your weight pulling you down. The scale would not push up and you and the weight reading would be zero. You would be "weightless." Roller coasters, like elevators, have sections where the acceleration is up or down. At these locations, people sense that they weigh more or less. When you design your roller coaster, you may want to take into consideration how large an acceleration your sample population would enjoy.

Physics To Go

1. The vector diagram shows a block of wood that can move up and down. The red vector represents the force pushing up on the block. The blue vector represents the weight of the block.
   a) Could the block be at rest?
   b) Could the block be moving up at a constant speed?
   c) Could the block be accelerating down?
   d) If you wrote "no" for any of the above questions, draw a force diagram for that description.

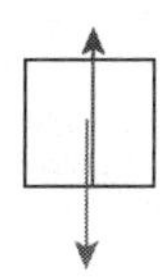

ANSWERS

Physics To Go

1. a) Since the forces are equal, the block could be at rest
   b) Since the forces are equal, the block could be moving up at a constant speed.
   c) Since the forces are equal, there is no net force down and therefore the block cannot be accelerating.
   d)

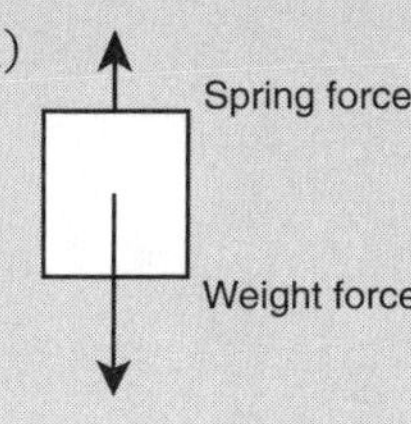

Chapter 4

ANSWERS

Physics To Go
(continued)

2. See chart below.
3. The elevator is accelerating down. The spring scale force (the reading on the scale) is less than the weight.
4. The acceleration will be up. The passenger will feel heavier and the scale will read a larger value.

2. Complete the following chart in your log. Some responses are provided.

	Acceleration (up, down, zero)	Scale reading (larger, smaller, equal to weight)
1. Elevator at rest on bottom floor	Zero	Equal
2. Elevator starts moving down		
3. Elevator moves down at constant speed		
4. Elevator comes to rest on bottom floor		
5. Elevator is at rest at bottom floor		
6. Elevator begins to move up	Up	Larger
7. Elevator moves up at constant speed		
8. Elevator comes to rest on top floor		
9. Elevator at rest on top floor	Zero	Equal

3. A student usually weighs 140 lbs. On an elevator, the person is surprised to find that the scale only reads 137 lbs. for a few moments. Describe the movement of the elevator.
4. A person in an elevator at rest weighs 600 N. The elevator is about to move from the 2nd floor to the 5th floor. When it first starts to move, what will the passenger observe about his weight?
5. An elevator at the top floor begins to descend with an acceleration of 1.5 m/s^2. A person is standing on a bathroom scale in the elevator.

2. Chart

	Acceleration (up, down, zero)	Scale reading (larger, smaller, equal to weight)
1. Elevator at rest on bottom floor	Zero	Equal
2. Elevator starts moving down	Down	Smaller
3. Elevator moves down at constant speed	Zero	Equal
4. Elevator comes to rest on bottom floor	Up	Larger
5. Elevator is at rest at bottom floor	Zero	Equal
6. Elevator begins to move up	Up	Larger
7. Elevator moves up at constant speed	Zero	Equal
8. Elevator comes to rest on top floor	Down	Smaller
9. Elevator at rest on top floor	Zero	Equal

a) Will the bathroom scale's reading increase or decrease once the elevator starts?

b) What will a bathroom scale read if a 50-kg person is standing on the scale during the acceleration?

6. A 50-kg student is on a scale in the elevator.

a) What will be the scale reading when the elevator is at rest?

b) What will be the scale reading when the elevator accelerates up at a rate of 2 m/s^2?

c) What will be the scale reading as the elevator travels up at a constant speed?

7. Explain the meanings of the three sketches below. Specifically, why is there a different scale reading for the same student in each elevator?

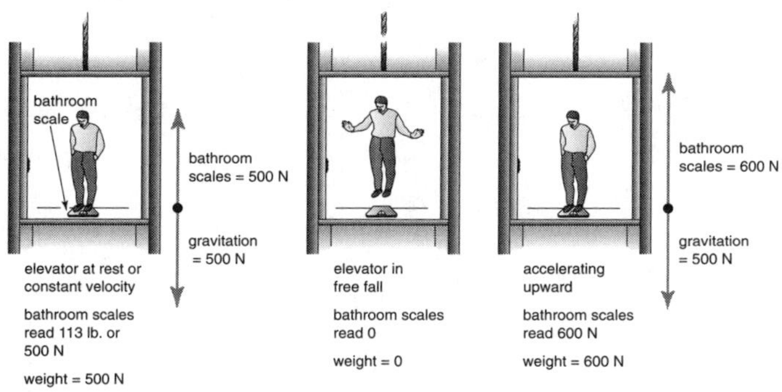

Stretching Exercise

Use a digital camera or projector and record some of the activities while riding in an elevator. Go with an adult to a place where there is an elevator that moves up and down several floors.

Answers

Physics To Go

(continued)

5. a) As the elevator begins to descend, the acceleration is down. The scale reading will decrease.

b) Weight – (force by spring) = ma

$(50.0 \text{ kg})(9.8 \text{ m/s}^2) - F_S = (50.0 \text{ kg})(1.5 \text{ m/s}^2)$

$490 \text{ N} - F_S = 75 \text{ N}$

$F_S = 415 \text{ N}$

6. a) Weight – force by spring = ma

$(50.0 \text{ kg})(9.8 \text{ m/s}^2) - F_S = (50.0 \text{ kg})(0 \text{ m/s}^2)$

$490 \text{ N} - F_S = 0 \text{ N}$

$F_S = 490 \text{ N}$

b) Weight – force by spring = ma

$(50.0 \text{ kg})(9.8 \text{ m/s}^2) - F_S = (50.0 \text{ kg})(2.0 \text{ m/s}^2)$

$490 \text{ N} - F_S = 100 \text{ N}$

$F_S = 390 \text{ N}$

c) Weight – force by spring = ma

$(50.0 \text{ kg})(9.8 \text{ m/s}^2) - F_S = (50.0 \text{ kg})(0 \text{ m/s}^2)$

$490 \text{ N} - F_S = 0 \text{ N}$

$F_S = 490 \text{ N}$

7. In the first diagram, there is no acceleration of the elevator and the forces are equal. The scale reading (the apparent weight) is 500 N.

In the middle diagram, the elevator is in free fall. There is no force up. Therefore, the spring scale (that typically provides the force up) reads zero.

In the last diagram, the elevator is accelerating up. To have an acceleration up, the scale must be providing a larger force up than in the first diagram. The scale will read 600 N.

Summary of Changes to Bathroom-Scale Readings on an Elevator

	Acceleration (up, down, zero)	Scale reading (larger, smaller, equal to weight)
1. Elevator at rest on bottom floor		
2. Elevator starts moving up		
3. Elevator moves up at constant speed		
4. Elevator comes to rest on top floor		
5. Elevator is at rest at top floor		
6. Elevator begins to move down		
7. Elevator moves down at constant speed		
8. Elevator comes to rest on bottom floor		
9. Elevator at rest on bottom floor		

For use with *Thrills and Chills*, Chapter 4, Activity 5: Weight on a Roller Coaster

NOTES

ACTIVITY 6
On the Curves

Background Information

It is suggested that you read the **For You To Read** and **Physics Talk** sections in the student text for **Activity 6** before proceeding in this section.

Any object moving in a circle requires a centripetal force–a force toward the center of the circle. This force may be a force of friction, a force of tension in a string, a force of gravitation, a normal force, a force of a spring. Regardless of what force (or combination of forces) produces the circular motion, the force is referred to as a centripetal force.

The centripetal force, according to Newton's Second Law, has a centripetal acceleration associated with it. This acceleration is always directed to the center of the circle. Although the speed of the object moving in a circle does not change, the direction of the velocity does change. The change in direction is the result of the acceleration.

The centripetal acceleration can be calculated from the equation (derived in the students text) $a = v^2/R$ where v is the velocity and R is the radius of the circle that the object is traveling. Applying Newton's Second Law, we arrive at the equation for centripetal force $F = ma = mv^2/R$.

When an object travels in a vertical circle, the weight of the object is always down. At the bottom of the vertical circle, a centripetal force vertically up is required if the object is to move in a circle. The downward weight cannot supply this force. The normal force of the surface is able to supply the required upward force. This upward force must be larger than the weight by the amount required for centripetal motion. The sum of the normal force and the weight must equal mv^2/R.

Planning for the Activity

Time Requirements

One class period is required to complete this activity.

Materials Needed

For each group:

- toy car, battery operated
- goggles, safety
- rubber stopper w/hole.
- batteries, AA

Advance Preparation and Setup

Check the toy car to make sure that the batteries will operate the car. You may wish to do this part of the activity with the entire class if supplies are limited.

You will need to provide each group with strings with rubber stoppers attached. One string should have one stopper, another string, two stoppers, and a third string, three stoppers. Ensure that the stoppers are attached to the string in such a way that they will not fly off when twirled.

Teaching Notes

Many students harbor a misconception that once an object is moving in a circular path, it will continue to move in a circular path irrespective of whether the force moving it in the original circle is removed. The first part of the activity has students confront this prior conception by having them observe a toy car with a string forcing it to move in a circle.

Another misconception that students harbor is that there is an outward force associated with circular motion. This probably comes from the sensation of being "pulled to the outside of the curve" when a car makes a fast turn. Students should learn that this "pull" is not a pull but rather the tendency of the passenger to continue moving in a straight line before the friction, the seat belt, or the wall of the car provides the force toward the center of the circle to keep the passenger moving with the car in a circle. It is best not to use the term "centrifugal" as this will confuse the students and is not necessary to describe circular motion.

Students should be asked about safety procedures before all activities. In this activity, students should realize that if the cork comes loose, it could fly into somebody's face. That is why safety glasses are required.

The concept of no change in speed requiring an acceleration is difficult for students. When they try to apply the equation $a = \Delta v/\Delta t$ they may think that with no change in speed, the numerator would be zero. Velocity is a vector and a change in velocity does occur when there is a change in direction. The students will need some help in following the description of the subtraction of velocity vectors.

You will also want to listen carefully to ensure that students do not begin to think that the centripetal force is an additional force. The centripetal force is the name given to string tension if the string tension keeps an object moving in a circle. If the object travels in a circle because of frictional forces, then friction is the centripetal force.

NOTES

Activity Overview

Students investigate circular motion. The analysis of centripetal forces and accelerations are applied to roller coaster design as students calculate the required centripetal forces for horizontal and vertical curves on the coaster. Students first become aware of some of the safety features required in roller coasters.

Student Objectives

Students will:

- Recognize that an object in motion remains in motion unless acted upon by a force–Newton's First Law.
- Explain how a force toward a fixed center will allow a car to travel in circular motion.
- Describe how the centripetal force is dependent on the speed and the radius of the curve and the mass of the object.
- Solve problems using the equation for centripetal force.
- Recognize that safety considerations limit the acceleration of a roller coaster to below 4 g.

Activity 6 On the Curves

GOALS

In this activity you will:

- Recognize that an object in motion remains in motion unless acted upon by a force – Newton's First Law.
- Explain how a force toward a fixed center will allow a car to travel in circular motion.
- Describe how the centripetal force is dependent on the speed and the radius of the curve and the mass of the object.
- Solve problems using the equation for centripetal force
- Recognize that safety considerations limit the acceleration of a roller coaster to below 4 g.

What Do You Think?

The first looping coaster was built in Paris, France. It had about a 4 m (13 ft.) wide loop. One of the largest loops today is about 35 m (120 ft.) wide.

- **Why don't you fall out of the roller coaster car when it goes upside down during a loop-the-loop?**

Record your ideas about this question in your *Active Physics* log. Be prepared to discuss your response with your small group and the class.

ANSWERS FOR THE TEACHER ONLY

What Do You Think?

It will be fascinating to listen to students describe their prior understandings. They all know that you don't fall out of the roller coaster when it is upside down, but few students will recognize that this is due to the circular motion. They will investigate this in the activity. You should not tell the students the answer in the **What Do You Think?** but rather ascertain their prior understandings.

The force of gravitation pulls you down, but this force is perpendicular to your large velocity at the top of the roller coaster and is only able to change the direction of your velocity to move you in a circular path. It is not great enough to change your velocity to a vertical direction and have you fall down.

For You To Do

In this activity, you will explore the behavior of the roller coaster on horizontal curves where you rip across the side and on vertical curves where you find yourself upside down. The more you understand about the requirements of curves on roller coasters, the more freedom you will have in designing your roller coaster.

Part A: Moving on Curves

1. A battery-operated car can move by turning on the switch. Investigate the toy car's motion under different circumstances.
   a) Let the car go. Describe its motion in your *Active Physics* log.

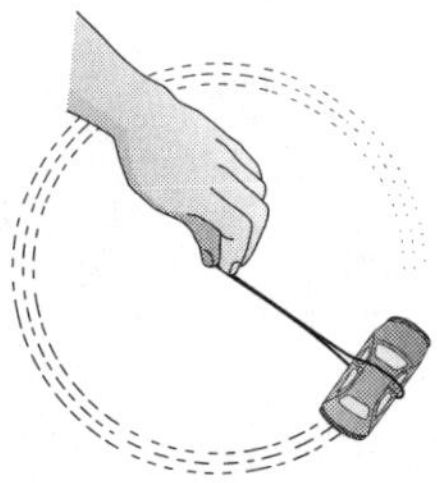

Be sure to pick up the car from the floor when it is not in use so it does not present a hazard for people walking in the room.

   b) Attach a string to the side of the car and hold the other end of the string fixed to a point on the floor. Describe the motion of the car.
   c) Predict what will happen to the motion of the car when your end of the string is let go. Provide a reason for your prediction.
   d) Test your prediction and record your observations in your log.

2. Your investigation with the toy car demonstrates that a force is needed for circular motion. This is big stuff. Whenever you see anything moving in a circle, you should remind yourself

ANSWERS

For You To Do

Part A: Moving on Curves

1. a) The car travels in a straight line.

   b) The car moves in a circle. The radius of the circle is the length of the string.

   c) If the string is let go, the car moves in a straight line. The line is tangent to the circle.

ANSWERS

For You To Do
(continued)

2. a) The tension in the string supplies the force.

   b) The force is toward the center of the circle.

3. a) The force of friction between the tires and the road keep a real car moving in a circle.

4. a)–c)

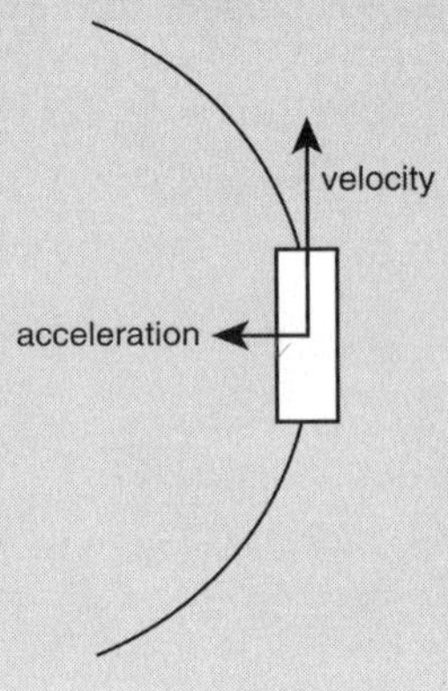

of the movement of the car. Without the force of the string, the car moves in a straight line. If something moves in a circle, there must be a force that keeps it moving in a circle.

a) What force kept the toy car moving in a circle?

b) In which direction must this force point?

3. There is no string that keeps a real car moving around a curve. However, if the car is to move around the curve, there must be a force pointing toward the center of the curve. Imagine the road surface covered with slick ice. The car would not "make the curve" but would keep moving in a straight line and go off the road. It wouldn't matter which way you pointed the wheels—no turning.

a) What is the force that keeps a real car moving around a curve?

4. In a roller coaster, there are horizontal curves similar to those on the road.

a) Sketch the coaster moving around such a curve.

b) Draw an arrow showing the velocity of the car. This is the direction it would go if there were no force.

c) Draw an arrow showing the direction of the force that keeps the coaster moving in a circle.

5. There are two orientations of the roller coaster car you will investigate as it travels in a horizontal circle. The passengers can be sitting up as they would in an automobile with the wheels of the roller coaster down. They could also be on their sides with the wheels of the roller coaster facing away from the center of the circle. In each of these orientations, the force moving the car in a circle will be toward the center.

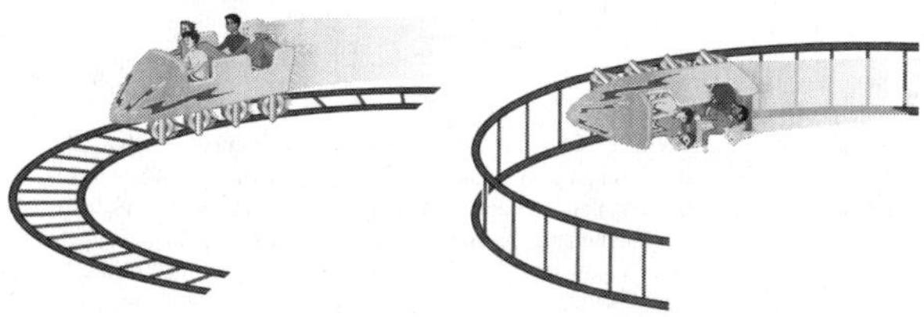

a) Is that the way that you drew it in **Step 4**? Make any changes necessary to your diagram.

b) Identify what force causes the roller coaster to move in the circle in each of these two cases. To help you identify the forces, you may want to imagine the circumstances with which the roller coaster would not "hold the turn" but would move off in a straight line.

6. The roller coaster may also do a loop-the-loop as it travels in a vertical circle. If the loop were a perfect circle, as illustrated below, there would always have to be a force toward the center of circle.

a) Make a sketch of the loop in your *Active Physics* log.

ANSWERS

For You To Do
(continued)

5. a) Students should modify their diagrams in **Step 4**, if necessary.

   b) When the car is flat on the ground, the force between the wheels and the width of the track holding the wheels provides the force. When the roller coaster is on its side, the force of the side of the track (the vertical track) provides the force.

6. a) You may wish to provide students with a diagram of the loop provided in the Blackline Master at the end of this activity.

ANSWERS

For You To Do
(continued)

6. b)

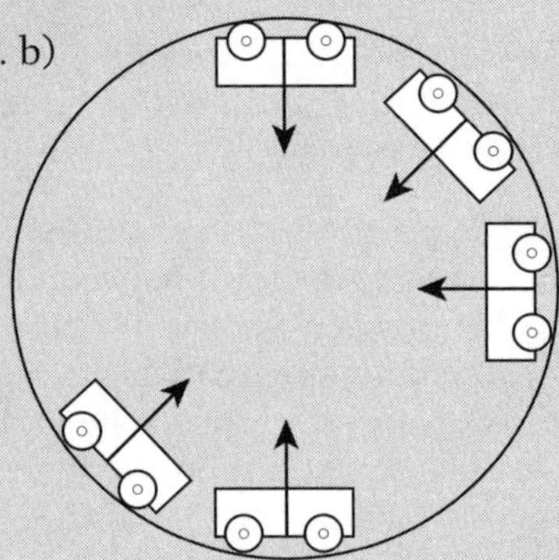

7. a)

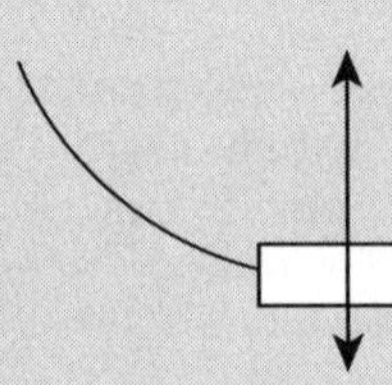

8. a)–c) Students check their diagrams.

9. a) Yes, the normal force could be zero if the gravitation force were large enough to move the cart in a circle.

b) Draw the force vectors showing the direction of the force for the positions of the roller coaster noted.

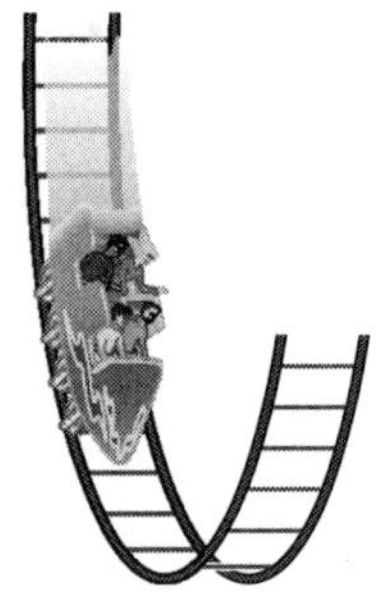

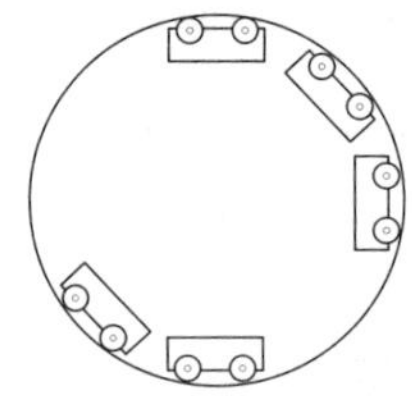

7. The gravitational force F_g is acting on the car at all times. To move in a circle, there must be a force toward the center of the circle. At the top of the circle, the gravitational force is toward the center of the circle. For a car at the bottom of the circle, the gravitational force is in the opposite direction to what is required for circular motion. The only other force at the bottom of the loop is the track pushing up on the car. This upward force must be responsible for the car moving in a circle.

a) Draw the gravitational force and the track's force on the car when the car is at the bottom of the loop and moving in a circle

8. Check your force diagram for the car at the bottom.

a) Is the gravitational force (or weight) vector pointing down?

b) Is the force of the track pointing up?

c) Is the force of the track pointing up larger than the weight vector pointing down?

9. The force of the track on the car is called the normal force F_N. (It is called the normal force because it is "normal" or "perpendicular" to the track.) This normal force must be present on the car when it rounds the loop at the bottom.

a) Could the normal force at the top of the loop be zero?

Part B: How Much Force is Required?

1. Your teacher will supply you with a rubber stopper and an attached string. You should wear safety glasses to protect your eyes in case the string should break or your partner accidentally loses grip of the string. You will use the stopper and string as a qualitative way to investigate the force needed to move something in a circle.

   Twirl the stopper at a slow speed in a horizontal circle. Increase the speed. Observe the force that your fingers are applying to the string as you increased the speed.

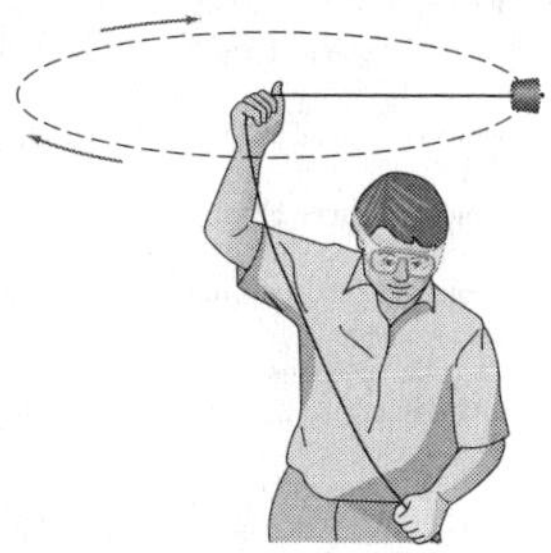

Everyone must be wearing safety glasses. Stand clear of anything that is glass.

a) Write down a description of what you observed concerning the speed of the stopper and the force on your fingers.

2. Now twirl a string with two or three rubber stoppers attached.

a) Compare the force that your fingers applied to a string with one rubber stopper and a string with more than one cork.

b) In **Steps 1** and **2** did you keep the speeds and radii (length of string) of the twirls identical? Why is this important?

c) Write down your observations about the force of your fingers and the mass of the stoppers.

Answers

For You To Do
(continued)

Part B: How Much Force is Required?

1. a) As you twirl the stopper at larger speeds, your fingers tighten up to supply a bigger force.

2. a) As the mass of the stoppers increase, the fingers tighten to supply a bigger force.

   b) You want to change only one variable, the mass. If you change more than one variable, you will not know what causes the change.

   c) As the mass increases, the force required also increases.

Chapter 4

ANSWERS

For You To Do
(continued)

3. a) As you twirl the stopper in a smaller circle, you tighten your fingers to supply a larger force.

   b) You must hold the mass and the speed of the stopper constant.

4. a) When the speed of the roller coaster increases, a larger force will be required.

   b) When the mass of the roller coaster increases, a larger force will be required.

   c) When the radius of the curve of the roller coaster *decreases*, a larger force will be required.

   d) You could strengthen the track by building it with stronger steel or reinforcing the steel around the curve.

5. a) An infinite speed would require an infinite force.

   b) If there were zero speed, there would be no force required. If you were going very, very slowly a very small force would be required.

3. Twirl one stopper, but this time change the length of the string.
   a) Write down a description of what you observed concerning the length of the string and the force on your fingers.
   b) What properties of the twirling stopper must you hold constant if you wish to compare only how changes in length affect the required force?

4. Transfer your observations about the cork on the string to your area of interest for the **Chapter Challenge**—roller coasters. To keep a roller coaster moving in a circle requires a force toward the center of the circle. The wheels on the track, the surface of the track, or the force of gravity can supply this force.
   a) How does the required force change when the speed of the roller coaster changes?
   b) How does the required force change when the mass of the roller coaster changes?
   c) How does the required force change when the radius of the curve changes?
   d) If the speed of a roller coaster were increased, how might you strengthen the track to provide the additional force required?

5. In physics, scientists often look at "limiting cases" to help them understand a concept better. A limiting case is the most extreme case that you may imagine. For instance, analyze the limiting cases for a roller coaster going around a horizontal curve that is not banked. If the car's speed got larger and larger and larger, the limiting case would be an infinite speed.
   a) If the cart were going at an infinite speed, how large a frictional force would be required by the track to keep the cart moving in the curve? Write your response down in your *Active Physics* log.
   b) The other limiting case is a car with zero speed. How large a frictional force would the track have to provide if the cart were moving very, very, very, slowly around a curve? Write your response down in your *Active Physics* log.

6. A roller coaster car in a vertical loop requires a force toward the center of the loop. At the top of the loop, the car requires a force toward the center of the loop, which is straight down. This force can be supplied by the force of gravity and by the normal force of the track.

a) In the limiting case, where the car is traveling as fast as possible, how much force will be required to keep it moving in the circle? Since the force of gravity is *mg* and doesn't change its value, what produces the very large force?

b) In the limiting case, where the car is moving as slowly as it can while still moving in a circle, the force of gravity is the only force acting on the car. Describe what would happen if the car moved faster than this. Where would the extra force come from? If the car moved slower than this, what would happen? (Hint: It would need a smaller force, but the force of gravity can't get smaller.) Would the car be able to travel in the circle?

c) Describe how the construction of a roller coaster track in a vertical loop is impacted by the speed of the roller coaster.

Part C: Is There an Equation?

1. The success of physics in describing the world is due to the discovery that mathematics can describe events precisely, accurately, and concisely. The equation for circular motion is

$$F_c = \frac{mv^2}{R}$$

where F_c is the centripetal force,

m is the mass,

v is the velocity, and

R is the radius.

This equation accurately describes your observations. (You can see how the equation is derived in the **Physics Talk** section.)

The centripetal force, F_c, on the left side of the equation is the force required to move something in a circle. It is always directed toward the center of the circle. (Reminder: when something moves in a circle a force is required. Remember

ANSWERS

For You To Do

(continued)

6. a) If it is going very fast, the force of gravitation will not be enough and the track itself will have to supply the remaining force.

b) If the car moved faster, the track itself would provide the required force. If the coaster car moved slower than this, the force of gravitation would be more than required to move the coaster in a circle and the coaster would move away from the track and toward the ground.

c) The faster the roller coaster speed at the top, the stronger the track must be at the top.

ANSWERS

For You To Do
(continued)

Part C: Is There an Equation?

1. a) If the mass increases, the force increases. This is what happened when we twirled two masses instead of one.

   b) If the velocity increases, the force increases. This is what happened when we twirled the cork at a higher speed.

   c) If the roller coaster doubles its speed, it requires four times the force ($2^2 = 4$).

   d) As the radius increases, the force decreases.

   e) The gentler the curve, the <u>smaller</u> the force required to keep the car moving along the curve. If the curve is tight (r is very small) then a <u>larger</u> force is required.

2. As we twirled the cork in a smaller circle, a larger force was required.

the toy car with the string attached. The string always supplied a force toward the center of the circle and the car moved in a circle.)

On the right side of the equation are variables that can change when objects move in circles.

a) If the mass increases on the right side of the equation, then the right side of the equation gets larger. What happens to the F_c? Describe in your log how this agrees with your observations.

b) If the velocity increases on the right side of the equation, what happens to the F_c? Describe in your log how this agrees with your observations.

c) The equation tells you more about how the change in velocity affects the force than you could determine from your qualitative exercise. The equation says that the force increases as the square of the velocity, v^2. If the velocity triples, then v^2 is nine times as large. Tripling the velocity requires nine times the force. If the velocity quadruples (4 times as large), then v^2 is sixteen (4×4) times as large. And, if the velocity increases by a factor of 10, then v^2 is 100 times as large.

A roller coaster car going with twice the speed around a banked curve needs a stronger track. Write down in your log how much stronger the track must be for a doubling of the speed?

d) If the radius of the curve increases on the right side of the equation, then the right side of the equation gets smaller since the R is in the denominator of the fraction. What happens to the F_c?

e) Complete the following sentence in your log: The gentler the curve, the ______ the force required to keep the car moving along the curve. If the curve is tight (R is very small) then a ______ force is required.

2. The limiting case of the large curve is where the curve's radius is so very large that the curve and a straight line are hardly distinguishable. On a straight path, no force is required.

a) Describe in your log how this agrees with your observations of the cork on a string.

PHYSICS TALK

Calculating Centripetal Force

Newton's First Law states that an object at rest stays at rest and an object in motion stays in motion unless acted upon by an unbalanced force. A car with no net force will be at rest or travel at a constant speed in a straight line. Newton's Second Law states that accelerations require forces: $F_{net} = ma$. Acceleration is defined as a change in velocity with respect to time.

$$a = \frac{\Delta v}{\Delta t}$$

This change in velocity can be a change in speed (the car can go faster or slower) or a change in direction (the car can move in a curve.) Newton's Second Law states that any acceleration requires a force.

Acceleration due to a change in speed is easy to calculate. When a car accelerates from 10 m/s north to 30 m/s north, its change in velocity is 20 m/s north. If the change occurred in 4 s, the acceleration is (20 m/s)/4 s = 5 m/s^2.

For acceleration due to a change in direction, the calculation is a bit more difficult. A car traveling at 10.0 m/s north that then travels 10.0 m/s east after 4.0 s also has acceleration. To determine the change in velocity, you have to calculate the change in the velocity by subtracting vectors $v_f - v_i$, where v_f is the final velocity and v_i is the initial velocity. This is mathematically equivalent to adding $v_f + (-v_i)$.

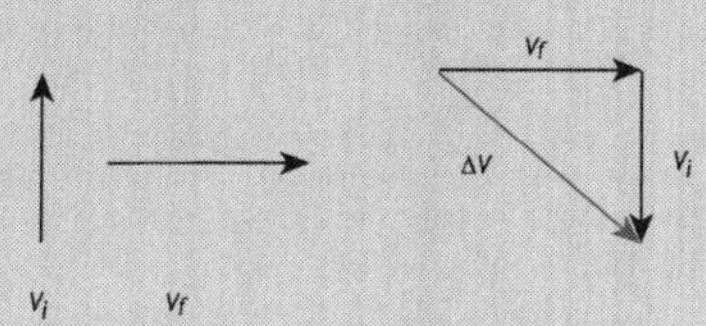

→

Chapter 4

The magnitude of this change in velocity Δv can be found with the Pythagorean Theorem and is equal to 14 m/s. You can see that the direction is 45º south of east. The acceleration is $\frac{\Delta v}{\Delta t}$ = (14 m/s)/4 s = 3.5 m/s^2 at an angle of 45º south of east.

Looking at the vector diagram, you can see that this makes sense. To change from moving north to moving east the car needs a push in the southeast direction. If these were two velocities of a car moving in a circle, this would be average acceleration during one-quarter revolution. The direction at the midpoint would be directly toward the center of the circle.

There is another way to calculate the acceleration without using vectors and without the need for the time. You can use an equation derived from an analysis of the circular motion.

$$a = \frac{v^2}{R}$$

If you know the speed of the roller coaster car and the radius of the circle, you can directly calculate the required acceleration. Knowing the mass, you can find the required force by using Newton's Second Law, $F = ma$.

Sample Problem

A roller coaster car moving at 12.0 m/s swings into a horizontal turn with a radius of curvature equal to 20.0 m.

a) What is the acceleration of the roller coaster?
b) If the mass of the passengers and car total 300 kg, what is the centripetal force required to keep the car on its tracks?

Strategy: Since you know the speed of the car and the radius of the circle, you can directly calculate the required acceleration. You can then use Newton's Second Law to calculate the force.

Activity 6 On the Curves

Givens:

$v = 12.0\text{ m/s}$

$R = 20.0\text{ m}$

$m = 300.0\text{ kg}$

Solution:

$$a = \frac{v^2}{R} = \frac{(12.0\text{ m/s})^2}{20.0\text{ m}} = 7.2\text{ m/s}^2$$

$$F = ma = (300.0\text{ kg})(7.2\text{ m/s}^2) = 2200\text{ N}$$

This force will have to be supplied by the track to the wheels of the coaster.

Physics Words

centripetal force: a force directed towards the center that causes an object to follow a circular path.

FOR YOU TO READ

Centripetal Force

The fun of a roller coaster is the fun of whipping around the turns and flipping upside down. All objects moving in circles require a force toward the center of the circle. In a roller coaster moving around the curve, this force is the track on the wheels of the car when the wheels are down during the curve.

In a roller coaster curve where the car tilts vertically and the wheels face the outside of the circle, the force toward the center is the force of the wall holding the track. This is called a normal force F_N because it is normal (perpendicular) to the wall.

In any circular motion, the force that keeps the object moving in a circle is called the centripetal force. The toy car moving in a circle had a **centripetal force** that was the force of tension in the attached string. A car moving around a curve has the force of friction between the tires and road as the centripetal force. The roller coaster car rounding a turn on its side may have the force of the track as the centripetal force. The centripetal force is not an additional

→

force. It is the name given to a force like friction, tension, gravity, or the normal force when that force causes an object to move in a circle.

When the roller coaster is in a vertical loop, the direction of the centripetal force is always changing to ensure that the force vector always points to the center of the circular track. Pay particular attention to how this is phrased. Although the force is always toward the center, the direction is always changing since in the circle, the force may be toward the left or the right or up, but still point toward the center.

In the vertical loop, this centripetal force can be either the gravitational force, the normal force of the track on the car or a combination of the two. When it is a combination of the two, you must add the forces as vectors. At the bottom of the circle, the normal force points toward the center of the circle (upward) while the gravitational force points downward. The vector sum of these two forces must be toward the center of the circle. You can therefore conclude that the normal force is larger than the gravitational force. The normal force corresponds to your apparent weight, as it did in the elevator activity. This is why you feel as if you weigh more at the bottom of the loop of the roller coaster.

At the top of the loop-the-loop, the gravitational force and the normal force both act downward, toward the center of the loop. The sum of these two vectors provides the required centripetal force. How much of the normal force is required will depend on the mass and velocity of the car.

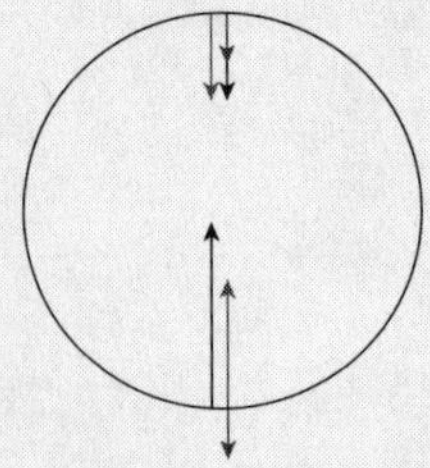

The blue force vectors show the required centripetal force to keep the car moving in a circular path at the top and bottom of the loop. Notice that the size of the vectors is different since the car has a larger speed at the bottom of the loop. The directions are different because the centripetal force must always point toward the center of the circle.

The red force vector represents the force of gravity or weight of the cars. Both weight vectors are identical because the weights of the roller coaster car are identical at the top and bottom.

The black vector represents the normal force of the track on the car. The sum of the normal force plus the weight must be equal to the required centripetal force. At the top, the normal force is small since the weight contributes to the centripetal force. If the speed decreases, the required centripetal force would be less and less. There comes a point where the gravitational force (weight) would be all that is required to keep the car moving in a circle. In that case, the normal force is zero. In this special situation where no normal force is required, you could actually have a small gap in the top of the track and the car would continue to move in a circle.

At the bottom of the roller coaster, the car would need a normal force of the track on the car that would be greater than the weight since the weight is not providing any help for the required centripetal force.

This is summarized in the following tables.

The force of gravity pulling you down is used to move you in a circle. You are moving down but you are also moving across. If the speed of the roller coaster were not sufficient, you would require a centripetal force less than the gravitational force to move you in a vertical circle. The gravitational force would cause you to fall. It's your large speed that keeps you from falling when you are upside down.

Apparent Weight and the Roller Coaster Ride

You discovered earlier that an elevator ride could give you a sense of weight changes during accelerations. In the roller coaster

Fast-moving roller coaster			
	Required centripetal force	**Force of gravity (weight)**	**Normal force (the force of the track on the car)**
At the top of the loop	5000 N	1000 N	4000 N
At the bottom of the loop	9000 N	1000 N	10,000 N

Slow-moving roller coaster			
	Required centripetal force	**Force of gravity (weight)**	**Normal force (the force of the track on the car)**
At the top of the loop	2100 N	1000 N	1100 N
At the bottom of the loop	6100 N	1000 N	7100 N

→

ANSWERS

For You To Read

The diagrams on this page are available as a Blackline Master at the end of this activity.

loop-the-loop, the passenger will also experience changes in apparent weight. The normal force is an indication of the apparent weight, as it was in the elevator. A passenger on the roller coaster feels lighter at the top of the roller coaster. This is similar to the elevator because in both cases you feel lighter because acceleration is down. A passenger on the roller coaster feels heavier at the bottom of the roller coaster. Once again, this is similar to the elevator because in both cases you feel heavier because acceleration is up.

In the slow-moving roller coaster in the chart, the apparent weight (normal force) at the top may only be 1100 N, while the apparent weight (normal force) at the bottom may be 7100 N.

There are three locations that you can use to summarize the discussion on forces and weight. On a level track with the cart moving at constant speed, the sum of the forces must be zero.

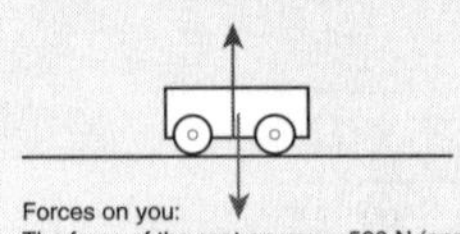

Forces on you:
The force of the seat on you = 500 N (apparent weight).
The force of gravity on you is 500 N (weight).

At the bottom of the hill or loop, there must be a net force toward the center of the circle to keep you moving in a circular path.

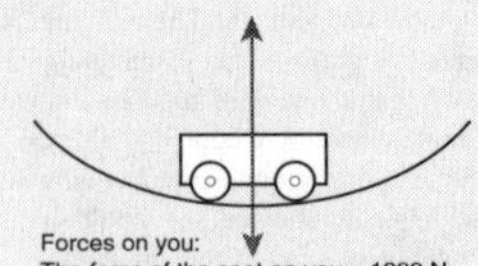

Forces on you:
The force of the seat on you = 1000 N.
The force of gravity on you is 500 N (weight).
You feel as if you weigh 1000 N (apparent weight).

At the top of the hill or loop, there must be a net force toward the center of the circle to keep you moving in a circular path.

Forces on you:
The force of the seat on you = 100 N.
The force of gravity on you is 500 N (weight).
You feel as if you weigh 100 N (apparent weight).

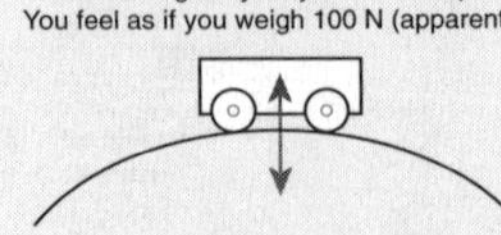

Roller coasters do not use loops that are circular. They use a clothoid loop (it has a big radius at the bottom and a small radius at the top). In this way, at the top of the loop the roller coaster is moving in a small circle, while at the bottom it is moving in a larger circle. This is done to ensure that the roller coaster car can make the turn at the top but not gain so much speed at the bottom that the person at the bottom would be pulling more than 4 g's.

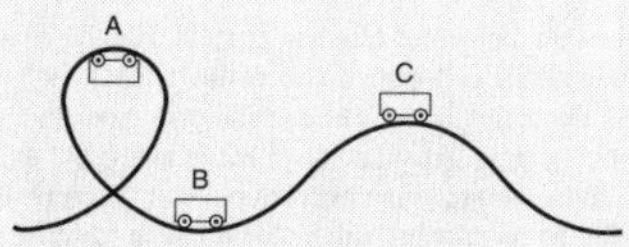

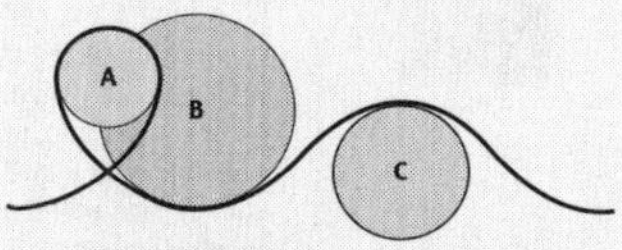

Each of these three situations is basically circular motion.

The type of loop the coaster designers use is called a clothoid. It has a large radius at the bottom,and a smaller radius at the top.

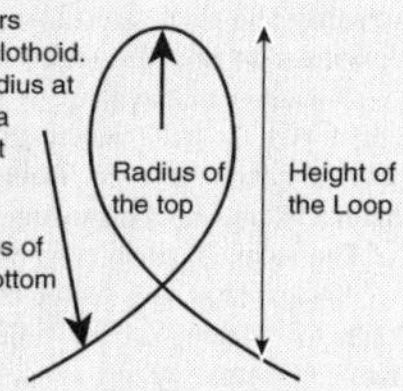

Safety on the Roller Coaster

Test pilots and astronauts experience lots of accelerations during their job performance. To prepare for this, they all go through physical training to see how much acceleration they can endure without getting sick or becoming unconscious. Experiencing an acceleration of more than nine times gravity for a sustained period will cause unconsciousness in most people. Since the acceleration due to gravity is 9.8 m/s^2 or approximately 10 m/s^2, you can refer to other accelerations in terms of 1 g. An acceleration of 2 g is approximately 20 m/s^2 while an acceleration of 8 g is approximately 80 m/s^2. Astronauts sometimes experience as much as 6 g's during liftoff.

Safety on a roller coaster requires that you stay below 4 g for the entire ride. You must never go beyond 4 g for even a short time. Changes in small accelerations may make a better ride than one big thrill from a single large acceleration.

Answers

Physics To Go

1. a) The car will travel in a circular path.

   b) If the string were to break, the cart will travel in a straight line tangent to the circle.

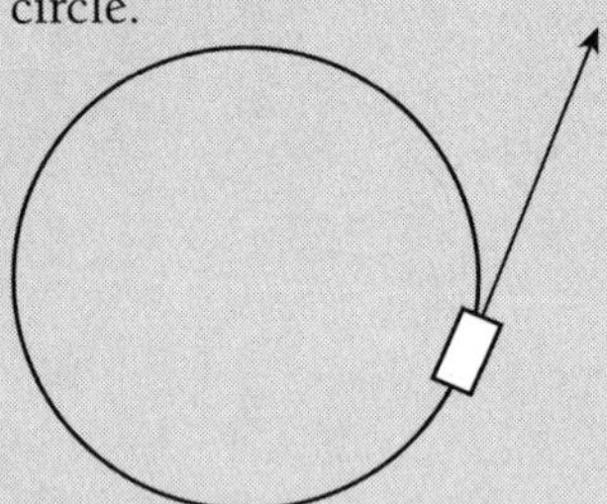

2. a) Friction between the road and the car.

   b) The car would travel in a straight line, tangent to the circle. (See diagram in **Question 1**.)

Reflecting on the Activity and the Challenge

The loop on a roller coaster is one of the big thrills. People are always nervous that they will fall out of the roller coaster when it is upside down. This does not happen because you arrive there with a large velocity. The gravitational force (weight) at the top of the roller coaster serves as the centripetal force that moves the roller coaster in a circular path. All objects moving in circles require a centripetal force toward the center. With a toy car attached to a string, the tension in the string is the centripetal force. A roller coaster car rounding a turn has the wheels in the track providing the centripetal force. A roller coaster making a turn on its side has the track's normal force as the centripetal force. The upside-down roller coaster has the gravity as the centripetal force. At the bottom of the loop, the normal force provides the centripetal force. This force must be larger than the gravitational force and passengers feel much heavier at the bottom of the loop. In designing your roller coaster, you will have to ensure that the roller coaster has enough speed to make the full circle. You will also have to ensure that it doesn't have so much speed at the bottom that the apparent weight is too great or the passengers may get injured.

Physics To Go

1. A battery-operated toy car is attached to a string.
   a) If the loose end of the string is held to the ground, draw the path of the car.
   b) If the string were to break, draw the path that the car would follow.

2. Consider a car on a road making a turn.
   a) What force has replaced the string of the toy car in **Question 1(a)**?
   b) If the car were to hit a section of ice, draw the path that the car would follow.

3. A person twirls a key chain in a circle. If she twirls it faster, she finds that she holds the chain tighter. Explain why this is necessary.

4. It's a cold night and the roads are icy. If your car is filled with friends, will it be easier or more difficult to make a turn? Explain why.

5. In the equation for circular motion $F_c = \frac{mv^2}{R}$ explain what each of the terms represents.

6. A roller coaster car is traveling east at 20 m/s. After 2 s, it is traveling north at 20 m/s.
   a) Did the speed of the roller coaster car change?
   b) Did the velocity of the roller coaster car change?
   c) How much did the velocity of the roller coaster car change?

7. A roller coaster car is traveling east at 20 m/s. After 16 s, it is traveling north at 20 m/s. The circular curve had a radius of 200 m. Calculate the acceleration of the car.

8. A roller coaster car is traveling in a loop. Identify the six force vectors in the diagram below.

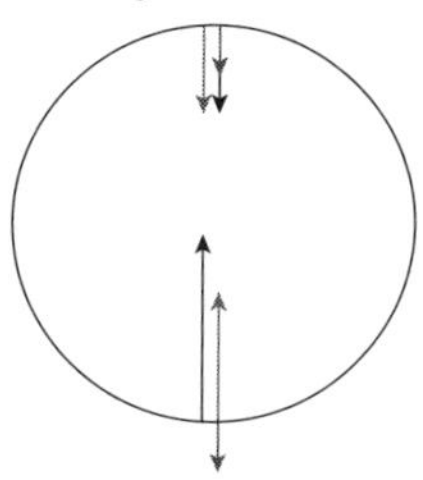

9. In explaining circular motion, someone correctly states that the centripetal force is a name for a force, but it is not an additional force. Explain what this means.

Answers

Physics To Go

(continued)

3. An object moving in a circle at a greater speed requires a larger centripetal force. That force is supplied by the grip of the fingers. To increase the grip, the person squeezes her fingers tighter.

4. It will be more difficult to make the turn because the mass has increased. A larger mass requires a larger centripetal force.

5. *F* is the centripetal force required for circular motion,

   m is the mass of the object moving in a circle,

   v is the speed of the object,

   R is the radius of the circular path that the object is moving.

6. a) No, the speed is still 20 m/s.

   b) Yes, the velocity changed because the direction changed.

   c) By constructing a vector diagram, we can see that the change is equal to 28 m/s northwest. (It increased its north speed by 20 m/s and decreased its east speed by 20 m/s which when added using the Pythagorean Theorem produces the change of 28 m/s NW.)

7. $a = v^2/R$

   $= (20 \text{ m/s})^2/200 \text{ m}$

   $= 4.0 \text{ m/s}^2$

8. At the top: The first down force vector (red) is the force of gravitation (the weight); the other down force vector (black) is the normal force of the track pushing down; the largest down force vector (blue) is the vector sum of the weight and the normal force. This vector sum of the two forces is the net force, the centripetal force.

   At the bottom: The down force vector (red) is the force of gravitation (the weight); the large up force vector (black) is the normal force of the track pushing up; the other up force vector (blue) is the vector sum of the weight and the normal force. This vector sum of the two forces is the net force, the centripetal force.

9. The force that moves something in a circle may be a tension in a string, weight, friction or a normal force. Whichever force (or combination of forces) keeps the object moving in a circle is called the centripetal force.

ANSWERS

Physics To Go
(continued)

10. See tables below.

11. At the bottom of the roller coaster loop, the normal force (supplied by the surface you are seated on) is larger and you would feel heaviest.

12. The speed and radius of the curve must be adjusted.

13. Answers will vary. The speed at the bottom should be greater than the speed at the top of the roller coaster.

10. Fill in the missing values in the tables that you created in your *Active Physics* log:

Fast-moving roller coaster			
	Required centripetal force	**Force of gravity (weight)**	**Normal force (the force of the track on the car)**
At the top of the loop	4000 N	500 N	
At the bottom of the loop	6000 N	500 N	

Slow-moving roller coaster			
	Required centripetal force	**Force of gravity (weight)**	**Normal force (the force of the track on the car)**
At the top of the loop	800 N	500 N	
At the bottom of the loop	2800 N	500 N	

11. At which section of a loop would the roller coaster passengers feel the heaviest? Why?

12. Safety requires the roller coaster to be able to make the complete loop and to keep the acceleration under 4 g. How can both of these safety features be accomplished at the same time?

13. Design a loop for your roller coaster. Calculate the speeds and accelerations at some key places in the loop.

14. Use the diagram of the Terminator Express roller coaster. Indicate at which of the following points the passengers will feel heavy, where they will feel light, and where it is uncertain.

Table 10.

Fast-moving roller coaster			
	Required centripetal force	**Force of gravity (weight)**	**Normal force (the force of the track on the car)**
At the top of the roller coaster	4000 N	2000 N	2000 N
At the bottom of the roller coaster	7500 N	2000 N	9500 N

Slow-moving roller coaster			
	Required centripetal force	**Force of gravity (weight)**	**Normal force (the force of the track on the car)**
At the top of the roller coaster	2500 N	2000 N	500 N
At the bottom of the roller coaster	3400 N	2000 N	5400 N

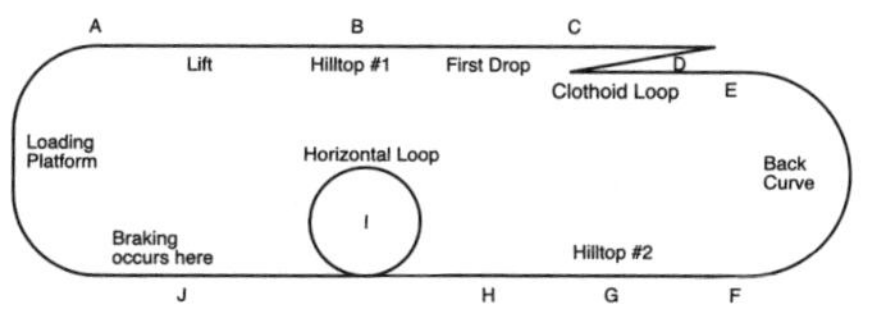

TOP VIEW

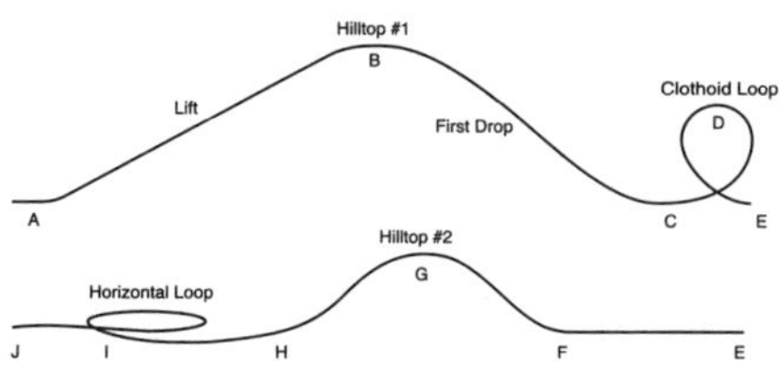

SIDE VIEW

a) C (bottom of hill #1)
b) D (top of the vertical loop)
c) E (bottom of the vertical loop)
d) F (bottom of hill #2)
e) lift hill (going up at constant speed)

15. Using the diagram of the Terminator Express, indicate at which of the following points the centripetal force is up, when it is down, when it is zero, and when it is sideways.
   a) C (bottom of hill #1)
   b) D (top of the vertical loop)
   c) E (bottom of the vertical loop)
   d) F (bottom of hill #2)
   e) lift hill (going up at constant speed)
   f) horizontal loop
   g) back curve

ANSWERS

Physics To Go

(continued)

14. a) C (bottom of hill #1): heavy
   b) D (top of the vertical loop): light
   c) E (bottom of the vertical loop): heavy
   d) F (bottom of hill #2): heavy
   e) lift hill (going up at constant speed): normal

15. a) C (bottom of hill #1): up
   b) D (top of the vertical loop): down
   c) E (bottom of the vertical loop): up
   d) F (bottom of hill #2): up
   e) lift hill (going up at constant speed): zero
   f) horizontal loop: sideways
   g) back curve: sideways

Diagram of Roller Coaster Car in Different Positions in a Loop

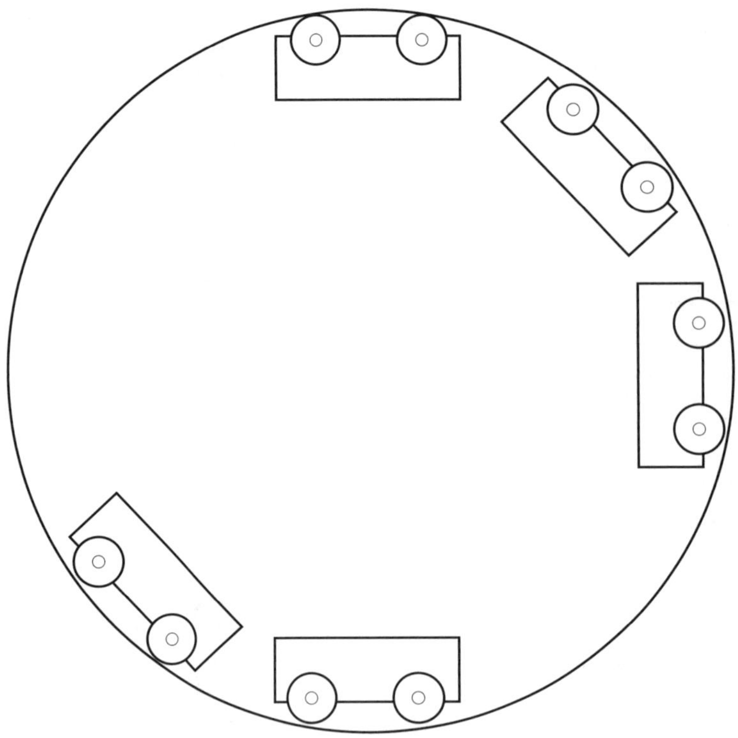

For use with *Thrills and Chills*, Chapter 4, Activity 6: On the Curves

Forces on a Roller Coaster Car on a Level Track, at the Bottom of a Loop, and at the Top of a Loop

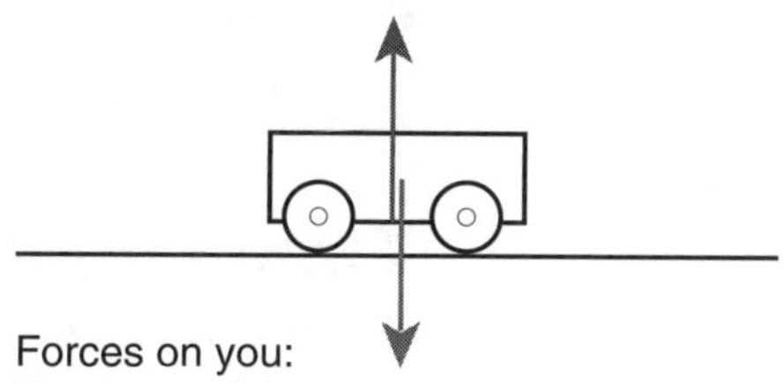

Forces on you:
The force of the seat on you = 500 N (apparent weight).
The force of gravity on you is 500 N (weight).

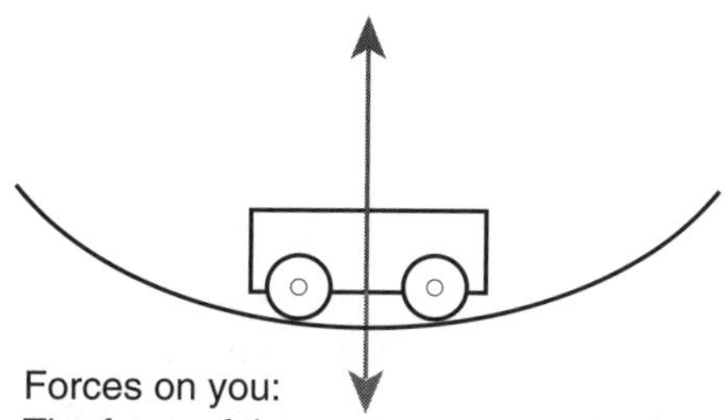

Forces on you:
The force of the seat on you = 1000 N.
The force of gravity on you is 500 N (weight).
You feel as if you weigh 1000 N (apparent weight).

Forces on you:
The force of the seat on you = 100 N.
The force of gravity on you is 500 N (weight).
You feel as if you weigh 100 N (apparent weight).

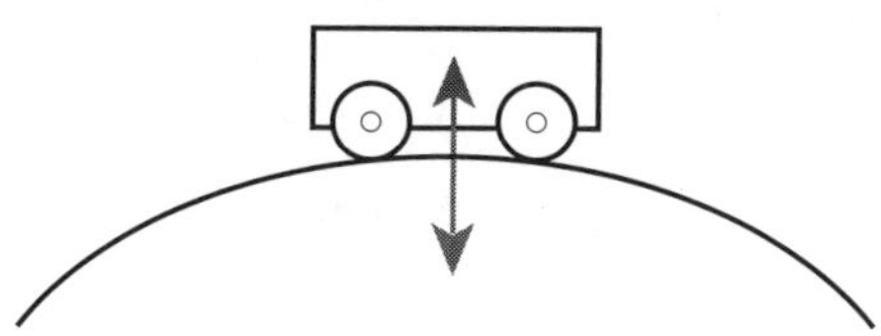

For use with *Thrills and Chills*, Chapter 4, Activity 6: On the Curves

ACTIVITY 7
Getting Work Done

Background Information

It is suggested that you read the **For You To Read** section in the student text for **Activity 7** before proceeding in this section.

Energy is often defined as the ability to do work. Work is defined as the product of force and displacement when they are both in the same direction. This is expressed mathematically as $W = F \cdot d$ where F is the force and d is the displacement and where the dot signifies a special kind of multiplication. The force and displacement can be multiplied if and *only* if they are in the same direction.

Both force and displacement are vector quantities. They both have direction. Work is a scalar quantity. It has no direction. Work is measured in joules, as is energy.

Work can be applied to an object to increase its kinetic energy. We can see that the force over a displacement will produce an acceleration (Newton's Second Law: $F=ma$) and will result in a larger velocity and therefore a larger kinetic energy ($KE=1/2\ mv^2$). Work can also be applied to an object to lift it in the air and increase the object's gravitational potential energy ($GPE = mgh$).

Work can be done on a spring and increase the spring's potential energy ($SPE = 1/2\ kx^2$).

In turn, the SPE can do work on an object and increase its KE or its GPE. When no work is done, the total energy is conserved. When work is done from an outside force, then the total energy of the system can increase or decrease.

In any closed system (i.e., no external forces), the total energy remains the same. This is an organizing principle of physics and is referred to as the conservation of energy.

Each type of energy can be calculated in joules.

Power, measured in watts, is defined as the rate of doing work ($W=P/t$).

Planning for the Activity

Time Requirements

One class period will be required to complete this activity.

Materials Needed

For each group:

- spring scale, 0-10 newton range
- stick, meter, 100 cm, hardwood
- cast iron right angle holder
- 1/2" x 9" steel rod (to act as crossarm)
- large ringstand
- protractor
- dynamics cart
- track for roller coaster demonstration
- track support

Advance Preparation and Setup

Ensure that friction will not play a large role in sliding the carts up the incline.

Teaching Notes

The first two steps in the **For You To Do** section are reviews of an earlier activity. If students seem unsure of the response, then there may be a need to review the prior activity about energy conservation. Work is intimately related to energy conservation.

Students should try to pull the carts up the ramp at constant speeds. The analysis of the inverse relationship between force and displacement

W	KE	GPE	SPE	Electrical	Heat	Solar	Sound	Tidal	Light
$W = F \cdot d$	$1/2\ mv^2$	mgh	$1/2\ kx^2$	VIt	$mc\Delta T$				

requires that students are familiar with inverse relationships from their math class. As a means to bridge the chasm between math and science class, students are asked to graph a simple inverse relationship. Students will be surprised to find out that the "dot" used in the $W = F \cdot d$ means more than simple multiplication. In early grades students were shown that for multiplication, they could use a dot, a cross, or just place variables next to one another. Once you learn about vectors, you find that there are two types of vector multiplication.

- The dot product, as in $W = F \cdot d$ where F is the force and d is the displacement and where the dot signifies a special kind of multiplication. The force and displacement can be multiplied if and *only* if they are in the same direction.
- The cross product, as in $F = qv \times B$ (to be learned in a later course) where the vectors can be multiplied only when they are perpendicular to one another. It is worth mentioning this other type of vector multiplication as a precursor of things to come and also to expand on the concept of the dot product.

Students are not required to calculate the components of the force that are in the same direction as the displacement but students who have had more familiarity with vectors or trigonometry can explore this as an additional topic.

NOTES

Activity Overview

To get the roller coaster started, work must be done to provide the coaster with gravitational potential energy. Students investigate the relationship between work, force, and displacement as they drag carts up different inclines. The work-energy theorem is then stated as a way to tie together this activity with the concept of the conservation of mechanical energy in earlier activities. Students also calculate the power required for a roller coaster as it ascends a hill.

Student Objectives

Students will:

- Measure and recognize that the product of force and distance is identical for lifting the object to the same height irrespective of the angle of the ramp.
- Define work $W = F \cdot d$ where F is the force and d is the displacement and where the dot signifies a special kind of multiplication.
- Explain the relationship between work and gravitational potential energy and spring potential energy.
- Define power as the rate of doing work and the units of power as watts.

Activity 7 Getting Work Done

GOALS

In this activity you will:

- **Measure and recognize that the product of force and distance is identical for lifting the object to the same height irrespective of the angle of the ramp.**
- Define work where F is the force **and d is the displacement and where** the **dot signifies a special kind** of **multiplication.**
- **Explain the relationship** between work **and gravitational potential energy** and **spring potential energy.**
- Define **power as** the **rate of** doing **work and** the units of power as watts.

What Do You Think?

The greatest drop for a roller coaster is 125 m (400 ft.). The roller coaster must be pulled up to that height to get the ride started.

- **Does it take more energy to slide the roller coaster up a steep incline than a gentle incline?**
- **Why is it more difficult to walk up a steep incline than a gentle incline?**

Record your ideas about these questions in your *Active Physics* log. Be prepared to discuss your response with your small group and the class.

ANSWERS FOR THE TEACHER ONLY

What Do You Think?

Although the total energy required is identical irrespective of the slope of the incline, it is common for students to think that a steep incline requires more energy. This is probably because they confuse energy with force. This is not the time to correct this misconception. The **What Do You Think?** question is here to elicit these prior understandings. After students complete the activity, it will then be appropriate to return to this question and find out "What Do You Think Now?"

A gentle incline does not require the same force. As the angle of the incline increases, a larger force is required to get to the same height because the distance traveled is so much less.

Activity 7 Getting Work Done

For You To Do

1. From the starting hill, the roller coaster is able to complete its entire trip without any motors, pushes, tires, or chains. The roller coaster car heads down a hill and has enough speed at the bottom to make it up to the next hill.
   a) Why do you think that the roller coaster cannot scale a higher hill than the one from which it began?

2. The roller coaster at the top of the hill is ready to go. It goes up and down the hills and around the curves without any energy input. The roller coaster is a closed system. No energy is added to the system and no energy leaves the system. Nobody adds energy to the roller coaster with motors. No energy is assumed lost by the system to friction or air resistance. The roller coaster as a closed system (an idealized, conceptual roller coaster) will keep on going back and forth forever. The cart will go from point A to B to C to D to E to F to G. It will then reverse and go from G to F to E to D to C to B to A. It will then begin the trip again.

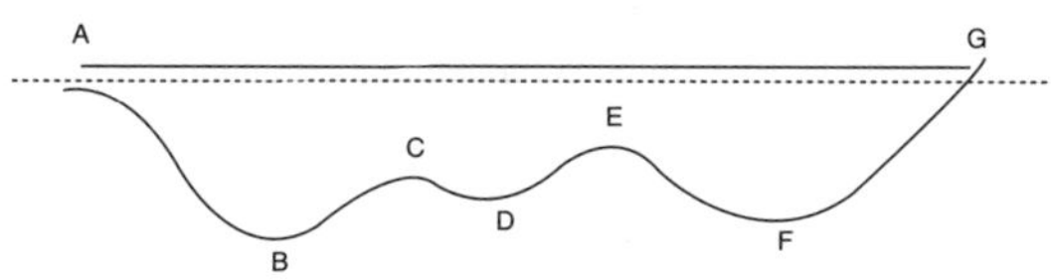

3. You will now investigate the force required to lift a roller coaster car to a certain height. You will use a cart and a track in your classroom. You can pull the cart to the top of the track with the use of a spring scale. The spring scale will indicate the force required to pull the cart. A meter stick can be used to record the distance that the cart moved along the track. You can then vary the length and angle of the track while keeping the height of the track constant.

ANSWERS

For You To Do

1. a) The roller coaster has a total energy defined by the height of the first hill. For the roller coaster to get to a larger height, it would need additional energy. A more sophisticated way of looking at this is to realize that if the final potential energy (GPE = mgh) were greater than the initial gravitational potential energy, then the kinetic energy would have to be less than the initial kinetic energy. Since the kinetic energy at the first hill was zero, then there is no lower kinetic energy. Kinetic energy can never be less than zero because it is a positive quantity. In the equation for kinetic energy, $KE=1/2\ mv^2$, the mass is always positive and the square of the velocity must also be positive.

ANSWERS

For You To Do
(continued)

3. a) Height of top of ramp = _____

Force to pull cart up ramp	Distance the cart is pulled

4. a) As the distance along the ramp increases (because of the change in angle of the ramp), the smaller the force required to pull the cart up the ramp.

5. a) See graph below

6. a) Yes, the graph should be similar to the hyperbola in **Step 5.**

   Students' answers will vary depending on their data. The product of force and distance should be a constant.

Pull the cart at a speed that keeps it on the track through the entire run. Before beginning to pull the cart, make sure that the way is clear for the person pulling to walk without obstruction.

a) Create a data table in which you can record the force required to pull the cart up four different tracks. (Reminder: You must always pull the cart to the same height and parallel to the track.)

4. Complete your investigation.

a) What conclusion can you reach about the distance along the track to attain a specific height and the force required to move the cart?

x	y	$xy = k$
1	12	12
2	6	12
3	4	12
4	3	12
6	2	12

5. When one quantity increases and a second quantity decreases, this is referred to as an inverse relation. If x is one quantity and y is the other quantity, one inverse relation can be described mathematically by the equation $xy = k$ where k is a constant. At left are some x and y values forming an inverse relation where $xy = 12$.

a) Make a graph to show the relationship for the inverse relation $xy = 12$.

6. Create a graph for the data from your experiment.

a) In the equation $xy = 12$, the product of the x and y values always equals 12. Does the product of the force and distance in your experiment always equal a certain value? Make the calculations and record the results on the side of your chart.

5. a)

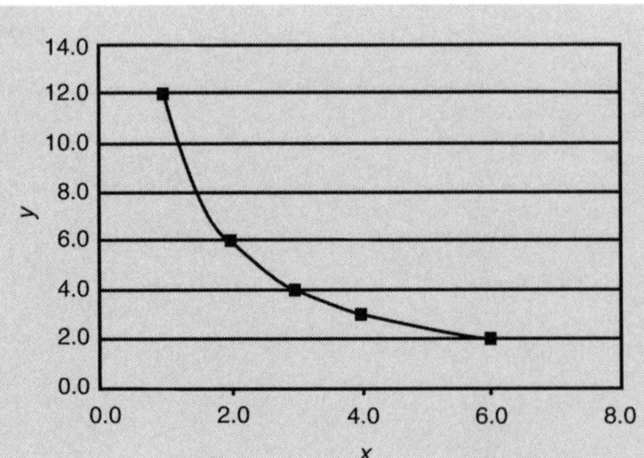

b) Why would the values in your experiment not be expected to be *exactly* the same?

7. Any time you take measurements, there is some uncertainty in the measurement. When you weigh yourself on a scale, the weight reading may be off by a little bit. If the scale reads 143 lb., you may actually weigh 143 lb. and a few ounces. The scale does not give you an exact measurement. No measurement is ever exact.

a) What are the uncertainties in your measurements of distance? Could your measurement of distance be off by as much as 3 cm? Could your measurement of distance be off by as much as 1 cm? What is the largest amount that your distance measurement may be off? Write down this value with the notation ± to signify that you may have been under or over by that amount. For instance, if you think that your distance measurement could have been off by 2 cm, you would write this as ± 2 cm.

b) Record the uncertainties in your measurements of force by noting the accuracy of your spring-scale reading.

8. Another way that you can get the cart to the top of the incline is to lift it vertically. Use the spring scale to lift the cart vertically.

a) Record the force required to lift cart vertically and the distance (height) that you lifted it.

b) Does this force and distance fit the same inverse relation as the track measurements?

9. There are other means by which the cart could be lifted to the top of the incline. For example, you could have had an electric motor pull the cart to the top. Brainstorm a list of at least three ways in which the cart could be brought to the top of the incline. (Brainstorming allows for all ideas to be included, even those that appear silly or impractical.)

a) Record your ideas in your *Active Physics* log.

ANSWERS

For You To Do
(continued)

6. b) Uncertainty in the values could account for some discrepancies. (Measurements always have uncertainties.)

7. a) Distance measurements certainly have uncertainties of at least 0.1 cm. They may be off by as much as 0.5 cm.

   b) Uncertainties in force could be as much as 1 N.

8. a) Answers will vary.

   b) It should fit the same inverse relationship.

9. a) Lift it by a spring pushing it.

   Lift it by a fan pushing on it.

   Lift it by battery power.

   Many, many other answers are acceptable.

Chapter 4

FOR YOU TO READ

Work

The roller coaster must get to the top of the first hill to begin the ride. In the activity, you moved a cart to the top of an inclined ramp by applying a force with the spring scale over a certain distance. In physics, this is called **work.** In physics work has a specific definition. Work is defined through the following equation:

$$W = F \cdot d$$

where F is the force and

d is the displacement.

The dot signifies a special kind of multiplication. The force and displacement can be multiplied if and only if they are partially in the same direction.

In the activity, the spring scale pulled the cart up the incline and the force was in the same direction as the displacement. You found that the work done was the same irrespective of the angle of the incline. The force was larger for a steeper incline, but the distance along the incline was smaller. The product of the force and distance was always the same. That quantity was the work that was done by the spring scale on the cart.

Sample Problem 1

A cart that weighs 300 N is lifted to the top of an incline 2 m above the ground.

a) What is the work done?

b) How much force would be required to lift the same cart to the same height using a 10-m track?

a) Strategy: The minimum force required to lift the cart is equal to its weight. The displacement is the height that the cart was lifted. The force and the displacement are both in the vertical direction.

Givens:

$F = 300$ N

$d = 2$ m

Solution:

$W = F \cdot d$

$W = (300 \text{ N})(2 \text{ m})$

$= 600$ Nm

$= 600$ J

b) Strategy: The work required to lift the cart would be identical since the cart began at the same height and ended at the same height. Since you know the new displacement, you can find the new force.

Givens:

$W = 600$ J

$d = 10$ m

Solution:

$W = F \cdot d$

$600 \text{ J} = F(10 \text{ m})$

$F = 60$ N

By using the track (ramp) only 60 N of force are required to slide the cart up the track instead of the 300 N to lift it. This is why the track is considered a simple machine. The same work is done, but with much less force. Of course, the force must be applied over a longer distance.

Direction of Force in Work Done

It may seem that the force would always be in the same direction as the displacement. This is not always the case. Consider a push lawn mower. The push lawn mower has no motor. It moves because someone pushes it.

The force is applied along the handle of the lawn mower. The displacement of the lawn mower is the distance along the ground. They are not in the same direction, but there is some work done.

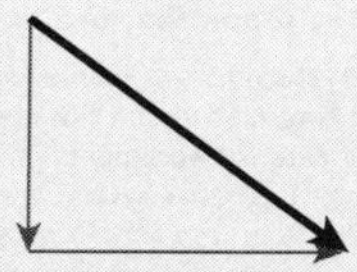

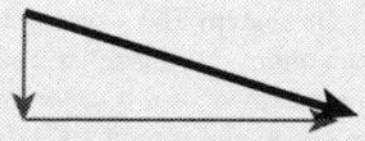

Although the entire force is not in the same direction as the displacement, some of the force is in the same direction of the displacement.

The force along the handle can be broken into its two vector components by finding the horizontal and vertical forces that, when added together, would be identical to the original vector. In the diagram at the top, it appears that the horizontal and vertical component forces are approximately equal in size. In the next diagram, the horizontal vector is much larger than the vertical component. Most of the force applied to the handle is now in the same direction as the displacement. Even though the total force is identical (note that the length of the force vector is the same in each diagram), the horizontal component is larger as the angle gets smaller. The same total force and the same displacement, but more work is done when the horizontal component is greater.

Why then don't you push the lawn mower with a small angle? Although more work would be done, it would hurt your back. Therefore, you sacrifice some work in order to make mowing more comfortable.

→

Chapter 4

Closed and Open Systems

A **closed system** has no external forces on it. With no external forces, a closed system also has no external work done. The ideal roller coaster is a closed system once the roller coaster is on top of the first big hill ready to go. It remains a closed system if friction or air-resistance forces do no work.

To start the roller coaster, work must be applied to the system. The work will increase the energy of the roller coaster system. The work to lift the roller coaster up the track is identical to the work to lift it vertically. The force required is equal to the weight of the car. The vertical displacement is the height that it must be lifted.

$$W = F \cdot d$$

$$W = \text{weight} \cdot \text{height}$$

$$W = mgh$$

The work done on the roller coaster is *mgh*. This is equal to the change in gravitational potential energy GPE of the roller coaster.

It would be difficult to raise a real roller coaster car by hand as you raised the cart in the activity. You could try to lift it with a large spring. The force of a spring that obeys Hooke's Law is $F = kx$. The force is not constant but changes as the stretch of the spring changes. If you stretch the spring a distance *x*, then the average force will be $\frac{1}{2}kx$. It is zero when the spring is not stretched at all and a maximum value of *kx* when the spring is stretched a distance *x*. The work done by a spring is

$$W = F \cdot d$$

$$W = (\tfrac{1}{2}kx)(x)$$

$$= \tfrac{1}{2}kx^2$$

The work done by a spring is equal to the potential energy of the spring SPE.

The roller coaster car is usually raised with electrical energy supplied by a motor. Electrical energy can be calculated by measuring the voltage, current, and time. Creating steam to push it up the incline could also have raised the roller coaster car. In this method, the heat energy can also be calculated. In all of these methods, work is done by the spring, by the electricity, or by the heat. The roller coaster system gains that amount of energy. The roller coaster has increased its GPE by exactly that amount.

In any closed system, the total energy remains the same. This is an organizing principle of physics and is referred to as the conservation of energy.

Although you treat the roller coaster as a closed system in your simplified analysis, there is work done by friction and work done by air resistance. This work removes energy from the roller coaster. The work done by friction, for instance, becomes heat energy that is dissipated to the air surrounding the roller coaster.

In an **open system**, you can add energy. You do work on the car to bring it to the top of the hill and in this way add gravitational potential energy GPE.

In an open system, you can lose energy. Friction can do work and the car will lose some of its kinetic energy and corresponding GPE and will not get up to the same height that it began.

In a closed system, no external forces act on the system and no energy is able to enter or leave the system. The total energy of the system will remain the same.

Power

A related concept to work is how fast the work is done. In the activity, you pulled the cart up the incline. You could have pulled it up with a variety of speeds. The rate of doing work is **power**.

$$P = \frac{W}{t}$$

Sample Problem

Tomas runs up the stairs in 24 s. His weight is 700 N and the height of the stairs is 10 m.

a) What is the work done by Tomas?

b) How much power must Tomas supply?

Givens:

$F = 700\text{ N}$

$d = 10\text{ m}$

Solution:

a) $W = F \cdot d$

$W = (700\text{ N})(10\text{ m})$

$= 7000\text{ J}$

b) $P = \frac{W}{t}$

$= \frac{7000\text{ J}}{24\text{ s}}$

$= 290\text{ W (watts)}$

Notice that the unit for power is joules per second, or watts. You are familiar with the power ratings of light bulbs in watts. You may have heard of horsepower as another unit for power. One horsepower is the power output of a horse over a specific time. One horsepower is approximately 750 W.

Physics Words

work: the product of the displacement and the force in the direction of the displacement; work is a scalar quantity.

closed system: a physical system on which no outside influences act; closed so that nothing gets in or out of the system and nothing from outside can influence the system's observable behavior or properties.

open system: a physical system on which outside influences are able to act; open so that energy can be added and/or lost from the system.

power: the time rate at which work is done and energy is transformed.

ANSWERS

Physics To Go

1. The work is identical in the two cases. The force is different since in the second case, most of the force is in the direction of the displacement.

Reflecting on the Activity and the Challenge

A roller coaster ride always begins with a slow, suspenseful ride to the top of the first hill. On the way up, the roller coaster is designed to shake a bit and to make a few extra noises in order to add to the drama. The roller coaster is gaining gravitational potential energy GPE on the way up. The motor is performing work on the roller coaster. Work is a precisely defined term in physics: $W = F \cdot d$. The work supplied by the motor increases the energy of the roller coaster. At the top of the incline, the motor is disengaged and the roller coaster is on its own. There is some work by friction with the air and track that removes energy from the roller coaster. In designing your roller coaster, you will have to include a motor to lift the roller coaster. You will have to decide on the slope of the track going up and the time you want the ride to take. Work and energy will be useful ways of describing what is needed in your design. You will also want to know how fast this work is done. For that you will use the concept of power where $P = W/t$.

Physics To Go

1. A student is asked to use the window pole to slide the window up. If the window moves the same distance up, is the work applied equal in the two cases shown? Is the force applied equal in the two cases?

2. A cart starts at the top of the incline. It slides down the incline a distance l and comes to rest after compressing a spring a distance x.
   a) Compare the GPE of the cart at the top of the incline and at the bottom.
   b) How much work was done on the cart by the force of gravity (the cart's weight)?
   c) How much work was done on the cart by the spring?
   d) What is the spring's SPE when it is compressed?
   e) Before hitting the spring, describe the total energy of the cart.
   f) At which point does the cart begin to slow down?

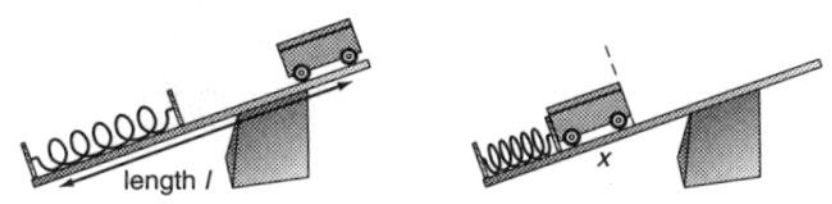

3. Calculate the work done in the following situations:
   a) A waiter applies a force of 150 N to hold a tray filled with plates on his shoulder. He then moves 7 m toward the kitchen door.
   b) A bowler lifts a 60 N bowling ball from the rack to his chest, a vertical distance of 0.5 m.
   c) A girl pulls her sled up a hill. The length of the hill is 40 m and the pulling force required was 75 N.
   d) The weight of an object is 500 N. It is lifted over a body-builder's head, a distance of 0.7 m.
4. Why are you told to conserve energy if the conservation of energy tells you that energy is always conserved? Create a better way of saying "conserve energy."
5. What is the difference between an open and closed system?
6. If you were to fill the cart you used in the activity with clay to represent the people in the roller coaster, what would have changed in the experiment?

ANSWERS

Physics To Go
(continued)

2. a) The GPE deceases as the cart descends the ramp.
   b) $W = Fd = mgh$
   c) The work by the spring $= 1/2\ kx^2$
   d) SPE $= 1/2\ kx^2$
   e) Before hitting the spring, the cart has some GPE, no SPE, and some KE.
   f) The cart begins to slow down when the force on the spring is greater than the force of gravity pulling the cart down the ramp. It is *not* when the cart first hits the spring because at that moment it is still gaining KE, but the spring is not yet applying a force.
3. a) No work done since the force is up and the displacement is horizontal. (The force and displacement are not in the same direction.)
   b) $W = Fd = mgh$ $= (60\ N)(0.5\ m) = 30\ J$
   c) $W = Fd = mgh$ $= (75\ N)(40\ m) = 3000\ J$
   d) $W = Fd = mgh$ $= (500\ N)(0.7\ m) = 350\ J$
4. Conserve energy implies "don't waste energy."
5. In a closed system, there is no energy loss or energy gain.
6. The force to lift the cart up the incline would have changed, the total work would have changed. The work in all trials would have been equal to each other.

Chapter 4

ANSWERS

Physics To Go
(continued)

7. a) $W = Fd = mgh$
$= (10{,}000 \text{ N})(20 \text{ m})$
$= 200{,}000 \text{ J}$

b) $P = W/t = 200{,}000 \text{ J}/150 \text{ s}$
$= 1333 \text{ W}$

8. The motor does positive work on the coaster to bring it to its highest point. Gravity does work to increase its kinetic energy until it reaches the bottom of the hill at which point gravity does negative work by decreasing the kinetic energy and increasing the gravitational potential energy.

The energy of the roller coaster is increased as the motor lifts the coaster to the top of the first hill. At all other points of the roller coaster, the total energy (GPE and KE) remains the same if we ignore losses due to friction.

7. A roller coaster car that weighs 10,000 N is lifted to the top of the first hill that is 20 m above the ground. To add suspense, the ride up takes 150 s.
   a) Calculate the work done by the motor.
   b) Calculate the power of the motor.

8. In the Terminator Express roller coaster, describe one trip in terms of work and energy.

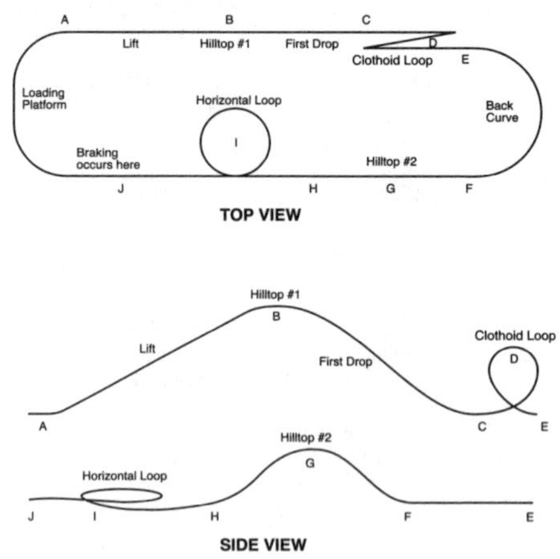

Stretching Exercise

Have the cart pulled up to the top of the incline with a motor. Measure the energy of the motor using voltage, current, and time. Compare the energy from the motor with the increase in GPE of the cart.

ACTIVITY 8
Vectors and Scalars

Background Information

It is suggested that you read the **For You To Read** section in the student text for **Activity 8** before proceeding in this section.

Roller coasters can be analyzed using forces or energy considerations.

As a scalar quantity, energy is a simpler way to calculate the velocities at all points on the roller coaster. To do this, we use the concept of conservation of energy and recognize that the sum of gravitational potential energy and kinetic energy will be identical for all points on the roller coaster. Although on straight-line slopes of the roller coaster, forces and accelerations can be used to calculate speeds. On curved parts of the track this is quite difficult because the vector sum of the gravitational force and the normal force is constantly changing.

Energy considerations do not tell us about how quickly the speeds change. These accelerations are what provide the thrills of a roller coaster. In this case, we need to look at the forces at each point on the roller coaster.

The concept map given at the end of this section brings together all of the enduring understandings illustrated in this chapter concerning energy and forces. The two maps are connected through the concept of work ($W = F \bullet d$).

Creating a roller coaster requires calculations of both speeds and accelerations.

Planning for the Activity

Time Requirements

One class period is required to complete this activity.

Materials Needed

For each group:

- Post-its®
- pennies
- tape, masking, 3/4 x 60 yds.

Advance Preparation and Setup

No advance preparation is required.

Teaching Notes

The creation of the concept map is a valuable way in which to review the concepts within the chapter. After each team creates their own concept maps, it would be interesting to have them share their concept maps. There is no one “correct” concept map. The different links that students create will reveal different ways of connecting the concepts. The concept map may also uncover some misconceptions or confusions and can serve as a formative assessment to guide review before students begin their roller coaster design projects.

Research on the differences between novices and experts point out that experts lean on general principles and patterns to analyze new situations while novices look for superficial similarities. This activity attempts to guide students in recognizing that roller coasters can be analyzed by both energy and force considerations. However, when the roller coaster ride has lots of turns and irregularities, an analysis of forces will be quite difficult. On the other hand, energy considerations cannot give any insights into how much time it takes for the roller coaster to get from one location to another. It would be worthwhile to have one set of students invent questions that are easier to answer with force considerations and another set of students invent questions that are easier to answer with energy considerations. They can then switch questions and try to answer with the two concepts (e.g., try answering the force question using energy first and then force concepts).

NOTES

Activity 8 Vectors and Scalars

GOALS

In this activity you will:

- Describe instances in which two cars will have the same speed but require different times to reach those speeds.
- Add vectors that are perpendicular to each other.
- Recognize that forces are vectors and energies are scalars.
- Explain how forces and energy considerations provide different insights into roller coaster rides.
- Choose whether energy or force considerations are more appropriate for analyzing aspects of roller coaster rides.

What Do You Think?

"The Snake" roller coaster stays at ground level throughout the ride. The passengers move left, then right, then left again.

- **Which parts of The Snake will be the most thrilling?**
- **If the speed of The Snake always remains the same, why will it still be fun?**

Record your ideas about these questions in your *Active Physics* log. Be prepared to discuss your response with your small group and the class.

Activity Overview

Students view a roller coaster as both an energy ride and as a force ride. The thrills of the roller coaster come from the accelerations. Energy as a scalar quantity cannot adequately describe the accelerations and students are introduced to vector addition to better appreciate how forces contribute to roller coaster fun.

Student Objectives

Students will:

- Describe instances in which two carts will have the same speed but require different times to reach those speeds.
- Add vectors that are perpendicular to each other.
- Recognize that forces are vectors and energies are scalars.
- Explain how forces and energy considerations provide different insights into roller coaster rides.
- Choose whether energy or force considerations are more appropriate for analyzing aspects of roller coaster rides.

ANSWERS FOR THE TEACHER ONLY

What Do You Think?

The parts that will be the most thrilling will be where the accelerations are the greatest. This will be around the tight turns or the quick left-right-left turns.

Speed does not provide the thrills of the roller coaster–it is the rapid changes in speed or large accelerations that provide the thrills.

ANSWERS

For You To Do

Part A: Energy and Forces in a Roller Coaster

1.–5. Concept lists and concept maps will differ among the students. Please see the end of this activity for a sample list of concepts and a sample concept map. Two concept maps are shown with the connection since the map would be too large to place on one sheet.

For You To Do

Part A: Energy and Forces in a Roller Coaster

1. Your study of roller coasters has actually taken two turns. You have investigated energy changes in roller coasters. You have also investigated forces and accelerations in roller coasters.

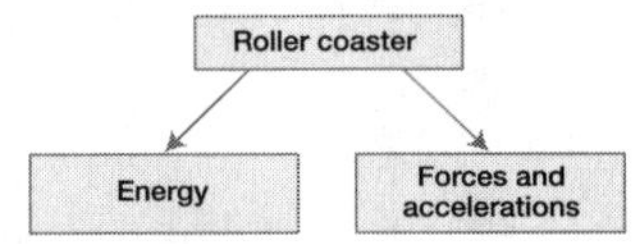

a) Copy this beginning of a concept map into your log. On a set of note-sized pieces of paper, write down some things you know about energy and how it relates to roller coasters. Each note should have one concept only.

2. Sort the concepts into a map.

a) Add these concepts to your log.

3. On a new set of notepaper, write down some things you know about forces and accelerations and how they relate to roller coasters. Each note should have one concept only.

4. Sort the concepts into a map.

a) Add these concepts to your log.

5. The left half of your map reminds you of the relationships between energy concepts. The right half of your map reminds you of the relationships between force and acceleration concepts.

a) Is there a bridge between these two sides of the map? Describe how energy is related to forces and accelerations.

6. You use both energy and force approaches to understand roller coasters because they both provide you with valuable information. Sometimes it is easier to look at a roller coaster as an energy ride, while other times it is best to look at a roller coaster as a force ride. As you become more comfortable with physics, you become better at matching what you want to know with the energy or force approach. Sometimes you need both and sometimes they are redundant.
7. In the roller coaster below, the initial height of the roller coaster is given.

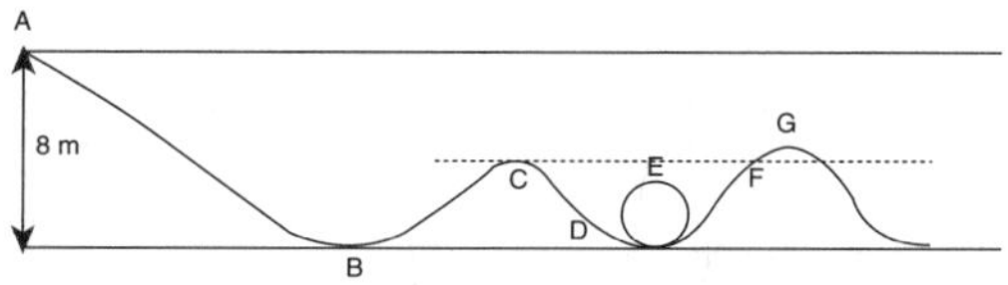

a) At which two points does the roller coaster have the same speed? (Neglect friction.)

b) How did you determine your answer? Write down your approach in your log.

8. Describe how the new roller coaster shown below is different from the roller coaster in **Step 7**.

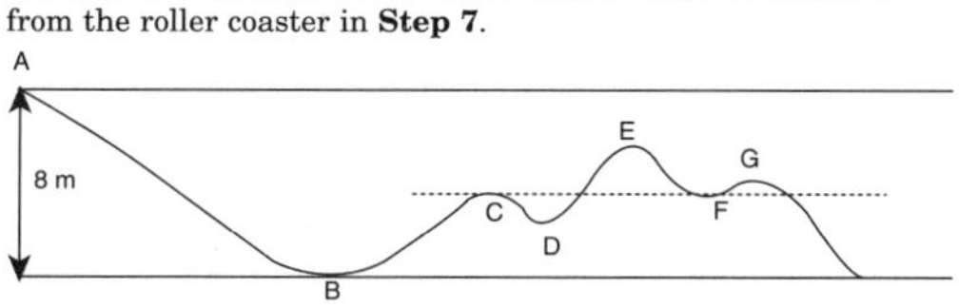

a) In this roller coaster, at which two points does the roller coaster have the same speed?

b) How did you determine your answer? Write down your approach in your log.

ANSWERS

For You To Do

(continued)

7. a)–b) The roller coaster has the same speed at points that have the same height. In this diagram, the roller coaster has the same speed at points C and F.

8. The roller coaster does not have a vertical loop.

   a) The roller coaster has the same speed at points C and F.

   b) The points C and F are at the same height and therefore have the same gravitational potential energy and the same kinetic energy.

ANSWERS

For You To Do
(continued)

9. The roller coaster would travel with a constant speed along the dotted line. Traveling at a constant speed is not nearly as much fun as changes in speed.

10. a) Both carts will have the same speed. Both carts have the same loss in gravitational potential energy and the same gain in kinetic energy. With the same kinetic energy and the same mass, they will have identical velocities ($KE = 1/2\ mv^2$)

    b) The cart moving along the steeper incline will get to the bottom in less time. If you take a look at the limiting case, an almost horizontal track, it will take a long, long time for the cart to descend.

Part B: Using Vectors to Describe a Path

1.–2. Students' hiding places and directions will vary. All directions should include the distances in meters or steps and the directions.

9. In either roller coaster on the previous page, the track could have been replaced with the dotted line.
   a) Why would the flat track be less fun than the roller coaster track?

10. Look at the following diagram.

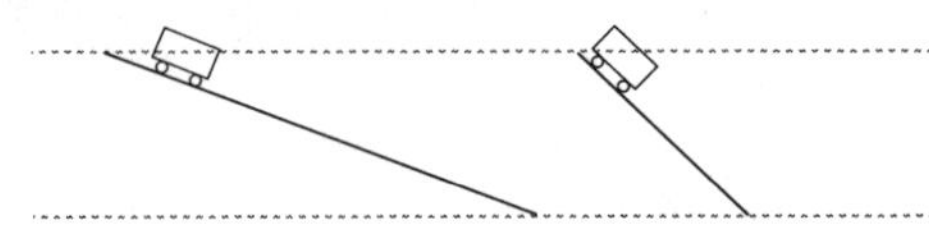

   a) Using energy principles, predict which cart would have the greatest speed when it reaches the bottom.
   b) Predict which cart will get to the bottom in the least time. On what did you base your response? Write down your explanation in your log.

Part B: Using Vectors to Describe a Path

1. Your teacher will give you a penny.
   a) Record the date of the penny. Hide the penny somewhere in the room.
   b) Provide a set of detailed instructions to allow another student to find your penny if they start at your desk.

2. Exchange directions and try to find your partner's penny.
   a) Did your instructions include how far they have to walk?
   b) Did your instructions include any changes in direction (left turns or right turns)?
   c) Did your instructions include reaching up or down?
   d) Rewrite the instructions so that each instruction describes how far the person should move in meters and in which direction.
   e) Compare this new set of directions with your first set. What advantages and disadvantages does each set have?

FOR YOU TO READ

Adding Scalars and Vectors

You can walk 30 m east. You can ride at 60 mph toward Mexico. Both descriptions include a number and a direction. Both are vectors. There are some descriptions that include a number, but no direction. There are 26 students in the classroom. The temperature is 18° C. Physicists have found that whether a number has a direction or not is an extremely important distinction. You can understand the world better if you recognize which quantities can have directions and deal with them accordingly.

It's fairly obvious that some quantities, like force always have directions. Some quantities, like your age, never have direction. There are some quantities, like how fast you are traveling, that can include direction. Your car can be traveling at 30 mph or you can describe the car traveling at 30 mph north.

A quantity with both a number (often referred to as magnitude) and a direction is called a vector. A quantity with a number and no direction is a scalar.

Scalars are easy to add, subtract, multiple, and divide. If you walk 15 km and then walk another 20 km, the total distance traveled is 35 km. After walking 35 km, you know how tired you will be and how worn your shoes will be. This scalar quantity is called distance. Traveling from New York to Florida, your average speed might be 50 mph. This takes into account the total distance traveled and the total time, but does not take into account any turns you made. Speed is also a scalar.

Displacement is a vector. You may walk 15 km north and then walk another 20 km east. Your total displacement is only 25 km. To add vectors, you must draw them and use vector addition. In this case, it is an application of the Pythagorean Theorem.

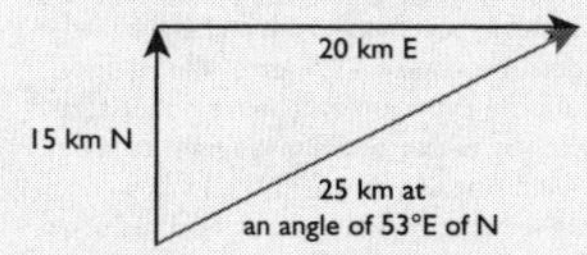

To add these two displacement vectors, you first drew the 15-km vector north. The vector was drawn to scale. To scale means that the length of the vector may be 15 cm or 1.5 cm or 0.15 cm. When you then draw the 20-km east vector you begin at the tip of the 15-km vector and then draw this vector using the same scale. If the 15-km

→

vector was 1.5 cm, then the 20-km vector will be 2.0 cm. You find the sum by drawing the resultant vector from the tail (or beginning) of the first vector to the end (or tip) of the second vector. This completes a triangle. You can then measure the length of the resultant with a ruler. In this case, the vector would be 2.5 cm that corresponds to a displacement of 25 km. You must measure the angle because all vectors have a direction. Using a protractor, you find that the angle is close to 53° east of north (east of the north direction).

Since the vectors are perpendicular, you can also use the Pythagorean Theorem. In any right triangle $a^2 + b^2 = c^2$ where c is the hypotenuse and a and b are the sides. The value $c = 25$ km is the solution from the Pythagorean Theorem that agrees with the vector addition diagram as it should. Just as you found the length, mathematically, you can also find the angle. The tangent button on the calculator, often labeled "tan" will tell you the angle if you know the lengths of the sides. Divide the side opposite the angle (= 20) by the side adjacent to the angle (= 15). By tapping "inverse tan" the calculator will provide the angle of 53°. Forces are vectors and add the same way as displacements.

Energy—A Scalar Quantity

Energy is a scalar and addition of scalars is simple. As you explored in earlier activities, the roller coaster ride may have GPE (gravitational potential energy) and KE (kinetic energy). It may have used electrical energy to lift the roller coaster to the top of the first hill. All energies can be calculated, but they are all measured in the same units, joules. And to find the total energy at any place or at any time, you just add up all the energies. This is what makes the roller coaster analysis using energies so powerful. After the roller coaster begins moving downhill, the total energy remains the same. The roller coaster begins with GPE and as the cart moves, the GPE converts to KE as the roller coaster picks up speed and then converts back to GPE as the cart goes higher and loses speed. Whatever the energy of the roller coaster is at the beginning of the ride, that is the energy at all times without friction. If two points on the roller coaster have the same height, then they must have the same GPE. If they have the same GPE, then they also have identical KE. It doesn't matter what the car did between the two points. It may have gone up, down, or in a loop-the-loop, but the KE will be the same at all points a specified distance above the ground.

In this activity, you looked at a roller coaster in **Step 7**. The speeds of the roller coasters are the same at points C and F. Both points C and F have the same height and therefore have the same GPE. Since *all* points on the roller coaster have the same total energy (GPE + KE) then both points must have the same KE. The same KE implies the same speed. (KE = $\frac{1}{2}mv^2$).

In the roller coaster in **Step 8** of this activity, the speeds were still the same at points C and F even though the track changed between C and F.

In roller coaster, energy considerations tell you three things:

- **The total energy (GPE + KE) is the same at every point (without friction or motors).**
- **The GPE depends only on the height (GPE = *mgh*) since the mass and the gravitational force remain the same.**
- **If two points on a roller coaster have the same height, the roller coaster is moving at the same speed at those two points.**

Energy considerations are path independent. You can look at the energy at one point and compare it to the energy at a later point. The energy will remain the same. It doesn't matter what happens between the places that are of interest.

In these four roller coaster sections, the cars begin at the top with zero KE and 20 J of GPE. When they reach the bottom, all will have the same KE (kinetic energy). This means that they will all have the same speed. To find this KE or speed, you had to only look at the beginning point and the final point. The path does not affect the final speed.

Force—A Vector Quantity

Although the roller coaster cars all get to the bottom with the same speed, they do not get there in the same time. To find the time, you would have to look at the forces and this becomes a vector problem. In all tracks, the force of gravity is always down. The normal force is always perpendicular to the track.

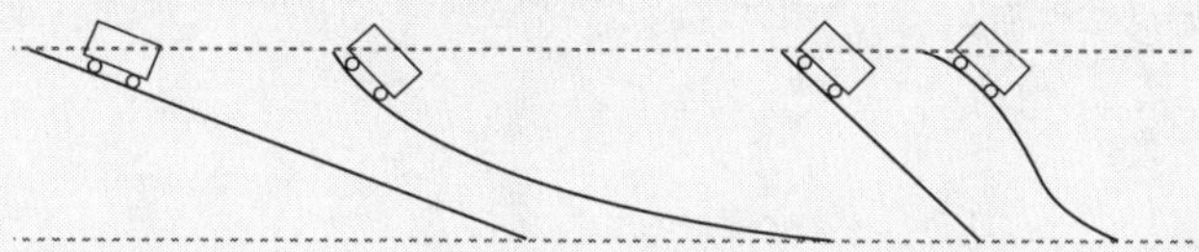

→

Thrills and Chills

The straight tracks are the easiest to analyze. The force of gravity and the normal force remain in fixed directions. You move down the incline and go faster and faster. The steeper the slope, the larger the gravitational force down the incline and the quicker you get to the bottom. It is a big acceleration for a short time and you reach the maximum speed. On a small incline, there is a small resultant force down the incline. It is a small acceleration for a long time and you reach the same maximum speed.

The inclines with shifting directions add to the thrill. Your speed changes as you move to different heights. As you move closer to the ground, your speed increases. The normal force (the force of the track on you) is always changing direction. This causes you to accelerate in lots of different directions. The changes in the acceleration (both in size and in direction) give you that bouncy feeling and the thrill of the roller coaster. The diagram below shows the gravitational force and normal forces at different points on a roller coaster.

- **On the straight incline, the gravitational force and the normal force remain in fixed directions. The car has a constant acceleration in magnitude and direction.**
- **On the curved incline, the normal force changes direction (it must be perpendicular to the incline) and changes in magnitude. The car has a changing acceleration in magnitude and direction. This provides big thrills.**

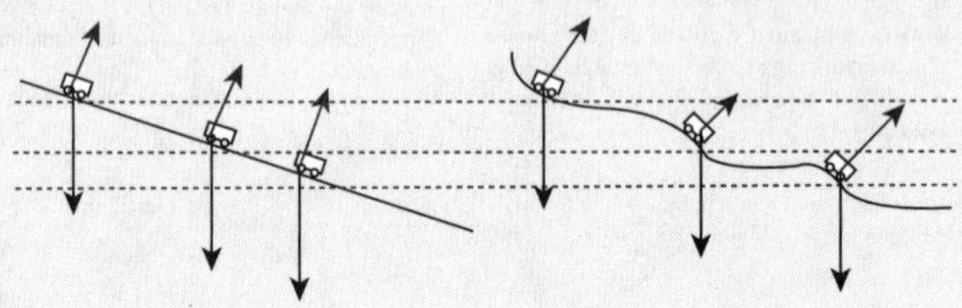

- **The speeds of the cars are identical on the two inclines. When the heights above the ground are the same, the GPE is the same. If the GPE is the same and the total energy is the same, the KE is the same. If the KE is the same, then the speed is the same.**

When to Consider Force or Energy

The mathematics of energy conservation requires simple addition. The mathematics of forces and accelerations requires vector addition. When the roller coaster looks complex, with lots of curves, physicists think of energy first because of the ease of using simple addition rather than vector addition.

When asked about how much time something will take, physicists think about forces and accelerations. Acceleration is the change in speed with respect to time.

Force and energy are related. The force of gravity does work on the roller coaster and increases its KE. Changes in energy always require work by a force. Work is a force applied over a distance ($W = F \cdot d$). The only external force doing work is gravity. When the roller coaster is moving down, the force is down and the displacement is down.

There is positive work on the roller coaster and it increases its KE. The normal force does no work since it is always perpendicular to the displacement. No part of the normal force is ever in the direction the roller coaster car is moving.

$W = F \cdot d$ where F is the force of gravity ($F = mg$) and d is the displacement and the dot tells you that the vector multiplication requires you to use only the distance in the same direction of gravity. This is the change in height (Δh).

The work done by gravity is $W = mg\Delta h$. The change in KE is $mg\Delta h$. This is equal to the loss in GPE. GPE = $mg\Delta h$.

Reflecting on the Activity and the Challenge

The thrill of the roller coaster comes from the changing velocities. You can analyze the changes in speed using energy considerations. Energy is a scalar. GPE can be easily calculated at every point on the roller coaster. Once you know the GPE, you can find the KE and then determine how fast the roller coaster moves. Understanding the mathematics of energy is as simple as $2 + 3 = 5$. Energies add with simple arithmetic just like all scalars.

You can also analyze the thrills of changing velocities by noting the forces acting on the roller coaster. Forces are vectors. They have both magnitude and direction. When more than one force acts on a roller coaster (e.g., the gravitational force and the normal force), you have to add them using vector arithmetic. You can always do this with a vector diagram. When the forces are perpendicular, you can also do this mathematically using the Pythagorean Theorem.

Designing a roller coaster requires you to know how fast it will be going at each point. You can use energy considerations to determine this.

You will also have to know how large the forces are because you will need to figure out the strength of the materials needed to provide the forces by the track. If too large a force is applied, the track may break. Adding the forces can provide you with this information.

You will also have to know accelerations of the passengers. Too large an acceleration or a change in acceleration and the riders may get sick or become unconscious. Newton's Second Law relating forces and accelerations ($F = ma$) can help you with this.

Making an exciting roller coaster requires changes in forces. The whips and turns and the ups and downs will change the speeds, the energy, and the forces on the passengers.

Activity 8 Vectors and Scalars

Physics To Go

1. A roller coaster makes a sharp right turn. The velocity of the roller coaster car is 5.0 m/s south before the turn and 5.0 m/s west after the turn.
   a) Determine the change in velocity of the roller coaster car using a vector diagram.
   b) Determine the change in velocity of the roller coaster car using the Pythagorean Theorem and the tangent button on your calculator.

2. A roller coaster makes a sharp right turn as it descends a hill. The velocity of the roller coaster is 5.0 m/s south before the turn. After the turn, the velocity of the roller coaster is 12.0 m/s west but it is also pointing downward at an angle of 25°.

   Ignore the downward angle.
   a) Determine the change in velocity of the roller coaster using a vector diagram.
   b) Determine the change in velocity of the roller coaster using the Pythagorean Theorem and the tangent button on your calculator.
   c) How would your answer change if you took into account the downward angle?

3. A roller coaster wall allows the riders to be on their side as they whip around a curve. The normal force (perpendicular to the track) is equal to 3000 N. The weight of the car with riders is 5000 N. What is the total force acting on the riders? (Be sure to include both magnitude and direction.)

4. All roller coasters that begin at the same height have the same speeds at the bottom.

   Explain why these two roller tracks provide the same change in speed.

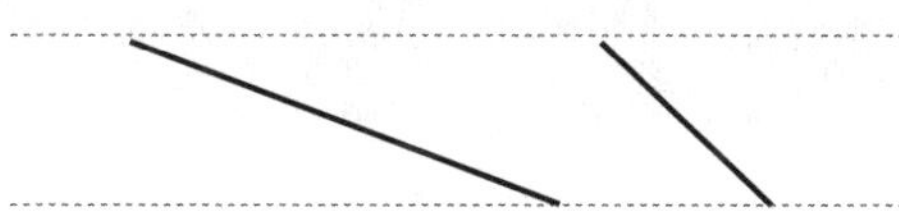

ANSWERS

Physics To Go

1. a) Using the Pythagorean Theorem, the change in velocity is 7.1 m/s northwest. The velocity changed by 5 m/s toward the west and 5 m/s north.

2. a)–b) Using the Pythagorean Theorem, the change in velocity is 13 m/s at 23° north of west.

   c) The 13 m/s and the downward velocity would have to be added as vectors.

3. The total force will be the vector sum of the forces which can be found using the Pythagorean Theorem and the tangent button.

   The force equals 5800 N. The angle would be 59° below the horizontal.

4. Both roller coasters begin with identical gravitational potential energies and zero kinetic energies. Their total energies will be identical at the bottom. Since their gravitational potential energies at the bottom are both zero, their kinetic energies must be identical.

ANSWERS

Physics To Go
(continued)

5. Identify the following as vectors (V) or scalars (S) .
   a) distance (S)
   b) displacement (V)
   c) speed (S)
   d) velocity (V)
   e) acceleration (V)
   f) force (V)
   g) kinetic energy (S)
   h) potential energy (S)
   i) work (S)
6. Which of the following are vectors (V) and which are scalars (S)?
   a) Traveling 30 km. (S)
   b) Weight (the force of gravity) is 600 N. (V) (direction is down).
   c) The roller-coaster car had a kinetic energy of 1200 J. (S)
   d) The car was traveling at 30 m/s toward the center of town. (V)

5. Identify the following as vectors or scalars:
   a) distance
   b) displacement
   c) speed
   d) velocity
   e) acceleration
   f) force
   g) kinetic energy
   h) potential energy
   i) work
6. Which of the following are vectors and which are scalars?
   a) Mark traveled 30 km.
   b) Maia's weight (the force of gravity on her) is 600 N.
   c) The roller coaster car had a kinetic energy of 1200 J.
   d) The car was traveling at 30 m/s toward the center of town.

Enduring Understandings (Concepts) For Roller Coaster Chapter of Active Pysical Science

- **1.** Energy
  - **1.1.** Different kinds of energy
    - **1.1.1.** Each energy can be calculated and measured
      - **1.1.1.1.** GPE mgh
      - **1.1.1.2.** KE = 1/2 mv^2
      - **1.1.1.3.** SPE = 1/2 kx^2
  - **1.2.** Conserved in closed systems
  - **1.3.** Not conserved in open systems
    - **1.3.1.** Work = change in energy
      - **1.3.1.1.** $W = F \cdot d$
    - **1.3.2.** Work can increase or decrease energy
  - **1.4.** Energy is a scalar
- **2.** Forces and accelerations
  - **2.1.** $a = \Delta v/\Delta t$
    - **2.1.1.** change in magnitude
    - **2.1.2.** change in direction
      - **2.1.2.1.** circular motion
        - **2.1.2.1.1.** $F = mv^2/R$
  - **2.2.** Accelerations require forces
    - **2.2.1.** $F = ma$
      - **2.2.1.1.** Newton's Second Law
    - **2.2.2.** Force of gravity on roller coaster
      - **2.2.2.1.** $F = mg$
        - **2.2.2.1.1.** weight
        - **2.2.2.1.2.** apparent weight
        - **2.2.2.1.3.** how a scale works
          - **2.2.2.1.3.1.** Hooke's Law
    - **2.2.3.** Normal force of track
      - **2.2.3.1.** Perpendicular
    - **2.2.4.** Forces add as vectors
      - **2.2.4.1.** Vector addition

Sample Concept Map I

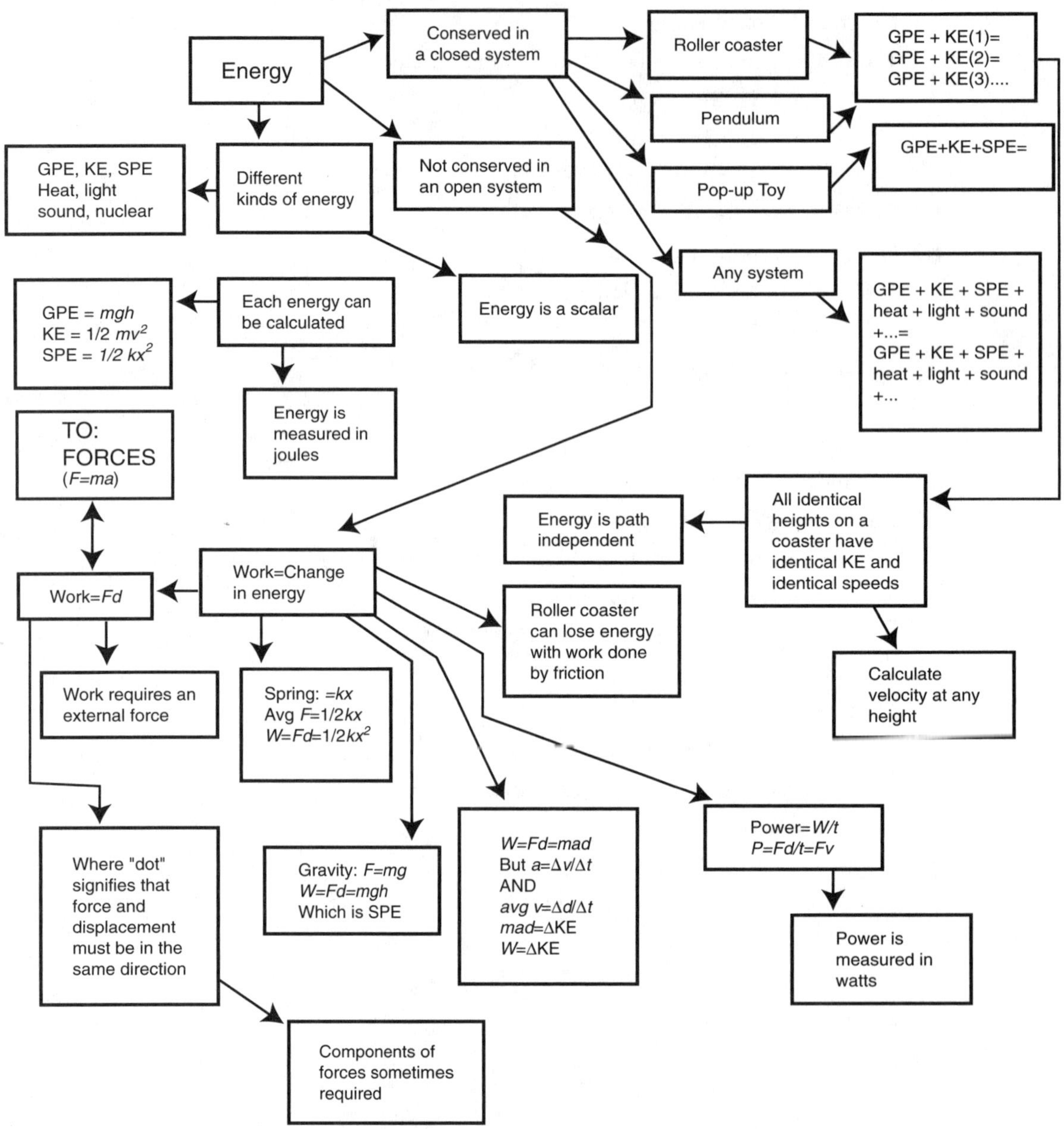

Sample Concept Map 2

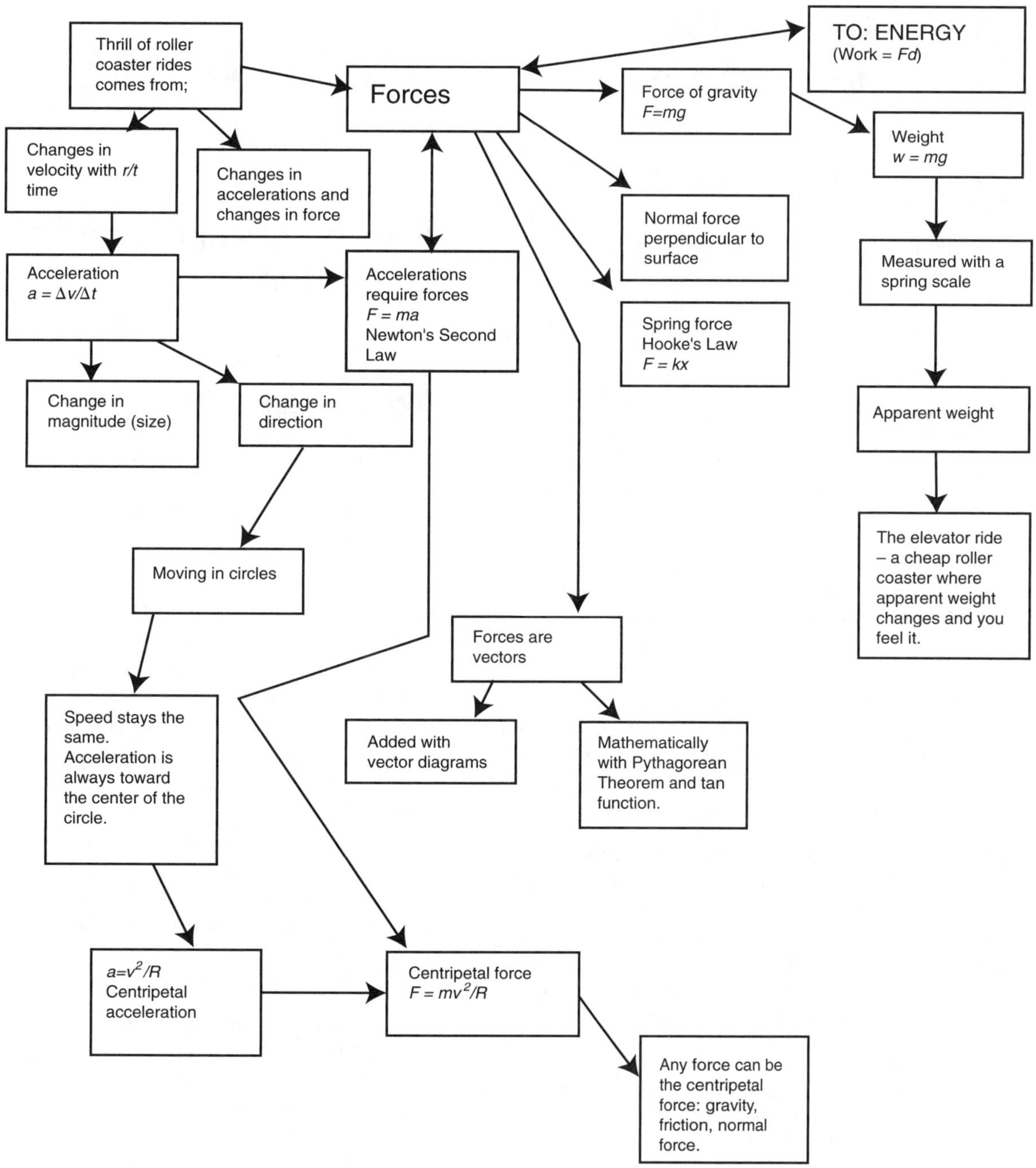

Chapter 4

ACTIVITY 9
Safety is Required but Thrills are Desired

Background Information

It is suggested that you read the **For You To Read** section in the student text for **Activity 9** before proceeding in this section.

The mathematics associated with roller coasters at this level of treatment include calculation of velocities at different heights of the roller coaster using the conservation of energy. These velocities can then be used to determine if the centripetal accelerations around curves on any radius are deemed too large for safety. We use a conservative value of 4 times the acceleration due to gravity ($4 \times 9.8\ m/s^2$ = approximately $40\ m/s^2$) as the maximum allowable acceleration.

At the top of a loop in the roller coaster, we know that the downward force will be at least 1 g—the acceleration due to gravity. This acceleration is the minimum centripetal acceleration at the top of the loop. From this minimum acceleration, we can calculate the minimum speed that a coaster must have to maintain circular motion and not leave the track. Any speed greater than this minimum speed will require a larger centripetal acceleration. This additional acceleration can be provided by the force of the track pushing down on the coaster.

The third safety concern of a roller coaster is that the track be strong enough to provide the forces required to hold up the track during the accelerations. This force is dependent on the mass of the roller coaster. It is therefore necessary to estimate the maximum weight of people in the roller coaster in designing a safe ride.

Roller coasters are made more fun by using optical illusions and sounds during the ride to heighten the thrill. For example, it may appear that the roller coaster is about to hit a wall or plunge into water. Sounds are used to make the roller coaster sound as if it may be broken during the long, slow uphill ride to the top.

Planning for the Activity

Time Requirements

One to two class periods will be required to complete this activity. The amount of time required will depend on the math skills of the students.

Materials Needed

- No materials needed

Advance Preparation and Setup

No advance preparation is required.

Teaching Notes

The discussions of safety features of the roller coaster should take place before you delve into the calculations with the students. On the board, you should list the three safety features that we are investigating:

- No accelerations can be greater than 4 g.
- The speed at the top must be greater than the speed required for circular motion with a centripetal acceleration of 1 g.
- The track and walls must be strong enough to support the weight of a roller coaster filled with passengers and to provide any centripetal forces on curves.

You may want to work through this activity with the students in a step by step fashion. You could begin with the class completing the sample problems in the **For You To Read** section. After assisting the students with this section, they can try their best to complete the **For You To Do** section.

You will have to determine the mathematics required for your students. This is dependent on their grade level and which math class they are enrolled in. The math teacher may provide some insight or assistance in the algebra required to complete these calculations. Try to remember that some of the students struggle in math class for months trying to isolate variables and manipulate equations. You will not be able to succeed in two days where the math teacher has struggled for months.

The students can have a clear, qualitative understanding of the safety features and be able to make simple calculations that do not require transposing equations. You should insist on this in that it will increase the confidence of the students when they do try the calculations.

The students will generate lots of good ideas for visual and sound effects that will make the roller coaster ride more thrilling. This part of the activity will allow all students to succeed and will prove to be a welcome respite from the math.

NOTES

NOTES

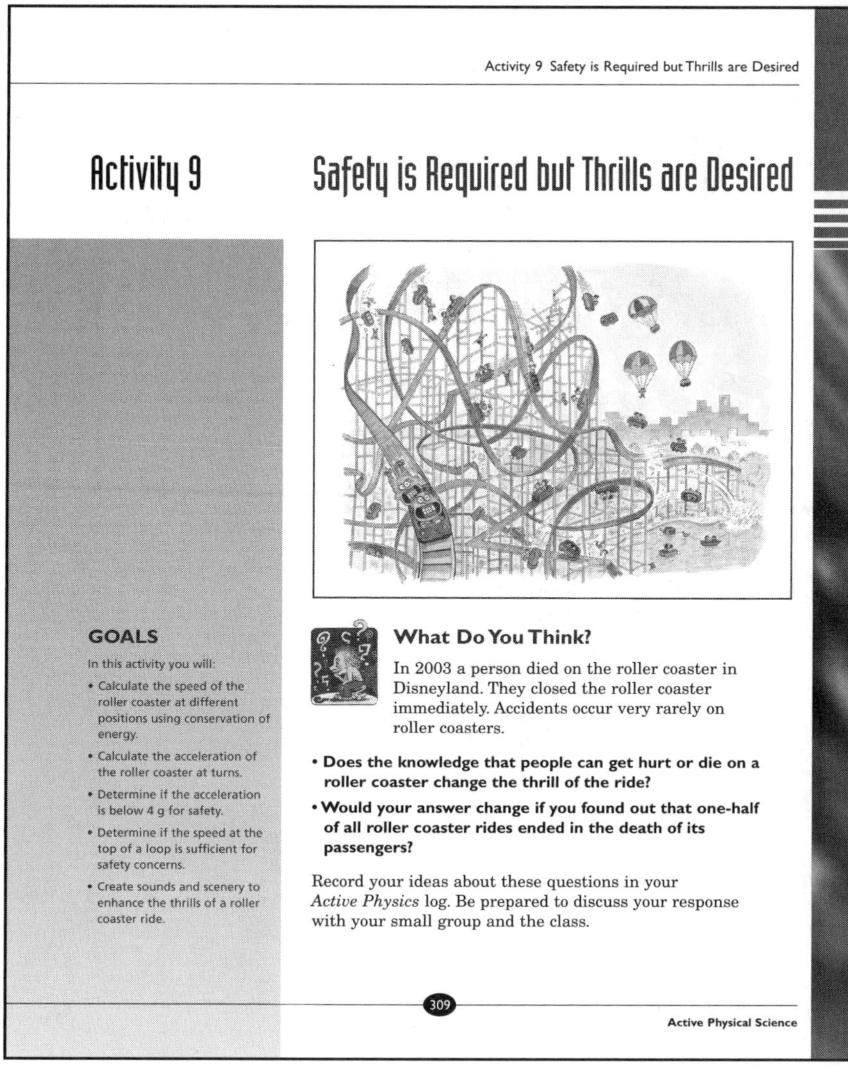

Activity 9 Safety is Required but Thrills are Desired

GOALS

In this activity you will:

- Calculate the speed of the roller coaster at different positions using conservation of energy.
- Calculate the acceleration of the roller coaster at turns.
- Determine if the acceleration is below 4 g for safety.
- Determine if the speed at the top of a loop is sufficient for safety concerns.
- Create sounds and scenery to enhance the thrills of a roller coaster ride.

What Do You Think?

In 2003 a person died on the roller coaster in Disneyland. They closed the roller coaster immediately. Accidents occur very rarely on roller coasters.

- **Does the knowledge that people can get hurt or die on a roller coaster change the thrill of the ride?**
- **Would your answer change if you found out that one-half of all roller coaster rides ended in the death of its passengers?**

Record your ideas about these questions in your *Active Physics* log. Be prepared to discuss your response with your small group and the class.

Activity Overview

Students investigate what may happen to passengers and the roller coaster if accelerations are too large. This leads to explorations of safety aspects of the roller coaster design including forces on the wheels, on the tracks and the need for strong materials. Students also discuss psychological means of making the roller coaster appear less safe and more fun.

Student Objectives

Students will:

- Calculate the speed of the roller coaster at different positions using conservation of energy.
- Calculate the acceleration of the roller coaster at turns.
- Determine if the acceleration is below 4 g for safety.
- Determine if the speed at the top of a loop is sufficient for safety concerns.
- Create sounds and scenery to enhance the thrills of a roller coaster ride.

ANSWERS FOR THE TEACHER ONLY

What Do You Think?

Students should distinguish between the thrill of the fear of being in danger and actually being in danger. Many people flock to horror movies but most people enjoy these films because they know it is not real. Films about war are different than war. Roller coaster rides where you fear you are out of control are different from airplanes in turbulence where you may actually be out of control.

ANSWERS

For You To Do

1. a) • Attendance—people will only go on rides they know are safe

   • Lawsuits—people will sue if they are hurt on a ride

   • Injury—nobody wants to be injured

   b) Students' answers will vary.

   Nausea—1 person per day and the ride should be closed

   Broken bones—1 injury every 3 years

   Unconsciousness—1 injury every 3 years

   Death—1 injury every 20 years if due to the ride and not negligence on the part of the rider.

2. a) Accelerations occur on the turns and loops as well as when the roller coaster goes down a hill.

   b) The acceleration in free fall is 9.8 m/s^2 which equals 1 g.

   c) It is not a safety concern with regard to consciousness. You will pick up speed quickly and will need a safe way to slow down.

Thrills and Chills

For You To Do

1. Safety is one of the criteria that you must meet in designing your roller coaster.

   a) List three reasons why safety is a major concern for roller coaster designers.

   b) How safe is safe? Your answer may depend on what injuries you describe for the roller coaster. Nausea and vomiting are one type of injury, broken bones are a second type of injury, becoming unconscious is a third type of injury, and death is the greatest injury. For the four types of injury listed, make an estimate of how many people could get injured on a roller coaster ride before it would be closed to the public. Be sure to include whether these are injuries in a day, a month, or a year.

2. Astronauts going into space had to withstand very large accelerations during rocket launch. After many tests of test pilots and race-car drivers, it was determined that people will become unconscious if the acceleration is greater than 9 g (or 9 times the acceleration due to gravity, that is, 9 x 9.8 m/s^2). Some people black out at 5 g or 6 g. This unconsciousness results from the blood leaving the brain during the high acceleration.

   The roller coaster manufacturer has indicated that the maximum acceleration at any place on the roller coaster can not exceed 4 g (or 4 times the acceleration due to gravity, that is, 4 x 9.8 m/s^2).

   a) At what locations on the roller coaster are there accelerations?

   b) If the roller coaster were to fall straight down, what would be the acceleration?

   c) Is this a safety concern?

3. A roller coaster car is traveling at 30.0 m/s at the bottom of a loop. The radius of the loop is 9.0 m.

310

Active Physics

Activity 9 Safety is Required but Thrills are Desired

a) Using the conservation of energy ($mgh + \frac{1}{2}mv^2$ = constant), calculate the initial height of a roller coaster to give it a speed of 30.0 m/s at the bottom of a loop at ground level. At the top of the roller coaster, the velocity is 0 m/s. At the bottom of the roller coaster, the height h equals 0 m.

b) Using the equation $a = v^2/R$, calculate the acceleration at the bottom of the loop.

c) Is this a safety concern?

d) At what speed would this loop with a radius of 9.0 m begin to be a safety concern?

e) At what speed would a loop with a smaller radius of 7.0 m begin to be a safety concern?

f) How fast would the roller coaster car be traveling at the top of the loop? (The top of the loop is 18.0 m above the ground since the loop has a radius of 9.0 m, and thus a diameter of 18.0 m.) You must use the initial height of the roller coaster that you calculated above to solve this problem.

g) Using the equation $a = v^2/R$, calculate the acceleration at the top of the loop.

h) Is this a safety concern?

i) There are two safety concerns regarding accelerations in a loop. The acceleration cannot exceed 4 g. This excessive acceleration would occur at the bottom of the loop, if at all. The acceleration at the top must be greater than 1 g (9.8 m/s^2). If the acceleration required for circular motion at the top of the loop is less than 1 g, the roller coaster car will leave the track and plummet to the ground. The speed at the top of the roller coaster must be large enough to require acceleration at least as great at 9.8 m/s^2.

4. The track must be strong enough to hold the roller coaster car without breaking. You can calculate the minimum strength of a track by assuming that the roller coaster car is filled with big football players or small Sumo wrestlers.

ANSWERS

For You To Do
(continued)

3. a) Top Bottom

GPE + KE = GPE + KE

$mgh + 0 = 0 + 1/2\ mv^2$

$h = v^2/2\ g$

$h = (30\ m/s)^2/2(9.8\ m/s^2)$

$h = 46$ m

b) $a = v^2/R$

$a = (30\ m/s)^2/(9\ m)$

$a = 100\ m/s^2$

c) Yes, this is more than 10 times the acceleration due to gravity or 10 g. People will lose consciousness.

d) If the acceleration were limited to 4 g = 40 m/s^2, you can find the corresponding velocity.

$a = v^2/R$

$v = \sqrt{aR}$

$v = \sqrt{(40\ m/s^2)(9\ m)}$

= 19 m/s

e) $a = v^2/R$

$v = \sqrt{aR}$

$v = \sqrt{(40\ m/s^2)(7\ m)} = 17$ m/s

f) Top Top of loop

GPE + KE = GPE + KE

$mgh + 0 = mgh + 1/2\ mv^2$

$gh_{top} = gh_{top\ of\ loop} + 1/2\ v^2$

$v^2 = 2g\ (h_{top} - h_{top\ of\ loop})$

$v^2 = 2\ (9.8\ m/s^2)(46\ m - 18\ m)$

$v = 23$ m/s

g) $a = v^2/R$

$a = (23\ m/s)^2/(9\ m) = 59\ m/s^2$

h) Yes, this is also a safety concern. The acceleration is greater than 4 g.

Chapter 4

ANSWERS

For You To Do
(continued)

4. a) $w = mg$

$w = (1000 \text{ kg})(9.8 \text{ m/s}^2)$

$= 98{,}000 \text{ N}$

5. a) • Sharp left turn—force is toward the center

• Sharp left turn with car on its side—force is toward center (identical to the first answer)

• Moving in a vertical circle—force is always toward center (up when car is at bottom of circle and down when car is at top of circle)

b) • Sharp left turn—the centripetal force is the force of the track that guides the wheels

• Sharp left turn with car on its side—the centripetal force is the wall that the track is attached to when the track is banked 90°

• Moving in a vertical circle—the centripetal force is a combination of the force of gravitation and the normal force of the track

6. a) $F_c = mv^2/R$

$F_c = (1000 \text{ kg})\ (15 \text{ m/s})^2/(12 \text{ m})$

$F_c = 18{,}750 \text{ N}$

7. a) $F_c = mv^2/R$

$F_c = (1000 \text{ kg})\ (15 \text{ m/s})^2/(12 \text{ m})$

$F_c = 18{,}750 \text{ N}$

a) What force would a roller coaster track have to supply to hold up a car filled with passengers if the total mass were 1000. kg on a horizontal section of track. (Remember the equation for weight is $W = mg$ where $g = 9.8 \text{ m/s}^2$.)

5. When the roller coaster makes turns, there must be a centripetal force pushing the roller coaster car toward the center of the circle. There are three types of turns that can be analyzed:

- **a sharp left turn on a flat track**
- **a sharp left turn where the car is turned on its side and the track is vertical**
- **a loop where the car is moving in a vertical circle**

a) For each of the three turns, identify the direction of the centripetal force (the force keeping the car moving in a circle).

b) For each of the three turns, identify the centripetal force. (The centripetal force could be the force of friction, the gravitational force, the normal force, the force of tension on a rope, the force of the wheels in the track or any combination of these forces.)

6. A roller coaster car has a maximum mass of 1000.0 kg and a sharp left turn where the car is turned on its side and the track is vertical with a radius of 12.0 m and the car is moving at 15.0 m/s.

a) Calculate the centripetal force required to keep the car moving in the curve.

7. A roller coaster car has a maximum mass of 1000.0 kg and a sharp left turn on a flat track with a radius of 12.0 m and the cart is moving at 15.0 m/s.

a) Calculate the centripetal force required to keep the car moving in the curve.

8. A roller coaster car has a maximum mass of 1000.0 kg and is about to enter a loop that has a radius of 12.0 m. The car is moving at 15.0 m/s.

Activity 9 Safety is Required but Thrills are Desired

a) Calculate the centripetal force required to keep the car moving in the curve.

b) This centripetal force that you calculated is the sum of the normal force of the track up toward the center of the roller coaster car and the force of gravity down to the ground. Copy the following diagram into your log:

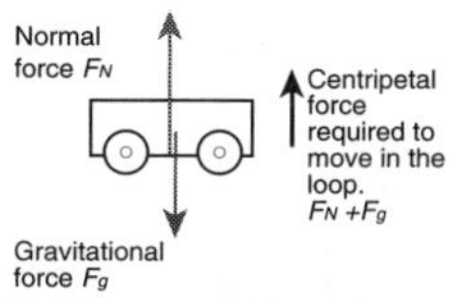

c) Calculate the force of gravity on the car.

d) Calculate the normal force on the car. This is the strength required of the steel that makes up the roller coaster.

9. The roller coaster designer can play all sorts of tricks to produce extra thrills on the same roller coaster.

a) Describe how you can add suspense to the trip up to the top of the first hill.

b) Describe how you can use sounds during the roller coaster ride to add to the thrills.

10. The choice of scenery surrounding the roller coaster can also add to the thrill. The roller coaster can look like it will dive into water when it is descending. The roller coaster can look like it will hit a building when it rounds a curve.

a) Describe three visual effects through the use of scenery that will add to the thrill of your roller coaster.

11. Thrills can come from big speeds. Thrills can also come from acceleration (changing velocity with respect to time). Change in velocity can be a change in speed or direction.

a) Describe three ways in which you can add thrills to a roller coaster design by having changes in velocity.

b) Describe three ways in which you can add thrills to a roller coaster design by having changes in acceleration.

ANSWERS

For You To Do
(continued)

8. a) $F_c = mv^2/R$

$F_c = (1000 \text{ kg})\ (15 \text{ m/s})^2/(12 \text{ m})$

$F_c = 18{,}750 \text{ N}$

b) Students copy diagram into their log.

c) $w = mg = (1000 \text{ kg})\ (9.8 \text{ m/s}^2) = 9800 \text{ N}$

d) *centripetal force required* $= F_N - F_g$

$F_N =$ *centripetal force required* $+ F_g$

$F_N = 18{,}750 \text{ N} + 9800 \text{ N}$

$F_N = 28{,}550 \text{ N}$

9.–11. For the following questions students' answers will vary. Encourage their creativity. Some sample answers are provided.

9. a) You can make the trip very slow; you can make it shake a bit.

b) You can make the roller coaster have some creaks and other noises to sound like it may be broken.

10. a) The roller coaster can come close to walls; the roller coaster can look like it is higher than it is; the roller coaster can look like the track will end.

11. a) The roller coaster can change directions; the roller coaster can come close to free fall; the roller coaster can make a sharp turn.

b) The roller coaster can change acceleration by having it zig-zag back and forth with different size turns; the loops provide changes in the acceleration; the roller coaster can turn and tilt at the same time.

Chapter 4

FOR YOU TO READ

Roller Coaster Safety

The roller coaster has to be safe in order to be fun. Analysis of the safety requirements of the roller coaster is a valuable way of reviewing some of the physics in earlier activities.

You know from research with test pilots that people will not be safe if their acceleration is greater than 4 g. A free fall provides an acceleration of 1g. Roller coasters may have steep inclines but they are generally less than free fall and therefore have an acceleration less than 1 g.

When the roller coaster rips around a corner or moves through the bottom of the loop, the acceleration can be much more than 1 g. Analyze the acceleration at the bottom of a loop. The acceleration can be computed by recognizing that the roller coaster at this location is moving in an arc of a circle. The centripetal acceleration must be toward the center of the circle and can be calculated by using the equation $a_c = v^2/R$. By varying the speed or the radius of the circle in the roller coaster design, you can limit the acceleration to less than 4 g.

Sample Problem 1

A roller coaster car with a mass of 800 kg is traveling at 15.0 m/s at the bottom of a loop. The loop has a radius of 5.0 m.

Activity 9 Safety is Required but Thrills are Desired

a) What is the required centripetal acceleration to keep the car moving in a circle?

a) Strategy: Use the equation for centripetal acceleration.

Givens:

$v = 15.0 \text{ m/s}$

$R = 5.0 \text{ m}$

Solution:

$$a = \frac{v^2}{R}$$

$$= \frac{(15 \text{ m/s})^2}{5.0 \text{ m}}$$

$$= 45 \text{ m/s}^2$$

This acceleration is greater than 4 g ($4 \times 9.8 \text{ m/s}^2 = 39.2 \text{ m/s}^2$) and is therefore unsafe.

b) One way to lower this acceleration would be to lower the velocity. Assume that the new design gives the car a velocity of 12.0 m/s. Calculate the required centripetal acceleration to keep the car moving in a circle.

b) Strategy: Use the equation for centripetal acceleration, again.

Givens:

$v = 12.0 \text{ m/s}$

$R = 5.0 \text{ m}$

$$a = \frac{v^2}{R}$$

$$= \frac{(12.0 \text{ m/s})^2}{5.0 \text{ m}}$$

$$= 29 \text{ m/s}^2$$

This acceleration is now less than 4 g ($4 \times 9.8 \text{ m/s}^2 = 39.2 \text{ m/s}^2$) and is therefore safe.

c) Another way to lower the acceleration is to make the loop larger. Using the original speed of 15.0 m/s, calculate the required centripetal acceleration if the radius of the loop were 7.0 m.

c) Strategy: Use the equation for centripetal acceleration, again.

Givens:

$v = 15.0 \text{ m/s}$

$R = 7.0 \text{ m}$

$$a = \frac{v^2}{R}$$

$$= \frac{(15.0 \text{ m/s})^2}{7.0 \text{ m}}$$

$$= 32 \text{ m/s}^2$$

→

This acceleration is now less than 4 g ($4 \times 9.8\ m/s^2 = 39.2\ m/s^2$) and is therefore safe. The largest centripetal acceleration (at the bottom of the loop) also requires the largest centripetal force. This maximum force will inform you as roller coaster designer of the strength of materials required to build this part of the roller coaster. The force moving the car in a circle is the normal force. The normal force required to move the car in a circle is much greater than the normal force that would support the car at rest at the bottom of the incline. The "at rest" car requires no net force. The normal force up (provided by the track) must equal the gravitational force down. This is shown with the vector diagram below. (Notice that the gravitational force acts on the center of the car while the normal force acts on the bottom of the car where it touches the surface.)

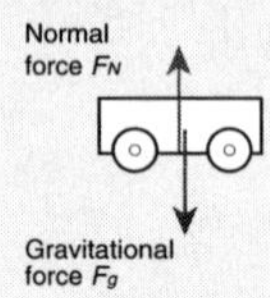

To move the car in a vertical circle, a centripetal force is required. The sum of the normal force and the gravitational force must equal the centripetal force required. Since the gravitational force is down, the normal force must be greater to provide the additional upward force needed. This is shown in the vector diagram below.

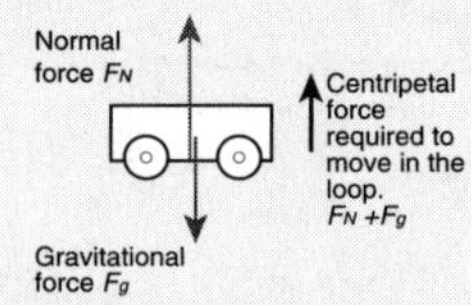

The centripetal force required can be calculated using Newton's Second Law: $F = ma$. In this case $F_c = \frac{mv^2}{R}$.

Sample Problem 2

A roller coaster car with a mass of 800.0 kg is traveling at 15.0 m/s at the bottom of a loop. The loop has a radius of 5.0 m. (**Part (a)** is a repetition of the calculation above.)

a) What is the required centripetal acceleration to keep the car moving in a circle?

b) What is the required centripetal force to keep the car moving in the circle?

c) What is the normal force by the track on the car?

a) Strategy: Use the equation for centripetal acceleration.

Givens:

$v = 15.0$ m/s

$R = 5.0$ m

Solution:

$$a = \frac{v^2}{R} = \frac{(15.0\ \text{m/s})^2}{5.0\ \text{m}} = 45\ \text{m/s}^2$$

This acceleration is greater than 4 g ($4 \times 9.8\ \text{m/s}^2 = 39.2\ \text{m/s}^2$) and is therefore unsafe.

b) *Strategy:* Use the equation for centripetal force.

Givens:

$v = 15.0$ m/s

$R = 5.0$ m

$m = 800.0$ kg

Solution:

$$F = \frac{mv^2}{R}$$

$$= \frac{(800.0\ \text{kg})(15.0\ \text{m/s})^2}{5.0\ \text{m}}$$

$$= 36{,}000\ \text{N}$$

This net force of 36,000 N up will allow the cart to move in the vertical circle.

c) *Strategy:* The normal force must be 36,000 N greater than the gravitational force to provide a net force of 36,000 N as required.

Solution: The gravitational force (weight) is:

$$w = mg = (800.0\ \text{kg})(9.8\ \text{m/s}^2)$$

$$= 7840\ \text{N (about 7800 N)}$$

Therefore, the normal force must equal 36,000 N + 7800 N = 43,800 N or about 44,000 N.

This indicates that the strength of the metal of the roller coaster must be at least 44,000 N or the track will break.

Similar calculations can be completed with the force required to make a turn on a horizontal part of the roller coaster and a turn where the roller coaster banks on its side as it whips around a turn.

Another safety feature requires that the speed at the top of the loop is great enough to complete the loop. A car that has too little speed will not make it to the top of the roller coaster and will not be able to move in the circle. It will find itself falling to the ground as the following diagram illustrates.

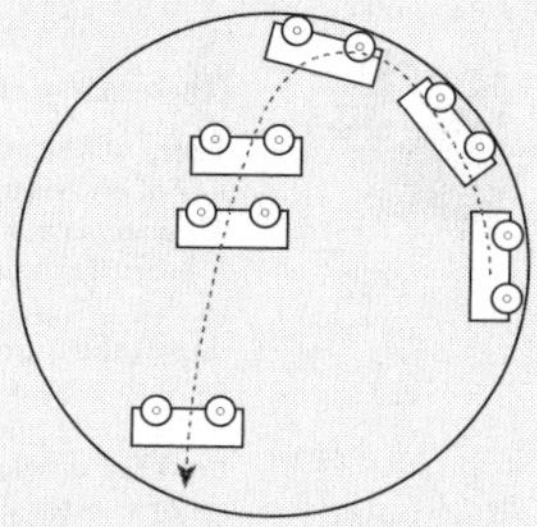

If gravity were the only force acting at the top of the roller coaster, then the car must require a centripetal acceleration equal to gravity.

→

Chapter 4

Sample Problem 3

What is the minimum speed required at the top of the loop to ensure that the car does not leave the track? The car has a mass of 800.0 kg. The loop has a radius of 5.0 m.

Strategy: The minimum speed pertains to the minimum centripetal acceleration of 9.8 m/s^2. Using the equation for centripetal acceleration, you can find the required speed.

Givens:

$a = 9.8\ \text{m/s}^2$

$R = 5.0\ \text{m}$

Solution:

$$a = \frac{v^2}{R}$$

$$v = \sqrt{aR}$$

$$= \sqrt{(9.8\ \text{m/s}^2)(5.0\ \text{m})}$$

$$= 7.0\ \text{m/s}$$

A speed of 7.0 m/s will be able to complete the upper part of the loop. A speed greater than 7.0 m/s will also be able to make the loop. The greater speed will require a larger centripetal force. The additional force will be provided by the track pushing down on the car.

Reflecting on the Activity and the Challenge

There is lots of creativity in designing a roller coaster. There is lots of creativity in designing a bridge, a building, and a table. All designs are constrained by the physics of the world. A beautiful bridge must also be a bridge that does not collapse. In this activity you learned about the safety features that you must take into account in your design for the roller coaster. You have to ensure that the accelerations are never above 4 g. This will require you to design the curves and loops with radii that limit the accelerations. You must also make sure that if your roller coaster does have a loop, the car will be able to complete the loop. You can vary the radius of any part of the track in the design. You can vary the velocity of the car by changing the launch height for the roller coaster. The higher the first hill, the more speed the coaster will have at the bottom. Safety is required, but thrills are desired. The activity also discussed ways in which you can use sound and scenery to improve the thrills of your design.

Physics To Go

1. An engineering company submits a plan for a roller coaster. What factors will you check to ensure that the roller coaster is safe?
2. A roller coaster car is traveling at 20.0 m/s at the bottom of a loop. The radius of the loop is 12.0 m.
   a) Using the conservation of energy ($mgh + \frac{1}{2}mv^2$ = constant), calculate the initial height of a roller coaster to give it a speed of 20.0 m/s at the bottom of a loop at ground level.
   b) Using the equation $a = v^2/R$, calculate the acceleration at the bottom of the loop.
   c) Is this a safety concern?
   d) At what speed would this loop with radius of 12.0 m begin to be a safety concern?
   e) At what speed would a loop with a smaller radius of 7.0 m begin to be a safety concern?
3. A roller coaster car is traveling at 25.0 m/s at the bottom of a loop. The radius of the loop is 10.0 m.
   a) Calculate the acceleration at the bottom of the loop.
   b) Is this a safety concern?
4. A roller coaster has an initial height of 50.0 m.
   a) What will be the speed of the roller coaster car at the bottom of the hill?
   b) The roller coaster car goes into a loop with a radius of 10.0 m. What is the acceleration required to move in the loop?
   c) What will be the speed of the roller coaster car at the top of the loop?
   d) What will be the required acceleration at the top of the loop to keep the car moving in a circle?
   e) Explain whether the roller coaster is safe at the bottom and the top of the loop.

ANSWERS

Physics To Go

1. To ensure that the roller coaster is safe, you should check to see if:
   - No acceleration is greater than 4 g;
   - The acceleration at the top of a loop is at least 1 g;
   - The tracks are strong enough to supply the forces required.

2. a) Top Bottom

$$GPE + KE = GPE + KE$$
$$mgh + 0 = 0 + 1/2\ mv^2$$
$$h = v^2/2g$$
$$h = (20\text{ m/s})^2/2(9.8\text{ m/s}^2)$$
$$h = 20\text{ m}$$

b) $a = v^2/R$

$$a = (20\text{ m/s})^2/(12\text{ m})$$
$$a = 33\text{ m/s}^2$$

c) There is no safety concern since the acceleration is less than 4 g (40 m/s²).

d) If the acceleration were limited to 40 m/s², you can find the corresponding velocity.

$$a = v^2/R$$
$$v = \sqrt{aR}$$
$$v = \sqrt{(40\text{ m/s}^2)(12\text{ m})}$$
$$v = 22\text{ m/s}$$

e) $a = v^2/R$

$$v = \sqrt{aR}$$
$$v = \sqrt{(40\text{ m/s}^2)(7\text{ m})}$$
$$v = 17\text{ m/s}$$

3. a) $a = v^2/R$

$$a = (25\text{ m/s})^2/(10\text{ m})$$
$$a = 63\text{ m/s}^2$$

b) There is a safety concern since the acceleration is more than 4 g (40 m/s²).

4. a) Top Bottom

$$GPE + KE = GPE + KE$$
$$mgh + 0 = 0 + 1/2\ mv^2$$
$$v^2 = 2gh$$
$$v = \sqrt{2gh}$$
$$v = \sqrt{2(9.8\text{ m/s}^2)(50\text{ m})}$$
$$v = 31\text{ m/s}$$

b) $a = v^2/R$

$$a = (31\text{ m/s})^2/(10\text{ m})$$
$$a = 98\text{ m/s}^2$$

c) Top Top of loop

$$GPE + KE = GPE + KE$$
$$mgh + 0 = mgh + 1/2\ mv^2$$
$$gh_{top} = gh_{top\ of\ loop} + 1/2\ v^2$$
$$v^2 = 2\ g(h_{top} - h_{top\ of\ loop})$$
$$v^2 = 2\ (9.8\text{ m/s}^2)(50\text{m}-20\text{m})$$
$$v = 24\text{ m/s}$$

d) $a = v^2/R$

$$a = (24\text{ m/s})^2/(10\text{ m})$$
$$= 58\text{ m/s}^2$$

e) Yes, there is a safety concern at both the top and bottom of the loop. The acceleration is greater than 4 g.

Chapter 4

ANSWERS

Physics To Go
(continued)

5. a) $a = v^2/R$

   $v = \sqrt{gR}$

   (The acceleration must equal the acceleration due to gravity $g = 9.8\ m/s^2$.)

   $v = \sqrt{(9.8\ m/s^2)(8\ m)}$

   $v = 8.9\ m/s$

   b) Top Top of loop

   GPE + KE = GPE + KE

   $mgh + 0 = mgh + 1/2\ mv^2$

   $gh_{top} = gh_{top\ of\ loop} + v^2$

   $h_{top} = h_{top\ of\ loop} + 1/2\ v^2/g$

   $h_{top} = 16\ m + 1/2\ (8.9\ m/s)^2/(9.8\ m/s^2)$

   $h_{top} = 16\ m + 4\ m$

   $h_{top} = 20\ m$

6. a) $a = v^2/R$

   $a = (12\ m/s)^2/(18\ m)$

   $a = 8.0\ m/s^2$

   b) $F_c = ma = mv^2/R$

   $F_c = (900\ kg)(8.0\ m/s^2)$

   $F_c = 7200\ N$

   c) The centripetal force will be provided by the strength of the track on the wheels.

7. a) $a = v^2/R$

   $a = (20\ m/s)^2/(15\ m)$

   $a = 26.7\ m/s^2$

   b) $F_c = ma = mv^2/R$

   $F_c = (900\ kg)(26.7\ m/s^2)$

   $F_c = 24{,}000\ N$

   c) Yes, the coaster is safe in that the wheels only have to supply a force of 24,000 N.

8. a) No, the acceleration is not dependent on the mass.

   b) The roller coaster will be traveling at the same speed.

   c) Yes, the roller coaster will require a stronger material. The force is proportional to the mass. ($F = ma$).

Thrills and Chills

5. A roller coaster has a loop with a radius of 8.0 m (diameter = 16.0 m).
   a) What speed must the roller coaster car have at the top of the loop if the only force at the top is the force of gravity and the centripetal acceleration is therefore $9.8\ m/s^2$?
   b) How high must the first hill be to provide this speed at the top of the loop?

6. A roller coaster car, when filled with people, has a mass of 900.0 kg. The roller coaster car rounds a curve on the ground with a radius of 18.0 m at a speed of 12.0 m/s.
   a) What is the centripetal acceleration of the car?
   b) What is the centripetal force on the car?
   c) What will provide this centripetal force?

7. A roller coaster car, when filled with people, has a mass of 900.0 kg. The roller coaster car rounds a curve on the ground with a radius of 15.0 m at a speed of 20.0 m/s.
   a) What is the centripetal acceleration of the car?
   b) What is the centripetal force on the car?
   c) The wheels in the tracks can provide a force of 25,000 N. Is the roller coaster safe?

8. A roller coaster is able to complete a loop when the car has two passengers. The car is loaded with six people.
   a) Will the centripetal acceleration change as a result of the change in mass?
   b) Will the roller coaster be going faster, slower, or the same speed at the bottom of the loop with the extra passengers?
   c) Will the roller coaster require a stronger material because of the increased number of riders?

320

Active Physics

PHYSICS AT WORK

Mark Rosenzweig

Not a job for the Queasy.

If you're scared of doing loop-the-loops at 100 mph or taking hairpin turns at 4 *g*'s, this is not the profession for you. But if you're one of those people who enjoys extreme excitement and quick thrills, Mark may be living your professional dream.

Mark works for Zamperla, one of the largest designers of roller coasters, spinners, and other stomach-turning amusement park rides in the world. "A lot of my time is spent at places like Disney World, Universal Studios, and other amusement parks like Legoland. I believe that visiting with customers and watching the reaction of park guests as they exit a ride is one of the best ways to judge its success."

Mark is not a scientist by trade. He did not go to graduate school or study engineering in college. After growing up in Long Island and earning an undergraduate degree in psychology at the State University of New York at Oneonta, he took jobs at small amusement parks in Long Island and then Michigan. Then Zamperla, an Italian company with an office in New Jersey, came calling, and the rest is hair-raising history.

As marketing and sales director, Mark is responsible for communicating the wishes and desires of potential riders to Zamperla's team of designers and engineers. And that means understanding many general principles of physics, engineering, and the human body. So, even though he is not a trained physicist, engineer, or human physiologist, he needs knowledge in all these areas to do his job effectively.

"One of our newest rides, called the Volari, will debut in several parks around the world this year," he says. "It's a flying roller coaster—something like hang-gliding. Riders are suspended in the air on their stomachs from an overhead track. Mark's team will work with the parks' staff to balance issues such as *g*-forces with turning angles and the strength of the track for a particular ride. "We can custom make a ride to fit in almost any space," says Mark. "If a customer wants their roller coaster to be indoors, weaving around columns—we can do that. We can do that because we combine an understanding of the physics with an appreciation for the consumers' need to be entertained."

Chapter 4

Chapter 4 Assessment

Now that you have finished this chapter, it is time to complete your challenge. Go back to the beginning of the chapter and read the challenge again. You may have already decided for whom you will be designing the roller coaster. Will it be for a young child, an adult, a physically challenged person, someone who is visually impaired, or a thrill-seeking daredevil?

Review the criteria. All designs must have certain components including:

- **at least two hills**
- **one horizontal curve**

In addition, you will have to provide evidence that the ride is safe. Safety data must include the height, speed, and acceleration of the roller coaster at five designated locations. Finally, you will have to calculate the energy required to get the roller coaster rolling.

Next, review the grading system that you and your classmates established before you started this chapter. Perhaps you may wish to change some of the criteria and the marking scheme now that you have more information about the topic. The more you know about what is expected of you for the **Chapter Challenge**, the better you will be at completing your assignment.

Physics You Learned

Velocity

Acceleration

Gravitational potential energy

Kinetic energy

Spring potential energy

Conservation of energy

Mass and weight

Hooke's Law

Force vectors

Weight changes during acceleration

Newton's First Law

Newton's Second Law

Circular motion

Centripetal acceleration

Centripetal forces

Normal forces

Scalars and vectors

Name: ______________________________

1) Work is being done when a force

A) of gravitational attraction acts on a person standing on the surface of the Earth

B) is exerted by one team in a tug-of-war when there is no movement

C) acts vertically on a cart that can only move horizontally

D) is exerted while pulling a wagon up a hill

2) A net force of 5.0 N moves a 2.0-kg object a distance of 3.0 m in 3.0 s. How much work is done on the object?

A) 1.0 J B) 10. J C) 30. J D) 15 J

3) A 0.50-kg sphere at the top of an incline has a potential energy of 6.0 J relative to the base of the incline. Rolling halfway down the incline will cause the sphere's potential energy to be

A) 6.0 J B) 3.0 J C) 0 J D) 12 J

4) Which mass has the *greatest* potential energy with respect to the floor?

A) 50-kg mass resting on the floor B) 6-kg mass 5 m above the floor

C) 2-kg mass 10 m above the floor D) 10-kg mass 2 m above the floor

5) Which graph *best* represents the relationship between potential energy (PE) and height above ground (*h*) for a freely falling object released from rest?

A) 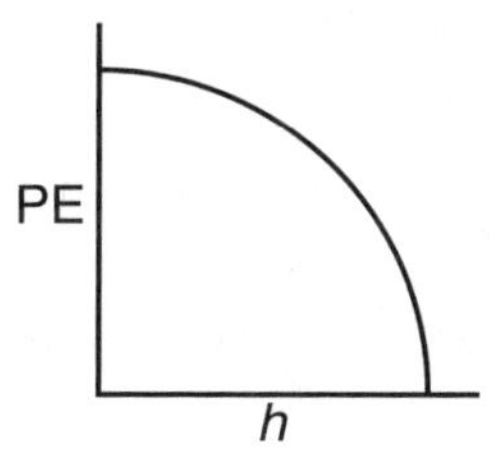

C) 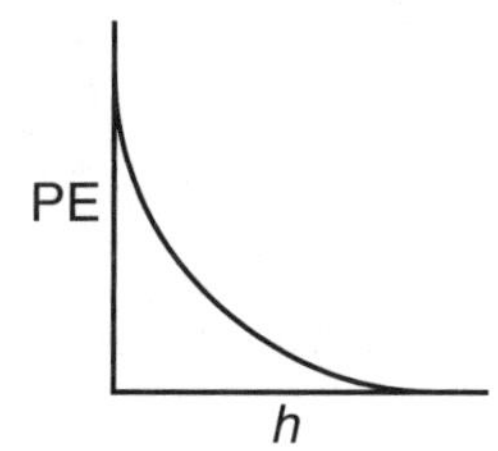

B 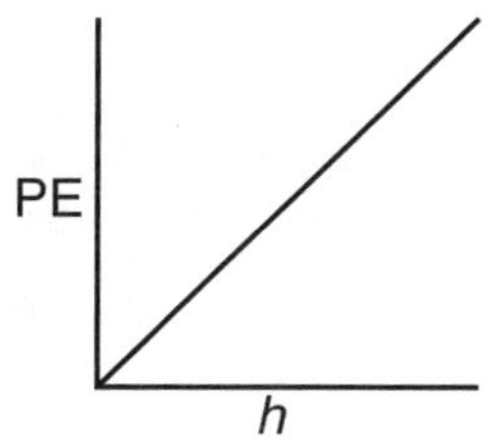

D) 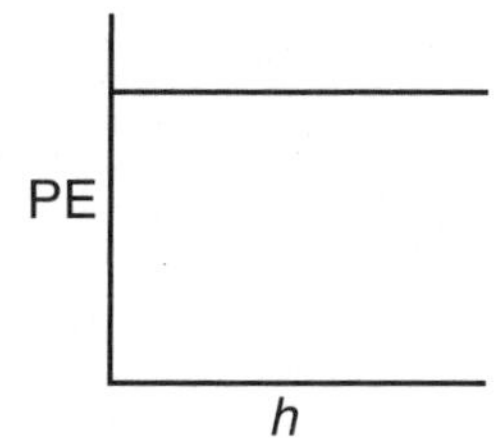

6) As an object falls *freely* in a vacuum, its total energy

A) increases B) decreases C) remains the same

7) As the speed of a bicycle moving along a level horizontal surface changes from 2 m/s to 4 m/s, the magnitude of the bicycle's gravitational potential energy

A) increases B) decreases C) remains the same

8) Three people of equal mass climb a mountain using paths *A*, *B*, and *C* shown in the diagram below.

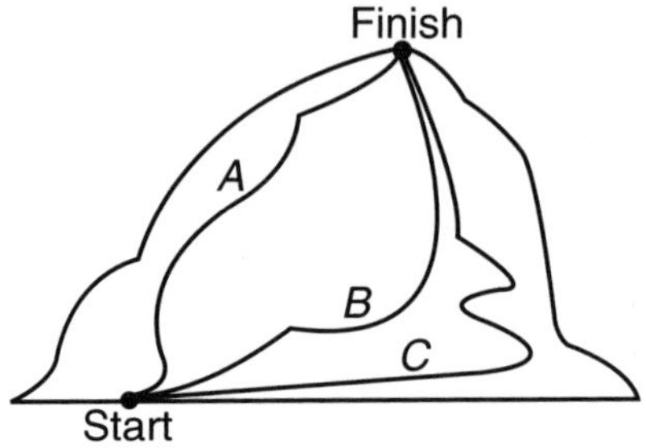

Along which path(s) does a person gain the *greatest* amount of gravitational potential energy from start to finish?

A) *C*, only
B) *B*, only
C) The gain is the same along all paths.
D) *A*, only

9) If the velocity of an automobile is doubled, its kinetic energy

A) decreases to one-half
B) quadruples
C) doubles
D) decreases to one-fourth

10) The diagram below represents a flat racetrack as viewed from above, with the radii of its two curves indicated. A car with a mass of 1000 kg moves counterclockwise around the track at a constant speed of 20 m/s.

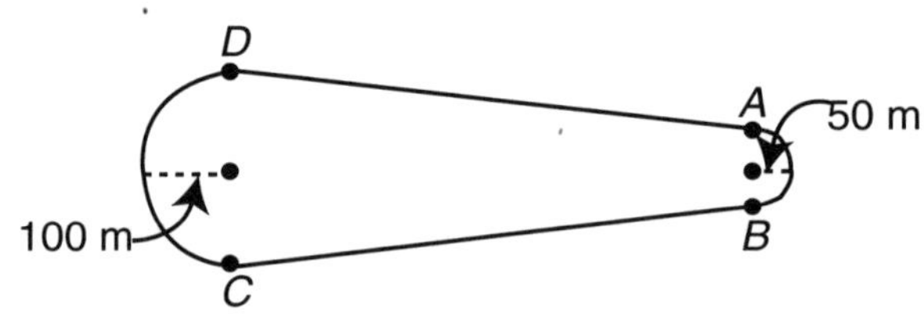

Compared to the kinetic energy of the car while moving from *A* to *D*, the kinetic energy of the car while moving from *D* to *C* is

A) greater
B) the same
C) less

11) A car of mass *M* on a frictionless track starts from rest at the top of a hill having height h_1, as shown in the diagram below.

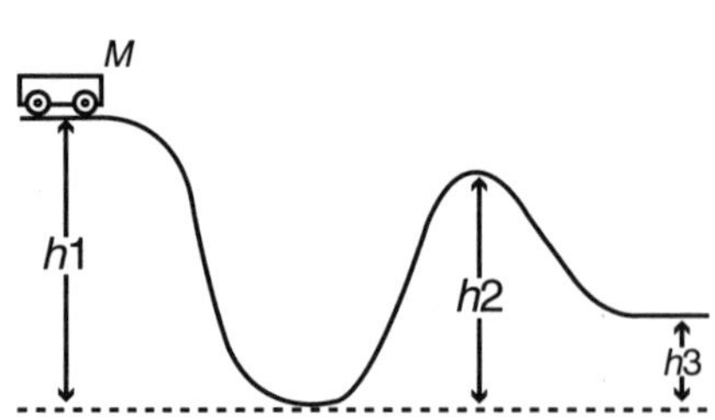

What is the kinetic energy of the train when it reaches the top of the next hill, having height h_2?

A) Mgh_1
B) 0
C) $Mg(h_1\text{-}h_2)$
D) $Mg(h_2h_3)$

12) Which cart shown below has the *greatest* kinetic energy?

A) v= 1 m/s B) v= 4 m/s C) v= 2 m/s D) v= 3 m/s

Questions 13 through 15 refer to the following:

The diagram below represents a 2.0-kg mass placed on a frictionless track at point *A* and released from rest.

Assume the gravitational potential energy of the system to be zero at point *E*.

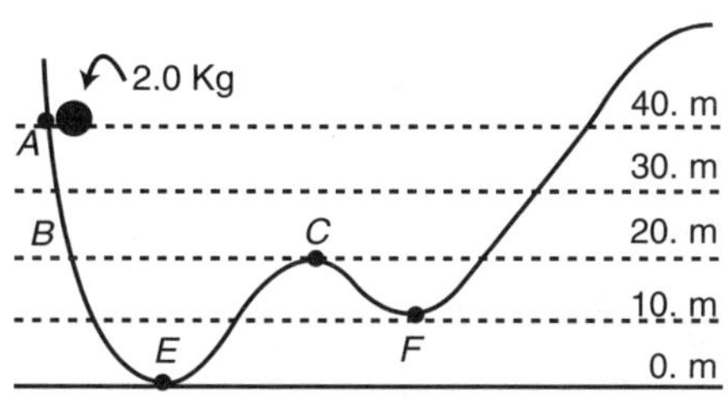

13) The gravitational potential energy of the system at point *A* is approximately

A) 7.0×10^2 J B) 80. J C) 8.0×10^2 J D) 20. J

14) Compared to the kinetic energy of the mass at point *B*, the kinetic energy of the mass at point *E* is

A) $\frac{1}{2}$ as great B) twice as great C) the same D) 4 times greater

15) If the mass were released from rest at point *B*, its speed at point *C* would be

A) 0 m/s B) 10. m/s C) 14 m/s D) 0.50 m/s

Questions 16 and 17 refer to the following:

The diagram below represents a 1.00-kg object being held at rest on a frictionless incline.

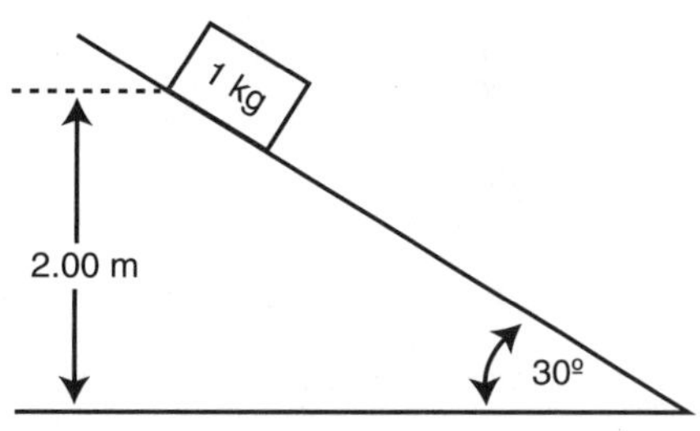

16) The object is released and slides the length of the incline. When it reaches the bottom of the incline, the object's kinetic energy will be *closest* to

A) 9.81 J B) 19.6 J C) 2.00 J D) 4.00 J

17) As the object slides down the incline, the sum of the gravitational potential energy and kinetic energy of the object will

A) decrease B) remain the same C) increase

18) Which graph *best* represents the relationship between the elongation of an ideal spring and the applied force?

A) 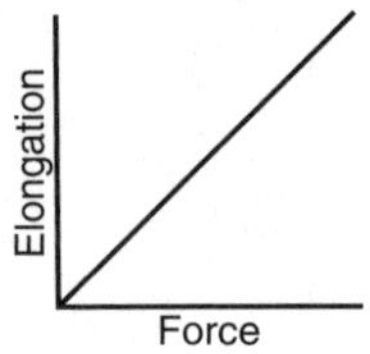

B) 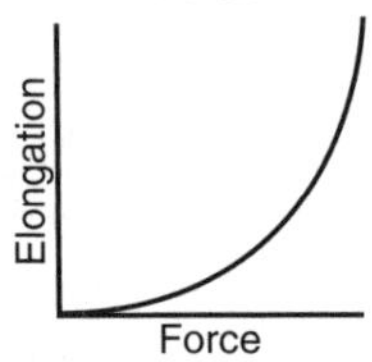

C) 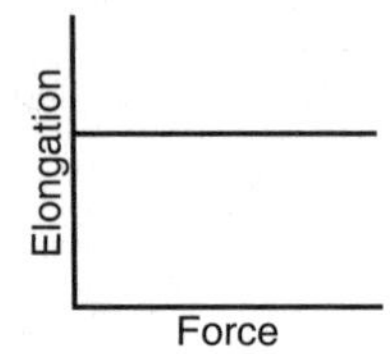

D) 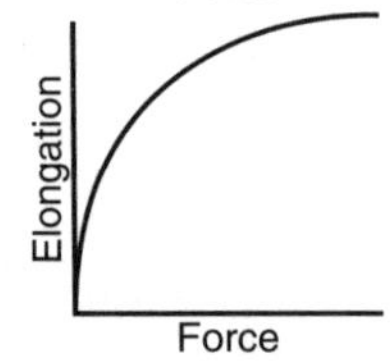

19) The graph below represents the relationship between the force applied to a spring and the elongation of the spring.

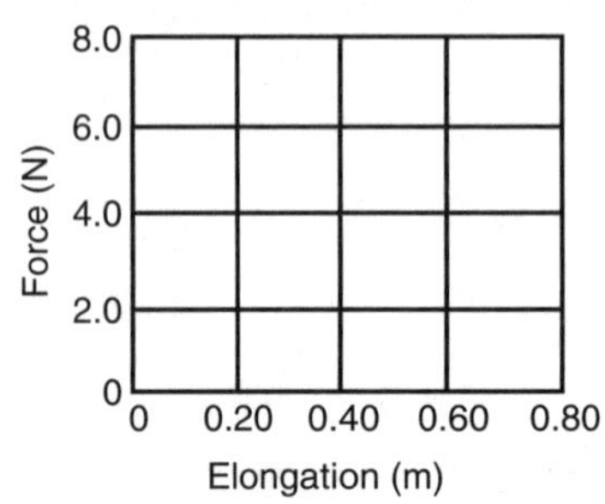

What is the spring constant?

A) 0.050 m/N B) 20. N/m C) 9.8 N/kg D) 0.80 N · m

20) The diagram below represents a block suspended from a spring. The spring is stretched .200 m. If the spring constant is 200. N/m, what is the weight of the block?

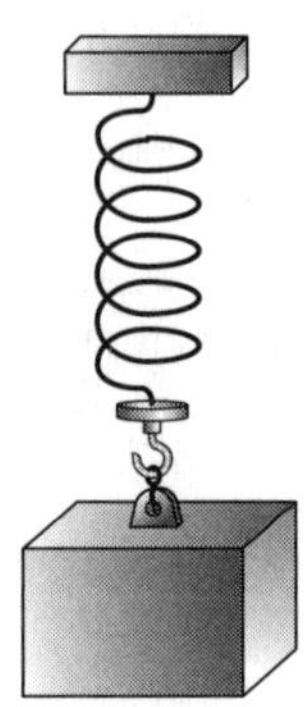

A) 4.00 N B) 40.0 N C) 8.00 N D) 20.0 N

21) A spring has a spring constant of 120 N/m. How much potential energy is stored in the spring as it is stretched 0.20 meter?

A) 12 J B) 4.8 J C) 24 J D) 2.4 J

22) Energy is measured in the same units as

A) momentum B) power C) force D) work

23) Which quantities are measured in the same units?

A) heat and temperature
B) work and energy
C) mass and weight
D) power and work

24) As an object is raised above the Earth's surface the gravitational potential energy of the object-Earth System

A) increases
B) decreases
C) remains the same

25) The diagram below represents a cart traveling from left to right along a frictionless surface with an initial speed of v. At which point is the gravitational potential energy of the cart *least*?

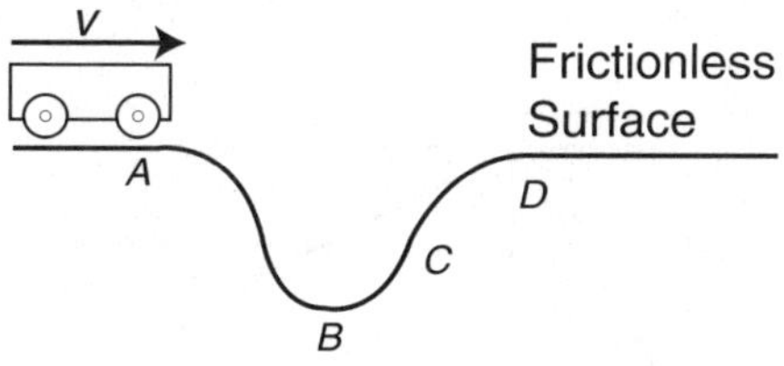

A) *A*
B) *B*
C) *C*
D) *D*

26) As an object falls freely near the Earth's surface, the loss in gravitational potential energy of the object is equal to its

A) loss of height
B) loss of mass
C) gain in kinetic energy
D) gain in velocity

27) As the pendulum swings freely from *A* to *B* as shown in the diagram below, the gravitational potential energy of the pendulum

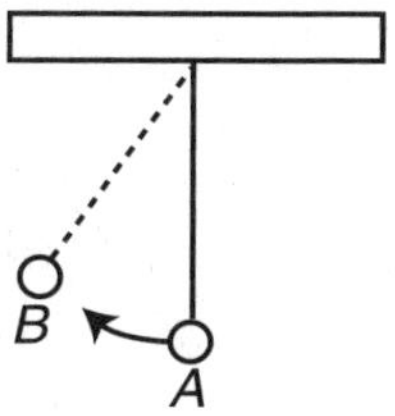

A) remains the same
B) increases
C) decreases

28) In the diagram below, an ideal pendulum released from point *A* swings freely through point *B*.

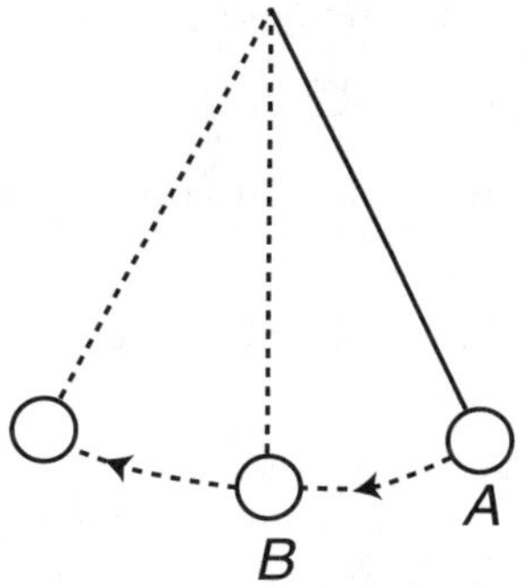

Compared to the pendulum's kinetic energy at *A*, its potential energy at *B* is

A) four times as great
B) twice as great
C) half as great
D) the same

29) As the pendulum swings from position *A* to position *B* as shown in the diagram below, what is the relationship of kinetic energy to potential energy? [*Neglect friction.*]

A) The kinetic energy decrease is more than the potential energy increase.

B) The kinetic energy increase is equal to the potential energy decrease.

C) The kinetic energy decrease is equal to the potential energy increase.

D) The kinetic energy increase is more than the potential energy decrease.

30) The diagram below represents a simple pendulum with a 2.0-kg bob and a length of 10. m. The pendulum is released from rest at position 1 and swings without friction through position 4. At position 3, its lowest point, the speed of the bob is 6.0 m/s.

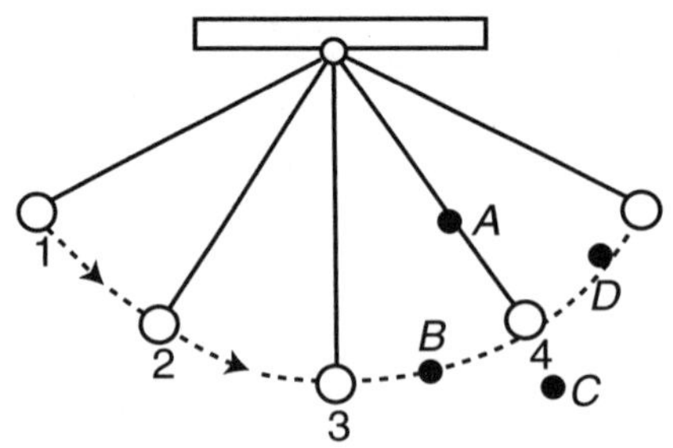

Compared to the sum of the kinetic and potential energies of the bob at position 1, the sum of the kinetic and potential energies of the bob at position 2 is

A) less B) the same C) greater

Questions 31 and 32 refer to the following:

The diagram below represents a 2.0-kg mass placed on a frictionless track at point *A* and released from rest. Assume the gravitational potential energy of the system to be zero at point *E*.

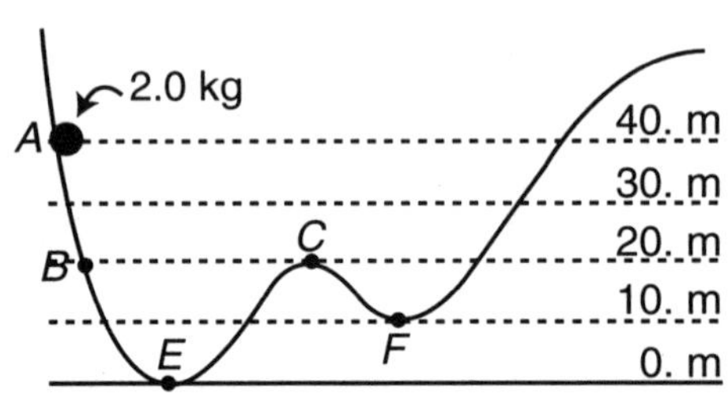

31) As the mass travels along the track, the maximum height it will reach above point E will be closest to

A) 10. m B) 40. m C) 20. m D) 30. m

32) Compared to the total mechanical energy of the system at point *A*, the total mechanical energy of the system at point *F* is

A) more B) less C) the same

33) Which is a scalar quantity?

A) distance B) force C) displacement D) acceleration

34) A student weighing 500. n stands on a spring scale in an elevator. If the scale reads 520. n, the elevator must be

A) moving downward at constant speed B) moving upward at constant speed

C) accelerating downward D) accelerating upward

35) The diagram represents a block sliding along a frictionless surface between points *A* and *G*.

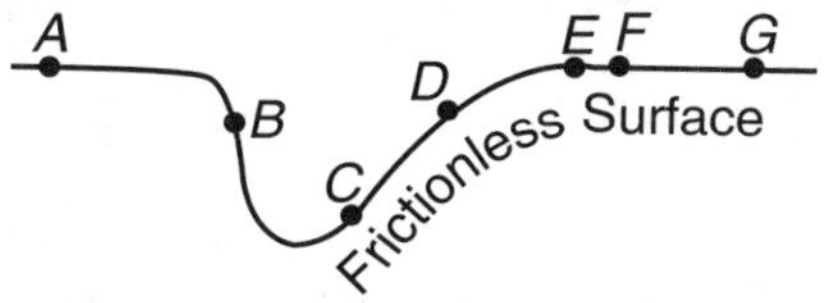

As the block moves from point *A* to point *B*, the speed of the block will be

A) constant, but not zero

B) increasing

C) zero

D) decreasing

Alternative Chapter Assessment Answers

1) D	8) C	15) A	22) D	29) B
2) D	9) B	16) B	23) B	30) B
3) B	10) B	17) B	24) A	31) B
4) B	11) C	18) A	25) B	32) C
5) B	12) D	19) B	26) C	33) A
6) C	13) C	20) B	27) B	34) D
7) C	14) B	21) D	28) D	35) B

NOTES

Chapter 4

NOTES

NOTES

NOTES